AF787241

Sustainable Food Systems in Sub-Saharan Africa

Bal Ram Singh · Liveness Jessica Banda ·
Andy Safalaoh · Agnes Mwangwela · Ruth Haug ·
Lars Olav Eik · Jennifer Joy West · Rattan Lal
Editors

Sustainable Food Systems in Sub-Saharan Africa

Plant Derived Nutritious Foods—Volume 1

Springer

Editors
Bal Ram Singh
Faculty of Environmental Sciences
and Natural Resource Management
Norwegian University of Life Sciences
Ås, Norway

Andy Safalaoh
Animal Science Department
Lilongwe University of Agriculture
and Natural Resources
Lilongwe, Malawi

Ruth Haug
Department of International Environment
and Development Studies, Faculty
of Landscape and Society
Norwegian University of Life Sciences
Ås, Norway

Jennifer Joy West
Department of International Environment
and Development Studies, Faculty
of Landscape and Society
Norwegian University of Life Sciences
Ås, Norway

Liveness Jessica Banda
Lilongwe University of Agriculture
and Natural Resources
Lilongwe, Malawi

Agnes Mwangwela
Lilongwe University of Agriculture
and Natural Resources
Lilongwe, Malawi

Lars Olav Eik
Department of International Environment
and Development Studies, Faculty
of Landscape and Society
Norwegian University of Life Sciences
Ås, Norway

Rattan Lal
4CFAES Rattan Lal Center for Carbon
Management and Sequestration,The Ohio
State University
Columbus, OH, USA

ISBN 978-3-032-11287-3 ISBN 978-3-032-11288-0 (eBook)
https://doi.org/10.1007/978-3-032-11288-0

© The Editor(s) (if applicable) and The Author(s) 2026. This book is an open access publication.

Open Access This book is licensed under the terms of the Creative Commons Attribution 4.0 International License (http://creativecommons.org/licenses/by/4.0/), which permits use, sharing, adaptation, distribution and reproduction in any medium or format, as long as you give appropriate credit to the original author(s) and the source, provide a link to the Creative Commons license and indicate if changes were made.
The images or other third party material in this book are included in the book's Creative Commons license, unless indicated otherwise in a credit line to the material. If material is not included in the book's Creative Commons license and your intended use is not permitted by statutory regulation or exceeds the permitted use, you will need to obtain permission directly from the copyright holder.
The use of general descriptive names, registered names, trademarks, service marks, etc. in this publication does not imply, even in the absence of a specific statement, that such names are exempt from the relevant protective laws and regulations and therefore free for general use.
The publisher, the authors and the editors are safe to assume that the advice and information in this book are believed to be true and accurate at the date of publication. Neither the publisher nor the authors or the editors give a warranty, expressed or implied, with respect to the material contained herein or for any errors or omissions that may have been made. The publisher remains neutral with regard to jurisdictional claims in published maps and institutional affiliations.

This Springer imprint is published by the registered company Springer Nature Switzerland AG
The registered company address is: Gewerbestrasse 11, 6330 Cham, Switzerland

If disposing of this product, please recycle the paper.

Foreword

Agriculture in Sub-Saharan Africa (SSA) is subsistence farming, dominated primarily by small land holders and resource-poor farmers. The use of modern agricultural machinery and external inputs (fertilizers, pesticides and irrigation) is rather limited. Indeed, the fertilizer rate in SSA is hardly 22 kg/ha compared with 146 kg/ha of average rate on global scale or one-seventh the rate of world average. Additionally, declining land resources due to soil degradation, harsh and uncertain climate and the rapidly increasing population have also contributed to declining crop productivity, food insecurity and nutritional quality of food. Thus, SSA is vulnerable to negative nutrient budget which is estimated at 22 kg/ha of N, 2.5 kg/ha of P and 15 kg/ha of K. The negative nutrient budget, being experienced on continental scale for several decades going back to 1960s, is aggravated with the widespread problem of soil degradation by a range of processes such as accelerated soil erosion and salinization. Yet, the population of SSA is increasing rapidly. In 2010, Africa's population was 800 million, which increased to 1.5 billion in 2024, and is projected to increase to 2.5 billion in 2050 and 4 billion in 2100.

The triple crisis of increase in population, severe soil degradation, and global warming is driven by mutually reinforcing feedback which is aggravated by poor governance and lack of will power to bring about the much-needed transformation. Increase in population aggravates emission of greenhouse gases (GHGs) which in turn exacerbate intensity and severity of soil degradation by erosion, drought and salinization. Thus, transforming agriculture is the most feasible option to break the vicious cycle by improving agronomic productivity and bridging the yield gap through protection, restoration and sustainable management of soil health and its life-support functions which generate ecosystem services for human wellbeing and nature conservancy.

Although increases in agronomic yield and annual growth rates have been observed in some SSA countries, sustaining such growth rates in future will be difficult because of rapidly growing population, unpredictable climate, limited water resources and irrigation facilities, accelerated soil erosion, soil pollution caused by organic and inorganic pollutants and increasing number of pervasive pests and pathogens due to warmer climate. According to IPCC, predicted 1.5 °C change in temperature in SSA will threaten the ecosystems, biodiversity, and human health to a greater extent than that in other regions. These threats not only will enhance food insecurity but may also create problems with food safety, undernutrition and malnutrition. Therefore, the need to enhance both agricultural productivity and nutrient-enriched foods in SSA is widely felt.

To extend and implement the already researched and proven pro-poor technologies and value-chains to small land-holder farmers, the Norwegian University of Life Sciences NMBU in collaboration with Lilongwe University of Agriculture and Natural Resources (LUANAR) started the Food Ma-Program. The aim of the program was to enhance production capacities with nutrient-rich food at smallholder farms and their economic and social resilience. The purpose of this book, accomplished through Food MA program, is to synthesize the knowledge generated on relevant and adaptive technologies for the betterment of farmer's livelihood and further to identify the most promising interventions and technologies to improve food production with high nutrition quality to safeguard human health of rural population.

The yield gap can be narrowed by integrated management of soil fertility and plant available water capacity. The anthropogenic climate change has a multiplier adverse effect on agronomic yield because of the increase in intensity and frequency of drought. With the global average of 20% of cropland area equipped for irrigation, that is SSA is estimated to range from 4 to 6%. Thus, soils of agroecoregions of SSA are strongly vulnerable to climate change with adverse effects on quality and quantity of food produced.

This book entitled *Sustainable Food Systems in Sub-Saharan Africa: Plant Derived Nutritious Foods—Volume 1* covers a wide range of themes to involve food production systems concerning sustainability and nutritional quality of food and adaptability to climate change, research-driven and proven technologies, innovative extension approaches and policy implementation.

We appreciate the contribution made by the successful authors, the editors and reviewers of different chapters of this book. We feel that the knowledge presented in the book is a crucial piece for tackling the challenges faced for increased and sustained food productivity and nutritional quality of food for better human health in SSA.

Prof. Siri Fjellheim
Rector
Norwegian University of Life Sciences
(NMBU)
Ås, Norway

Prof. Emmanuel Kaunda
Vice Chancellor
Lilongwe University of Agriculture
and Natural Resources (LUANAR)
Lilongwe, Malawi

Preface

Agriculture is the backbone for economic growth and livelihoods of most households, especially rural smallholder farmers in Sub-Saharan Africa (SSA) as > 60% of population of SSA is engaged in agriculture, dominated by smallholders (< 2 ha), and they constitute approximately 80% of all farms and employ about 175 million people directly. In many SSA countries, women make at least half of the labor force. Declining land resources caused by rapid soil degradation and harsh and uncertain climate coupled with continuous cropping with minimum investment in soil improvement make the agricultural production unsustainable and food insecure for most of the rural population in SSA. Rural sector in SSA is not only the most food insecure but also the most malnourished among developing regions. Malnutrition and micronutrient deficiencies remain widespread and have significant functional implications and poor human health because more than one third of all stunted children under 5 live in Africa. In SSA, 50% of people suffer due to Zn deficiency, 57% of pregnant women in Africa suffer from iron deficiency anemia and nearly 30% are deficient in zinc and iodine.

Agricultural production and food insecurity and food safety have been further aggravated by harsh and changing climate and thus marginalizing the livelihood opportunities of rural population. Changing climate has made the SSA region as one of the hot spots leading to severe soil degradation, drastic depletion of nutrients and soil organic matter stocks, contamination and scarcity of water, and reduction of the above and below ground biodiversity. The need to test research-driven and proven technologies for their sustainability and adaptability and to extend innovative and transformative technologies including extension approach, participatory research, subsidies, integrated seed systems and gender equality to smallholders is widely felt.

The Food Ma-Program was initiated by the Norwegian University of Life Sciences NMBU in collaboration with Lilongwe University of Agriculture and Natural Resources (LUANAR) with an aim to extend and implement the already researched pro-poor technologies and value-chains to small land-holder farmers. Therefore, enhancing production capacities with nutrient-rich food at smallholder farms and their economic and social resilience may improve food and nutrition security at different levels. Hence, this book provided the platform to contributors to identify the most promising interventions and technologies to improve food production with high nutrition quality to safeguard human health of rural population.

The book containing 28 chapters includes 5 thematic parts, viz. (i) Plant Food Systems and Innovative Technologies, (ii) Nutrient-Dense and Biofortified Food, (iii) Soils, Climate Change and Carbon Sequestration, (iv) Policy Formulation and Implementation for Smallholders, and (v) Prospects of Green Revolution and Research Perspectives. Although the focus of the book is on SSA, > 110 scientists representing the four continents of Africa, Asia, Europe and North America have contributed to this book.

All authors of the book deserve our appreciation and thanks for their outstanding contributions. We also thank the staff at Springer for following the proposed publication schedule and bringing out the publication on time

Ås, Norway Bal Ram Singh
Lilongwe, Malawi Liveness Jessica Banda
Lilongwe, Malawi Andy Safalaoh
Lilongwe, Malawi Agnes Mwangwela
Ås, Norway Ruth Haug
Ås, Norway Lars Olav Eik
Ås, Norway Jennifer Joy West
Columbus, USA Rattan Lal

Contents

About the Editors

Bal Ram Singh, Ph.D. is Professor Emeritus at the Norwegian University of Life Sciences. He earned his M.Sc. from the Indian Agricultural Research Institute (IARI), New Delhi, and Ph.D. degree from G. B. Pant University of Agriculture and Technology, Pant Nagar, India. His research program focuses on bioavailability and mobility of heavy metals in the soil and plant system, fertility management and agricultural sustainability in soils of the tropics and on carbon sequestration in soils. Prof. Singh is Fellow of ASA and SSSA and Recipient of several awards from the Soil Science Society of America, such as International Award in Soil Science, Soil Science Applied Research Award and Distinguished Service in Soil Science Award.

He has served as Chairman of the program board "Soils and Plants" of the Research Council of Norway and as Deputy Head of the Department, in addition to many national and international committees. He chaired the Cost Action FA0905 (EU) on "Mineral Improved Crop Production for Healthy Food and Feed", in which > 200 scientists from 31 countries participated. He has supervised 76 graduate students and 16 visiting fellows/scientists from 20 countries and published **496** articles, including **240** in peer reviewed journals, **49** book chapters, **13** edited/coedited books and journals, and **162** invited seminars/lectures. He is Chief Editor of the journal *Soils and Human Health of the Frontiers in Soil Science*, in the editorial boards of **4** journals, and Reviewer of > **30** journals. He has served as Chair of Division 3 of the International Union of Soil Science (2016–2022) and is Vice President of the

Norwegian Society of Soil Science and Member of the Geomedicine committee—Food, Environment and Health of the Norwegian Academy of Science and Letters.

Dr. Andy Safalaoh is Associate Professor of Monogastric Nutrition and Production at the Lilongwe University of Agriculture and Natural Resources (LUANAR). He is Science, Technology, and Innovation Scholar and Fellow of the Leadership in Environment and Development (LEAD) Programme at Imperial College London (Cohort 10, 2004). He holds a Ph.D. in Science and Technology Studies (University of Nottingham, UK), an M.Sc. in Animal Science (Oklahoma State University, USA), and a B.Sc. Degree and Diploma in General Agriculture (Bunda College of Agriculture, University of Malawi).

Andy has also participated in various professional programs, including the Higher Education Solutions Network Tech Con (Massachusetts Institute of Technology, USA, 2016), the International Ph.D. School on Innovation and Economic Development (Brazil, 2012), and the Pathways Approach Program (University of Sussex, 2012). Additionally, he attended the ISDA Symposium on Innovation and Sustainable Development in Agriculture and Food (CIRAD/INRA, France, 2010) and specialized training in monitoring and evaluation, facilitation, and baseline survey study techniques (SCF/Inc., 1986).

From 2007 to 2010, Dr. Safalaoh served as Regional Expert and Consulting Facilitator for the Agricultural Science and Technology Innovation Systems Programme for the African, Caribbean, and Pacific region at the Centre for Agricultural and Rural Development (CTA), Wageningen, The Netherlands. As LUANAR Programmes Coordinator, Andy also played a pivotal role in designing and securing funding for the FooDMa and TRANSFORM programs, supported by the Norwegian Government in Malawi. In 2020, Andy co-edited the Springer-published book *Climate Impacts on Agricultural and Natural Resource Sustainability in Africa*. At LUANAR, he has held several leadership roles, including Founder and Editor-in-Chief of the *Bunda Journal of Agriculture and Natural Resources*, Editor of the *Bunda Newsletter* (2002–2009), University

Programmes Coordinator, Deputy Head of the Animal Science Department, and Innovation Hub Coordinator.

Andy's research focuses on evaluating non-conventional and climate-smart feedstuffs such as insects, millets and agro-byproducts. He has supervised over 20 M.Sc. students and five Ph.D. students. His academic and professional contributions continue to influence agricultural and environmental sustainability in Malawi and beyond.

Dr. Agnes Mwangwela is Associate Professor in Food Science and Deputy Vice Chancellor at Lilongwe University of Agriculture and Natural Resources (LUANAR), Malawi, where she has worked for 28 years. Dr. Mwangwela is Holder of a Ph.D. in Food Science from the University of Pretoria, an M.Sc. and B.Sc. from the University of Malawi. Dr. Mwangwela currently leads the Malawi Food Data System (MAFOODS) which is entrusted with Food Composition data compilation and management for Malawi. Dr. Mwangwela served as Dean of Food and Human Sciences and Director of Bunda College. A Seasoned Researcher and Academician is involved in teaching food science and supervision of postgraduate students and has been awarded research grants from McKnight foundation, International Institute for Tropical Agriculture (IITA) and Africa Rising; International Foundation for Science (IFS), South African NRF, USAID.

Dr. Mwangwela is a recipient of The Norman E. Borlaug International Agricultural Science and Technology fellowship. She is passionate about mentorship and fostering resilient food systems to improve lives through the career development programs under the African Women in Agriculture Research and Development (AWARD) and the SARUA-UKRI Food System Network. Dr. Mwangwela contributes to government policy formulation and implementation in her area of expertise through the National Nutrition Coordinating committee and the Nutrition Policy Advisory Team for DNHA.

Dr. Liveness Jessica Banda is Associate Professor in Animal Physiology and Programs Coordinator at Lilongwe University of Agriculture and Natural Resources (LUANAR), Malawi, where she has worked for over 22 years. She holds a Ph.D. in Animal Genomics and Disease Resistance from the University of Edinburgh, an M.Sc. in Animal Science, and a B.Sc. in Agriculture from the University of Malawi. Her work spans teaching, research, outreach, and consultancy in animal science and community development, with research interests in livestock reproduction, physiology, production systems, and sustainable food systems in the context of climate change. She has served as Head of the Animal Science Department, sits on governance boards of various organizations, serves as an external academic examiner, and reviews journal articles and academic programs. As Programs Coordinator, she leads the implementation of several initiatives funded by partners such as the Royal Norwegian Embassy, World Bank, USAID, NORAD, the Flanders Government, and Wellcome Trust, while also coordinating resource mobilization and building strategic institutional partnerships. Liveness is Alumnus of the African Women in Agricultural Research and Development (AWARD), Borlaug Fellowship, and Fulbright Visiting Scholar Program.She is passionate about mentorship, youth empowerment, and fostering resilient food systems to improve lives.

Professor Ruth Haug is Professor of International Development Studies at the Norwegian University of Life Sciences (NMBU). She received her Ph.D. from the University of Maryland and began her professional career working nationally and internationally within the field of food security and rural development. At the Norwegian University of Life Sciences, Ruth Haug was Deputy Vice Chancellor in the period 2008–20013, and before that she was heading the Department of International Environment and Development Studies (Noragric) for seven years. Her field of expertise is within food systems, food security and international food policy; conflicts and humanitarian action; climate change and management of natural resources; agricultural advisory services; and gender issues in development. Ruth Haug has research, teaching and consultancy experience in Africa and Asia and has worked with NORAD, the

World Bank, and the Norwegian Ministry of Agriculture. She has supervised many M.Sc. and Ph.D. students and has a good publication record. She likes teaching and has taught many university courses. Ruth Haug has been Member of numerous national and international boards and committees; and she has been involved in several international policy processes in relation to food and agriculture. In particular, she has been representing Norway in CGIAR meetings in the period 1995 to 2012 and has been Member of the CGIAR Fund Council. She was Member of the International Food Policy Research Institute (IFPRI) Board of Trustees until the end of 2017 and the Agricultural Research in the Dry Areas (ICARDA) board until the end of 2020. She has also been representing Norway in CFS and FAO meetings and was Member of the Svalbard Seed Vault Council for eight years.

Author photo: By Tommy Normann/NMBU

Lars Olav Eik is Professor Emeritus at The Faculty of Landscape and Society (LANDSAM) at the Norwegian University of Life Sciences (NMBU). LANDSAM is an interdisciplinary faculty working with national and global sustainable development. His field of research specialisation is in animal nutrition and small ruminant production systems. After graduating from NMBU in 1982, he joined the Sokoine University of Agriculture (SUA), Tanzania, working on dry season feeding of ruminants and introducing dairy goat keeping in Tanzania. After this assignment, he returned to NMBU and completed his Ph.D. based on work with dairy goats in Norway.

Since 1990, he has worked on research programs in Norway, the European Union (EU), Tanzania, Malawi, Ethiopia and South Africa. His work is concentrated with farmers and the private sector. He is keenly interested in developing multifunctional production systems and value chains for small ruminants in tropical and temperate regions. He has supervised several M.Sc. and Ph.D. students, especially those coming from East Africa. His teaching covers small ruminant nutrition, production systems, tropical animal husbandry, and aquaculture.

In the past several years, he has coordinated two collaborative programs between NMBU and the Lilongwe University of Agriculture and Natural Resources (LUANAR) in Malawi. He initiated and participated in a new young staff exchange program between institutions in Norway, LUANAR and several universities in Tanzania, dealing with agricultural value chains. He was instrumental in introducing Norwegian Dairy Goats to the country in Tanzania. They soon gained popularity and multiplied to the estimated number of 500,000 milking goats. This venture has uplifted the livelihood of many small farmers in East African countries.

In Norway, he initiated the introduction of cashmere goats. He currently participates in a project developing quality yarn based on a blend of cashmere and wool from Norwegian short-tailed sheep. The work on wool is part of a broader ongoing initiative for the multipurpose use of small ruminants in the country.

Dr. Rattan Lal, Ph.D. is Distinguished University Professor of Soil Science and Director of the CFAES Rattan Lal Center for Carbon Management and Sequestration (Lal Carbon Center) at The Ohio State University (OSU); Adjunct Professor of the University of Iceland and the Indian Agricultural Research Institute (IARI), India; and Visiting Professor at Pontifical Catholic University of Valparaíso (PUCV), Chile. Dr. Lal is widely recognized globally as Pioneer in soil-centric agricultural management to improve food security globally and develop climate-resilient agriculture through soil carbon sequestration, sustainable intensification, enhancing use efficiency of agroecosystems, sustainable management of soils, and soil health.

Born on September 5th, 1944, in Punjab, India, Rattan Lal completed his education from Punjab Agricultural University, Ludhiana (B.Sc., 1963), Indian Agricultural Research Institute, New Delhi (M.Sc., 1965), and the Ohio State University, Columbus (Ph.D., 1968). He served as Sr. Research Fellow with the University of Sydney, Australia (1968–69), Soil Physicist at IITA, Ibadan, Nigeria (1969–1987), and Professor of Soil Science at OSU (1987–to date).

He is Fellow of the American Society of Agronomy (ASA, 1985), Soil Science Society of America (SSSA, 1986), Third World Academy of Sciences (1992), American Association for the Advancement of Sciences

(1996), Soil and Water Conservation Society (SWCS, 1997), Indian Natl. Academy of Agricultural Sciences (1998), and Rothamsted, UK (2013). He received the Hugh Hammond Bennett Award of the SWCS (2005), Borlaug Award (2005), Liebig Award (2006), M. S. Swaminathan Award in India (2009), COMLAND Award in Germany (2009), and the Sustained Achievement Award from the Renewable Natural Resources Foundation (2017). He has also received the Medal of Honor from UIMP, Santander, Spain (2018); the Distinguished Service Award from the International Union of Soil Sciences in Rio, Brazil (2018); as well as the Alumni Medalist Award (2019) and 422nd Commencement Medal (2019) from OSU, Columbus, Ohio, USA. He served as President of the World Association of the Soil and Water Conservation (1987–1990), the International Soil Tillage Research Organization (1988–1991), the Soil Science Society of America (2005–2007), and the International Union of Soil Sciences, Vienna, Austria (2017–2018). He has served as Member of the Federal Advisory Committee on National Assessment of Climate Change-NCADAC (2010–2014) and as Senior Science Advisor to the Global Soil Forum of IASS, Potsdam, Germany (2010–2012). He was also member of Steering Committee of the Global Soil Week, IASS, Potsdam, Germany (2012–2016); member of Advisory Board of FACCE-JPI of the European Council (2013–2016); Chair of Advisory Committee to UNU-FLORES, Dresden, Germany (2013–2019); member of the SPI of UNCCD, Bonn, Germany (2018–2022); and member of United Nations Food Systems Summit Science Committee and Action Track 3 (UNFSS) (2021). He is Chair in Soil Science and Goodwill Ambassador for Sustainable Development Issues of the Inter-American Institute for Cooperation on Agriculture (IICA), Costa Rica, San José (2020). **He serves as Member and Chair of SERDP Scientific Advisory Board of the Department of Defense (2011–2018; 2023–2025), and he was appointed as Member to the Board for International Food and Agricultural Development (BIFAD) by President Joe Biden (2022–2025).**

He has received honorary degrees of Doctor of Science from Punjab Agricultural University in

Ludhiana, India (2001); the Norwegian University of Life Sciences, Ås, Norway (2005); Alecu Russo Balti State University, Moldova (2010); Technical University of Dresden, Germany (2015); University of Lleida, Spain (2017); Gustavus Adolphus College in Saint Peter, Minnesota (2018); Pontifical Catholic University of Valparaíso, Chile (2019); Amity University, Uttar Pradesh, India (2020); and IARI, New Delhi, India (2021).

He is ranked #1 globally and in the U.S. among Agricultural Scientists (Plant Scientists and Agronomists) of the world by Research.com (2023, 2024). Clarivate/Thomson Reuters has included Dr. Lal on the list of the World's Most Influential Scientific Minds (2014–2016) and that of Highly Cited Researchers (2014–2023). Professor Lal was Lead Author for the IPCC (1998–2000) and was awarded Nobel Peace Prize Certificate by the IPCC in 2007.

He has mentored 125 graduate students, 15 undergraduate honors students, 82 postdoctoral researchers and research scientists, as well as hosted 188 visiting scholars from around the world. He has authored/co-authored more than 3000 research publications overall, including more than 1150 refereed journal articles and 580 book chapters, and he has written 24 books and edited/co-edited 84 books.

Professor Lal is Laureate of the Gulbenkian Prize for Humanity (2024), Mahatma Ghandi Award for Environmental Protection (2022), Padma Shri Award (2021), World Food Prize (2020), Arrell Global Food Innovation Award (2020), Japan Prize (2019), U.S. Awasthi IFFCO Prize (2019), GCHERA World Agriculture Prize (2018), and Glinka World Soil Prize (2018).

Jennifer Joy West is Head of the Department of International Environment and Development Studies (Noragric) at NMBU. As Environmental Social Scientist and Associate Professor at the department, she has several decades of interdisciplinary research experience working at the interface between climate change adaptation, rural and agricultural development and transformations toward sustainability in Norway and internationally. She holds a master's in management of Natural Resources and Sustainable Agriculture (MNRSA) and Ph.D. in international environment and development studies, both from Noragric. In her role as head of department, she leads academic and strategic development of research and education across Noragric's five study programmes and four research clusters dealing with rights, accountability and power in development, environmental governance, global orders, conflict, and security and climate change and agricultural development. West's primary academic expertise lies at the interface between climate adaptation, agricultural development and climate services in support of sustainable development, with a particular focus on East Africa and on inter/transdisciplinary methodologies integrating natural and social sciences and qualitative and quantitative approaches. She has conducted extensive fieldwork in Tanzania and has also contributed to research on enhancing the relevance of climate information and services for adaptation planning in Norway/Europe and to initiatives aimed at enhancing social science contributions to climate change and climate services research internationally.

Introduction

Plant Derived Nutritious Foods Under Changing Climate in Sub-Sahara Africa

Bal Ram Singh, Agnes Mwangwela, Ruth Haug, and Rattan Lal

Abstract Africa is one of the most food-insecure regions of the world, hosting about 282 million undernourished people, accounting for 38% of the global total of 735 million; and hence the access to adequate and nutritious food remains a big challenge. Degraded land resources, low agronomic productivity, and low nutrient dense foods are the major causes of malnourishment including micronutrient deficiencies in Sub Saharan Africa (SSA). More than 50% of people in SSA suffer from Zn deficiency, and 57% of pregnant women suffer from Fe deficiency anemia. Soil and water are fundamental resources for livelihoods and food security in Africa as 60% of people in SSA derive their livelihoods from these resources. Promotion of nutrition-sensitive agriculture including biofortification, the breeding of staple crops that are richer in essential micronutrients than traditional varieties, and enhancing soil health can be a feasible and cost-effective approach to addressing deficiencies in vitamin A, Fe, and Zn. Mitigation of climate change by adoption of recommended management practices has numerous co-benefits, such as increase in agronomic productivity, improvement in nutritional quality of food, increase in renewability and quality of water resources, and improvement in activity and species diversity of soil organisms. Small landholders in SSA face numerous challenges in securing sufficient and nutritious food. In order to address these challenges, it is essential to implement policies such as redistribution of land, subsidizing inputs such as seed and fertilizer, strengthening agricultural advisory services, establishing saving and credit groups, mechanization efforts, floor prices to enhance decent farm gate prices, insurance

B. R. Singh (✉)
Faculty of Environmental Sciences and Natural Resource Management, Norwegian University of Life Sciences, Ås, Norway
e-mail: balram.singh@nmbu.no

A. Mwangwela
Lilongwe University of Agriculture and Natural Resources, Bunda Campus, Lilongwe, Malawi

R. Haug
Department of International Environment and Development Studies, Faculty of Landscape and Society, Norwegian University of Life Sciences, Ås, Norway

R. Lal
CFAES Rattan Lal Center for Carbon Management and Sequestration, The Ohio State University, Columbus, OH, USA

© The Author(s) 2026

B. R. Singh et al. (eds.), *Sustainable Food Systems in Sub-Saharan Africa*,
https://doi.org/10.1007/978-3-032-11288-0_1

schemes to deal with risks and failures, climate adaptation measures etc. Furthermore, the large yield gap in SSA requires policies which are pro-farmer and pro agriculture, so that farmers are encouraged to adopt modern and proven innovations for sustainable management of soil resources and for ushering the soil-centric Green Revolution in SSA.

1 Introduction

Africa's population is the fastest growing in the world and the current population of 1.09 billion (B) may reach to 2.2 B by 2050 and more than 4 B by 2100 (FAO 2017). Smallholder farms constitute approximately 80% of all farms in Sub-Saharan Africa (SSA) and women comprise at least half of the labor force (FAO 2017). Agriculture plays a dominant role in income generation for most households in SSA region, and the agricultural productivity of small land holder farmers is seen as the key in alleviating poverty and ensuring food security. Land resources in SSA are experiencing increasing trends in degradation due to overexploitation and poor resource use. During the decade of 2010s, SSA experienced the worst land degradation globally, accounting for 22% of the total global annual cost of land degradation amounting to US$ 300 B (Nkonya et al. 2018). Climate-related factors, particularly rainfall variability, accelerated soil erosion and poor farming practices, including continuous cropping with little to no inputs (nutrient mining), low soil organic matter (SOM) content due to residue removal, overgrazing along with biomass burning, excessive tillage and cultivating marginal land, have been identified as major causes for the degradation of agricultural lands (Tully et al. 2015; Nezomba et al. 2017). These factors, especially frequent tillage, and other perturbations, disintegrate soil aggregates, redistribute crop residues, and alter soil aeration, moisture, and temperature regimes. This trend accelerates microbial decomposition and oxidation of SOM that releases CO_2 or CH_4, which is emitted into the atmosphere (Were et al. 2015), thus resulting in reduced carbon stock in soils.

Soil and water resources play a critical role in achieving food security. Unfortunately, Africa is one of the most food-insecure regions of the world, hosting about 282 million undernourished people, accounting for 38% of the global total of 735 million (FAO et al. 2023). Degraded land resources, low agronomic productivity, and low nutrient dense foods are the major causes for malnourishment in the rural population of SSA. Malnourishment, including micronutrient deficiencies, is widespread in SSA. Micronutrients are needed in the human diet in minute quantities, yet it is estimated that over 3 B people suffer from micronutrient deficiencies globally, resulting in deleterious impacts on human health (Cakmak and Kutman 2018). Furthermore, 50% of people in SSA suffer from Zn deficiency and 57% of pregnant women suffer from Fe deficiency anemia, and nearly 30% of people globally are deficient in Zn and I.

In most developing countries including SSA, access to adequate and nutritious food is a big challenge. Many countries are making efforts to make interventions that could help combat malnutrition and hidden hunger. Among these interventions, biofortification of food crops is the most sustainable solution to malnutrition and hidden hunger. Agronomic biofortification, through application of mineral fertilizers to soil or directly on crops to increase the content and bioaccessibility of specific nutrients in food crops (Jha and Warkentin 2020), is one of the approaches that have been used to combat malnourishment. The other approach is genetic biofortification through plant breeding to improve the concentration and bioaccessibility of minerals in crops, by utilizing the genetic differences between crops of similar species (Marques et al. 2021). In addition, an alternative school of thinking promotes diversification of diets to become more nutritious through increased consumption of e.g., legumes, vegetables, and fruits.

Small land holders face many challenges in meeting food security, which include small land size, unaffordable inputs, unavailability of services (e.g., extension and credit), shortage of labor, lack of market access and low farm gate prices. There is also high risk and uncertainty such as climate change, as well as social and gender inequality (Leach et al. 2020; Giller et al. 2021). These challenges need to be urgently addressed.

This chapter introduces the issues pertaining to food systems, soil, and water resources for crop production. It also addresses issues of nutrient-dense and biofortified food, dietary diversification, climate change impacts on food production and soil carbon sequestration, policy implementation on smallholder's farms; and it concludes by discussing ways of bringing the Green Revolution and agroecological approaches to farmers in Africa.

2 Plant Food Systems and Food Demands

As of 2023, SSA is not on course to meeting the Sustainable Development Goal (SDG) 2 targets to end hunger and ensure access by all people to safe, nutritious, and sufficient food all year round; and the need to end all forms of malnutrition remains high (FAO 2023). Plant food crops, mostly cereal grains, roots, and tubers, form the staple food of the SSA people. African countries have not succeeded in providing access to a healthy and adequate diet for all. Fruits, vegetables, plant oils and animal proteins, although available, are unaffordable to the 430 million Africans who live in extreme poverty, and produce their own food crops or buy food from the local open markets, where they face relatively high food costs in comparison to other regions in the world of a similar level of development (FAO 2023). The task of eradicating hunger is dependent on ensuring that bottlenecks in the food system are addressed to facilitate affordability and availability of a variety of nutritious, quality, and safe foods for all people irrespective of their location i.e., rural, peri-urban, or urban. In SSA, the agro-food system, from food production, processing, distribution, and procurement to consumer behavior is anchored by a diversity of small and large farmers, artisanal processors, and small and medium enterprise (Sirdas et al. 2021).

The traditional rural–urban dichotomy does not exist any longer; it has been replaced with a rural–urban continuum, characterized by a growing connectivity and interlinkage across urban, peri-urban, and rural areas. With almost 70% of the population in SSA projected to live in cities by 2050, urbanization is driving changes in agrifood systems along the rural–urban continuum, with increasingly more people relying on the market as their source of food (FAO et al. 2023). Furthermore, the shift is accompanied by increasing demand for cheaper, convenient, and processed foods, some of which are energy dense, high in fats, sugars and/or salt, and can contribute to the triple burden of malnutrition—where undernutrition, overnutrition and micronutrient deficiencies coexist in the population. These changing food preferences present an opportunity for innovations for healthy food product development, food preservation and technologies that boost the capacity of urban and peri urban agriculture to adapt to climate change mediated draughts and the declining agricultural lands due to expansion of towns and cities (Ruhweza 2020). The urbanization transition also means that the food system will have to develop by creating formal and complex food value chains that expand income-generating activities in off-farm employment for women and youth and increase the variety of health and nutritious food products, especially from legumes, oil seeds, fruits, and vegetables.

Nevertheless, it is important to recognize that with the projected urbanization, supply chains are likely to elongate and consumer incomes to increase, and this will lead to a decline in consumption of home-grown food. Therefore, it is important to ensure that consumers and business actors are aware of the food safety and quality standard required in the food system. In addition, strengthening of the food safety and quality assurance regulatory system is also irreplaceable. The growth of informal food processing and food service sectors requires capacity building to ensure that all people have access to safe and quality food. Food safety and quality issues facing the expanding crop food system in sub–Saharan Africa include mycotoxins, production of horticultural produce in potentially polluted sites, inappropriate use of pesticides, nutrient, and aesthetic loss due to poor handling and storage, as well as food poisoning, among others.

Apart from addressing the supply side of the food system, it is also necessary to consider the issue of food loss and waste, both in quantity and quality. It is estimated that the amount of food loss in SSA is as high as 37% (Ruhweza 2020). Experts agree that most leguminous grains and cereals are lost during post-harvest handling and storage on-farm, while loss of fresh produce is incurred in processing, packaging, and distribution (Sheahana and Barrett 2017a). Poor distribution infrastructure, preservation and consumer behavior greatly contribute to these alarmingly high losses.

3 Soil and Water Resources for Plant Production

Soil and water are fundamental resources for livelihoods and food security in Africa. Sixty percent (60%) of people in SSA derive their livelihoods from soil and its resources (Motaroki et al. 2020). Soil and water resources therefore play a critical role in ensuring food security and human wellbeing. Africa remains the world's most food-insecure region, with almost one-fourth of people—over 282 million—being undernourished (FAO et al. 2023). Both degradation and unsustainable use of soil and water resources undermine the vital role these resources can play in agriculture production and rural development across the continent. About 22% of Africa's total land area is under forests and woodlands, while 66% is covered by arid zones and deserts. Arable land constitutes about 630 million(M) ha, representing about 50% of the world's arable land (Motaroki et al. 2020). Only 10.6% of African soils are generally of good quality and support the largest proportion of the population, while majority of soils are of poor quality, highly acidic and frequently degraded and waterlogged, on which more than 200 M people depend (Motaroki et al. 2020).

4 Promoting Nutrition-Sensitive Agriculture

Nutrition-specific interventions or programs are those that address the immediate determinants of fetal and child nutrition and development—adequate food and nutrient intake, feeding, caregiving and parenting practices, and low burden of infectious diseases (Ruel et al. 2018). It requires a new way of thinking, planning, implementing, and partnering, as well as the active engagement of a variety of stakeholders from multiple sectors. It also requires identifying critical entry points where nutrition goals can be incorporated into agro-food systems (Jaenicke and Virchow 2013). Nutrition-sensitive agricultural practices include biofortification, livestock transfer programs, value chains for nutritious foods, and irrigation programs. Biofortification, the breeding of staple crops that are richer in essential micronutrients than traditional varieties, has been shown to be a feasible and cost-effective approach to addressing deficiencies in vitamin A, Fe, and Zn (Bouis and Saltzman 2017). At the same time, agronomic biofortification (i.e., use of mineral fertilizers through soil and foliar application to increase the content and bioaccessibility of specific nutrients in food crops) has also proven successful in many countries (Cakmak and Kutman 2018). Diversification of diet is another way through which crop production may contribute towards better nutrition and health e.g., through producing and consuming more legumes and vegetables, including neglected and underutilized crops (Haug et al. 2023).

The Harvest Plus Program initiated in 2003 was aimed at improving the nutritional value of food crops through biofortification, especially in Asia and African countries, to ensure the availability and accessibility of high-quality biofortified varieties of staples and the bioavailability of nutrients after consumption (Bouis 2018). Common examples of biofortification include golden rice, yellow and orange maize biofortified

Table 1 Biofortified crops

Biofortified	Crop Target micronutrient	Countries where crop has been tested	References
Cassava	Vitamin A	Nigeria, Democratic Republic of Congo	Talsma et al. (2016)
Beans	Iron	Uganda, Zimbabwe	Finkelstein et al. (2017)
Maize	Vitamin A	Nigeria, Democratic Republic of Congo	Gannon et al. (2014)
Wheat	Zinc	India	Signirell et al. (2019)
Rice	Zinc	Bangladesh	Sanjeeva et al. (2020)

Modified and adapted from Lowe (2021)

with vitamin A, Zn and Fe biofortified-rice and wheat, and beans (Lowe 2021). Some examples of plant-based biofortified food crops are presented in Table 1

5 Climate Change, Food Production and Soil Carbon Sequestration

World soils, the major reservoir of C among the terrestrial pools, contain two types of C: soil organic carbon (SOC) and soil inorganic C (SIC). To 1-m depth, world soils store 1550 Pg of SOC and 950 Pg of SIC stock (Lal 2018b). Thus, total soil C stock of 2500 Pg to 1-m depth is about 4 times that of the biotic stock (620 Pg) and 3 times that of the atmospheric stock (880 Pg). The atmospheric stock is increasing at the rate of about 5 Pg C /yr because of the ever-increasing emissions from fossil fuel combustion and land use conversion. Since the on-set of agriculture about 10–12 millennia ago, global soil C stock may have decreased by about 135 Pg because of deforestation, land use conversion from native to managed ecosystems and emission of greenhouse gases (GHGs) such as CO_2, CH_4, and N_2O (Lal et al. 2018). The magnitude of SOC depletion is aggravated by soil degradation such as that by wind and water erosion, salinization, acidification, elemental imbalance, loss of soil biodiversity etc. Therefore, degraded soils of agroecosystems have lower C stocks than their counterparts under natural ecosystems. This historic loss of soil C stock from the agroecosystem has created soil C sink capacity that can be refilled by conversion to a restorative land use and adoption of recommended management practices (RMPs) for soil specific conditions. Important among these are system-based conservation agriculture (Lal 2015a) and growing of a cover crop during the off-season (Lal 2015a, b). These and other RMPs can also advance the goals of 4 per Thousand initiative launched at COP21 in Paris (Lal 2019).

The process of re-carbonization of soil through adoption of RMPs is called soil C sequestration. Similar to two components of soil C stock, C sequestration also involves that of SOC and SIC components. The strategy of SOC sequestration is to create a positive soil/ecosystem C budget through increasing the input of biomass-C into the soil, at a rate which exceeds that of its depletion by pedological processes such as erosion, decomposition and leaching (Lal 2004).The process of SIC sequestration on the other hand works through formation of secondary carbonates and leaching of bicarbonates (Monger et al. 2015).The rate of SOC sequestration varies according to climate, soil type, land use, and farming/cropping systems and their management, and may range from negative (depletion) to as much as 1.5 Mg C/ha yr. And the process of SIC sequestration may be biotic and abiotic, and varies according to climate, soil and farming systems.

In addition to adaptation and mitigation of anthropogenic climate change, re-carbonization of world soils has numerous additional benefits. Important among these are: increase in agronomic productivity, improvement in nutritional quality of food, increase in renewability and quality of water resources, improvement in activity and species diversity of soil organisms, and advancement of the Sustainable Development Goals (SDGs) of the U.N. Agenda 2030 (U.N. 2015).

The technical potential of SOC sequestration in world soils is about 2.5 Pg C/yr or ~ 25% of the emissions from combustion of fossil fuel (Lal 2018a). With identification of non-carbon fuel sources, sequestration of SOC and SIC is a promising option to create a drawdown of atmospheric CO_2 and keep global warming beneath 2^0C. However, there is a need for development and implementation of the Soil Health Act, which has the provision to incentivize farmers through payments for ecosystem services for adoption of RMPs.

6 Implementation of Policies on Smallholder's Farms

Small land holder men and women farmers in SSA face numerous challenges in securing sufficient production and income from the farm to provide for themselves and their families. These challenges include small land area, unaffordable inputs, unavailability of services like extension and credit, shortage of labor, lack of access to markets, low farm gate prices, high risk and uncertainty such as climate change, and social and gender inequality (Leach et al. 2020; Giller et al. 2021; Diao et al. 2021). At the national levels, policies are made to address these challenges such as redistribution of land, subsidizing inputs such as seed and fertilizer, strengthening agricultural advisory services, establishing saving and credit groups or micro finance mechanisms, mechanization efforts, floor prices to enhance farm gate prices, insurance schemes to deal with risks and failures, climate adaptation measures, and efforts towards promoting gender equality. However, setting goals and approving policies is no guarantee for achieving the indicated outputs and results (Badiane et al. 2018; Haug et al. 2021).

Food systems analysis, as well as the efforts towards transforming current food systems to become more sustainable and equitable, are a global priority spearheaded by the U.N. (UNFSS 2021). Small land holder men and women farmers are important actors in African food systems. According to Béné et al. (2019), four different narratives explain the areas in which food systems are failing; namely closing the yield gap, improving nutrition, reducing inequality and controlling climate change. In 2021, UN organized the first *Food System Summit* (UNFSS) with the main objective of transforming food systems in accordance with the SDGs (UNFSS 2021). UNFSS (2021) and *UNFSS + 2 stocktaking* came up with country pathways and reporting systems with regard to transforming national food systems in relation to the four action areas below, and including an additional fifth action, namely *means of implementation* (UNFSS-Secretary General 2021; UNFSS + 2 2023):

- Nourish All People
- Boost Nature-Based Solutions of Production
- Advance Equitable Livelihoods and Decent Work and Empowered Communities
- Build Resilience to Vulnerabilities and Shocks and Stresses
- Means of implementation

Means of implementation is an important part of the national pathways' stocktaking process focusing on implementation, and is expected to aid results regarding the need for transformation. The most frequently mention *means of implementation* identified in the country pathways stocktaking UNFSS + 2 is *policy and regulations,* followed by innovation and knowledge, human resources capacities and finance and investment (UNFSS + 2 2023). However, although changing the policy and revising regulatory frameworks might be an appropriate means of implementation, the challenge still exists as to how to make sure that there will be successful implementation of the new policy and regulations.

Getting the policy right to ensure conducive conditions that enable smallholder men and women farmers to perform well, i.e. by moving out of poverty and adapting to climate change, have been on African governments and development agencies' agenda for a long time. However, the so-called "right policy" might have limited impact at smallholder farm level as implementing policies demands political will, efficient governance including ability to enforce rules, as well as sufficient financial, institutional, and human resources. Succeeding in reaching what is often called the last miles of smallholder men and women farmers are a tall task for governments in African countries where most of the population is rural. Implementing policies is difficult, but there are some success stories of policies being implemented and smallholder men and women farmers being reached e.g., by subsidized inputs, relevant agricultural advice and improved market access in ways that increase smallholders' production and income. Still, there is the challenge to move beyond policy formulation and decisions to a more dynamic implementation that makes a fundamental difference on-farm for men and women smallholders in Africa.

7 Bringing Green Revolution and Agroecological Approaches to Farmers

The Green Revolution (GR) of the 1960s by-passed SSA because soil restoration and its sustainable management were not part of the original package, which was based on use of in-put responsive varieties grown with high use of chemical fertilizers on land equipped for irrigation. In the meanwhile, soil degradation in SSA has been an important factor of malnutrition (Lal 2009). Therefore, improvement and sustainable management of soil health is critical to improving agronomic productivity. Soils of SSA are prone to accelerated erosion by water and wind, soil crusting and compaction caused by decline in soil structure, acidification aggravated by an indiscriminate use of chemical fertilizers (e.g., ammonium sulfate), salinization through irrigation by poor quality of water, and elemental imbalance by none or inappropriate use of fertilizers (only N and no or little P, K, Ca, Mg, and micronutrients).

The present state of soils of agroecosystems necessitates an urgent action plan. The rate of fertilizer use in SSA is less than 20 kg/ha compared with the global average use of 135 kg/ha (Sheahan and Barrett 2017b). On the whole continent, including Northern Africa, only about 13 million hectares of arable land are under irrigation, which is equal to 6 per cent of the total cultivated area (compared to 37 per cent in Asia, 14 per cent in Latin America) (Scheumann et al. 2017). Thus, agronomic yield of cereals in SSA is about 1.5 Mg/ha compared with the global average of 4 Mg/ha. The large yield gap in SSA requires policies which are pro-farmer and pro agriculture so that farmers are encouraged to adopt modern and proven innovations for sustainable management of soil resources.

There exist several plans for how African countries could improve soil health in agroecosystems. Obviously, such plans need to take the various national and local contexts into account. Below, we include an example of a possible plan that is suggested by Lal (2019):

1. Soil health/quality map at 1:10,000 scale with regards to key properties such as bulk density, texture, aggregation, plant available water holding capacity, pH, CEC, C, N, P, K.
2. Increase land area equipped for irrigation by 5% of the cropland area every 5 years, with specific targets for 2025, 2030, 2035, 2040, 2045 and 2050.
3. Increase rate of application of fertilizer use by 25 kg/ha every 5 years, so that the rate (kg/ha) increases from 20 now to 25 by 2025, 40 by 2030, 50 by 2035, 70 by 2040,90 by 2045 and 100 by 2050.
4. Land area under conservation agriculture must be increased from 1 M ha at present to 50 M ha by 2050.

To implement these plans, there is a need for implementation of the Soil Health Act at national and continental levels. The Soil Health Act must have the provisions to reward farmers for ecosystem services. Payment to farmers for sequestration of carbon in land-based sinks (soil and trees) should be made at the rate of US$50/ credit.

Implementation of these recommendations is critical to advancing SDGs of the Agenda 2030 of the United Nations (UN 2015), improving human health and well-being, and restoring degraded soils and desertified ecosystems. According to UN (2023), only around 15% of the SDGs are expected to be met by 2030.

The Declaration of COP28 in Dubai, with clear focus on agriculture and soil health, may bring attention to Africa and the need to restore and sustainably manage soil health. The COP 28 Declaration was signed by 159 nations, covering about 80% of world's land, with clear focus on sustainable agriculture, resilient food systems and climate action through management of soil health. The COP28 Declaration is widely viewed as good news for the world's small landholder farmers.

8 Conclusions

The soil-centric Green Revolution is long overdue in SSA. The COP28 Declaration, which focused on agriculture with specific attention to soil health and its management, is good news for 500 million small land holder farmers, many of whom are in SSA. Focus on traditional crops (e.g. sorghum, millet, teff, root crops) rather than rice and wheat may also help small land holder farmers. Traditional crops and the small holder family farms can play a critical role in achieving food security and advancing Sustainable Development Goals of the U.N. Agenda 2030. Implementation of the COP28 Declaration and whatever may emerge from COP30 in Brazil may require additional financial support to reward farmers for strengthening of ecosystem services such as carbon sequestration in land-based sinks. Protection and enhancement of soil health through re-carbonization of depleted soils of the agro-ecosystems may enable farmers to adapt better to anthropogenic climate change. Strengthening of extension services may also be important to in translation of proven science into action in SSA.

References

Badiane O, Collins J, Dimaranan B, Ulimwengu J (2018) An assessment of the new alliance for food security and nutrition; African Union: Addis Ababa, Ethiopia. Available online: https://au.int/sites/default/files/documents/34472-doc-nafsn20full20report20with20annexes.pdf.

Béné C, Oosterveer P, Lamotte L, Brouwer ID, de Haan S, Prager SD, Talsma EF, Khoury N (2019) When food systems meet sustainability—current narratives and implications for actions. World Dev 113:116–130. https://doi.org/10.1016/j.worlddev.2018.08.011

Bouis HE, Saltzman A (2017) Improving nutrition through biofortification: a review of evidence from Harvest Plus, 2003 through 2016. Glob Food Sec 12:49–58. https://doi.org/10.1016/j.gfs.2017.01.009

Bouis H (2018) Reducing mineral and vitamin deficiencies through biofortification: progress under harvest plus. World Rev Nutr Diet 118:112–122. https://doi.org/10.1159/000484342

Cakmak I, Kutman UB (2018) Agronomic biofortification of cereals with zinc: a review. Eur J Soil Sci 69:172–180

Diao X, Reardon T, Kennedy A, DeFries RS, Koo J, Minten B, Takeshima H, Thornton P (2021) The future of small farms: innovations for inclusive transformation, food systems summit brief prepared by research partners of the scientific group for the food systems summit. Bonn: Center for Development Research (ZEF) in cooperation with the Scientific Group for the UN Food System Summit. https://doi.org/10.48565/scfss2021-qq40
FAO (2017) The future of food and agriculture–trends and challenges. Rome. ISBN 978–92–5–109551–5
FAO, IFAD, UNICEF, WFP and WHO (2023) The State of Food Security and Nutrition in the World 2023. Urbanization, agrifood systems transformation and healthy diets across the rural–urban continuum. Rome, FAO. https://doi.org/10.4060/cc3017en
Finkelstein J, Haas J, Mehta S (2017) Iron-biofortified staple food crops for improving iron status: a review of the current evidence. Curr Opin Biotechnol 44:138–145. https://doi.org/10.1016/j.copbio.2017.01.003
Gannon B, Kaliwile C, Arscott SA, Schmaelzle S, Chileshe J, Kalungwana N et al (2014) Biofortified orange maize is as efficacious as a vitamin A supplement. in Zambian children even in the presence of high liver reserves of vitamin A: a community-based, randomized placebo-controlled trial. Am J Clin Nutr 100:1541–1550. https://doi.org/10.3945/ajcn.114.087379
Giller KE, Delaune T, Silva JV, van Wijk M, Hammond J, Descheemaeker K, van de Ven G, Schut AGT, Taylya G, Chihowo R, Anderson JA (2021) Small farms and development in sub-Saharan Africa: farming for food, for income or for lack of better options? Food Secur 13:1431–1454. https://doi.org/10.1007/s12571-021-01209-0
Haug R, Hella JP, Mulesa TH, Kakwera MN, Westengen OT (2023) Seed systems development to navigate multiple expectations in Ethiopia, Malawi and Tanzania. World Develop Sustain (Elsevier). 3. https://doi.org/10.1016/j.wds.2023.100092
Haug R, Nchimbi-Msolla S, Murage A, Moeletsi M, Magalasi M, Mutimura M, Hundessa F, Cacchiarelli L, Westengen OT (2021) From policy promises to result through innovation in african agriculture? World 2(2):253–266. https://doi.org/10.3390/world2020016
Jaenicke H, Virchow D (2013) Entry points into a nutrition-sensitive agriculture. Food Secur 5:679–692. https://doi.org/10.1007/s12571-013-0293-5
Jha AB, Warkentin TD (2020) Biofortification of pulse crops: status and future perspectives. Plants 9:73. https://doi.org/10.3390/plants9010073
Lal, R (2004) Soil carbon sequestration impacts on global climate change and food security. Science 304 (80):1623–1627. https://doi.org/10.1126/science.1097396
Lal R (2009) Soil degradation as a reason for inadequate human nutrition. Food Secur 1(1):45–57. https://doi.org/10.1007/s12571-009-0009-z
Lal R (2015a) Sequestering carbon and increasing productivity by conservation agriculture. J Soil Water Conserv 70(3):55A-62A. https://doi.org/10.2489/jswc.70.3.55A
Lal R (2015b) Soil carbon sequestration and aggregation by cover cropping. J Soil Water Conservation 70(6):329–339. https://doi.org/10.2489/jswc.70.6.329
Lal R (2018a) Saving global land resources by enhancing coefficient of agroecosystems. J Soil Water Conserv 73(4):100A–105A. https://doi.org/10.2489/jswc.73.4.105A
Lal R, Smith P, Jungkunst HF, Mitsch WJ, Lehmann J, Nai PKR, McBratney AB, Sá JCdM, Schneider J, Yuri LZ, Alba Z, Skorupa LA, Zhang HL, Minasny B, Srinivasrao C, Ravindranath NH (2018b) The carbon sequestration potential of terrestrial ecosystems. Soil Water Conserv 73(6):145A-152A. https://doi.org/10.2489/jswc.73.6.145A
Lal R (2019) Promoting "4 Per Thousand" and "Adapting African Agriculture" by south-south cooperation: Conservation agriculture and sustainable intensification. Soil Tillage Res 188:27–34. https://doi.org/10.1016/j.still.2017.12.015
Leach M, Nisbett N, Cabral L, Harris J, Hossain N, Thompson J (2020) Food politics and development. World Dev 134(2020):1–19. https://doi.org/10.1016/j.worlddev.2020.105024
Lowe NM (2021) The global challenge of hidden hunger: perspectives from the field. Proc Nutr Soc 80:283–289. https://doi.org/10.1017/S0029665121000902

Marques E, Darby HM, Kraft J (2021) Benefits and limitations of non-transgenic micronutrient biofortification approaches. Agronomy 11:464. https://doi.org/10.3390/agronomy11030464
Monger HC, Kraimer RA, Khresat S, Cole DR, Wang X, Want J (2015) Sequestration of inorganic carbon in soil and groundwater. Geology 43(5):375–378. https://doi.org/10.1130/G36449.1
Motaroki L, Omollo E, Karim A, Joshua M, Wamukoya G (2020) Sustainable land and water management under a changing climate to ensure food security in Africa. Policy Brief No. 5. AGNES (African Group of Negotiators Experts Support). www.agnes-africa.org
Nezomba H, Mtambanengwe F, Tittonell P, Mapfumo P (2017) Practical assessment of soil degradation on smallholder farmers' fields in Zimbabwe: integrating local knowledge and scientific diagnostic indicators. CATENA 156:216–227
Nkonya EM, Koo J, Kato E, Johnson T (2018) Climate risk management through sustainable land and water management in Sub-Saharan Africa. In: Leslie L, Nancy M, David Z, Solomon A, Giacomo B (eds) Climate smart agriculture: Building resilience to climate change, Part V: Case studies: Farm level response to improving adaptation and adaptive capacity. Food and Agricultural Organization of the United Nations (FAO) and Springer, pp 445–476
Ruel MT, Quisumbing AR, Balagamwala M (2018) Nutrition-sensitive agriculture: what have we learned so far? Glob Food Sec 17:128–153
Ruhweza A (2020) How can innovations in urban agriculture strengthen Africa's food systems and deliver benefits for nature and the environment? https://www.iwmi.cgiar.org/archive/cosai/news/how-can-innovations-urban-agriculture-strengthen-africa's-food-systems-and-deliver-benefits/index.html
Sanjeeva Rao D, Neeraja CN, Madhu Babu P, Nirmala B, Suman K, Rao LVS, Surekha, K, Raghu P, Longvah T, Surendra P, Kumar R, Babu VR, Volet SR (2020) Zinc biofortified rice varieties: challenges, possibilities, and progress in India. Front Nutr 7:26. https://doi.org/10.3389/fnut.2020.00026
Scheumann W, Houdret A, Bruntrup M (2017) Unlocking the irrigation potential in Sub-Sahara Africa: are public-private partnerships the way forward? Briefing Paper 7/2017. German Development Institute / Deutsches Institut für Entwicklungspolitik (DIE) Bonn, Germany
Sheahana, M, Barrett, CB (2017a) Review: food loss and waste in Sub-Saharan Africa. Food Policy 70:1–12. https://doi.org/10.1016/j.foodpol.2017.03.012
Sirdey N, Bricas N, Camara AD (2021) Food systems in sub saharan Africa: overview and specific features. Grain De Sel 81:6–7 https://www.inter-reseaux.org/wp-content/uploads/Inter-Reseaux-GDS81-GB-p6-7.pdf
Signorell C, Zimmermann MB, Cakmak I, Wegmüller R, Zeder C, Hurrell R et al (2019) Zinc absorption from agronomically biofortified wheat is similar to post-harvest fortified wheat and is a substantial source of bioavailable zinc in humans. J Nutr 149:840–846. https://doi.org/10.1093/jn/nxy328
Talsma EF, Brouwer ID, Verhoef H, Mbera GN, Mwangi AM, Demir AY, Maziya-Dixon, B,8 Boy E, Zimmermann MB, Melse-Boonstra A (2016) Biofortified yellow cassava and vitamin A status of Kenyan children: a randomized controlled trial. Am J Clin Nutr. 103:258–267. https://doi.org/10.3945/ajcn.114.100164
Tully K, Sullivan C, Weil R, Sanchez P (2015) The state of soil degradation in Sub-Saharan Africa: baselines, trajectories, and solutions. Sustainability 7:6523–6552
UNDP Sustainable Development Goals (2015) UNDP. https://www.undp.org/sustainable-development-goals.
UNFSS+2(2023) National Pathways Analysis Dashboard | UN Food Systems Coordination Hub (unfoodsystemshub.org.
UNFSS (2021) Food Systems Summit | United Nations
UNFSS-Secretary General (2021) Secretary-General's Chair Summary, Statement of Action on United Nations Food Systems Summit|UN Press
UN (2015) What are the sustainable development goals. United Nations Plaza, New York
U.N. (2023) Meeting coverage and press releases. https://press.un.org/en/2023/gashc4372.doc.htm

Were KO, Singh BR, Dick ØB (2015) Effects of land cover changes on soil organic carbon and total nitrogen stocks in the Eastern Mau Forest Reserve, Kenya. In: Lal R, Singh BR, Mwaseba DL, Kraybill D, D Hansen, Olav Eik L (eds) Sustainable intensification to advance food security and enhance climate resilience in Africa Springer New York, USA, pp 113–136

Open Access This chapter is licensed under the terms of the Creative Commons Attribution 4.0 International License (http://creativecommons.org/licenses/by/4.0/), which permits use, sharing, adaptation, distribution and reproduction in any medium or format, as long as you give appropriate credit to the original author(s) and the source, provide a link to the Creative Commons license and indicate if changes were made.

The images or other third party material in this chapter are included in the chapter's Creative Commons license, unless indicated otherwise in a credit line to the material. If material is not included in the chapter's Creative Commons license and your intended use is not permitted by statutory regulation or exceeds the permitted use, you will need to obtain permission directly from the copyright holder.

Plant Food Systems and Innovative Technologies

Integrated Nutrients Management: An Approach for Drought Management and Food Security Under Climate Change

R. S. Antil, R. P. Narwal, and B. R. Singh

Abstract The world is passing through a time when food and nutrition security has become a serious issue because of increasing population, increasing incidence of drought due to changing climate, shrinking natural resource base, and ever-increasing management constraints. The increasing food demands of a growing human population and the need for an environmentally friendly strategy for sustainable agricultural development require significant attention when addressing the issue of enhancing crop productivity. This chapter discusses the role of different components of integrated nutrient management (INM) in addressing drought and resolving these concerns, as INM is a promising strategy for addressing such challenges. The results of various long-term experiments (LTEs) conducted in Sub-Saharan Africa (SSA) and India have shown *that neither organic nor mineral fertilizers alone can achieve sustainability in crop production.* Strong and convincing evidence indicates that INM practice could be an innovative and environmentally friendly strategy for sustainable agriculture worldwide. INM holds great promise in meeting the growing nutrient demands of intensive agriculture. Comprehensive literature research has revealed that INM increases crop yields by 8–150% as compared with conventional practices; it also increases water and nutrients use efficiency and enhances economic returns to farmers. Inorganic fertilizers, organic manures, crop residues, legumes/green manure and bio-fertilizers are the main components of INM system. The use of INM increases crop productivity and grain quality, ensuring food security and environmental and soil quality in sustainable manner. INM is an approach that optimizes water use efficiency during drought, maintains soil health, plant nutrient availability

R. S. Antil
Amity Food and Agriculture Foundation, Amity University Uttar Pradesh, Noida, India
e-mail: rsantil@amity.edu

R. P. Narwal
Chaudhary Charan Singh Haryana Agricultural University, Hisar, Haryana, India

B. R. Singh (✉)
Faculty of Environmental Sciences and Natural Resource Management, Norwegian University of Life Sciences, Ås, Norway
e-mail: balram.singh@nmbu.no

© The Author(s) 2026

B. R. Singh et al. (eds.), *Sustainable Food Systems in Sub-Saharan Africa*,
https://doi.org/10.1007/978-3-032-11288-0_2

and crop productivity through optimization of the benefits from all possible sources of plant nutrients.

Keywords Crop productivity · Organic manures · Fertilizers · Soil health · Climate change

1 Introduction

The current world population is 8 billion (B) and is expected to reach 9.6 B by 2050 (https://www.un.org/2020). India is the most populous country in the world and its population is projected to reach 1.66 B by 2050 (https://www.pewresearch.org). The population of SSA was 179 M in 1950, 1.09 B in 2020 and may reach 2.12 B by 2050 and 3.78 B by 2100 (https://www.un.org/2020). More than 60% of the population of SSA is engaged in agriculture, dominated by smallholders and family farmers (< 2 ha), and they constitute approximately 80% of all farms in SSA and employ about 175 million people directly. As population increases, there is pressure on the existing cultivable lands to produce more food to meet increasing demands. However, continuous cultivation of land results in reduction of inherent soil nutrients. India is one of the major players in the agriculture sector worldwide and it is the primary source of livelihood for about 58% of India's population. India also has the world's largest area planted to wheat, rice, and cotton, and is the second-largest producer of fruit, vegetables, tea, farmed fish, cotton, sugarcane, wheat, rice, cotton, and sugar in the world (https://www.fao.org›india›india-at-a-glance). But India faces several challenges related to agriculture and food security that include fragmentation of land-holdings, degradation of resource quality, increasing incidence of drought and water scarcity due to rapid change of climate and its induced adverse impacts, large scale migration of agriculturists to non-farming sectors, etc. The food grain production of India has increased from 51 (1950–51) to 323 million tons (Mt) (2022–2023). A six-fold increase in food grain production during the last 72 years, combined with inadequate and imbalanced use of nutrients, has led to extra mining of all the essential nutrients, which undermines the soil's health. And yet, soil health is important to ensure food and nutritional security of the country. For efficient use of fertilizers, all nutrients must be applied in balanced proportions. The NPK consumption ratio (2022–2023) of India was 7.7:3.1:1 against ideal ratio of 4:2:1. The problem of imbalanced fertilizer use is worse in the case of secondary and micronutrients, wherein the use is much less compared to crop requirement. Hence, there is a need to narrow down the NPK consumption ratio to sustain crop productivity and soil health. The Green Revolution of 1965–66 helped India, for the first time since Independence, to gain self-sufficiency in food production. This was achieved through the adaptation of modern technologies such as the use of high yielding varieties, seeds, mechanized farm tools, irrigation facilities, and use of pesticides and fertilizers. However, during the last decades, the compounded growth rates for production and productivity of major crops generally declined or stagnated compared to 1995s (Kumari et al. 2020;

Antil et al. 2022). Limited by soil degradation, yield increases from improved crop varieties are estimated at only 28% in Africa as compared to 88% in Asia. The average yield across Africa is half that of India and one-fifth of the yields in the US. The large yield gap could be caused by a decrease in the amount of organic matter (OM) in the soil, an uneven fertilizer application, excessive nutrient mining, and a lack of secondary and micronutrients (Antil et al. 2022). Antil and Raj (2020) reported that farmers have to add larger amounts of fertilizer each year to maintain the same level of yield as the year before. The use of organic manures along with chemical fertilizers may be an effective alternative approach for further improving crop yields and sustaining soil health (Antil et al. 2022). Thus, the objective of this review article is to examine the concepts and principles of INM, and to assess its role in improving agricultural productivity by management of drought on the one hand, but also in improving crop quality and soil health under changing climate on the other.

2 Overview of INM

2.1 Nutrient Use Efficiency (NUE)

The persistent decline in NUE, soil fertility status, and environmental quality are the key constraints that get in the way of achieving sustainability in cropland soils of Asia and SSA. The NUE (yield per unit fertilizer input or in terms of recovery of fertilizer applied) is also relatively low in soils of India and SSA. In soils of India, NUE varies from 30 to 50% for N, 15 to 20% for P, 60 to 70% for K, 8 to 10% for S, and 1 to 2% for micronutrients. The crop response to fertilizers has been declining continuously. The average crop response to fertilizer application (kg grain produced per kg of NPK applied) decreased from 12.1:1 (during the 1960s) to 3:1 (in the 2020s). With the decrease in crop response to fertilizers, the profitability of the farmers has declined year after year. Current fertilizer application rates in SSA average only about 16 kg ha^{-1} $year^{-1}$, compared with over 100 kg ha^{-1} in Europe and North America, and over 150 kg ha^{-1} in China and other regions of Asia. The average NUE in SSA is about 10% (which is very low), whereas it has reached 22.2% in Asia. It has been demonstrated that the efficiencies of nutrient vary between 24 and 38% for total N, 34 and 38% for P and 18 and 34% for K in SSA (Tittonell et al. 2010).

2.2 Fertility Status of Soils

In many ecoregions of India and SSA, inadequate and unbalanced fertilizer use has resulted in widespread nutritional deficiencies and decline in soil health. The soils of

SSA are poor in nutrients, with low soil organic carbon (SOC), clay contents, and ion exchange capacity. About 80% of the agricultural land in SSA is degraded, mainly due to soil erosion, soil nutrient depletion, soil organic matter decline, and soil biodiversity loss, low nutrient application, and soil acidification (Zingore et al. 2015). The soils of SSA are largely unhealthy due to years of nutrient mining and limited input of organic or inorganic amendments. As such, crop yields have stagnated leading to high levels of food insecurity and poverty. Around 40% of cropland soils of Africa have low nutrient capital reserves, 25% have aluminum toxicity, 18% have high leaching potential, and 9% have high P-fixation capacity (Bado and Bationo 2018). A large proportion of soils of SSA (56%) are characterized as "acid infertile soils" (Oxisols and Ultisols in USDA Soil Taxonomy), 16% as "very infertile sandy soils" (Psamments), 12% as "poorly drained soils" (Aquepts), and 12% as "moderately fertile and well-drained soils" (Alfisols, Vertisols, Mollisols, Andepts, Tropepts, and Fluvents) (Bado and Bationo 2018). Similarly, soils of India are deficient in N, P, K, and S by 63, 42, 13, and 40%, respectively. Soils of India are also deficient in micro-nutrients such as Zn, Fe, Cu Mn, B and Mo by 37, 13, 4, 7, 23, and 13%, respectively (Shukla et al. 2018).

2.3 *Nutrient Balance in Soils*

Most soils in SSA and India record negative nutrient levels, and they continue to do so because crops remove significantly more nutrients than are added by manures and fertilizers. The farmers in SSA harvested about 4.4, 0.5, and 3.0 Mt ha^{-1} yr^{-1} of N, P and K; while returning only 0.8, 0.3, and 0.2 Mt, of these nutrients. Hence, soil nutrient mining demands five times nutrient application every year (Amede and Diallo 2022). In most farms in SSA, negative nutrient balances have been reported because the use of nutrients has been below 10 kg ha^{-1} (Gram et al. 2020), as compared to over 100 kg ha^{-1} in Europe, North America, and China (Rurinda et al. 2020). Nutrient balance sheets in most soils of India have been deficient and continue to be. This is primarily because nutrient removals by crops far exceed the nutrient additions through manures and fertilizers. For the past 50 years the gap between removals and additions has been estimated at 10 Mt NPK per year, and it is expected to rise to 15 Mt in 2022–23. Thus, nutrient removal may increase by a factor of two as compared to the present situation (Antil et al. 2022).

2.4 *Drought Management by INM*

Under recent climatic changes, both biotic and abiotic stresses are a serious threat for global food security and plant production sustainability. Among the abiotic stresses, drought stress is gaining attention due to its adverse effect on plant growth and development and significant reduction in plant yield and biomass causing global food insecurity. Certain physiological, metabolic, and biochemical processes are affected by drought stress, and this hampers plant productivity. Adequate nutrient supply under INM optimizes water use efficiency during drought and creates higher osmotic pressure in plant cells, which results in a better resistance to drought. To tackle the adverse effect of the drought stress on plants, INM can play an important role in enhancing drought tolerance (Singh et al. 2015).

3 Components of INM and Their Effect on Crop Productivity

3.1 Balanced Fertilization

Application of plant nutrients at an ideal ratio and in sufficient amounts is referred to as balanced fertilization. Balanced fertilization is the proper supply of all nutrients (macros and micros) throughout the growth cycle of a crop. As stated by Majumdar et al. (2013), balanced fertilization works based on 4R concept: the right fertilizer, right rate of usage, at the right time, and in the right place to manage fertilizer application and achieve stakeholder-set economic, social, and environmental objectives. The benefit of balanced fertilization is toincrease yield and crop quality, maintain soil fertility and improve the environment quality. At present, the consumption ratio of the NPK (2022–23) in India is about 7.7:3.1:1 against the ideal ratio of 4:2:1. The NPK use ratio is rather wide and application of K, S and micronutrients is often ignored. Balanced fertilization, using chemical fertilizers of macro and micronutrients, not only has the potential to improve crop yields, NUE and farmer profits but can also lead to positive impacts on environment footprint of fertilizer (Kumar et al. 2020). Imbalanced fertilizer application is one of the reasons for decline in soil health (Antil and Raj 2020). Singh et al. (2012a) evaluated the effect of balanced and imbalanced application of plant nutrients on crop productivity and soil health in rice–wheat cropping system and concluded that the rice and wheat grain yield decreased significantly in all the treatments where imbalanced application of fertilizer nutrient was made. The rate of rice and wheat yield decline was the lowest in control (0.055 Mg ha^{-1} yr^{-1} and 0.038 Mg ha^{-1} yr^{-1}) and the highest decline was observed in N_{150} $P_{75}K_{75}$ (0.134 Mg ha^{-1} yr^{-1} and 0.116 Mg ha^{-1} yr^{-1}) treatment, respectively. Results indicated that balanced fertilization with N, P and K doubled both maize and rice yields (i.e., from 2 to 4 Mg ha^{-1}) in Africa compared to that for the current practice. The main constraints in promoting balanced use of fertilizers include inadequate and ill-equipped soil testing facilities, lack of organic manures, inadequate extension services, wide gap in dissemination of knowledge between research institutions, and lack of awareness among farmers about the benefits of balanced fertilization.

3.2 *Use of Organic Resources (Organic Manures, Composts Etc.)*

3.2.1 Organic Manures

In the intensive cropping system, the sole use of inorganic fertilizers results in infertility as well as unfavorable soil physical and chemical conditions and biological processes. Soil health deteriorates year after year. Organic manures can be used in conjunction with fertilizers to combat soil degradation and maintain healthy crop growth. Organic manures are tried-and-true tools for boosting soil fertility and productivity. In addition to providing macro- and micronutrients, organic manures also assist in boosting the soil's physical, chemical, and biological qualities. It also helps to preserve the soil's C:N ratio. The use of organic manures in Indian agriculture has been declining steadily. Based on results of 29 years of LTEs, it could be concluded that the maximum grain yield of 4.7 Mg ha^{-1} in wheat was obtained when 100% recommended dose of fertilizers (RDF) was applied through mineral fertilizers after 50% N of RDF being substituted with farmyard manure (FYM) in rice. Grain yield of wheat declined under control and sub-optimal fertilizer inputs (50 or 75% RD of NPK fertilizer), whereas positive yield increment was observed under treatments receiving organic supplements (Kumari et al. 2017). A significant reduction in growth and yield attributes of basmati rice–wheat systems in Upper Indo-Gangetic Plains was recorded under natural farming (NF), as compared to integrated crop management (ICM) and organic management. Grain yield of basmati rice was reduced by 15.2, 51.2 and 53.9% under organic management, NF and control as compared to ICM, respectively. Similarly, 49.5, 58.6, and 63.8% reduction in grain yield of wheat was recorded under organic management, NF and control as compared to ICM, respectively (Ghasal et al. 2023). Zhou et al. (2022) found that addition of organic manures with chemical fertilizers increased yield of wheat by 26.4 to 44.6%, and that of maize by 12.5 to 40.8%, compared to those for recommended rate of fertilizers.

Ejigu et al. (2021) reported that the addition of compost did not boost maize yields compared to application of sole mineral fertilizers in north-western Ethiopia. Padbhushan et al. (2021) collected data from 1989–2018, based on different research studies in India, mainly on nutrient management in rice and wheat crops. The results showed that the yield of rice and wheat was 1.4 and 4.9% more in INM than that in 100% NPK, and that respective yields were comparatively higher in loamy soils (2.8%) and clayey soils (1.0%) than those in sandy soils. Most of the research results clearly demonstrated that INM enhances crop yields by 8–150%, and improves soil fertility as compared with conventional practices (Table 1).

Applications of inorganic and organic nutrients in combination, had a significant ($p < 0.001$ and $p < 0.01$) effect on grain yield of wheat under small holder farmers in SSA. Higher wheat grain yield and straw yields were obtained from the application of organic and inorganic plant nutrients under small holder farmers in SSA (GeremewBiramo 2018). Yield increases were over 100%, owing to improvement

Table 1 Integrated impact of nutrient management on soil fertility and crop productivity

Crops	Response to integrated nutrient management	References
Wheat–Maize	Significantly increased SOC and total N, enzymatic activities	Liang et al. (2014)
Maize	RDF + vermicompost (VC) enhance NPK availability and microbial activity	Kannan et al. (2013)
Pea–French bean	FYM level improved SOC, microbial colonization	Mahanta et al. (2013)
Wheat–Corn	Enhance SOC availability by FYM and municipal solid waste compost	Hemmata et al. (2010)
Chilli	Highest available NPK and micronutrients, higher yield with INM	Naidu et al. (2009)
Wheat	Increase the availability of OC, NPK	Jat et al. (2013)
Rice-Rice	Rice-Rice Significantly increased in NUE and SOM	Xu et al. 2008
Rice-Barley	Significant improvement in enzymatic activity	Liang et al. (2003)
Rice	Enhance available NPK in post-harvest soil	Cho et al. (2008)
Maize	Revealed that highest grain yield was recorded with 3/4th RDF + VC	Kumar et al. (2012)
Green	Available NPK and humic substances were higher with INM	Lakshmi et al. (2014)
Cotton	INM significantly increased NPK uptake and sustain soil fertility	Marimuthu et al. (2014)
Maize	VC + RDF enhances 100 seed weight, grain yield	Kannan et al. (2013)
Pea-Wheat	Significantly higher yield with manure and NP	Liu et al. (2013)
Cereal-Legume	Green manure with mineral fertilizer ensures higher crop productivity, soil fertility	Rahman et al. (2013)

in soil fertility (Table 2). Using organic fertilizers in combination with inorganic fertilizers also shortened the time to maturity, which is a good strategy to enable the plant to escape terminal moisture stress which occurs under rain-fed crop production systems. Thus, considering the poor soil fertility management by resource poor smallholder farmers and the high cost of mineral fertilizers, combined use of organic and mineral fertilizers at appropriate rates is central to enhance productive capacity of soil and to improve crop yield and its productivity.

An LTE was initiated in 1995 to study the effects of different combinations of organic manures (FYM, pressmud, poultry manure) and fertilizers on yield, soil fertility and C-Sequestration under pearl millet—wheat cropping system. The lowest grain yields of pearl millet and wheat were recorded when either 15 Mg FYM or 5 Mg poultry manure or 7.5 Mg pressmud ha^{-1} was applied alone. However, the crops' yield increased significantly when NP fertilizer was applied in conjunction with organic manures, and it was on par with the RD of applied NP alone

Table 2 Response of wheat grain yield and straw yield to integrated soil fertility management treatments. under small farmers in SSA

Treatments	Grain yield (Mg ha^{-1})	Straw yield (Mg ha^{-1})
Control	1.26e	2.39d
Recommended NP rate (60/20)	3.16a	4.51a
50% recommended NP rate (30/10) + 50% compost/ manure (3.25 Mg ha^{-1})	2.88b	4.19a
CD (p = 0.05)	0.23	0.56
CV (%)	9.4	15.4

*, *** = significant at $p < 0.05$ and $p < 0.001$, respectively;
NS, Not significant. Means in a column with the same letter are not significantly different ($p < 0.05$)
Source Extracted and modified from GeremewBiramo (2018)

or applied in conjunction with organic manures (Kumara et al. 2013). This trend in crop yield suggests that the amount of N released by organic manures was not enough to meet the remaining N requirement of the crop. The application of 15 Mg FYM ha^{-1} $year^{-1}$ along with RDF increased the productivity of diverse cropping systems (mustard (*Brassica juncea* L.)-sorghum (*Sorghum bicolor* L.) (fodder), mustard-cotton (*Gossypium arboretum* L.), wheat-cotton, rice (*Oryza sativa*)-wheat) compared to that for the application of RDF alone, according to the results of three years of field demonstrations in farmer's fields (Antil and Raj 2020).

3.2.2 Composts

Modern agriculture uses compost and mulch on annual crops, perennials, orchards, vineyards, and grasslands. Compost improves soil properties, provides nutrients in a stable organic form, increases plant growth and health, and conserves water. Antil et al. (2013) reported that grain and straw yields of wheat were significantly higher when compost made from the agro-industrial waste (sewage sludge, distillery effluent, press mud, and poultry waste composts) was used, as compared to their raw materials. When compared to unfertilized control, the yields of compost-fertilized wheat grain increased by 118% by using poultry waste compost, followed by press mud compost with the RD of NPK fertilizer. Apart from distillery effluent compost, composts from agro-industrial waste with RD of NK fertilizers produced wheat grain yields like those achieved with RD of NPK fertilizers, suggesting a net saving of 100% of P fertilizer (Antil et al. 2013). Application of enriched phospho-compost resulted in a 52% increase in groundnut pod yield as compared to ordinary compost applied in a groundnut-wheat cropping system (Antil and Narwal 2007). Kabato et al. (2022) reported that combined application of the NP fertilizer (55.2%:46%), along with 6 Mg ha^{-1} of compost, produced the highest (8.16 Mg ha^{-1}) wheat grain yield, increasing wheat yield by 27.45% over that obtained by the sole application of the chemical fertilizer because of significant improvement in the physico-chemical

Table 3 Effect of crop residue management in rice–wheat system on rice and wheat yield (pooled data of 4 years)

Yield	Grain yield under crop residue management practices (Mg ha^{-1})				CD ($p = 0.05$)
	Removed	Burned	Incorporated	Incorporated + green manuring	
Rice grain	2.2	2.2	3.4	3.6	0.072
Rice straw	3.2	3.2	4.5	4.4	0.048
Wheat grain	2.4	2.4	3.7	3.8	0.004
Wheat straw	5.2	5.2	7.9	8.1	0.037

*, *** = significant at $p < 0.05$ and $p < 0.001$, respectively
Source Extracted and modified from Singh et al. (2019)

properties of the soil. Therefore, integrated use of compost and fertilizer can be more efficient and sustainable for wheat productivity than using chemical fertilizers alone.

3.3 Recycling of Crop Residues

Crop residues are good sources of plant nutrients and are important components of INM. Crop residues, besides supplying nutrients to the current crop, also have a sustainable residual effect on crops succeeding in the system. About 30% N, 60–70% P and 75% K contained in crop residues can be available to the first crop, and the rest to subsequent crop (Sarkar et al. 2020). Thus, addition of crop residues can improve crop productivity, SOC content, NUE, soil physical properties and microbial and enzymatic activity (Antil and Raj 2020). Data from a 4-year study shown in Table 3 shows that the crop residue management practices involving incorporation of residues can significantly increase the grain yields of rice and wheat crop in the system over those in practices involving removal or burning of crop residues (Singh et al. 2019). The incorporation of crop residues with or without green manuring is promising for the environmentally friendly and effective utilization of crop residues under the rice–wheat system.

3.4 Legumes and Green Manure

Incorporation and decomposition of residues of legumes and green manure can solubilize N, P, K, and micronutrients (Zn, Mn, Fe, and Cu) in the soil, which mitigates nutrient deficiencies by recycling nutrients, decreasing leaching, and reducing gaseous losses of N, while also improving the effectiveness of applied plant fertilizers. Sunnhemp (*Crotolaria juncea*) and *dhaincha* (*Sesbania aculeata*) are the most important common green manure crops. When cultivated for grain or fodder in a cropping system or when added as green manure, legumes may be a crucial component of INM. Antil and Raj (2020) observed that the productivity of the rice–wheat cropping system was increased by using legumes as green manure, forage, or grain crops, which also rejuvenated soil fertility. In a rice–wheat cropping sequence, incorporation of mungbean (*Vigna radiata L.*) residues, after picking of the pods, significantly increased yield over that of the fallow treatment. Intercropping with legumes can significantly improve soil fertility, rhizobacteria community and diversity, and nutrient availability that are determinants of increased crop growth performance (Gogoi et al. 2018). Results from an All India Coordinated Research Project on Cropping Systems on Integrated Farming Systems indicated that about 25 to 50% N can be saved under rice–wheat, rice-rice, rice-maize, maize-wheat, pearl millet-wheat and sorghum-wheat cropping systems by growing moongbean as a catch crop (Chamkhi et al. 2022). Singh et al. (2012a) reported that application of 50% of RDF ($N_{75}P_{37.5}K_{37..5}Zn_{25}$) with *Sesbania* green manuring produced rice grain yield comparable to that obtained with 100% of RDF ($N_{150}P_{75}K_{75}Zn_{25}$), indicating a saving of approximately 50% of fertilizers.

3.5 *Bio-Fertilizers*

Use of biofertilizers is one of the important components of INM, as they are a cost effective and renewable source of plant nutrients to supplement the chemical fertilizers for sustainable agriculture. The use of biofertilizers (e.g., *Azotobacter*, *Azospirillum, Rhizobium* for N, and Phosphate and potassium solubilizing microorganisms, Vesicular–Arbuscular Mycorrhiza for other nutrients availability in crop cultivation) can improve crop yields and quality. Soil inoculation with *Azotobacter*, *Azospirillum,* and phosphate solubilizing bacteria produced crop yields by 5–10% over farmers' practice (Mahmud et al. 2021). In comparison to control and application of only FYM (@ 10 Mg ha^{-1}) + *Azotobactor spp.* + *Pseudomonas striata* + 50% RDF, the combined application of 20 Mg FYM ha^{-1} + 100% RDF + *Azotobactor* spp. + *Pseudomonas striata* produced significantly higher shelling percentage, protein content, oil yield, pod and haulm yield (Chaudhary et al. 2022). The results of various LTEs indicated that use of biofertilizers along with chemical fertilizers played a significant role in improving crop yield and soil health (Table 4).

3.6 *Improvement in Soil Water Retention and Drought Tolerance by INM*

Drought stress is the principal global threat to crop productivity. Approximately one third of the global cultivated area is prone to perpetual drought. INM significantly

Table 4 Integrated impact mineral fertilization and biofertilizer on crop performance and soil health

Crops	Response of biofertilizers with mineral fertilization	References
Apple	Biofertilizers play a significant role in the crop production, help to build up the lost micro flora and improve the crop yield and soil health	Mosa et al. (2015)
Corn	Bioinoculants with ½ RDF obtained a significant result in terms of ear length, ear weight and grain yield of corn	Sumagaysay (2014)
Lentil	FYM and biofertilizer improve soil health	Moraditochaee et al. (2014)
Soybean	RDF with biofertilizers resulted in improved productivity, soil fertility and nutrient balance	Gharpinde et al. (2014)
Review	Results indicate that liquid biofertilizer enhance and restore soil health	Pindi et al. (2012)
Sunflower	Sunflower yielded a significantly higher grain and biological yield with biofertilizer –N fertilization-FYM	Akbari et al. (2011)
Wheat	Significant response of biofertilizers on growth and crop productivity	Singh and Prasad (2011)
Mungbean	Significantly enhanced crop productivity and soil fertility status with bioinoculants and mineral fertilization	Rana et al. (2011)

improves the growth and yield of rainfed chickpea (*Cicer arietinum* L.) grown on sandy clay loam soil (Dewangan et al. 2017). The highest seed/stover yield was found under 100% RDF + 2% urea spray and significantly superior to 100% RDF + 25% N (VC), 100% RDF + 5 kg Zn ha^{-1} and 100% RDF (Table 5). This envisages a direct role of N to seed growth and a guided help in minimizing osmotic imbalance present during final stage in seed filling (Kachot et al. 2001). Among the drought mitigating practices, application of 2% KCl + 0.4% sodium selenite spray was found significantly superior to water spray (Table 6). According to Hussain et al. (2017), higher rates of application of K with Zn are needed to enhance maize production, increase net income, and alleviate the adverse effects of drought stress.

4 Effect of Components of INM on Soil Health

4.1 *Soil Physical Properties*

Among the different physical properties of soil, bulk density (BD) is a key metric for assessing soil health due to its correlations with other soil state (strength and porosity) and rate (moisture retention and flow characteristics) variables. Soil aggregation, a physical property related to soil structure, is significantly improved by the addition of organic materials. Application of NPK fertilizers in combination with organic manures significantly improved soil aggregation, soil water retention, micro porosity,

Table 5 Yield and economics of chickpea as influenced by integrated nutrients and drought mitigating practices

Treatment	Seed yield (Mg ha^{-1})	Stover yield (Mg ha^{-1})
INM		
100% RDF + 2% Urea spray	1.35	3.26
100% RDF + 25% RDN (VC)	1.32	3.22
100% RDF + 5 kg Zn ha^{-1}	1.22	3.02
CD ($p = 0.05$)	0.032	0.071
Control	1.17	2.90
Water spray	1.21	2.97
2% KCl + 0.4% sodium selenite	1.32	3.22
CD ($p = 0.05$)	0.026	0.061

Source Extracted and modified fromDewangan et al. (2017)

Table 6 Effect of crop residue management in rice–wheat system on soil properties (pooled data of 4 years)

Indicators	Crop residue management practices				CD (p = 0.05)
	Removed	Burned	Incorporated	Incorporated + green manuring	
BD (Mg m^{-3})	1.57	1.59	1.48	1.46	0.02
Infiltration rate (cm h^{-1})	0.32	0.32	0.38	0.41	0.01
Aggregate stability (%)	9	10	14	14	0.04

Source Extracted and modified from Singh et al. (2019)

and available water capacity and reduced BD of the soil in different cropping systems (maize-wheat-cowpea (fodder), rice–wheat) (Das et al. 2014; Antil and Raj 2020). The integration of 50% RDF, along with either 5 Mg ha^{-1} poultry manure or FYM or both, resulted in significant improvement of soil physical properties at par with the result where 100% RDF was used. Manured soils also showed lower BD and higher porosity, hydraulic conductivity (HC) and aggregate stability, relative to fertilized soils (Wailare and Kesarwani 2017).

The 4-year results revealed that the incorporation of crop residue (Table 6) significantly improved the soil properties such as infiltration rate, aggregate stability, and decreased BD in rice–wheat system over removal or burning of crop residue practices (Singh et al. 2019). Comprehensive literature research revealed integrated impact of nutrient management on soil physical properties (Table 7).

Table 7 Integrated impact of nutrient management on soil physical properties

Crops	Response to integrated nutrient management	References
Soybean	Significantly decreased in BD	Aziz et al. (2015)
Maize	Significantly improved total porosity, HC, soil moisture content	Nwite et al. (2014)
Wheat	RDF + FYM @ 5 Mg ha^{-1} significantly reduced BD	Kusro et al. (2014)
Maize	BD, particle density reduced and pore space %, significantly increased	Kannan et al. (2013)
Cotton-Wheat	Significantly improved in BD, total porosity, water stable aggregates (WSA), MWD	Hassan et al. (2013)
French bean	Integrated treatments improved BD, water holding capacity (WHC)	Datt et al. (2013)
Pea-Wheat	Soil moisture conserves, significantly increased WSA, mean weighted diameter (MWD)	Liu et al. (2013)
Soybean	BD significantly lowered with the application of FYM	Aziz et al. (2015)
Tomato	Soil temperature, BD significantly reduced	PA and Salami (2012)
Maize–Mustard	Significantly increased WSA, WHC, and decreased BD	Saha et al. (2010)
Wheat-Soybean	Significantly reduced BD, increased MWD, SOC	Bhattacharyya et al. (2007)
Rice–Wheat	Increased MWD, total porosity, WHC	Rasool et al. (2007)

4.2 Soil Chemical Properties

Several studies have shown that the application of organic manure in conjunction with fertilizers increased the SOC and available N contents and their fractions more effectively than the application of fertilizers alone (Meena et al. 2015). Application of 50% of RDF NP fertilizers for 16 years in pearl millet—wheat cropping system decreased the SOC content from its initial level. However, the application of organic manures alone or in combination with NP fertilizers for 16 years of pearl millet—wheat cropping system increased the SOC significantly. Application of organic manures with or without NP fertilizers could not sustain the initial level of N. However, soil fertility with respect to P, K, Zn, Mn, Fe and Cu can be maintained with the application of organic manures (15 Mg FYM or 7.5 Mg pressmud or 5 Mg poultry manure ha^{-1} yr^{-1}) in conjunction with RD of N under pearl millet—wheat cropping systems (Antil et al. 2011; Kumara et al. 2013).

The C sequestration rate showed a negative trend with greater magnitude (-18.8 kg ha^{-1} yr^{-1}) with the application of 75 kg N + 30 kg P_2O_5 ha^{-1} for 16 years, but it increased to 387.5 kg ha^{-1} yr^{-1} with 15 Mg FYM ha^{-1}. A combined application of fertilizers with FYM, poultry manure and pressmud for 16 years increased the C sequestration rate of soil to 456, 119. and 225 kg ha^{-1} yr^{-1}, respectively (Kumara et al. 2014). The increase in C sequestration rate of soil, with the addition of organic manures plus NP fertilizers, might be due to better crop growth with accompanying higher root biomass generation and higher return of plant residues on surface. The SOC stock was increased by 23.8% in rice, 15.1% in wheat, 25.3% in loamy soils, and 14.4% in clayey soils in INM over 100% NPK. INM increased SOC sequestration by 2.3 Mg ha^{-1} as compared to using 100% NPK (Padbhushan et al. 2021). Addition of compost increased soil pH, SOC, total N, available P, S, and exchangeable bases compared to application of sole mineral fertilizers in north-western Ethiopia (Ejigu et al. 2021). Conversely, mineral fertilizers addition alone aggravated soil acidity, depletion of SOC and soil nutrients although it provided higher maize yields over plots amended with sole compost and the control treatments. The 4-year results revealed that the incorporation of crop residues alone or in combination with green manure significantly improved the soil chemical properties such as electrical conductivity, SOC, available N, P, and K in the soil in the rice–wheat system over removal or burning of crop residue practices (Singh et al. 2019). The SOC content increased by 63 and 53%, respectively, under incorporation of crop residues with green manuring and incorporation of crop residues alone, in comparison to the practice of burning of crop residues (Table 8).

Integrated application of organic and inorganic fertilizers improves fertility status of the soil (GeremewBiramo 2018). Soils fertilized with manure had higher contents of OM and numbers of micro-fauna than soils fertilized with chemical fertilizers, and were more enriched in P, K, Ca and Mg in topsoil and nitrate N, Ca and Mg in sub soils in Sub-Saharan Africa (Tewodros and Belay 2015). Agegnehu et al. (2014) also reported that the soil pH, total N and SOM content, cation exchange capacity and exchangeable cations were significantly improved due to the application of FYM in SSA.

Application of FYM as a component of INM increased SOC content of the soil, and the increase ranged from 0.03 to 0.06% $year^{-1}$ under different cropping systems (mustard (*Brassica juncea* L.)-sorghum (*Sorghum bicolor* L.) (fodder),

Table 8 Effect of crop residue management in rice–wheat system on soil chemical properties (pooled data of 4 years)

Indicators	Crop residue management practices				CD ($p = 0.05$)
	Removed	Burned	Incorporated	Incorporated + green manuring	
pH (1:2.5)	6.7	6.7	6.8	6.8	NS
EC (d Sm^{-1})	0.20	0.21	0.27	0.28	0.01
OC (%)	0.40	0.38	0.58	0.62	0.14
Available N (Kg ha^{-1})	175	178	205	230	2.82
Available P (Kg ha^{-1})	20	18	32	34	0.46
Available K (Kg ha^{-1})	190	188	264	265	0.84

Source Extracted and modified from Singh et al. (2019)

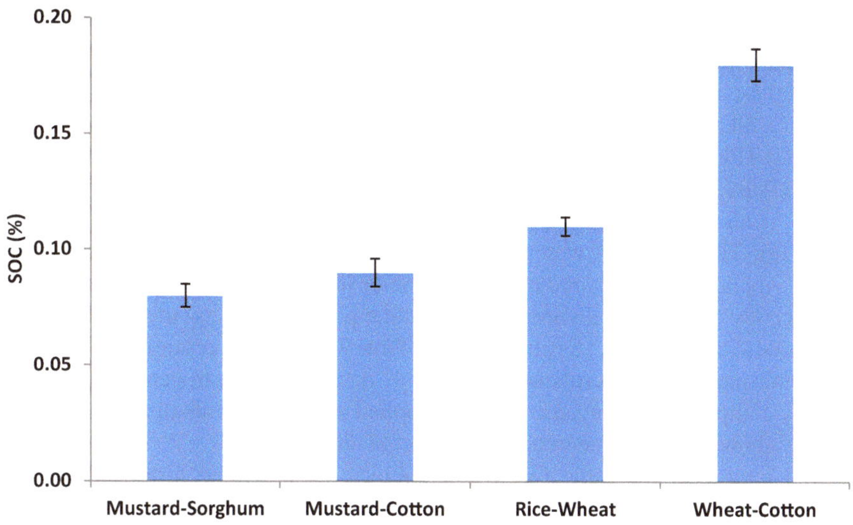

Fig. 1 Influence of FYM on buildup of SOC in soil as a component of INM under various cropping systems. Values are reported ± standard deviation. *Source* Extracted and modified from Antil and Raj (2020)

mustard-cotton (*Gossypium arboretum* L.*)*, wheat-cotton, rice–wheat). The cumulative increase in SOC content under different cropping systems (Mustard-sorghum (fodder), mustard-cotton, wheat-cotton, rice–wheat) as shown in Fig. 1.

Highest SOC, available N, P, K and Fe in soil were found under organic farming package and ICM and least under NF treatments. Lower SOC content under NF treatments indicated that the quantity of organic matter added through different inputs was not sufficient to sustain the microbial population added through different management practices (Ghasal et al. 2023). Improvement in SOC content and nutrient status under organic management and INM was also reported by Moe et al. (2019). Bijay-Singh (2018) did a meta-analysis of the data published on crop yields and soil parameters in different cropping systems (maize-wheat, rice-rice, and rice–wheat) from LTEs. They found that the buildup of SOC in upland maize-wheat system was less compared to the rice-based cropping systems, and that the application of optimum rate of N, P and K fertilizers resulted in build-up of SOC over no-fertilizer control in all the three cropping systems.

4.3 Soil Biological Properties

A knowledge of soil biological properties is important for maintaining soil quality, plant health, soil resilience, soil fertility and soil sustainability. Application of organic manure improves the microbial condition of soil (Verma et al. 2019). There are reports suggesting that combined use of organic manure with chemical fertilizers improves the population of microbes and enzymatic activities (Mallikarjun and Maity 2018). The highest microbial biomass C (218.2 μg g^{-1}), urease activity (221.9 μg urea g^{-1} h^{-1}), alkaline phosphatase (121.5 μg PNP g^{-1} h^{-1}) and dehydrogenase activity (44.5 μg TPF $g^{-1}d^{-1}$) recorded, with treatment, 70% RDF with 30% N by FYM and pressmud equally. Significantly lower value of urease activity (169.8 μg urea g^{-1} h^{-1}), alkaline phosphatase (81.83 μg PNP g^{-1} h^{-1}) and dehydrogenase activity (27.07 μg TPF $g^{-1}d^{-1}$) were observed with unfertilized control (Jat and Singh 2017). Application of organic manures alone or in combination with NP fertilizers significantly increased dehydrogenase, alkaline phosphatase and urease activity as compared to application of fertilizers alone (Table 9). However, combined application of organic manures and fertilizers significantly decreased dehydrogenase and urease activity compared with organic manures applied alone, while alkaline phosphatase activity showed the reverse trend (Sheoran et al. 2020).

The long-term inorganic fertilizer combined with organic fertilizer incorporation practices significantly enhanced soil sucrose (30.1–51.9%), urease (28.4–38.3%), and β-1,4-glucosidase (34.6–122.4%) activity. However, nitrite reductase, polyphenol oxidase, and catalase were significantly lower (27.3–49.9%, 8.5–26.3% and 23.3–34.3%) than single applications of inorganic N fertilizer groups. The organic manure addition improved soil fauna feeding activity by 35.2%–42.5%, and the excessive application of inorganic N fertilizer reduced the activity level of soil fauna (Zhou et al. 2022). Incorporation of residues significantly increased bacterial and fungal population, enzyme dehydrogenase and phosphatase activity in the soil in rice–wheat system over removal or burning of crop residue practices (Singh et al. 2019). Ghasal et al. (2023) reported that among the different management practices (organic farming package, ICM and NF), the highest population of bacteria, fungi and actinomycetes and glomalin content were found under organic management.

Table 9 Effect of long-term application of organic manures and fertilizers on soil enzymes activity

Treatment	Dose (Mg ha^{-1})	Fertilizer (kg ha^{-1})		Dehydrogenase activity (μg TPF g^{-1} 24 h^{-1})	Alkaline phosphatase activity (μg PNP g^{-1} h^{-1})	Urease activity (μgNH_4^+–N g^{-1} h^{-1})
		N	P_2O_5			
Fertilizer	0	150	60	32.3	581	65.0
FYM	15	0	0	63.7	684	87.7
	15	150	30	50.5	734	71.3
Poultry manure	5	0	0	48.6	705	85.2
	5	150	30	40.1	756	75.8
Pressmud	7.5	0	0	58.1	665	97.6
	7.5	150	30	39.4	673	71.3
CD (p = 0.05)				5.7	18	7.9

Source Extracted and modified from Sheoran et al. (2020)

5 Effect of INM Components on NUE and Nutritional Quality of Food Crops

Long-term studies indicated that application of organic manures along with fertilizers enhanced the use efficiency of N at different locations (Dwivedi et al. 2016). In Inceptisols (Ludhiana), the N use efficiency in maize recorded under 100% N (alone) was 16.7% which increased to 23.5, 36.4, 40.2% on integration with P, PK and FYM, respectively. A similar trend was noted in Mollisols (Pantnagar) under the rice–wheat system. INM led to an increase in use efficiency of P and K at different locations (Ludhiana, Palampur and Pantnagar) under the All India Coordinated Research Project on LTEs (Singh et al. 2012b).

Several factors can directly or indirectly affect the nutritional quality of crops. Application of organic manures (VC, FYM) alone or in combinations improved the quality of bread wheat, protein, starch, wet and dry gluten and gluten index of bread wheat, as compared to control (Ali et al. 2020). Pooled data of protein and nutrient content revealed that the organic manures applied alone resulted in lower protein, P and K contents in wheat grains. However, when organic manures were applied in combination with N or NP fertilizers, then only protein content improved significantly. The protein, P and K contents in wheat grain ranged from 15.0 to 15.9, 0.44 to 0.53 and 0.25 to 0.30%, respectively under different treatment combinations (Table 10). Among all treatments, the highest (15.9%) and the lowest (15.0%) protein content in wheat grain was observed with pressmud$_{7.5}$N$_{150}$P$_{30}$ and poultry manure$_5$, respectively (Sheoran et al. 2017). The INM increases economic returns to farmers compared with conventional practices, while improving grain quality and sustainability under small-holder farmers in SSA (Wei and Ma 2015).

Table 10 Effect of long-term application of organic manures and fertilizers on protein content in wheat grain (pooled data of 2013 and 2014 years)

Treatment	Dose (Mg ha^{-1})	Fertilizer (kg ha^{-1})		Protein (%)
		N	P_2O_5	
Fertilizer	0	150	60	15.6 cd
FYM	15	0	0	15.1a
	15	150	30	15.8d
Poultry manure	5	0	0	15.0a
	5	150	30	15.7d
Pressmud	7.5	0	0	15.ab
	7.5	150	30	15.9d
CD ($p = 0.05$)				0.38

Note: Different lower-case letters within columns indicate significance at $p < 0.05$
Source: Extracted and modified from Sheoran et al. (2017)

6 Constraints in Adoption of INM Technology

- Lack of knowledge: Farmers often have inadequate knowledge on the use of fertilizers in balanced proportions.
- Funding: Farmers lack access to credit, especially in rural areas. They have insufficient funds to buy manure and fertilizer that are important for INM.
- Land degradation: Degradation of lands due to intensive cropping and over-exploitation by the enormous pressure of the ever-increasing population is a big constraint to INM.
- Monsoon vagaries: Indian and SSA agriculture is dependent on monsoon. The risk of water deficit during drought prone periods is considered the most important deterrent to fertilizer use and to INM. During monsoon, water erosion is a serious threat to soil fertility and productivity.
- Limitation of small holdings: Land is fragmented in India and Africa, and most of the farmers hold small holdings. This prevents the application of INM at commercial level.
- Limitation of biofertilizer: There are several constraints to the effective utilization and popular use of biofertilizer e.g., use of the biofertilizer is crop and location specific, low shelf life of the microorganisms, need for careful handling etc. Moreover, poor-quality bio-fertilizers that reach the farmers are ineffective and their marketing becomes difficult because the products contain living or latent organisms.
- Lack of knowledge and poor advisory services.

7 Conclusions

To manage long term soil fertility, productivity, drought as well as environment quality, integration of organic and inorganic fertilizer can be the most sustainable practices to adopt. Several nutrient management practices have been attempted; however, current studies on LTEs indicate that INM was found to be the best strategy for agricultural sustainability, as well as for minimizing environmental pollution and the detrimental impacts of drought stress on plants. Use of organic, biofertilizers and green manures along with fertilizers has been found effective in improving and maintaining soil fertility, increasing NUE so maximizing productivity in different cropping systems. The success of INM depends on how precisely the different components of INM have been used. In addition, INM helps to increase the activity of soil microorganisms and soil C sequestration, and thus improves the soil's physical, chemical, and biological properties. Adequate nutrition delivery under INM optimizes water usage efficiency during droughts and increases osmotic pressure in plant cells, which results in a superior resilience to droughts. INM can be very helpful in boosting drought resistance and reducing the negative effects of drought stress on plants. Sprays of 2% KCl + 0.4% sodium selenite spray and potassium with zinc enhance crop productivity, increase net income and lessen the negative impacts of drought stress. The interest of agriculture specialists and farmers must be turned towards the technique of INM, which is an acceptable option, a cost-effective practice, and an eco-friendly approach. This can be easily followed by farmers and can result in reducing the need of fertilizers and in providing greater yield productivity with better crop quality. INM can therefore be used as a part of global strategy to ensure food security, protect the environment, and address drought under climate change.

References

Agegnehu G, vanBeek C, Bird MI (2014) Influence of integrated soil fertility management in wheat and tef productivity and soil chemical properties in the highland tropical environment. J Soil Sci Plant Nutr14(3):Temuco set 2014 Epub 02-Ago-2014. https://doi.org/10.4067/S0718-95162014005000042

Akbari P, Ghalavand A, Modarres SAM, Alikhani MA (2011) The effect of biofertilizers, nitrogen fertilizer and farmyard manure on grain yield and seed quality of sunflower (*Helianthus annus L.*). J Agric Sci Technol 7 (1):173–184

Ali N, Khan MN, Ashraf MS et al (2020) Influence of different organic manures and their combinations on productivity and quality of bread wheat. J Soil Sci Plant Nutr 20:1949–1960

Amede T, Diallo A (2022) Enhancing agronomic efficiency of fertilizers in SubSaharan Africa: evidence from the field. Grow Africa 1(1):18–22

Antil RS, Raj D (2020) Integrated nutrient management for sustainable crop production and improving soil health. In: Meena RS et al (eds) Nutrient dynamics for sustainable crop production, Springer Nature Singapore Pte Ltd. Singapore, pp 67–102. https://doi.org/10.1007/978-981-13-8660-23

Antil RS, Narwal RP (2007) Integrated nutrient management for sustainable soil health and crop productivity. Indian J Fert 3(9):111–112

Antil RS, Narwal RP, Singh B, Singh JP (2011) Long-term effects of FYM and N application on soil health and productivity under pearl millet–wheat cropping system. Indian J Fert 7:14–32

Antil RS, Raj D, Narwal RP, Singh JP (2013) Evaluation of maturity and stability parameters of composts prepared from a wide range of organic wastes and their response to wheat. Waste Biomass Valor 4:95–104

Antil RS, Singh BR, Narwal RP (2022) Raising soil organic matter to improve productivity and nutritional quality of food crops in India. In: Rattan Lal (ed) Soil organic matter and feeding the future: environmental and economic impacts, CRC Press/Taylor and Francis, Boca Raton, FL, pp 235–257. https://doi.org/10.1201/9781003102762-10

Aziz MA, Mushtaq T, Ali T, Islam T, Rai AP (2015) Effect of integrated nutrient management on soil physical properties using soybean (*Glycine max* (L.) Merill) as indicator crop under temperate conditions. J Enviorn Sci Comput Sci Engg Technol 4:56–64

Bado VB, Bationo A (2018) Integrated management of soil fertility and land resources in Sub-Saharan Africa: involving local communities. Adv Agron 150:1–33

Bhattacharyya R, Chandra S, Singh RD, Kundu S, Srivastva AK, Gupta HS (2007) Long term farmyard manure application effects on properties of a silty clay loam soil under irrigated wheat–soybean rotation. Soil Tillage Res 94:386–396

Bijay-Singh S (2018) Optimizing fertilizer use to maintain soil Health. In: Reicosky D (ed) Managing soil health for sustainable agriculture vol 2 Monitoring and management, Burleigh Dodds Science Publishing, Cambridge, UK, (ISBN: 9781786761927). www.bdspublishing.com

Chamkhi P, Cheto S, Geistlinger J, Zeroual Y, Kouisni L, Bargaz A, Ghoulam C (2022) Legume-based intercropping systems promote beneficial rhizobacterial community and crop yield under stressing conditions. Ind Crops Prod 183(9):114958

Chaudhary P, Singh S, Chaudhary A, Sharma A, Kumar G (2022) Overview of biofertilizers in crop production and stress management for sustainable agriculture. Front Plant Sci 13:930340. https://doi.org/10.3389/fpls.2022.930340

Cho BJY, Son JG, Song CH, Hwang SA, Lee YM, Jeong SY, Chung BY (2008) Integrated nutrient management for environmental-friendly rice production in salt-affected rice paddy fields of Saemangeum reclaimed land of South Korea. Paddy Water Environ 6:263–273

Das BD, Chakraborty VK, Singh P, Aggarwal R, Singh R, Dwivedi BS (2014) Effect of integrated nutrient management practice on soil aggregate properties, its stability and aggregate-associated carbon content in an intensive rice-wheat system. Soil Tillage Res 136:9–18

Datt N, Dubey YP, Chaudhary R (2013) Studies on impact of organic, inorganic and integrated use of nutrients on symbiotic parameters, yield, quality of french-bean (*Phaseolus vulgaris* l.) vis-à-vis soil properties of an acid Alfisol. African J Agric Res 8(22):2645–2654

Dewangan S, Singh RP, Singh MK, Singh S (2017) Effect of integrated nutrient management and drought mitigating practices on performance of rainfed chickpea (*Cicer arietinum*). Indian J Agric Sci 87(3):301–305

Dwivedi BS, Singh VK, Meena MC, Dey A, Datta SP (2016) Integrated nutrient management for enhancing nutrient use efficiency. Indian J Fert 12(4):62–71

Ejigu W, Selassie YG, Elias E, Damte M (2021). Integrated fertilizer application improves soil properties and maize (*Zea mays* L.) yield on Nitisols in Northwestern Ethiopia. Heliyon 7:e06074. https://doi.org/10.1016/j.heliyon.2021.e06074

Geremew B (2018) The role of integrated nutrient management system for improving crop yield and enhancing soil Fertility under small holder farmers in Sub-Saharan Africa: a review. Mod Con Develop Agronom 2(5):MCDA.000547. https://doi.org/10.31031/MCDA.2018.02.000547

Gharpinde B, Gabhane VV, Nagdeve MB, Sonune BA, Ganvir MM (2014) Effect of integrated nutrient management on soil fertility, nutrient balance, productivity and economics of soybean in an Inceptisol of semi-arid region of Maharashtra. Karnataka J Agric Sci 27(3):303–307

Ghasal PC, Mishra RP, Choudhary J, Dutta D, Bhanu C, Meena AL, Ravisankar N, Kumar A, Panwar AS (2023) Evaluation of integrated crop management, organic management and natural farming in basmati rice-wheat system under Upper Indo-Gangetic Plains. J Plant Nutr. https://doi.org/10.1080/01904167.2023.2209117

Gogoi N, Baruah KK, Meena RS (2018) Grain legumes: impact on soil health and agroecosystem. In: Meena RS et al (eds) Legumes for soil health and sustainable management, Springer Nature Singapore Pte Ltd. pp 511–530

Gram G, Roobroeck D, Pypers P, Six J, Merckx R, Vanlauwe B (2020) Combining organic and mineral fertilizers as a climate-smart integrated soil fertility management practice in sub-Saharan Africa: a meta-analysis. PLoS ONE 15(9):e0239552. https://doi.org/10.1371/journal.pone.0239552

Hassan M, Rafique E, Rashid A (2013) Physical and hydraulic properties of Aridisols as affected by nutrient and crop-residue management in a cotton-wheat system. Acta Scientiarum Agron 35(1):127–137

Hemmata A, Aghilinategh N, Rezainejad Y, Sadeghi M (2010) Long-term impacts of municipal solid waste compost, sewage sludge and farmyard manure application on organic carbon, bulk density and consistency limits of a calcareous soil in central Iran. Soil Tillage Res 108:43–50

Hussain S, Maqsood M, Lal R, Hussain M, Sarwar MA, Bashair M, Ullah A, Haq IU (2017) Integrated nutrient management strategies to alleviate drought stress in hybrid maize in Punjab, Pakistan. Romanian Agricult Res 34:1–10. First Online. DII 2067–5720 RAR 2017–44

Jat LK, Singh YN (2017) Short-term effects of organic and inorganic fertilizers on soil properties and enzyme activities in rice production. Int J Curr Microbiol Appl Sci 6(2):185–194

Jat LK, Singh SK, Latare AM, Singh RS, Patel CB (2013) Effect of dates of sowing and fertilizer on growth and yield of wheat (*Triticum aestivum*) in an Inceptisol of Varanasi. Indian J Agron 58(4):611–614

Kabato W, Ergudo T, Mutum M et al (2022) Response of wheat to combined application of nitrogen and phosphorus along with compost. J Crop Sci Biotechnol 25:557–564. https://doi.org/10.1007/s12892-022-00151-7

Kachot NA, Malvia DD, Solanki RM, Sagarka BK (2001) Integrated nutrient management in rainy season groundnut (*Arachis hypogea*). Indian J Agron 46(3):516–522

Kannan RL, Dhivya M, Abinaya D, Krishna R, Krishnakumar S (2013) Effect of integrated nutrient management on soil fertility and productivity in maize. Bull Environ Pharmacol Life Sci 2(8):61–67

Kumar R, Lal M, Naresh RK, Kumar M, Yadav S, Kumar R, Chandra MS, Rajput P (2020) Influence of balanced fertilization on productivity, nutrient use efficiency and profitability of rice in Inceptisol: a review. Int J Curr Microbiol Appl Sci 9(1):568–590

Kumar M, Baishaya LK, Ghosh DC, Gupta VK, Dubey SK, Das A, Patel DP (2012) Productivity and soil health of potato (*Solanum tuberosum* L.) field as influenced by organic manures, inorganic fertilizers and biofertilizers under high altitudes of eastern Himalayas. J Agric Sci 4(5):223–234

Kumara BH, Antil RS, Raj D (2013) Long-term effects of nutrient management on soil health and crop productivity under pearl millet—wheat cropping system. Indian J Fert 9(12):86–97

Kumara BH, Antil RS, Raj D (2014) Impact of long-term manures and fertilizer application on carbon sequestration and its efficiency under pearl millet—wheat cropping sequence. Int J Farm Sci 4(1):21–26

Kumari N, Mehta VP, Bhatia JK (2020) Foodgrains production in India: trend and decompositions analysis. Econ Aff 65(3):333–342

Kumari R, Kumar S, Kumar R, Das A, Kumari R, Choudhary CD, Sharma RP (2017) Effect of long-term integrated nutrient management on crop yield, nutrition and soil fertility under rice-wheat system. J Appl Nat Sci 9(3):1801–1807

Kusro PS, Singh DP, Paikra MS, Kumar D (2014) Effect of organic and inorganic additions on physico-chemical properties in Vertisol. Am Int J Res Form Appl Nat Sci 5(1):51–53

Lakshmi CSR, Rao PC, Sreelatha T, Padmaja G, Madhavi M, Rao PV, Sireesha A (2014) Residual effects of INM on humus fractions, micronutrient content and their uptake by *rabi* greengram under rice-pulse cropping system. Res Crops 15(1):96–104

Liang Q, Chen H, Gong Y, Yang H, Fan M, Kuzyakov Y (2014) Effects of 15 years of manure and mineral fertilizers on enzyme activities in particle-size fractions in a North China Plain soil. Euro J Soil Biol 60:112–119

Liang Y, Yang Y, Yang C, Shen Q, Zhou J, Yang L (2003) Soil enzymatic activity and growth of rice and barley as influenced by organic manure in an anthropogenic soil. Geoderma 115:149–160

Liu CA, Li FR, Zhou LM, Zhang RH, Jia Y, Lin SL, Wang LJ, Siddique HM, Li FM (2013) Effect of organic manure and fertilizer on soil, water and crop yields in newly built terraces with loess soils in a semi-arid environment. Agric Water Manag 117:123–132

Mahanta D, Bhattacharyya R, Gopinathc KA, Tuti MD, Jeevanandan K, Chandrashekara C, Arunkumar R, Mina BL, Pandey BM, Mishra PK, Bisht JK, Srivastva AK, Bhatt JC (2013) Influence of farmyard manure application and mineral fertilization on yield sustainability, carbon sequestration potential and soil property of Garden pea–French bean cropping system in the Indian Himalayas. Sci Hortic 164:414–427

Mahmud AH, Upadhyay SK, Srivastava AK, Bhojiya AA (2021) Biofertilizers: a Nexus between soil fertility and crop productivity under abiotic stress. Curr Res Environ Sustain 3:100063. https://doi.org/10.1016/j.crsust.2021.100063

Majumdar K, Johnston AM, Dutt S, Satyanarayana T, Roberts TL (2013) Fertilizer best management practices concept, global perspectives and application. Indian J Fert 9(4):14–31

Mallikarjun M, Maity SK (2018) Effect of integrated nutrient management on soil biological properties in kharif rice. Int J Curr Microbiol App Sci 7(11):1531–1537

Marimuthu S, Surendran U, Subbian P (2014) Productivity, nutrient uptake and postharvest soil fertility as influenced by cotton-based cropping system with integrated nutrient management practices in semi-arid tropics. Archives Agron Soil Sci 60:87–101

Meena RS, Dhakal Y, Bohra JS, Singh SP, Singh MK, Sanodiya P (2015) Influence of bioinorganic combinations on yield, quality and economics of mungbean. Am J Exp Agric 8(3):159–166
Moe K, Htwe AZ, Thu TTP, Kajihara Y, Yamakawa T (2019) Effects on NPK status, growth, dry matter and yield of rice (Oryza sativa) by organic fertilizers applied in field condition. Agriculture 9(5):109. https://doi.org/10.3390/agriculture9050109
Moraditochaee M, Azarpour E, Bozorgi HR (2014) Study effects of bio-fertilizers, nitrogen fertilizer and farmyard manure on yield and physiochemical properties of soil in lentil farming. Int J Biosci 4(4):41–48
Mosa WFA, El-Gleel Paszt LS, Frc1 M, TrzciDski P (2015) The role of biofertilization in improving apple productivity: a review. Adv Microbiol 5:21–27
Naidu DK, Radder BM, Patil PL, Hebsur NS, Alagundagi SC (2009) Effect of integrated nutrient management on nutrient uptake and residual fertility of chili (*CV. Byadgi dabbi*) in a Vertisol. Karnataka J Agric Sci 22:306–309
Nwite JN, Igboji PO, Okonkwo GI (2014) Effect of integrated nutrient management on selected soil physical properties and grain yield of maize in Abakaliki, South Eastern Nigeria. J Agric Econ Exten Rural Develop 2(11):211–217
P A, Salami TB (2012) Effect of integrated nutrient management approach on soil chemical and physical properties and performance of tomato (*Lycopersicon lycopersicum*) under mildly-acidic Alfisol conditions. Int J Appl Agric Apic Res 8(1):91–98
Padbhushan R, Sharma S, Kumar U, Rana DS, Kohli A, Kaviraj M, Parmar B, Kumar R, Annapurna K, Sinha AK, Gupta VSR (2021) Meta-analysis approach to measure the effect of integrated nutrient management on crop performance, microbial activity, and carbon stocks in Indian soils. Front Environ Sci Sect Soil Processes 9. https://doi.org/10.3389/fenvs.2021.724702
Pindi PK, Satyanarayana SDV (2012) Liquid microbial consortium—a potential tool for sustainable soil health. J. Biofertili. Biopestic. 3(4):1–9
Rahman MH, Islam MR, Jahiruddin M, Rafii MY, Hanafi MM, Malek MA (2013) Integrated nutrient management in maize-legume-rice cropping pattern and its impact on soil fertility. J Food Agric Environ 11:648–652 (2013)
Rana MM, Chowdhury AK, Bhuiya MSU (2011) Effects of plant population and biofertilizer on the growth parameters of three summer mungbean (*Vigna radiata* L.) cultivars. Bangladesh J Agric Res 36(3):537–542
Rasool R, Kukal SS, Hira GS (2007) Soil physical fertility and crop performance as affected by long term application of FYM and inorganic fertilizers in rice–wheat system. Soil Tillage Res 96:64–72
Rurinda J, Zingore S, Jibrin MJ, Balemi T, Masuki K, Andersson JA, Pampolino MF, Mohammed I, Mutegi J, Kamara AY, Vanlauwe B, Craufurd PQ (2020) Science-based decision support for formulating crop fertilizer recommendations in sub-Saharan Africa. Agric Syst 180:102790. https://doi.org/10.1016/j.agsy.2020.102790
Saha R, Mishra VK, Majumdar B, Laxminarayana K, Ghosh PK (2010) Effect of integrated nutrient management on soil physical properties and crop productivity under a maize (*Zea mays*)-mustard (*Brassica campestris*) cropping sequence in acidic soils of northeast India. Commun Soil Sci Plant Anal 41:2187–2200
Sarkar S, Skalicky M, Hossain A, Brestic M, Saha S, Garai S, Ray K, Brahmachari K (2020) Management of crop residues for improving input use efficiency and agricultural sustainability. Sustainability 12(9808):1–24. https://doi.org/10.3390/su12239808
Sheoran S, Raj D, Antil RS, Mor VS, Dahiya DS (2017) Productivity, seed quality and nutrient use efficiency of wheat (*Triticum Aestivum*) under organic, inorganic and integrated nutrient management practices after twenty years of fertilization. Cereal Res Commun 45(2):315–325
Sheoran S, Raj D, Antil RS, Mor VS, Grewal KS (2020) Soil microbial properties influenced with long term application of manures and fertilizers. Indian J Agric Sci 90:178–182
Shukla AK, Behera SK, Pakhre A, Chaudhari SK (2018) Micronutrients in soils, plants, animals and humans. Indian J Fert 14(4):30–54

Singh B, Ryan J, Campbell C, Kröbel R (2015) Nutrient management and water use efficiency for sustainable production of rain-fed crops in the World's dry areas. In: Drechsel P, Heffer P, Magen H, Mikkelsen R, Wichelns D (eds) Managing water and fertilizer for sustainable agricultural intensification, international fertilizer industry association, international water management institute, international plant nutrition institute, international potash institute, pp 139–167
Singh JP, Kaur J, Mehla DS, Narwal RP (2012a) Long-term effects of nutrient management on soil health and crop productivity under rice-wheat cropping system. Indian J. Fert. 8(8):28–48
Singh M, Dwivedi BS, Datta SP (2012b) Integrated nutrient management for enhancing productivity, nutrient use efficiency and environmental quality. In: Soil science in the service of nation, 55–67, ISSS, New Delhi
Singh RK, Sharma GK, Kumar P, Singh SK, Singh R (2019) Effect of crop residues management on soil properties and crop productivity of rice-wheat system in Inceptisols of Seemanchal*r*egion of Bihar. Current J Appl Sci Technol 37(6):1–6
Singh RR, Prasad K (2011) Effect of bio-fertilizers on growth and productivity of wheat (*Triticum aestivum*). J Farm Sci 1(1):1–8
Sumagaysay CL (2014) Bio N fertilization on corn. Int J Biotechnol 3(6):85–90
Tewodros M, Belay Y (2015) Review on integrated soil fertility management for better crop production in Ethopia. Sky J Agric Res 4(1):21–32
Tittonell P, Rufino MC, Janssen BH, Giller KE (2010) Carbon and nutrient losses during manure storage under traditional and improved practices in smallholder crop-livestock systems—evidence from Kenya. Plant Soil 328:253–269. https://doi.org/10.1007/s11104-009-0107-x
Verma S, Mahapatra P, Kumar SB, Sinha AK, Minz A, Deep KP, Yadav S, Pal J, Palatty AE (2019) Long term effect of nutrient management on biological properties of acid soil under maize-wheat cropping system. J Pharmacogn Phytochem 5:208–211
Wailare AT, Kesarwani A (2017) Effect of integrated nutrient management on growth and yield parameters of maize (*Zea mays* L.) as well as soil physico-chemical properties. Biom J Sci Tech Res 1(2):294–299
Wei W, Ma B (2015) Integrated nutrient management for sustaining crop productivity and reducing environmental impact: a review. Sci Total Environ 512–513:415–427
Xu MG, Li DC, Li JM, Qin DZ, Kazuyuki Y, Yasukazu H (2008) Effects of organic manure application with chemical fertilizers on nutrient absorption and yield of rice in Hunan of Southern China. Agric Sci China 7(10):1245–1252
Zhou Z, Zhang S, Jiang N, Xiu W, Zhao J, Yang D (2022) Effects of organic fertilizer incorporation practices on crops yield, soil quality, and soil fauna feeding activity in the wheat-maize rotation system. Front Environ Sci 10:1058071. https://doi.org/10.3389/fenvs.2022.1058071
Zingore S, Mutegi J, Agesa B, Tamene L, Kihara J (2015) Soil degradation in sub-Saharan Africa and crop production options for soil rehabilitation. Better Crops 99(1):24–26

Open Access This chapter is licensed under the terms of the Creative Commons Attribution 4.0 International License (http://creativecommons.org/licenses/by/4.0/), which permits use, sharing, adaptation, distribution and reproduction in any medium or format, as long as you give appropriate credit to the original author(s) and the source, provide a link to the Creative Commons license and indicate if changes were made.

The images or other third party material in this chapter are included in the chapter's Creative Commons license, unless indicated otherwise in a credit line to the material. If material is not included in the chapter's Creative Commons license and your intended use is not permitted by statutory regulation or exceeds the permitted use, you will need to obtain permission directly from the copyright holder.

Biological Nitrogen Fixation and Yields of Bambara Groundnut Cultivars on Alfisols in Central Malawi

Keston O. W. Njira, Eliza Mankhwazi, and Elton Panulo

Abstract Bambara groundnut (*Vigna subterranean*), also referred to as Bambara nut or Ground beans, is considered as one of the most important legumes in African countries where it is grown. However, it is one of the neglected crops that needs to be promoted in the current situation as it is a crop that offers many benefits, including being a nutrient-dense grain. It also potentially resilient to challenges that arise due to climate change and soil degradation as it biologically fixes nitrogen (N). This chapter presents a case study of an evaluation of biological nitrogen fixation (BNF) and yields of six Bambara groundnut cultivars. The six cultivars include two released varieties of Makata and Kayera plus four landraces namely Red, Black, Chikope cha Nyani and Khaki. The study was implemented at the Lilongwe University of Agriculture and Natural Resources research farm in the two cropping seasons of 2015/16 and 2016/17. It was implemented in a randomized complete block design (RCBD), with three replicates in each of the two seasons. Results showed that Khaki consistently led to significantly higher ($P < 0.05$) two year mean biologically fixed N (72.2 kg ha^{-1}), total N yield, number of pods per plant, grain yield and total dry matter yield than the other landraces and varieties except Red that had similar higher amounts of biologically fixed N (71.2 kg ha^{-1}). On the other hand, the other four cultivars produced a range of 45.0 to 52.5 kg ha^{-1} biologically fixed N. Grain yield ranged from 808 kg ha^{-1} (Kayera) to 1259 kg ha^{-1} (Khaki). This study has brought about some new information about the Bambara landraces that exist in Malawi and will be beneficial for breeders in improving varieties of Bambara groundnuts. Further studies should be conducted in different agroecological zones and on different soil types that exist in Malawi.

K. O. W. Njira (✉) · E. Mankhwazi · E. Panulo
Department of Crop and Soil Sciences, Lilongwe University of Agriculture and Natural Resources, Lilongwe, Malawi
e-mail: knjira@luanar.ac.mw

© The Author(s) 2026

B. R. Singh et al. (eds.), *Sustainable Food Systems in Sub-Saharan Africa*,
https://doi.org/10.1007/978-3-032-11288-0_3

1 Introduction

Bambara groundnut (*Vigna subterranea*), also referred to as Bambara nut or Ground beans is considered as one of the most important legumes in African countries where it is grown (Olanrewaju et al. 2021; Yalaukani 2023). Basu et al. (2007) noted that the species *subterranea* is further differentiated at the subspecies level into two groups, which are var. *spontanea*, comprising the wild forms that are mostly found in parts of West Africa and var *subterrranea*, common in many parts of sub-Saharan Africa. This crop is a legume that helps to improve soil fertility through biological nitrogen fixation (BNF). Substantial amounts of nitrogen are reported to be fixed by Bambara cultivars (Gerrano et al. 2021). Mussa et al. (2016) reported a range of 31 to 83 kg ha^{-1} biologically fixed nitrogen. Bambara is resilient and resistant to high temperature and can withstand marginal soils where many other legume species may face productivity challenges (Gerrano et al. 2021). This is an indicator that Bambara is one of the crops that can easily adapt to climate change. Additionally, Bambara is one of the nutrient-dense legumes, implying that is a good crop for food and human nutritional security. Bambara groundnut is reported to contain 64.4% carbohydrate, 23.6% protein, 6.5% fat, and 5.5% fibre (Halimi et al. 2019; Tan et al. 2020). It contains high contents of minerals that vary by landrace and environment. Amarteifio et al. (2006) reported ranges for macro-minerals (mg/100 g dry matter) indicate Ca 37–128, K 1545–2200, Mg 159–335, Na 16–25 and P 313–563 and ranges of micro-nutrients (ppm) of Cu 3.0–13.2, Fe 23.0–150, and Zn 13.9–77.

Surprisingly, it is one of the neglected crops in Malawi and other parts of the sub-Saharan Africa. However, it is one of the crops that need to be promoted in the current situation as it is a crop that offers many benefits and has a lot of potential resilience against climate change and soil degradation challenges. On the other hand, it is noted that most of Bambara cultivars exist as landraces, and information on their productivity is scanty. This chapter presents results of a two-year case study on evaluation of biological nitrogen fixation and yield of the different Bambara cultivars conducted at a research farm in Central Malawi.

2 Materials and Methods

2.1 Site Identification and Baseline Soil Characterization

The study was conducted at the Lilongwe University of Agriculture and Natural Resources (LUANAR) research farm (14° 12′ S, 33° 45′ E) within the Mkwinda Extension Planning Area (EPA). The soil of the site is characterized as Chromic Luvisols (Typic Hapludalfs, Alfisols) (Chilimba et al. 2011; Mutegi et al. 2015).

Sampling of soils was conducted using an auger from depths ranges of 0–20 cm and 20–40 cm, based on the simple random sampling plan (Anderson and Ingram 1993). Preplant and post-harvest soil sampling was conducted for determination of the fertility parameters. The samples were air-dried and analysed for a number of parameters such as soil texture by the hydrometer method, pH in water by the potentiometric method, total nitrogen (N%) by micro-Kjedahl method, and plant extractable phosphorus (P) by Mehlich-3 method as outlined by Chilimba (2007). Organic carbon (C) was determined using wet oxidation by dichromate and exchangeable basic cations by Mehlich-3 method, according to Chilimba (2007). Post-harvest samples were collected from each plot for analysis of mineral N, which is normally required when using the modified nitrogen difference method for estimation of BNF (Unkovich et al. 2008).

2.2 *Bambara Groundnut Cultivars*

Bambara groundnut cultivars were sourced from the Genetic Resources Centre at Chitedze Agriculture Research Station. These included landraces obtained from different areas of Malawi. Two of the six cultivars were released varieties (Kayera and Makata), while the other four were landraces (Black, Chikope cha Nyani/chikope, Khaki and Red) (Fig. 1).

2.3 *Field Experimental Design and Plot Layout*

The experiment was laid out in the randomized complete block design (RCBD) with three replicates. The blocking factor was the slope of the land. Treatments included the six Bambara cultivars Black, Chikope cha nyani, Kayera, Khaki, Makata and Red. Plant spacing was 15 cm between planting station along the ridge/row. Seeds were planted one seed per station and ridges were 75 cm apart. The plant and ridge spacings aforementioned were implemented based on the conventional procedures (MoAFS 2012). A maize plot was included on each replicate, and this was required because it was used as a reference crop in the determination biologically fixed N. Plot sizes were 10 by 10 m.

2.4 *Nodulation Assessment, Plant Sampling and Analysis*

Nodule numbers were assessed at 50% flowering of the Bambara groundnut plant. The procedure included sampling of 10 plants per plot (each treatment). Nodules were counted for each of the ten plants sampled from each treatment before computing averages. Plant samples were collected from the fields at 50% flowering from the Bambara groundnut plants and at tasseling stage for maize for determination of % N. Sampled plant tissues were air-dried and ground into powdered material and were subjected to chemical analysis according to Chilimba (2007).

Fig. 1 Bambara groundnut cultivars evaluated for biological nitrogen fixation and yields at LUANAR research farm, central Malawi

2.5 Determination of Biologically Fixed N and Nitrogen Percent Derived from the Atmosphere (%N_{dfa})

This included determination of N yield by multiplying the plant's total N (in percentage) by total dry matter yield. The BFN was determined using the modified Nitrogen-Difference method (Unkovich et al. 2008), Eq. 1, which is achieved by computing the difference between total plant nitrogen (N uptake or N yield) of an N_2-fixing legume and a control or reference crop (non-N_2-fixing). The modification of this technique is intended to reduce errors of measurements when a non-nodulating legume is not available and the two plant species being used are not botanically well matched. Hence, the difference of post-harvest soil N in the legume and control plots is added to the original formula.

$$Q = [\text{N yield (legume)} - \text{N yield (control)}] + [\text{N soil (legume)} - \text{N soil (control)}] \quad (1)$$

where

Q (kg ha^{-1}) = Quantity of the biologically fixed N.

N yield [legume] (kg ha^{-1}) = N yield of a legume.

N yield [control] (kg ha^{-1}) = N yield of a non-fixing plant.

N soil (kg ha^{-1}) = Post-harvest soil N in legume or control.

plot.

The percentage of N derived from the atmosphere was estimated using Eq. 2

$$\%\ N_{dfa} = \frac{N_2 \text{ fixed}}{\text{N yield}} \times 100 \quad (2)$$

where:

N_2 fixed is the biologically fixed N; N yield is the total amount of N uptake.

The nitrogen harvest index (NHI) and N balance were determined according to McDonald (1989) as outlined in the following equation

$$Nbalance = (Nleg \times Ndfa) - (Nleg \times NHI)$$

Nleg = legum N.

Ndfa = percent of nitrogen derived from the atmosphere/proportion of N from BNF.

NHI = proportion of crop N harvested in grain.

2.6 Determination of Total Dry Matter Yields

The total dry matter yield (TDMY)—total above ground biomass plus the pods—was determined. At harvesting, the Bambara groundnut crop in the net plot of 8 by 8 m was dug out and soils attached to the roots were shaken off. This was followed by getting fresh weight of the net plot biomass. Randomly sampled 500 g (from each et plot) of fresh biomass were taken to the laboratory for oven drying and determination of dry matter yield, and this was extrapolated to dry matter yield per hectare, according to Mloza-Banda (1994).

2.7 Statistical and Data Analysis

All data were subjected to analysis of variance (ANOVA) using GenStat 17th edition statistical package. Means were separated using the least significant difference (LSD) test at 5% significant level.

Table 1 Soil properties of the study site

Soil property	0–20 cm	20–40 cm
pH $_{water}$ 1:2.5	6.05	6.2
Organic C (%)	1.11	0.9
SOM (%)	1.80	1.6
Total N (%)	0.05	0.05
Available P (mg/kg)	37.0	28.4
Mg $Cmol_{(+)}$/kg	1.24	1.18
K $Cmol_{(+)}$/kg	0.36	0.23
Fe mg/kg	19.12	15.7
Zn mg/kg	2.57	4.1
Mn mg/kg	10.3	11.49
Bulk density (g/cm^3)	1.54	1.55
Particle size distribution		
Sand	46.6	47.5
Silt	22.7	29.0
Clay	30.8	23.2
Textural class	Sand clay loam	Loam

3 Results and Discussions

3.1 *Baseline Soil Properties*

The baseline soil analysis results indicated low contents of nitrogen and soil organic matter (Table 1). However, available P, Mg, K and Zn indicated adequate levels. Soil pH was slightly acidic in a range that is tolerated by most crops and bacteria, including rhizobia. The indication of low N levels is not healthy for the growth of most crops. On the other hand, the modified N difference method for estimating biologically fixed N used in this study give reasonably reliable results in soils of low N levels (Danso 1995).

3.2 *Nodulation and Biological Nitrogen Fixation*

Results (Table 2) shows that in year 1, mean nodule numbers ranged from 21 per plant by Khaki to 36 per plant by Makata; while in year 2, it ranged from 19.0 nodules per plant by Red to 38 nodules per plant by Makata. On the other hand, the two-year mean nodule numbers per plant ranged from 19.8 to 37 by Makata, which consistently showed significantly higher ($P < 0.05$) number of nodules than the rest of the cultivars. Puozaa et al. (2021) in a study conducted in Ghana on certain

Table 2 Nodulation and biological nitrogen fixation in different Bambara groundnut cultivars grown on Alfisols in central Malawi

Bambara groundnut cultivar	Year 1		Year 2		2 year mean
	Nodule numbers	BNF (kg N ha^{-1})	Nodule numbers	BNF (kg N ha^{-1})	Nodule numbers
Black	29.0^{abc}	57.1^{a}	19.7^{a}	42.0^{a}	24.3^{a}
Chikope	26.0^{ab}	53.7^{a}	24.0^{ab}	51.3^{a}	25.2^{a}
Kayera	22.3^{a}	47.7^{a}	17.3^{a}	42.3^{a}	19.8^{a}
Khaki	21.0^{a}	72.9^{b}	32.7^{bc}	71.5^{b}	26.8^{a}
Makata	36.0^{c}	59.1^{ab}	38.0^{c}	45.3^{a}	37.0^{b}
Red	32.3^{bc}	88.2^{c}	19.3^{a}	54^{a}	25.8^{a}
F. pr	0.019	0.001	0.008	0.045	0.012
LSD (0.05)	8.45	15.2	10.64	16.2	7.77
CV%	16.7	13.2	23.2	37.2	16.1

Key: *F.pr* F probability, *LSD (0.05)* Least significant difference at 5% significant level, *CV* coefficient of variation, Means with different superscript letters are statistically diiferent

Bambara landraces also noted variations in nodulations where black seedcoat was considered a marker for enhanced nodulation. However, in this study the black-seed coated Bambara landrace never showed highest nodule numbers. Amounts of N due to BNF ranged from 47.7 to 88.2 kg N ha^{-1} by Kayera and Red, respectively, in the first season and ranged from 42 to 71 kg N ha^{-1} for Black and Khaki, respectively, in the second season. The two-year mean amounts obtained from BNF ranged from 45.0 to 72.2 by Kayera and Khaki, respectively (Fig. 2). Khaki and Red biologically fixed higher amounts, which were significantly higher ($P < 0.05$) than the other varieties. Khaki biologically fixed 37.7% more N than Kayera. Results obtained in this study have shown mean number of nodules and amounts of biologically fixed N, which vary by landrace or cultivar but are in similar ranges to other studies, including a study with different Bambara landraces on a tropical acidic soil of Malaysia by Mussa et al. (2016). Mussa et al. (2016) reported biologically fixed N in a range of 49.3 to 83.2 $kgNha^{-1}$ from the different landraces of Bambara. The variations show the differences in genetic potential and adaptability of the different landraces to a given environment.

3.3 *Yields, Ndfa, Nitrogen Harvest Index and N Balance*

Results in Table 3 shows that there were significant differences on a number of parameters, including a two-year mean grain yield, total dry matter yield, number of pods per plant and total N yield as influenced by the differences in the Bambara cultivars. Grain yield ranged from 808 to 1259 kg ha^{-1} for Kayera and Khaki, respectively, and grain yield for Khaki was significantly higher ($P < 0.05$) than the rest of the cultivars.

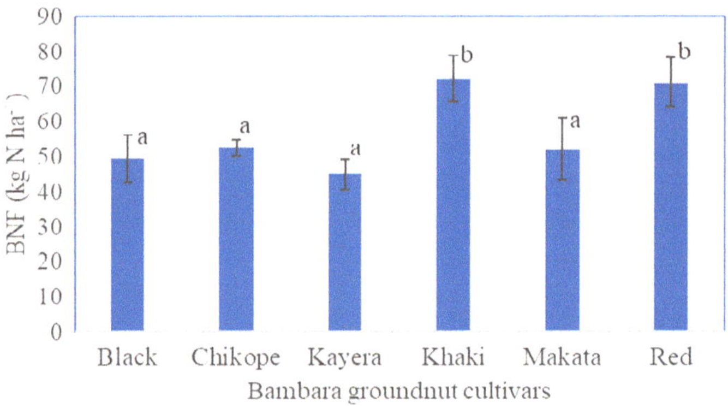

Fig. 2 Two year mean biologically fixed nitrogen (kg N ha^{-1}) by different Bambara groundnut cultivars grown on Alfisols in Central Malawi

It was also noted that TDMY, number of pods per plant and total N yield were also highest for Khaki, and were significantly higher than the rest of the cultivars except Red that had similarly higher amounts of total N yield. However, two-year mean amounts of NHI, Ndfa and N balance did not show significant differences among the different cultivars. The consistently high performance in terms of grain yield, TDMY, number of pods and total N yield for Khaki can be attributed to its genetic potential to produce large amount of biomass, fix nitrogen while partitioning enough nutrients for the grain. Similarly, Pungulani et al. (2012) reported Bambara grain yields in a range of 841 to 1322 kgh^{a-1} from a study conducted in a number of sites in Malawi. On the other hand, Mussa et al. (2016) in a study on Bambara landraces reported grain yield ranges of 703 to 2256 kg ha^{-1}. Variations in a number of phenotypic traits, including growth parameters, were also reported in a study conducted in Nigeria on different Bambara landraces (Olanrewaju et al. 2021). Khan et al. (2021) noted that genetic variation is an important factor that is beneficial in cross breeding.

4 Concluding Remarks

This study included two released varieties of Makata and Kayera and four landraces namely Red, Khaki, Chikope cha Nyani and Black. From this study, it was noted that these cultivars showed significant differences on nodulation, biological nitrogen fixation, total N yield, number of pods per plant, grain yield and total dry matter yield. This study has exposed additional information that can help breeders to conduct genetic analysis and selection of these Bambara landraces. Though conducted over two growing seasons, this study had a limitation of number of sites. Therefore, it is recommended that further studies should include evaluation of the landraces in different agroecological zones and in different soil types. The two cultivars, Khaki and

Table 3 Two-year mean yields, pods, N harvest indices (NHI), Ndfa and N balance by different Bambara groundnut cultivars grown on Alfisols in Central Malawi

Bambara groundnut cultivar	Grain yield (kg ha^{-1})	TDMY (kg ha^{-1})	Number pods/plant	Total N yield (kg N ha^{-1})	NHI	Ndfa (%)	N balance (kg N ha^{-1})
Black	984^{b}	2546^{ab}	27.8^{a}	74.3^{a}	42.1	65.1	19.0
Chikope	1009^{b}	2749^{b}	30.3^{a}	78.9^{a}	41.1	66.4	20.2
Kayera	808^{a}	2211^{a}	31.2^{a}	72.2^{a}	34.8	61.9	20.0
Khaki	1259^{c}	3272^{c}	48.3^{b}	98.7^{b}	42.8	72.5	30.2
Makata	999^{b}	2695^{b}	29.3^{a}	78.7^{a}	44.5	64.8	17.4
Red	993^{b}	2552^{ab}	42.7^{b}	95.7^{b}	37.8	71.9	33.9
F. pr	< 0.001	0.004	< 0.001	0.017	0.163	0.100	0.094
LSD (0.05)	102.8	407.0	5.89	16.15	7.96	8.3	13.4
CV%	5.6	8.4	9.3	10.7	10.8	6.9	31.4

Key: *F.pr* F probability, *LSD (0.05)* Least significant difference at 5% significant level, *CV* coefficient of variation, *TDMY* total dry matter yield, Means with different superscript letters are statistically different

Red may require special attention as the produced higher two year mean biological fixed N than the rest.

Acknowledgements The authors would like to thank the Chitedze Research Station Gene Bank Management for providing Bambara landraces that were used in this study.

References

Amarteifio JO, Tibe O, Njogu RM (2006) The mineral composition of bambara groundnut (Vigna subterranea (L) Verdc) grown in Southern Africa. Afr J Biotech 5(23):2408–2411

Anderson JM, Ingram JSI (1993) Tropical soil biology and fertility: a handbook of methods of analysis, 2nd edn. CAB International, Wallingford, UK

Basu S, Roberts JA, Azam-Ali SN, Mayes S (2007) Bambara groundnut. In: Pulses, sugar and tuber crops. Springer Berlin Heidelberg, Berlin, Heidelberg, pp 159–173

Chilimba ADC (2007) Methods of soils, plants, fertilizers and Misclaneous analyses: a working manual for Chitedze and Bvumbwe research stations soils laboratories. Ministry of Agriculture. Lilongwe, Malawi

Chilimba ADC, Young SD, Black CR, Rogerson KB, Ander EL, Watts MJ, Lammel J, Broadly MR (2011) Maize grain and soil surveys reveal suboptimal dietary selenium intake is widespread in Malawi. Scientific Rep 1:72. https://doi.org/10.1038/srep00072

Danso S (1995) Assessment of biological nitrogen fixation. Fertil Res 42:33–41

Gerrano A, Eifediyi EK, Labuschagne M, Ogedegbe FO, Hassen AI (2021) Production practices of bambara groundnut. Food Potent Industr Appl Bambara Groundnut 7–25. https://doi.org/10.1007/978-3-030-73920-1_2

Halimi RA, Barkla BJ, Mayes S, King GJ (2019) The potential of the underutilized pulse Bambara groundnut (Vigna subterranea (L.) Verdc.) for nutritional food security. J Food Compos Anal 77:47–59. https://doi.org/10.1016/j.jfca.2018.12.008

Khan MMH, Rafii MY, Ramlee SI, Jusoh M, Al Mamun M (2021) Genetic analysis and selection of Bambara groundnut (Vigna subterranea [L.] Verdc.) landraces for high yield revealed by qualitative and quantitative traits. Sci Rep 11(1):7597. https://doi.org/10.1038/s41598-021-87039-8

McDonald GK (1989) The contribution of nitrogen fertiliser to the nitrogen nutrition of rainfed wheat crops in Australia: a review. Aust J Exp Agric 29(3):455–481

Ministry of Agriculture and Food Security (2012). Guide to agricultural production and natural resource management in Malawi. Agricultural Communications Branch. Lilongwe. Malawi, pp 366

Mloza-Banda HR (1994) Principles and practices of crop management: field study guide. Crop Science Department. Bunda College of Agriculture, pp 137

Musa M, Massawe F, Mayes S, Alshareef I, Singh A (2016) Nitrogen fixation and N-balance studies on Bambara groundnut (Vigna subterranea L. Verdc) landraces grown on tropical acidic soils of Malaysia. Commun Soil Sci Plant Anal 47(4):533–542. https://doi.org/10.1080/00103624.2016.1141928

Mutegi J, Kabambe V, Zingore S, Harawa R, Wairegi L (2015) The fertilizer reccomendation issues in Malawi: gaps, challenges, opportunities and guidelines. Soil Health Consort Malawi 54

Olanrewaju OS, Oyatomi O, Babalola OO, Abberton M (2021) Genetic diversity and environmental influence on growth and yield parameters of Bambara groundnut. Front Plant Sci 12:796352. https://doi.org/10.3389/fpls.2021.796352

Peoples MB, Faizah AW, Rerkasem B, Herritedge DF (1989) Methods for evaluating nitrogen fixation by nodulating legumes in the field. ACIAR Canberra 89

Pungulani L, Kadyampakeni D, Nsapato L, Kachapila M (2012) Selection of high yielding and farmers' preferred genotypes of bambara nut (Vigna subterranea (L.) Verdc) in Malawi. Am J Plant Sci. https://doi.org/10.4236/ajps.2012.312A221

Puozaa DK, Jaiswal SK, Dakora FD (2021) Black seedcoat pigmentation is a marker for enhanced nodulation and N2 fixation in Bambara groundnut (Vigna subterranea L. Verdc.) landraces. Front Agron 3:692238. https://doi.org/10.3389/fagro.2021.692238

Tan XL, Azam-Ali S, Goh EV, Mustafa M, Chai HH, Ho WK,... Massawe F (2020) Bambara groundnut: an underutilized leguminous crop for global food security and nutrition. Front Nutri 7:601496. https://doi.org/10.3389/fnut.2020.601496

Unkovich M, Herridge D, Peoples M, Cadisch G, Boddey B, Giller K, … Chalk P (2008) Measuring plant-associated nitrogen fixation in agricultural systems. Australian Centre for International Agricultural Research (ACIAR)

Yalaukani L (2023) Phenotypic variation of finger millet (*Eleusine coracana* (L.) Gaertn) and Bambara groundnut (*Vigna subterranea* (L.) Verdc) germplasm for farmer preferred traits. MSc Thesis. Lilongwe University of Agriculture and Natural Resources. Malawi

Open Access This chapter is licensed under the terms of the Creative Commons Attribution 4.0 International License (http://creativecommons.org/licenses/by/4.0/), which permits use, sharing, adaptation, distribution and reproduction in any medium or format, as long as you give appropriate credit to the original author(s) and the source, provide a link to the Creative Commons license and indicate if changes were made.

The images or other third party material in this chapter are included in the chapter's Creative Commons license, unless indicated otherwise in a credit line to the material. If material is not included in the chapter's Creative Commons license and your intended use is not permitted by statutory regulation or exceeds the permitted use, you will need to obtain permission directly from the copyright holder.

Enhancing Soil Nitrogen Management for Lowland Rice Production in Sub-Saharan Africa

Twaha Ali Basamba, Boshwenda Chuma Andre, Mambani Banda Pierre, Ann Tumushabe, and Kassim Ssekabira

Abstract Many long-grain rice varieties have been introduced in the tropics, including Rwanda. However, farmers prefer the local varieties to these new ones. The aim of this study was to evaluate how the new rice varieties respond to different nitrogen levels, determine the best nitrogen rate and location for enhancing rice grain yield, and assess the economic viability of nitrogen fertilizer rates and their effective application. Five nitrogen levels (0, 40, 80, 120 and 160 kg N ha^{-1}) and two methods of nitrogen application (broadcast and deep placement) were tested on two rice varieties (WAB 880-1-38-20-28-P1-HB and IR 28). The optimum fertilizer nitrogen rates were 106 and 98 kg N ha^{-1} for WAB 880-1-38-20-28-P1-HB and IR 28, respectively. There was no significant difference in rice grain yield between the two methods of placement; the maximum rice grain yield was obtained at 120 kg N ha^{-1} for both varieties. The recommended fertilizer nitrogen rates were 106 and 98 kg ha^{-1} for WAB and IR28 varieties, respectively.

Keywords Soil nitrogen · Rice varieties · Grain yield · Nitrogen fertilizers · Rwanda

---

T. A. Basamba (✉)
College of Agricultural and Environmental Sciences, School of Agricultural Sciences, Department of Soil Science and Land Use Management, Makerere University, Kampala, Uganda
e-mail: twaha.basamba@mak.ac.ug

B. C. Andre
Rwanda Agriculture and Animal Development Board, Kigali, Rwanda

M. B. Pierre
Department of Soils and Environmental Management, National University of Rwanda, Butare, Rwanda

A. Tumushabe · K. Ssekabira
Department of Environmental and Biological Sciences, College of Engineering and Applied Sciences, Kampala International University, Kampala, Uganda

© The Author(s) 2026
B. R. Singh et al. (eds.), *Sustainable Food Systems in Sub-Saharan Africa*,
https://doi.org/10.1007/978-3-032-11288-0_4

1 Introduction

Rice production has become a vital component of agricultural production in sub-Saharan Africa and in Rwanda in particular (World Bank 2020). The country currently has to import about 30% of its rice needs from countries such as Tanzania, Vietnam and Thailand (Senthilkumar 2020). To meet the local demand from cultivated areas and face regional competition, the Rwandan government, through its strategic "vision 2030" program expects to reach a national average yield of 7 t ha^{-1} (Cyamweshi et al. 2017). Rice production is strongly linked to the amount of nitrogen fertilizer applied (Chuma et al. 2020). Unfortunately, in Rwanda, blanket recommendations (80 N, 34 P, 34 K kg ha^{-1}), developed three decades ago, are still being used regardless of the crop species, variety and the array of soil, climatic and cropping conditions existing in the field. This is not likely to result into optimal efficiency and benefits (Elrys et al. 2021). There is urgent need to update and tailor fertilizer needs to a range of production conditions. Rwanda's population is predicted to be 20,000,000 in 2030 (AU 2020). Providing food for the growing human population from limited arable land poses serious challenges. In response to these challenges, the Government of Rwanda has identified rice as a priority crop since 2002. Subsequently, the country has invested tremendous amounts of resources into the development of the rice sub-sector. Rising incomes, growing urban population and changing lifestyles have further increased the demand for rice in the country (FAO 2018). In response to these concerns, the Rwandan Government, through the Rwanda Agriculture Board (RAB), conducted a Participatory Varietal Selection exercise in different marshlands under rice cultivation. In each marshland, 3–4 long grain varieties were selected by farmers according to the following criteria: resistance to diseases, tillering capacity, early maturity, number of panicles and number of grains per panicle (CARD 2018). Soil infertility is a constant threat to sustainability of rice cultivation in marshlands (Lambrecht et al. 2016). The low input intensive mono-cropping pattern in the marshlands constantly depletes the soil and water reserves. Lack of suitable fertilizer recommendations and high fertilizer costs are two major reasons for the poor nutrient management in rice fields (Zenna et al. 2017). Since rice production is strongly linked to the amount of N fertilizer applied to soils, it is important to improve the management of this soil nutrient in order to increase yields and profitability of rice (Massoa et al. 2016). Hence, the overall objective of this study was to contribute to the enhancement of soil nitrogen fertilizer management under rice cultivation in tropical lowland Eastern Rwanda. The specific objectives of the study were: to determine the response of WAB and IR 28 rice varieties to different nitrogen rates for increased rice production; to determine the optimum N rate and placement method for lowland rice, and; to determine the use efficiency of nitrogen fertilizer rates.

2 Materials and Methods

This study was conducted in Cyaruhogo lowland which is located in the Eastern Province of Rwanda, at 1579 m altitude above sea level. The average annual temperature ranges between 16 and 30 °C without significant variations, with average rainfall of about 1250 mm per annum. The initial physico-chemical properties of the soils are presented in Table 2. The experiment was conducted in a factorial design with three factors i.e. nitrogen placements, nitrogen fertilizer rates and rice varieties. The first and third factors had two levels, while the second factor had five levels (Table 1).

The field experiment was laid out in a split-split-plot arrangement with 20 treatments replicated three times. The experiment was conducted in two seasons. The land was ploughed once with a mould board plough. Plots were marked and bunds made in heavy clay soil around each plot; this facilitated uniform irrigation and minimized plot contamination. A fertilizer dose of 34 kg ha^{-1} for P and K was applied in form of Triple Super Phosphate and Muriate of Potash. Phosphorus and potash were applied on the day of transplanting. N fertilizer was applied either at 10–15 cm depth using urea briquettes (deep placement) or on surface (broadcasting application) in three splits at 14 Days After Transplanting (DAT), at 28 DAT and at panicle initiation. Two rice varieties were used in this study i.e. WAB 880-1-1-20-28-P1-HB (WAB) and IR 28, both having African geographical origin, from Ivory Coast and Egypt, respectively. They were introduced in Rwanda by RAB's rice program in 2007 and released in 2010 after being tested under Rwandan conditions among other 408 varieties in the Eastern province. The potential yields for these varieties are 9.3 and 9.2 tons ha^{-1} for WAB and IR 28, respectively (CARD 2018). Seeds obtained from RAB were treated with fungicide before being sown in nurseries. The 21 days old seedlings were transplanted in plots of 2.5 × 3.5 m at 20 × 20 cm spacing. Five

Table 1 Treatments and treatment levels

Main plot Nitrogen placement	Sub plot Nitrogen rates (kg ha^{-1})	Sub-sub-plot Rice varieties
Broadcast placement (1)	0 (1)	WAB 880-1-1-20-28-P1-HB (1)
Deep placement (2)	40 (2)	IR 28 (2)
	80 (3)	
	120 (4)	
	160 (5)	

Table 2 Initial physico-chemical properties of the experimental site (Cyaruhogo marshland, Eastern Rwanda)

pH	Exchangeable P (mg kg^{-1})	Exchangeable K(cmol(+)kg^{-1})	Available N	Organic matter	Clay ← (%) →	Sand	Silt
5.68	8.26	6.8	0.26	4.18	49.54	29.32	21.14

hills were randomly selected and marked in plot; these hills were used for all data collection and the average from the five measurements was calculated for data analysis. Plant height measurements were taken before panicle initiation stage; the mean from five plants per treatment was determined. Leaf area data was generated at the end of vegetative growth. Leaf area was determined by the non-destructive length x width method using the relation Leaf area (cm^2) = K × length (cm) × width (cm), where K is a constant (Rachman et al. 2022). A constant (K) = 0.8 was used in this study. Leaf width and length were measured using a tailor's tape. The measurement was taken from the same five plants selected in each plot and four leaves in each of the five plants were measured. After calculation of leaf area for each plant, the average value was calculated for each plot. Panicle numbers were recorded at harvest by visual counting of panicle numbers of each of the five plants selected in the plots. The average number of productive tillers per plant in each plot was determined. Rice grains were harvested at maturity. At harvest, the rice grain from each of the five plants selected for growth and yield parameters were collected separately and labeled. After drying, one hundred grains were counted from each sample and weighed using a precision balance. The record for 100 grains was multiplied by 10 to get the 1000-grain weight. The grain weight for each treatment was recorded after sun-drying to a constant weight. The yields obtained per plot were converted to yield per hectare. Nitrogen Use Efficiency for cereals is defined as the total quantity of nutrient uptake in grains over the control per unit of nutrient applied. NUE (%) = (Gf-Gu/Na) * 100 where Gf is the total quantity of nutrient in grains of fertilized plot (kg), Gu is the total quantity of nutrient in grains of the unfertilized plot (kg) and Na is the quantity of nutrient applied (kg). Before transplanting, soil samples were drawn from the 0–30 cm depth by a core sampler of 8 cm diameter at six places in the experimental field in each block's location. The six samples were thoroughly mixed, bulked and a representative composite sample drawn for determination of organic C using the Walkley and Black method, total N using the Kjeldahl method, while extractable P was determined by the Olsen Method (in this method, P is extracted from soils using $NaHCO_3$ and then P is determined calorimetrically) (Okalebo et al. 1993). K was determined using the Mehlich III multi-nutrient extraction method (in this method, the extractant is a dilute acid-fluoride-EDTA solution of pH 2.5) (Mehlich 1978). At harvest, representative grain and straw samples of rice were drawn from the bulk produce of the net plot harvest for determination of N. For N determination, plant material was digested separately and analyzed by the modified Kjeldahl method (Okalebo et al. 1993). In this study, linear, quadratic, square root and Mitscherlich models were tried. The relationship between yield and fertilizer dosage was found to be best defined by the quadratic model as follows: $Y = a + bx + cx^2$ where Y = grain yield (kg ha^{-1}); X = the dosage of the nitrogen applied (kg ha^{-1}) and a, b and c are model parameters. Estimated nutrient efficiency rate values were obtained from the model by finding "x" by equating the first derivatives of the reaction equations to zero. Data collected from the different growth stages was subjected to statistical analysis. The Genstat 14th edition software and ASSISTAT Version 7.6 beta tools were used to carry out analysis of variance and means separation. Ms Excel software was used for regression analyses. The level of significance used in the "F" test was 5% ($\alpha = 0.05$).

3 Results and Discussion

3.1 Effect of N Rates, Modes of N Placement and Rice Varieties on Soil Nitrogen Content and Nitrogen Uptake by High Yielding Rice Varieties

3.1.1 Experimental Site

The experimental site soil was clay loam in texture. A composite sample was analyzed for physico-chemical properties (Table 2). Analysis of soil samples revealed that Organic matter content was above the critical level for Rwandan soils (>2%) and the N content in soil was 0.26%.

3.1.2 Effect of N Rates, Modes of N Placement and Rice Varieties on Soil N Concentration

The highest N rate (160 kg ha^{-1}) gave a soil N content higher than the initial one. Soil N content increased with increase in N rates; at 40, 80 and 120 kg N ha^{-1}, the soil N content after the experiment was set up was almost the same as the initial one; however, it increased abruptly at 160 kg ha^{-1} (Fig. 1).

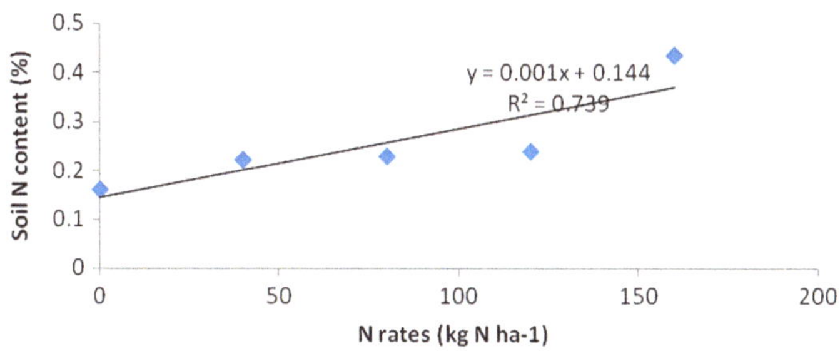

Fig. 1 Effect of N rates on soil N content

3.1.3 Effect of N Rates, Modes of N Placement and Rice Varieties on N Concentration in Plant Tissue by High Yielding Rice Varieties

Both the WAB 880-1-38-20-28-P1-HB and IR 28 rice cultivars' N content in plant tissue was unaffected by the mode of N placement (Table 3).

There was a strong correlation between the quantity of N applied and its concentration in grains. Grain N concentration increased with increase in N rates up to 120 kg ha^{-1}; then it decreased at 160 kg N ha^{-1} (Fig. 2).

Except for N placement, all other treatments (N rates, varieties and seasons) had significant effects on N concentration in grains. Nitrogen concentration in plant tissue was lower at low N rates than at high N rates; but for both varieties, the N concentration in plant tissue was not significantly different at 80, 120 and 160 kg N ha^{-1} rates. These results are supported by Singh (2020) who observed that application of nitrogen fertilizers promotes the concentration of nitrogen in plants. During

Table 3 Effect of N placement, N rates, rice varieties and seasons on N concentration (%) in plant tissue

	N placement*	N rates**	Rice varieties***	Seasons****
1	2.241 a	1.728 d	2.162 b	2.183 b
2	2.231 a	1.970 c	2.310 a	2.289 a
3		2.414 b		
4		2.768 a		
5		2.301 b		
Smd	0.0827	0.1308	0.0827	0.0827

Smd Significant minimum difference; *N Placement: 1. Broadcast placement, 2. Deep placement; **N Rates: 1: 0 kg ha^{-1} 2. 40 kg ha^{-1} 3. 80 kg ha $^{-1}$ 4. 120 kg ha^{-1} 5. 160 kg ha^{-1};***Varieties: 1. WAB 880-1-1-20-28-P1-HB, 2. IR 28; ****Seasons: 1. Wet season (January-June), 2. Dry season (July-December)

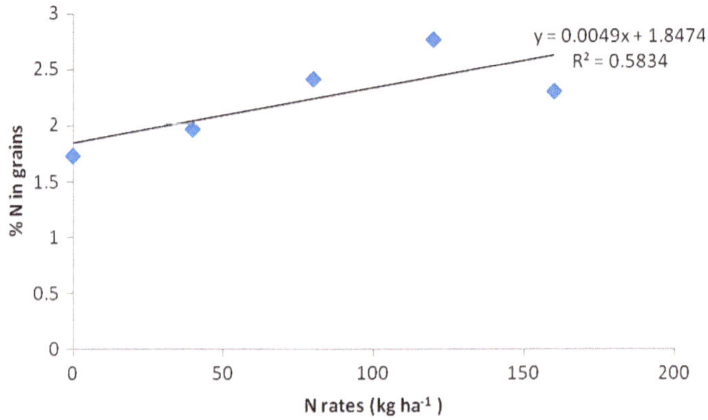

Fig. 2 Relationship between N rates and rice plant N concentration

Table 4 Effect of N rates, modes of N placement and rice varieties on plant height (cm)

	N placement	N rates	Varieties	Seasons
1	85.5 b	79.8 d	100.3 a	87.8 a
2	87.9 a	85.1 c	73.1 b	85.6 b
3		87.4 b		
4		89.0 b		
5		92.3 a		
Smd	0.9	1.9	0.9	0.9

Smd Significant minimum difference; N placements, N rates, varieties and seasons are as explained in Table 3

the productive stage, deep placement produced almost the same effect as broadcast placement; this might be attributed to the greater synchrony between crop demand and nutrient supply. This is because of split applications during the productive stage rather than a single dose as in the case with deep placement application (Suriyagoda 2022). However, these results are in contrast with those by Saito et al. (2019) who found out that urea deep placement had a significant effect on growth and yield parameters.

3.2 Response of High Yielding Rice Varieties to Various Nitrogen Levels

3.2.1 Effect of Applied N on Rice Growth Parameters

Plant Height

There was an increasing trend in rice plant height with increase in each successive level of fertilizers for both the WAB 880-1-38-20-28-P1-HB and the IR 28 varieties (Table 4). Plant height increased for about 3 cm for each 40 kg added. However, for the IR 28 variety, the plant height doesn't change significantly when N rates are increased from 40 to 120 kg ha^{-1} (Table 5).

Leaf Area

The mode of N placement significantly affected the leaf area of WAB 880-1-38-20-28-P1-HB and IR 28 varieties; deep placement gave bigger leaf areas than broadcast placement (Table 6). Leaf area increased with increasing N rates and plant height. Our results are similar to those obtained by Drinkwater and Snapp (2022) who found that with provision of adequate quantities of Nitrogen during the vegetative stage, rice plants become vigorous and plant height increases.

Table 5 Interaction effects between N rates and rice varieties and seasons on plant height (cm)

Rates(kg N ha^{-1})	Varieties		Seasons	
	WAB 880-1-1-20-28-P1-HB	IR 28	WET	DRY
0	92.1 eA	80.1 cA	79.5 dA	80.1 cA
40	97.8 dA	84.7 bA	85.4 cA	84.7 bA
80	100.8 cA	86.9 abA	87.9 cA	86.9 abA
120	103.8 bA	86.5 bB	91.5 bA	86.5 bB
160	107.3 aA	89.7 aB	94.9 aA	89.7 aB
Smd	Columns: 2.8	Rows: 2.0	Columns: 2.8	Rows: 2.0

Small letters: row comparison; Capital letters: column comparison; Smd: Significant minimum difference

Table 6 Effect of N rates, modes of N placement and rice varieties on plant leaf area (cm^2)

	N fertilizer placement	N rates	Varieties	Seasons
1	38.3 b	21.9 e	43.7 a	1.0 b
2	41.9 a	33.5 d	36.6 b	1.2 a
3		39.7 c		
4		48.6 b		
5		56.9 a		
Smd	0.9	1.9	0.9	0.9

Smd: Significant minimum difference; N placements, N rates, varieties and seasons are as explained in Table 3

The WAB 880-1-38-20-28-P1-HB rice variety showed greater plant heights and leaf areas as compared to IR 28; this might be due to genetic differences. Upland and lowland rice cultivars have different morphological features; the tillering ability, leaf area expansion and root development are quite different and all these differences are associated with dry matter production (Uwizeyimana et al. 2018).

3.2.2 Effect of Applied N on Rice Productive Parameters

Productive Tillers

The variety IR28 is shorter and less vigorous but has a higher number of productive tillers (Table 7). The response of both WAB 880-1-38-20-28-P1-HB and IR 28 varieties productive tillers to N rates was impressive; there was an abrupt increase in the number of productive tillers between 80 and 120 kg N ha^{-1} for both varieties. A greater number of productive tillers were observed in the dry rather than in the wet season.

Table 7 Interaction effects between N fertilizer rates and rice varieties and seasons on productive tillers (tillers per hill)

N rates (kg N ha^{-1})	Varieties	Seasons		
	WAB 880-1-38-20-28-P1-HB	IR 28	WET	DRY
0	5.2 cB	10.4 dA	7.2 dA	8.4 dA
40	6.8 bcB	15.2 cA	10.8 cA	11.2 cA
80	8.4 bB	20.4 bA	13.2 bB	15.6 bA
120	12.8 aB	30.0 aA	18.0 aB	24.8 aA
160	7.0 bcB	15.4 cA	10.2 cB	12.2 cA
Smd	Columns: 2.3	Rows: 1.6	Columns: 2.3	Rows: 1.6

Small letters: row comparison; Capital letters: column comparison; Smd: Significant minimum difference

Table 8 Interaction effect of N fertilizer placement and seasons on productive tillers (tillers per hill)

Placements	Seasons	
	DRY	WET
Broadcast	14.6 aA	11.0 bB
Deep	14.3 aA	12.8 aB
Smd	Columns: 1.0	Rows: 1.0

Small letters: row comparison; Capital letters: column comparison; Smd: Significant minimum difference

Table 9 Interaction effect of varieties and seasons on productive tillers

Varieties	Seasons	
	WET	DRY
WAB 880-1-38-20-28-P1-HB	7.4 bA	8.6 bB
IR 28	16.4 aA	20.4 aB
Smd	Columns: 1.0	Rows: 1.0

Small letters: row comparison; Capital letters: column comparison; Smd: Significant minimum difference

Deep placement gave a bigger number of productive tillers in the wet season than in the dry season although no significant difference in the number of productive tillers was observed between deep and broadcast placements in the dry season (Table 8). Both varieties produced a bigger number of productive tillers in the dry than in the wet season (Table 9).

The number of tillers increased with increase in nitrogen levels. Johnson et al. (2023) also reported increase in the number of tillers with increase in nitrogen rates. The IR 28 variety produced many productive tillers as compared to the WAB variety at the same nitrogen levels. Similar results were reported by Kabirigi et al. (2017) who found out that tiller productivity of rice depends on genetic potential.

Table 10 Effect of N rates, modes of N placement and rice varieties on 1000 grain- weight (g)

	N placement	N rates	Rice varieties	Seasons
1	27.0 a	26.6 bc	30.8 a	25.6 b
2	26.6 a	26.0 bc	22.9 b	27.7 a
3		27.3 b		
4		28.9 a		
5		25.2 c		
Smd	0.6	1.4	0.6	0.6

Smd: Significant minimum difference; N placements, N rates, varieties and seasons are as explained in Table 3

Table 11 Interaction effect of N fertilizer placement and rice varieties on 1000 grain-weight (g)

N placement	Rice varieties	
	WAB 880-1-38-20-28-P1-HB	IR 28
Broadcast	30.7 aA	23.4 aB
Deep	30.8 aA	22.3 bB
Smd	Columns: 0.9	Rows: 0.9

Small letters: row comparison; Capital letters: column comparison; Smd: Significant minimum difference

Table 12 Interaction effect of varieties and seasons on 1000 grain-weight (g)

Varieties	Seasons	
	WET	DRY
WAB	30.2 aB	31.3 aA
IR 28	21.6 bB	24.1 bA
Smd	Columns: 0.9	Rows: 0.9

Small letters: row comparison; Capital letters: column comparison; Smd: Significant minimum difference

1000-Grain Weight

N fertilizer placements and rates had no influence on the 1000-grain weight. The rate of 160 kg ha^{-1} (the highest) gave the least value (Tables 10 and 11). Both IR 28 and WAB 880-1-38-20-28-P1-HB varieties gave high 1000-grain values in the dry season (Table 10).

The 1000-grain weight was not influenced by increasing nitrogen rates, probably because it is a genetic trait; the same results were reported by Harerimana et al. (2023) who found out that nitrogen fertilization had no effect on 1000-grain weight in Eastern African conditions. The 1000-grain weight might have been influenced by other factors such as solar radiation and temperature rather than nitrogen fertilizer in Rwandan conditions. Other studies, for example, showed that 1000-grain weight decreased with incremental nitrogen fertilizers (Tuyishime 2023). However, there were significant differences in 1000-grain weight among thc two varieties; the WAB variety had greater 1000-grain weight values; this may not necessarily have been induced by nitrogen fertilization. As reported by Jiao et al. (2018), the individual grain weight is usually a stable varietal character and management practice has less effect on its variation (Table 12).

3.2.3 Effect of Applied N Fertilizer on Grain Yield

Grain yield increased with increase in nitrogen rates up to 120 kg ha^{-1}for both the WAB 880-1-38-20-28-P1-HB and IR 28 rice varieties; further increase of N fertilizer beyond 120 kg ha^{-1} resulted in rice grain yield decrease (Fig. 3).

Fertilizer application method had no influence on yield as shown in Table 13; the seasons and varieties had significant influence on yield, especially for the WAB variety.

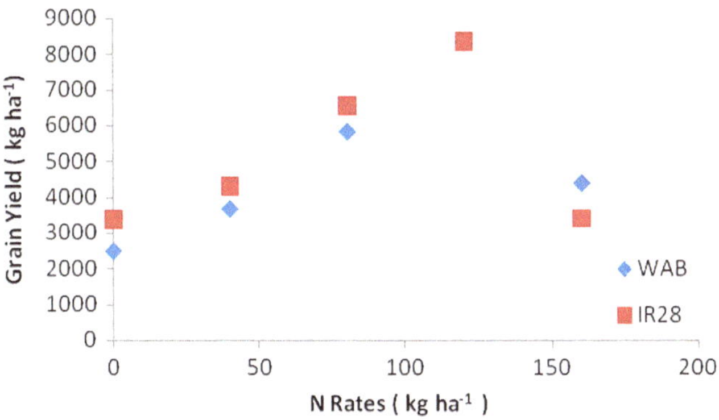

Fig. 3 Grain yield of WAB and IR 28 rice varieties at different N rates

Table 13 Effect of N rates, modes of N placement and rice varieties on grain yield (tons ha^{-1})

	N placement	N rates	Rice varieties	Seasons
1	4.986 a	2.933 d	4.939 b	4.875 b
2	5.161 a	3.986 c	5.207 a	5.271 a
3		6.195 b		
4		8.353 a		
5		3.899 c		
Smd	0.231	0.512	0.231	0.231

Smd: Significant minimum difference; N placements, N rates, varieties and seasons are as explained in Table 3

The WAB and IR28 varieties performed differently under the same climatic conditions; the IR 28 variety performed better in the wet season than in the dry season; in fact, the results from Table 14 show that during the wet season, at the same rates, there is a significant difference in grain yield between the two varieties, with IR 28 having the higher grain yield.

Grain yield was strongly influenced by the number of productive tillers (Fig. 4).

N rates increased grain yield through increasing the number of productive tillers. The productive tillers number had a quadratic association with grain yield and the influence was significant ($Y = 0.4165x - 0.4212$, $R^2 = 0.97$) (Fig. 5). Rautaray et al. (2020) reported a similar trend, with optimum grain yield obtained at 120 kg N ha $^{-1}$ for upland rice. Even at 40 kg N ha^{-1}, there was a significantly increased grain yield of the rice varieties, suggesting that N is a major limiting nutrient in the study area. This finding agrees with results obtained by Setiawati et al. (2020). With the current rate applied in Rwanda (80 kg N ha^{-1}), it is not possible to get good WAB 880-1-38-20-28-P1-HB variety performance (Khechba et al. 2021). Contrary to previous studies by Harerimana et al. (2021), nitrogen placements had no effect on yield, be it for the WAB 880-1-38-20-28-P1-HB or the IR 28 variety.

Table 14 Interaction effect of rice varieties and seasons on grain yield (tons ha^{-1})

Varieties	Seasons	
	WET	DRY
WAB 880-1-38-20-28-P1-HB	4.8587 aA	5.0200 bA
IR 28	4.8920 aB	5.5217 aA
Smd	Columns: 0.3261	Rows: 0.3261

Small letters: row comparison; Capital letters: column comparison; Smd: Significant minimum difference

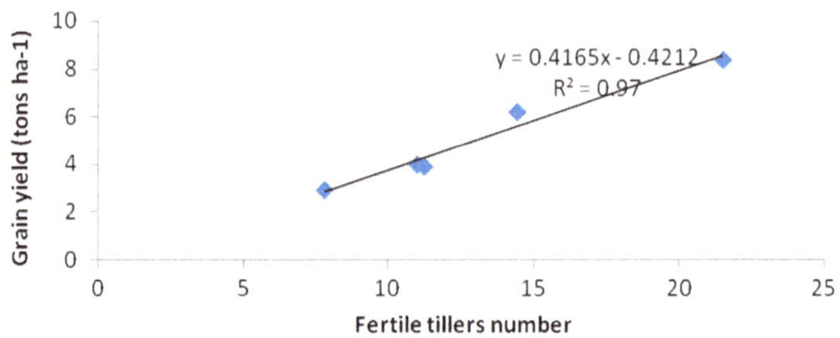

Fig. 4 Relationship between rice grain yield and productive tillers

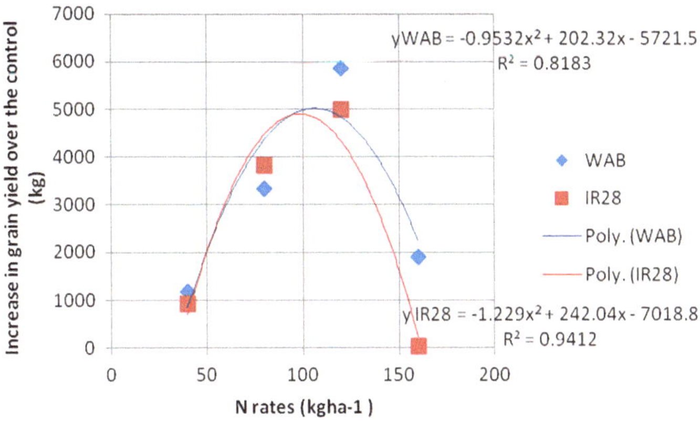

Fig. 5 Regression analysis between the increase in grain yield over the control and N rates for the WAB and IR28 rice varieties

3.3 Optimum N Rate Determination for the High Yielding Rice Varieties

The maximum point, which is the efficient or optimum rate, was determined from the quadratic relationship between the increase in grain yield and N rates (Fig. 5).

The WAB 880-1-38-20-28-P1-HB variety showed higher increase in grain yield over the control for all nitrogen rates as compared to the IR 28 variety. This might be attributed to genetic traits. Sesiawati et al. (2020) reported that some genotypes have potential to produce better grain yield than others at the same optimum rate. R^2 values (0.81 and 0.94, for WAB 880-1-38-20-28-P1-HB and IR 28, respectively) from the regression analysis of economic grain yield and nitrogen rates prove that nitrogen is the most limiting nutrient for getting high yields for rice production in lowland Eastern Rwanda.

3.4 Nitrogen Use Efficiency (NUE-In Percentage) of WAB and IR28 Varieties

WAB variety had higher NUE values compared to IR28 variety at all N rates (Table 15). This shows that WAB variety utilized nitrogen more efficiently than IR28 variety. These results show that variety IR28 failed to effectively utilize the supplied nutrients and this consequently negatively affected its reproductive growth and led to the crop's failure. The NUE values increased with increase in N rates up to 120 kg ha^{-1}.

Table 15 Nitrogen use efficiency of the WAB and IR 28 varieties under different N rates

N rates	WAB variety		IR 28 variety	
	*N uptake in grains (kg ha^{-1})	**Nitrogen use efficiency (%)	N uptake in grains (kg ha^{-1})	Nitrogen use efficiency (%)
0	27.8 d	–	34.6 d	
40	36.4 c	21.5	39.6 c	12.5
80	49.1 b	26.7	51.2 b	20.8
120	65.9 a	31.8	67.6 a	27.5
160	48.9 b	13.2	35.8 d	0.8

*N uptake was calculated from N concentration in grains and grain yield. N content in grains was determined by multiplying N concentration in grains with the yield, **N use efficiency was calculated using the formula (Gf-Gu/Na)* 100 where Gf is the total quantity of nutrient in grains of a fertilized plot (kg), Gu is the total quantity of nutrient in grains of the unfertilized plot (kg) and Na is the quantity of nutrient applied (kg)

4 Conclusion and Recommendations

Plant height, leaf area and number of tillers of both WAB and IR28 rice varieties increased with increased N rates but not the 1000-grain weight. Deep placement was more efficient during the vegetative than productive phases for both the WAB and the IR28 varieties. Nitrogen placement had no effect on the grain yield of the two varieties. WAB and IR28 varieties gave higher yields in the dry season but IR 28 performed better in the wet season. Both varieties show great performance between the 80 and 120 kg ha^{-1} N rates and therefore, the blanket rate (80 kg N ha^{-1}) applied in the Eastern Rwandan conditions is not optimum and efficient. These results should be extended to other areas of Sub-Saharan Africa to test the performance of these varieties in different agro-ecological zones of the tropics.

Acknowledgements The authors are grateful to the Alliance for a Green Revolution in Africa (AGRA) and the Regional Universities Forum for Capacity Building in Agriculture (RUFORUM) for funding this research.

References

AU (African Union) (2020) Second biennial review report of the African Union Commission on the implementation of the Malabo declaration on accelerated Africa agricultural growth and transformation for shared prosperity and improved livelihoods. Assembly of the union Thirty-Third

CARD (2018) NRDS (National Rice Development Strategy) task force reports. coalition for African rice development. https://riceforafrica.net/index.php/nrds-page

Chuma BA, Cotter M, Kalisa A, Rajaona A, Senthilkumar K, Stuerz S, Vincent I, Asch F (2020) Altitude, temperature and N-Management effects on yield and yield components of contrasting lowland rice cultivars. J Agron Crop Sci 206:456–465. https://doi.org/10.1111/jac.12420

Cyamweshi AR, Kayumba J, Nabahungu NL (2017) Optimizing fertilizer use within the context of integrated soil fertility management in Rwanda. Fertilizer use optimization in sub-Saharan Africa. CABI, GB, pp 164–175

Drinkwater LE, Snapp SS (2022) Advancing the science and practice of ecological nutrient management for smallholder farmers. Front Sust Food Syst 6:921216

Elrys AS, Desoky ESM, Alnaimy MA, Zhang H, Zhang JB, Cai ZC (2021) The food nitrogen footprint for African countries under fertilized and unfertilized farms. J Environ Manag 279:111599. https://doi.org/10.1016/j.jenvman.2020.111599

FAO (2018) Rwanda: hunger and food insecurity. Available at: http://www.fao.org/faostat/en/#country/184

Harerimana B, Zhou M, Shaaban M, Zhu B (2021) Estimating nitrogen flows and nitrogen footprint for agro-food system of Rwanda over the last five decades: challenges and measures. Front Environ Sci 9:778699

Harerimana B, Zhou M, Zhu B, Xu P (2023) Regional estimates of nitrogen budgets for agricultural systems in the East African Community over the last five decades. Agron Sust Dev 43(2):27

Jiao J, Shi K, Li P, Sun Z, Chang D, Shen X, Wu D, Song X, Liu M, Li H (2018) Assessing of an irrigation and fertilization practice for improving rice production in the Taihu Lake region (China). Agric Water Manag 201:91–98

Johnson JM, Ibrahim A, Dossou-Yovo ER, Senthilkumar K, Tsujimoto Y, Asai H, Saito K (2023) Inorganic fertilizer use and its association with rice yield gaps in sub-Saharan Africa. Glob Food Sec 38:100708

Kabirigi M, Ngetich FK, Mwetu KK, Rushemuka P, Wasige EJ, Ruganzu VM, Nabahungu NL (2017) Implications of tillage practices, management of soil surface and fertilizer application on sustainable dryland agriculture: a case study of Eastern Rwanda

Khechba K, Laamrani A, Dhiba D, Misbah K, Chehbouni A (2021) Monitoring and analyzing yield gap in Africa through soil attribute best management using remote sensing approaches: a review. Rem Sens 13(22):4602

Lambrecht I, Vanlauwe B, Maertens M (2016) Integrated soil fertility management: from concept to practice in Eastern DR Congo. Int J Agric Sustain 14(1):100–118

Massoa C, Ebanyat P, Baijukya F, Bekunda M, Bouaziz S, Wendt J, Vanlauwea B (2016) Nitrogen management for future food security: Sub-Saharan African case-study

Mehlich A (1978) New extractant for soil test evaluation of phosphorus, potassium, calcium, sodium, manganese and zinc. Commun Soil Sci Plant Anal 9:477–492

Okalebo JR, Gathua KW, Woomer PL (1993) Laboratory methods of soil and plant analysis. UNESCO-ROSTA, tropical soil biology and fertility programme. In: Soil science society of East Africa technical publication No1, p 88

Rachman B, Ariningsih E, Sudaryanto T, Ariani M, Septanti KS, Adawiyah CR, Yuniarti E (2022) Sustainability status, sensitive and key factors for increasing rice production: a case study in West Java Indonesia. Plos One 17(12):e0274689

Rautaray SK, Pradhan S, Mohanty S, Dubey R, Raychaudhuri S, Mohanty RK, Ambast SK (2020) Energy efficiency, productivity and profitability of rice farming using Sesbania as green manure-cum-cover crop. Nutri Cycl Agroecosys 116:83–101

Saito K, Vandamme E, Johnson J, Tanaka A, Senthilkumar K, Dieng I, Wopereis MCS (2019) Yield-limiting macronutrients for rice in sub-Saharan Africa. Geoderma 338:546–554. https://doi.org/10.1016/j.geoderma.2018.11.036

Senthilkumar K, Rodenburg J, Dieng I, Vandamme E, Sillo FS, Jean-Martial Johnson JM, Rajaona A, Ramarolahy JA, Abera BB, Kajiru GJ, Mghase J, Lamo J, Rabeson R, Kazuki Saito K (2020) Quantifying rice yield gaps and their causes in Eastern and Southern Africa. J Agron Crop Sci 206:478–490

Setiawati MR, Prayoga MK, Stöber S, Adinata K, Simarmata T (2020) Performance of rice paddy varieties under various organic soil fertility strategies. Open Agricult 5(1):509–515

Singh S (2020) Effect of nitrogen application through urea and Azolla on yield, nutrient uptake of rice and soil acidity indices in acidic soil of Meghalaya. J Env Biol 41(1):139–146

Suriyagoda LBD (2022) Rice production in nutrient-limited soils: Strategies for improving crop productivity and land sustainability. J Nat Sci Found Sri Lanka 50(3)

Tuyishime O (2023) Drainage intensity in paddy rice fields: nitrogen flows, salinity, rice yield and greenhouse gas emissions. Acta Universitatis Agriculturae Sueciae (58)

Uwizeyimana D, Mureithi SM, Karuku G, Kironchi G (2018) Effect of water conservation measures on soil moisture and maize yield under drought prone agro-ecological zones in Rwanda. Int Soil Water Cons Res 6(3):214–221

World Bank (2020) Rwanda food smart country diagnostic. International Bank for Reconstruction and Development, Washington, DC

Zenna N, Senthilkumar K, Sie M (2017) Rice production in Africa. In: Bhagirath S, Jabran CK, Mahajan G (eds) Rice production worldwide. Cham, Springer International Publishing, pp 117–135. https://doi.org/10.1007/978-3-319-47516-5_5.

Open Access This chapter is licensed under the terms of the Creative Commons Attribution 4.0 International License (http://creativecommons.org/licenses/by/4.0/), which permits use, sharing, adaptation, distribution and reproduction in any medium or format, as long as you give appropriate credit to the original author(s) and the source, provide a link to the Creative Commons license and indicate if changes were made.

The images or other third party material in this chapter are included in the chapter's Creative Commons license, unless indicated otherwise in a credit line to the material. If material is not included in the chapter's Creative Commons license and your intended use is not permitted by statutory regulation or exceeds the permitted use, you will need to obtain permission directly from the copyright holder.

Suitability Assessment of Arable Ferralitic Soils for Maize and Cassava Production in Uganda

Twaha Ali Basamba Ateenyi, Sam Komakech, Boshuwenda Chuma Andre, Ann Tumushabe, and Kassim Ssekabira

Abstract In Uganda, crop yields continue to decline due to soil degradation. This has led to large scale and continuous conversions of natural habitats into land for crop growing to feed the increasing human and livestock populations. The suitability of available arable soils for crop growth is critical for conservation of natural habitats. The study was aimed at ascertaining the suitability of Makerere University Agricultural Research Institute, Kabanyolo (MUARIK) arable soils for Maize (*Zea mays L.*) and Cassava (*Manihot esculenta*) production. A total of 30 composite samples were collected from 6 blocks and taken to Makerere University School of Agricultural Sciences (Soil, Water and Plant analytical laboratory) where soil physical and chemical properties were determined. Results show that the soils were generally acidic, with pH of 5.1 to 5.8. Soil Organic Carbon (SOC) and Nitrogen (N) were medium, while available P (Phosphorus) and Calcium (Ca) were deficient in the soil. Potassium (K) was low to high in content. The soil texture was clay. The soils at MUARIK are Ferrallitic (Oxisols). Analysis of variance showed that available P and soil pH were significantly different between the study blocks ($p < 0.05$). The rest of the soil properties were not significantly different across the site. The MUARIK soil's modest suitability for producing maize and cassava was determined through comparisons to the agronomic growth requirements for those crops. This soil's limited availability of P and K for growing maize is a severe constraint and the clayey texture and low calcium levels of the soils are a constraint for cassava growing.

Keywords Ferralsols · Arable land · Land degradation · Maize · Cassava

T. A. B. Ateenyi (✉) · S. Komakech
Department of Soil Science and Land Use Management, College of Agricultural and Environmental Sciences, Makerere University, Kampala, Uganda
e-mail: twaha.basamba@mak.ac.ug

B. C. Andre
Agriculture and Animal Development Board, Kigali, Rwanda

A. Tumushabe · K. Ssekabira
Department of Environmental and Biological Sciences, College of Engineering and Applied Sciences, Kampala International University, Kampala, Uganda

© The Author(s) 2026

B. R. Singh et al. (eds.), *Sustainable Food Systems in Sub-Saharan Africa*,
https://doi.org/10.1007/978-3-032-11288-0_5

1 Introduction

About 11% of the earth's surface is currently cultivated (1406 Mha) and about 24% is potentially arable, most of which (2500 Mha) is composed of acid soils, with 1700 Mha located in the humid tropics (Anusha et al. 2023). Thus, tropical acid soils, mainly Ultisols and Oxisols, represent the largest remaining soils potential for future agriculture. Acidity is a major determinant of soil fertility across the globe (Bharti et al. 2017; Ismaili et al. 2023). Given the effect of soil acidity on crop growth, such as aluminium toxicity, nutrient availability and microbial activities, a great challenge is likely to be encountered in future agriculture.

Almost 50% of the world's arable land is in Sub-Saharan Africa (SSA), yet only a small percentage of it meets the requirements for agricultural crop production. Present crop yields in Sub-Saharan Africa are low and frequently much below the potential yields due to water limitations and climate change effects (Badapalli et al. 2023). In many smallholder farms in the savannah zone of West Africa, for example, decline in soil fertility, resulting in low crop production, is a major reason affecting food security (Keson et al. 2023) because sustainable crop intensification cannot be achieved if farmers cultivate inherently infertile soils that are deficient in essential nutrients such as nitrogen and phosphorus (Soropa et al. 2019; Bhular et al. 2023). This under performance is due to several factors. FAO & ITPS (2015) pointed out ten of the threats to soil functions and they include: soil erosion, Soil Organic Carbon (SOC) loss, nutrient imbalance, soil acidification, soil contamination, waterlogging, soil compaction, soil sealing, salinization and loss of soil biodiversity. In Sub-Saharan Africa, soil erosion is regarded as one of the primary causes of stagnant or declining agricultural output, the lack of proper supply of additional ecosystem services, as well as the depletion of soil organic matter, soil nutrients and soil biodiversity (Tamene and Le 2015). The effects of soil degradation are felt most keenly in developing nations where a sizeable section of the population depends directly on the land for their livelihoods, as is the situation in SSA (Tully et al. 2015). Ssali (2000) revealed that Uganda's soils were initially considered fertile and little was done to improve productivity in a systematic way. However, confirmatory research on land suitability evaluation frameworks with special focus on the significance of soils in crop cultivation is missing (Sharififar et al. 2016). Yet knowledge of the kind of soils and the extent of their distribution are pre-requisites in developing rationale and local level land use plans for agriculture, forestry and other land uses (Radocaj and Jurisic 2022). Semalulu and Kaizi (2015) noted the difficulties facing soil resources in general, including land degradation, a lack of understanding about good soil or land use practises, population pressure and soil pollution, among other issues. Continued land degradation is due to lack of environmental awareness among farmers, traditional agricultural practices, minimal inorganic fertilizer use and little to no use of improved crop varieties (Minai 2015). Proper interventions need to be undertaken to ensure that crop production is increased to match the rapidly increasing population of Uganda. One of such interventions is **'soil suitability evaluation'**. Soil suitability is a pre-requisite for sustainable agriculture and plays a pivotal role in crop production, especially in degraded lands (Mustafa et al. 2016; Li et al. 2023). In this study, the suitability of arable ferralitic soils for the growth of Maizc (*Zea mays* L.) and Cassava (*Manihot esculanta*) was assessed at Makerere University Agricultural Research Institute, Kabanyolo (MUARIK) by identifying the various physical and chemical characteristics of the soils and then comparing them with agronomic growth requirements for maize and cassava.

2 Materials and Methods

2.1 Description of the Study Area

MUARIK is located 21 km north of Kampala, the capital city of Uganda, along the Gayaza-Namulonge road on spatial coordinates 0°27′60"N, 32°36′24"E at an altitude range of 1250 to 1320 m above sea level. It receives an average annual precipitation of 1218 mm and slightly drier periods from June to July and December to February. The average annual temperature is 21.5 °C. Bananas are the main food crop and Robusta coffee is the main cash crop. Maize is the secondary cash crop and sweet potatoes are a secondary food crop (UBOS 2022). The vegetation is mainly forest savannah with pastures suitable for intensive livestock production.

2.2 Soil Sampling

Composite soil samples were collected following the standard procedures as explained in Okalebo et al. (2002). Sub-samples were collected to a depth of about 30 cm (effective rooting depth) from randomly selected locations and thoroughly mixed to obtain 5 representative composite samples from each block so that analysis of these samples gives average values for the entire block. Soil samples were collected using a hand hoe and trowel, placed in labelled plastic containers and transported to the laboratory for analysis. A total of 30 samples were collected from the 6 blocks in five (5) replicates (Table 1).

2.3 Laboratory Analyses

The soil samples were air-dried at room temperature for three consecutive days, sieved through a 2 mm sieve and then analysed for pH, organic carbon, available P, total N, exchangeable cations (Ca, Na and K) and texture. Soil pH was determined in a soil: water solution at a ratio of 1:2.5 using a pH meter. Available Phosphorus (P) was determined by using the Bray and Kurtz No. 1 method. The soil was extracted

Table 1 Soil samples collected at Makerere University Agricultural Research Institute Kabanyolo (MUARIK)

Block (s)	Soil samples				
A (Maize)	A1	A2	A3	A4	A5
B (Banana plantation)	B1	B2	B3	B4	B5
C (Cassava plot)	C1	C2	C3	C4	C5
S (Sorghum plot)	S1	S2	S3	S4	S5
W (Water melon plot)	W1	W2	W3	W4	W5
CB (Cassava-banana intercrop)	CB1	CB2	CB3	CB4	CB5

using Bray 1 solution and the P determined by the calorimetric procedure using a spectrophotometer. Soil texture was determined using the hydrometer method as described by Okalebo et al. (2002). The percentages of sand, silt and clay were obtained and the soil textural class ascertained using the textural triangle. Exchangeable K, Ca and Na were obtained by treating the soil samples with excess 1 M ammonium acetate solution. Later, the concentrations of exchangeable Na and K in the extract were measured by flame photometer and the concentration of Ca was measured by atomic absorption spectrophotometry. The Kjedhal method of determining nitrogen was used due to its convenience and provision of fast results, while the quantitative Black and Walkley method of determining Total Organic Carbon (TOC) was used because of its accuracy and quick generation of results as outlined by Okalebo et al. (2002).

2.4 Data Analysis

Laboratory data on soil physical and chemical properties was entered into MS Excel to generate frequencies, means, standard errors of the means, coefficients of variation, percentages and correlation coefficients as explained by Tashayo et al. (2020). The data was fed into GenStat® (12th edition) statistical software and analysis done for one-way analysis of variance (ANOVA) (in randomised blocks) at 5% significance level to ascertain the significance of variations in soil properties between the blocks. Correlations were run to ascertain the strengths and directions of association between soil physical and chemical parameters.

3 Results and Discussion

3.1 Soil Chemical Properties

3.1.1 Soil pH

Soil pH ranged from 4.9 to 6.2 (Table 2). The average soil pH for the blocks was 5.5. Analysis of variance showed significant differences in soil pH between blocks (Table 3). Soil pH was generally moderately acidic. The results on soil pH concur with those of Okiror et al. (2017) and Majaliwa (1998) who reported MUARIK soils to be acidic; this is probably caused by one or a combination of factors such as crop removal (the depletion of basic cations in crop harvest), continuous cultivation practices, leaching of basic cations, continuous use of only two acidifying fertilizers (urea and di-ammonium phosphate) coupled with the decomposition of organic residues that produce organic acids such as carbonic acid (Radocaj and Jurisic 2022; Badapalli et al. 2023). Base-rich nutrients get absorbed by plants rapidly, while leaving acid concentration behind. Frequent cultivation may result in rapid decomposition of organic matter and weakening of soil structure, which later results into lowering of soil pH (Vasu et al. 2018). According to FAO (1998), moderately acidic soils may be deficient in phosphorus, calcium, magnesium and molybdenum; this may be the reason for low P and Ca in our study blocks. The low pH can also be attributed to rainfall that is naturally acidic because of the presence of carbonic acid, which results from the atmosphere's carbon dioxide and moisture; MUARIK is located near Kampala Capital City where the atmosphere has more sulphuric and nitric acids from burning fossil fuels and acid rains from the highly polluted city (UBOS 2022).

Table 2 Means of SOC, N and P in the different blocks at MUARIK

BLOCKS		SOC	N	P
	pH	%		ppm
A	5.57 ± 0.105	1.71 ± 0.010	0.20 ± 0.006	4.21 ± 2.060
B	5.43 ± 0.188	1.69 ± 0.218	0.20 ± 0.028	3.39 ± 1.172
C	5.83 ± 0.128	1.69 ± 0.102	0.21 ± 0.020	1.42 ± 0.338
S	5.76 ± 0.079	1.58 ± 0.214	0.21 ± 0.014	2.64 ± 0841
W	5.51 ± 0.147	1.60 ± 0.110	0.19 ± 0.010	5.21 ± 1.485
CB	5.14 ± 0.131	1.42 ± 0.198	0.17 ± 0.005	12.11 ± 3.721
Grand mean	5.54	1.62	0.20	4.83
lsd (5%)	0.424	0.474	0.049	5.165
p-value	0.045	0.786	0.520	0.005
Critical levels*	5.5–6.5	1.5–3	0.12–0.25	15

*Okalebo et al. (2002)

Table 3 Summary of ANOVA for soil physical and chemical properties at MUARIK

Source of variation	df	pH	SOC	N	P	K	Na	Ca	sand	silt	clay
Sample	4	*	ns	ns	**	ns	ns	ns	ns	ns	*
Blocks	5										
Residual	20										
Total	29										

* and ** show significant differences at $\alpha = 0.05$ and $\alpha = 0.01$, respectively, and ns means not significant and df means degrees of freedom

According to Wanyama et al. (2021), maize cultivation may be feasible at a pH range of 5.0 to 8.0. The 5.1 to 5.8 pH range found in the soils at MUARIK indicates that these soils are excellent for producing maize. Yet according to Ogunde and Olanbo (2017) and Nanganoa et al. (2020), this soil pH is only moderately favorable for cassava production.

3.1.2 Soil Organic Carbon (SOC)

SOC ranged from 0.8 to 2.9% (Table 2). Block A had the largest mean SOC (1.7%), followed by Blocks B, W, S and CB which had the lowest SOC (1.4%). There were no significant differences in SOC between any of the sample blocks (Table 3). SOC was generally medium in content at MUARIK, according to classification of soil properties by Abegunde et al. (2015) (Table 4). SOC pool is the most reliable indicator for monitoring soil degradation, especially that caused by accelerated erosion (Lal 2015). According to Purnamasari et al. (2022), increase in the amount of clay increases the amount of carbon retained in soil because clay constitutes organo-mineral complexes. It combines with SOC in soil and helps to retain carbon within the soil for long periods. In this study, %clay and SOC had a significant negative correlation, meaning that the observed SOC cannot be attributed to the high clay content. Crop residues are normally used as animal feed during the dry season and also as a source of fuelwood; this has led to massive removal of crop residues (Bhullar et al. 2023). Climate also affects SOC content; according to Choudhary et al. (2023), SOC stocks increase based on annual precipitation and biomass amounts and decrease relative to temperature. Furthermore, high precipitation causes temperatures to decrease to some extent. SOC content decreases with increases in the annual temperature as reported by Sanchez et al. (2014) and Everest and Sungur (2021). Abagyeh et al. (2016) indicated that SOC level from 0.8% and above may be suitable for maize production. Since the SOC in this study ranges from 1.4 to 1.7%, it means that MUARIK soils are highly suitable for maize growing (Abegunde et al. 2015). However, this observed SOC is moderately suitable for cassava production according to Imakumbili et al. (2019).

Table 4 Average exchangeable bases (K, Na and Ca) at MUARIK

Block (s)	K	Na	Ca
	Cmol(+)/kg		
A	0.15 ± 0.025	0.05 ± 0.012	0.08 ± 0.004
B	0.21 ± 0.056	0.03 ± 0.002	0.08 ± 0.015
C	0.25 ± 0.025	0.04 ± 0.004	0.11 ± 0.016
S	0.15 ± 0.037	0.44 ± 0.409	0.25 ± 0.155
W	0.24 ± 0.054	0.04 ± 0.001	0.08 ± 0.005
CB	0.12 ± 0.011	0.04 ± 0.004	0.07 ± 0.009
Grand mean	0.18	0.11	0.11
lsd (5%)	0.117	0.490	0.192
p-value	0.148	0.445	0.390
Critical levels	2.0	< 1	0.8

3.1.3 Soil Nitrogen (N)

There was no significant difference in soil N content among the blocks (Tables 2 and 3). Soil N content varied from 0.12 to 0.28%. Nitrogen is the most critical element obtained by plants from the soil and is a bottleneck in plant growth. N content of soils generally ranges from less than 0.02% in sub-soils to more than 2.5% in peat soils (Tashoya et al. 2020). The medium to slightly high total N content in top soils in this study could be attributed to the poor drainage conditions of the soil that reduces the rate of organic matter decomposition as well as the addition of farmyard manure and plant residues that are rich in carbon and nitrogen (Choudhary et al. 2023). The observed nitrogen content, according to Nuga and Akinbola (2015), may probably be due to the medium level of SOC content and the main reason for this is high precipitation. Based on these soil nitrogen results, maize can be grown with ease because the nitrogen content highly suits maize production (Wanyama et al. 2021). However, there is need to increase the soil nitrogen content for cassava production because the observed nitrogen level in this study is a bit low for cassava production according to Purnamasari et al. (2022).

3.1.4 Available Phosphorus (P)

The overall mean for P was 4.83 ppm (below the critical level) (Table 2). Available P was generally low in the study area (grand mean is 4.83 ppm). Low P in soil causes lack of phosphorus in crops; this may prevent other nutrients from being acquired because P has a stimulating effect on root growth, since it is usually concentrated in the root tips of most plants (Tashayo et al. 2020). Low P might be due to soil erosion that washes away the top soils with relatively higher amounts of available P. Tillage has significant effects on the content of total, organic and available P (Everest and Sungur 2021). Therefore, low available P at MUARIK may probably be due to the tillage activities that are performed every year on the arable land. Low available P may also be due to the soil inherent low P content, high P-fixation by Al and Fe oxides and hydroxides that are predominant in the Oxisols at MUARIK and complete removal of biomass from the fields (Nigussie and Kissi 2012). Phosphorus sorption to Fe and Al oxides generally predominates in acid ($pH < 7.0$) to neutral soils (Nanganoa et al. 2020). Other studies have also demonstrated that acidic soils are generally deficient in available P (Vasu et al. 2018). These soils would most likely respond to judicious phosphorus-containing fertilizer application (Imakumbili et al. 2019). The available P recorded in this study is marginally suitable for maize production but is in moderate levels for cassava production (Li et al. 2023).

3.1.5 Exchangeable Cations (K, Na and Ca)

K values in the different blocks did not significantly differ from one another (Table 4). These results show that K, just like P, is a deficient nutrient at MUARIK for production of maize i.e. the soil K in this area is marginally suitable for maize growing (Anusha et al. 2023). Imakumbili et al. (2019) indicated that soil K levels from 0.15 to 0.25 Cmol(+)/kg are high for cassava production. Therefore, the soil K levels in this study are high for cassava production. K is essential for the general vigour of plants, supports development of a strong root system and increases resistance to certain diseases (Nanganoa et al. 2020). Analysis of variance showed no significant differences in soil Na levels between different blocks (Table 4). Soils are considered healthy and productive when they contain < 1 Cmol(+)/kg of Na and this renders MUARIK soil Na content suitable for majority of the crops, including maize. According to Keson et al. (2023), Na level < 2.0 Cmol(+)/kg is considered medium for cassava production and therefore the Na level in this study is suitable for cassava growing as well. Ca levels were far below the critical levels and not significantly different between blocks (Table 4). Ca was very low, much as Ca is present in adequate amounts in most soils (Li et al. 2023); this is because it is a component of several primary and secondary minerals in the soil, which are essentially insoluble. If the soil is made from limestone or another high calcium material, it will usually have more calcium; and if the soil is made from sandstone or shale, it will usually have lower levels of Ca (Choudhary et al. 2023). According to McLaughlin (2014), acid soils have less Ca and high pH soils normally have more. Therefore, the low Ca levels in MUARIK soils may be attributed to their acidic pH levels.

3.2 *Soil Physical Properties*

3.2.1 Texture

While there was no significant difference in the percentages of sand and silt between the blocks, there was a significant difference in the proportion of clay. Clay content

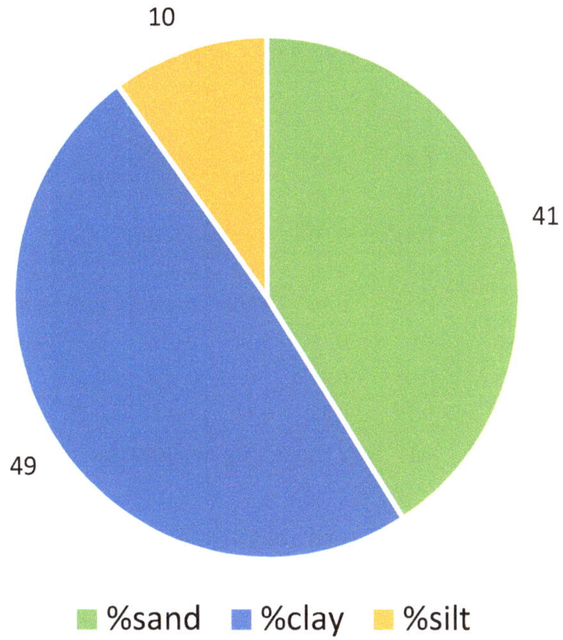

Fig. 1 Soil texture at MUARIK

was the highest (grand mean: 49%), followed by sand content (grand mean: 41%) and silt content (grand mean: 10%). For all the blocks, the soil's textural classification was clay (Fig. 1).

The general clay texture recorded in this study was also previously reported by Okiror et al. (2017). According to Nanganoa et al. (2020), clay percentage gives a clear indication of the problem nature of soils. A 10% clay threshold has been used as an indicator of the expansion potential, whilst 32% clay content exhibits extreme expansion potential (Bhullar et al. 2023). Such clay content has implications for the shrink-swell properties of the soil. Ismaili et al. (2023) agree that fine-textured clayey soils have small pores and release water gradually, which renders them susceptible to water logging because of their high-water retention capacity. Such soils may not be suitable or may be marginally suitable (depending on the type of clay) for cassava production which requires well drained soils for the cassava root tubers growth, but it is highly suitable for maize production (Nanganoa et al. 2020).

3.3 Results of Correlation Analysis

Sand and clay had the highest negative significant correlation ($r = -0.94$, $p < 0.01$), while Na and Ca had the highest positive significant correlation ($r = 0.99$, $p < 0.01$) (Table 5). pH had a positive significant correlation with SOC and K ($p < 0.05$). OC had a positive significant relationship with N and K ($p < 0.01$), positive significant

Table 5 Spearman's Correlations matrix of soil properties

Parameter	pH	%OC	%N	P	K	Na	Ca	%sand	%clay
pH									
%OC	0.37*	–							
%N	0.31	0.47**	–						
P	−0.31	0.18	−0.23	–					
K	0.41*	0.50**	0.41*	−0.10	–				
Na	0.17	0.25	0.07	0.002	0.12	–			
Ca	0.24	0.29	0.09	0.004	0.20	0.99**	–		
%sand	−0.02	0.38*	0.14	0.016	0.20	−0.06	−0.03	–	
%clay	−0.06	−0.44*	−0.17	0.030	−0.21	0.05	0.04	−0.94**	–
%silt	0.19	0.35	0.14	−0.12	0.15	−0.01	−0.03	0.32	−0.63**

Values with * and ** were significantly correlated at $\alpha = 0.05$ and $\alpha = 0.01$, respectively

correlation with sand and a negative significant correlation with clay ($p < 0.05$). Clay and silt had a negative significant correlation ($p < 0.01$). A high negative significant correlation between sand and clay was also reported by Okiror et al. (2017). This result is not uncommon because these two soil components are complementary to each other (Badapalli et al. 2023). The high positive correlation between Ca and Na can be attributed to the possibility of the two minerals having similar parent material mineralogy. The positive significant correlation between SOC and N concur with findings by Wanyama et al. (2021) where it was found out that increase in SOC increased total nitrogen content in soil.

3.4 Comparison of MUARIK Soil with Requirements for Maize and Cassava Production

Soil characteristics of the MUARIK region were compared to the norms for growing cassava and maize. The soil pH, SOC, OM, N and texture were found to be highly favourable for maize cultivation, while P and K were only marginally favourable for maize production, according to studies by Abagyeh et al. (2016) and Mustafah et al. (2016) (Table 6). Cassava could be grown in soil with moderate/medium pH, P and Na levels. Cassava development required low SOC, OM and N levels. For the production of cassava, soil K level was high and Ca concentration was quite low. If the soil clay is structurally dominant, the texture is marginally to moderately acceptable for cassava development or not suitable.

Table 6 Comparisons of MUARIK soil with requirements for maize and cassava production

Parameter (s)	MUARIK soil	Grand mean	Comparisons	
			Requirements for Maize	Requirements for Cassava
pH	5.14–5.83	5.54	*S1	**3
OC (%)	1.42–1.71	1.62	S1	2
OM (%)	2.44–2.95	2.79	S1	2
N (%)	0.17–0.21	0.20	S1	2
P (ppm)	1.42–12.11	4.83	S3	3
K (Cmol(+)/Kg)	0.12–0.24	0.18	S3	4
Na (Cmol(+)/Kg)	0.03–0.44	0.11	S2	3
Ca (Cmol(+)/Kg)	0.07–0.25	0.11	S3	1
Textural class	Clay	Clay	S1	S2–S3 N

**1 = Very low; 2 = Low; 3 = Moderate; 4 = High; 5 = Very high for cassava production. *S1 = highly suitable; S2 = Moderately suitable; S3 = Marginally suitable; N = Not suitable for maize or cassava production

4 Conclusion

MUARIK soils were highly and moderately suitable for maize and cassava production, respectively. For successful cassava production at MUARIK, Ca levels should be increased in the soil, preferably by applying gypsum soil amendment (calcium sulphate) to raise the soil pH. Given the low Ca concentration of the soil and the need for fertiliser, Ca-NPK would be a preferable inorganic fertilizer choice. Since most of Uganda is covered by acidic soils, mostly Ferralsols, the significance of our findings is that maize and cassava production should be promoted on a commercial scale so that our farmers can engage in massive production of these two commodities for urban markets and agro-industrial development.

References

Abagyeh SOI, Iodoga S, Agber PI (2016) Land suitability evaluation for maize (Zea mays) production in selected sites of the Mio-Benue valley, Nigeria. Int J Agric Policy Res 4(3):46–51

Abegunde L, Adedeji O, Tope-Ajayi O (2015) Land suitability analysis for maize production in Egbeda Local Government Area of Oyo State Using GIS Techniques. Int J Environ Ecol Eng 9(3):78–91

Anusha BN, Raghu KB, Kumar BP, Sree PP, Veeraswamy G, Swarnapriya C, Rajasekhar M (2023) Integrated studies for land suitability analysis towards sustainable agricultural development in semi-arid regions of AP India. Geosyst Geo-Environ 2(2):100–131. https://doi.org/10.1016/j.geogeo.2022.100131

Badapalli PK, Kottala RB, Madiga R (2023) Land suitability analysis and water resources for agriculture in semi-arid regions of Andhra Pradesh, South India using remote sensing and GIS techniques. Int J Energ Water Res 7:205–220. https://doi.org/10.1007/s42108-021-00151-3

Bharti VS, Dotaniya ML, Shukla SP, Yadav VK (2017) Managing soil fertility through microbes: prospects, challenges and future strategies. In: Agro-environmental sustainability. Springer, Cham, pp 81–111

Bhullar A, Nadeem K, Ali RA (2023) Simultaneous multi-crop land suitability prediction from remote sensing data using semi-supervised learning. Sci Rep 13:6823. https://doi.org/10.1038/s41598-023-33840-6

Choudhary K, Boori MS, Shi W, Valiev A, Kupriyanov A (2023) Agricultural land suitability assessment for sustainable development using remote sensing techniques with analytic hierarchy process. Soc Environ 32:101051. https://doi.org/10.1016/j.rsase.2023.101051

Everest T, Sungur A (2021) Özcan, H (2021) Determination of agricultural land suitability with a multiple-criteria decision-making method in Northwestern Turkey. Int J Environ Sci Technol 18:1073–1088. https://doi.org/10.1007/s13762-020-02869-9

FAO (Food and Agricultural Organisation of the United Nations) and ITPS (Intergovernmental Technical Panel on Soils) (2015) Status of the World's Soil Resources (SWSR)–technical summary. food and agriculture organization of the United Nations and intergovernmental technical panel on soils, Rome, Italy

FAO (Food and Agricultural Organisation of the United Nations) (1998) Guidelines on Land Evaluation for Rain fed Agriculture. Soils Bulletin No.52, FAO, Rome, Italy

Imakumbili ML, Semu E, Semoka JM, Abass A, Mkamilo G (2019) Soil nutrient adequacy for optimal cassava growth, implications on cyanogenic glucoside production: a case of konzo-affected Mtwara region Tanzania. PLoS ONE 14(5):e0216708

Ismaili M, Krimissa S, Namous M, Htitiou A, Abdelrahman K, Fnais MS, Lhissou R, Eloudi H, Faouzi E, Benabdelouahab T (2023) Assessment of soil suitability using machine learning in arid and semi-arid regions. Agronomy 13:165. https://doi.org/10.3390/agronomy13010165

Keson J, Silalertruksa T, Shabbir HG (2023) Land suitability class and implications to Land-Water-Food Nexus: a case of rice cultivation in Thailand. Energy Nexus10:100205, ISSN 2772–4271. https://doi.org/10.1016/j.nexus.2023.100205

Lal R (2015) Restoring soil quality to mitigate soil degradation. Sustainability 7(5):5875–5895

Li X, Wu K, Hao S, Yue Z, Ran Z, Ma J (2023) Mapping cropland suitability in China using optimized MaxEnt model. Field Crops Res 302:109064, ISSN 0378–4290, https://doi.org/10.1016/j.fcr.2023.109064

Majaliwa MJG (1998) Effect vegetation cover and biomass development on soil loss from maize-based cropping systems. Unpublished M. Sc. Thesis, Makerere University, Department of Agricultural Production, School of Agricultural Sciences, pp 79

McLaughlin JW (2014) Forest soil calcium dynamics and water quality: implications for forest management planning. Soil Sci Soc Am J 78(3):1003–1020

Minai JO (2015) Assessing the spatial variability of soils in Uganda, open access theses. http://docs.lib.purdue.edu/open_access_theses/581

Mustafa AA, Singh M, Ahmed N, Sahoo RN, Khanna M, Sarangi A, Mishra AK (2016) Soil suitability analysis for crop planning in Kheragarah Tehsil of Agra district, Uttar Pradesh. J Indian Soc Soil Sci 64(4):311–318

Nanganoa LT, Ngome FA, Suh C, Basga SD (2020) Assessing soil nutrients variability and adequacy for the cultivation of maize, cassava, and sorghum in selected agroecological zones of Cameroon. Int J Agronom 4:1–20

Nigussie A, Kissi E (2012) Physico-chemical characterization of a Nitisol in Southwestern Ethiopia and its fertilizer recommendation using NuMaSS. Glob Adv Res J Agric Sci 1(4):66–73

Nuga BO, Akinbola GE (2015) Land suitability evaluation of soils on two mapping units for some arable crops in south eastern Nigeria. J Agric Vet Sci 8(3):66–77

Ogunde A, Olanbo AR (2017) A web-based decision support system for evaluating soil suitability for cassava cultivation. ASTESJ 2(1):42–50

Okalebo JR, Gathua KW, Woomer PL (2002) Laboratory methods of plant and soil analysis: a working manual. TSBF-UNESCO, Nairobi, Kenya, pp 104
Okiror P, Lejju JB, Bahati J, Rukunda GK, Sebuuwufu CI, Mulindwa P, Ocan JJ (2017) Suitability of Kabanyolo soils for fruit and vegetable production. Open J Soil Sci 7:19–33. https://doi.org/10.4236/ojss.2017.72002
Purnamasari RA, Noguchi R, Ahamed T (2022) Land suitability assessment for cassava production in Indonesia using GIS, remote sensing, and multi-criteria analysis. In: Ahamed T (eds) Remote sensing application. new frontiers in regional science: Asian perspectives, vol 59. Springer, Singapore. https://doi.org/10.1007/978-981-19-0213-0_4
Radočaj D, Jurišić M (2022) GIS-based cropland suitability prediction using machine learning: a novel approach to sustainable agricultural production. Agronomy 12:2210
Sánchez B, Rasmussen A, Porter JR (2014) Temperatures and the growth and development of maize and rice: a review. Global Change Biol 20(2):408–417
Semalula O, Kaizi CK (2015) Overview of the status of soil resources in Uganda and the needs for and priorities for its sustainable management, NARO-Kawanda, pp 34
Sharififar A, Ghorbani H, Sarmadian F (2016) Soil suitability evaluation for crop selection using fuzzy sets methodology. Acta Agriculturae Slovenica 107(1):159–174
Soropa G, Nyamangara J, Nyakatawa EZ (2019) Nutrient status of sandy soils in smallholder areas of Zimbabwe and the need to develop site-specific fertiliser recommendations for sustainable crop intensification. South African J Plant Soil 36(2):149–151
Ssali H (2000) A century of soils research and development in Uganda. Uganda J Agric Sci 5:22–27
Tamene L, Le QB (2015) Estimating soil erosion in Sub-Saharan Africa based on landscape similarity mapping and using the revised Universal Soil Loss Equation (RUSLE). Nutr Cycl Agroecosyst 102(1):17–31
Tashayo B, Honarbakhsh A, Akbari M, Eftekhari M (2020) Land suitability assessment for maize farming using a GIS-AHP method for a semi-arid region, Iran. J Saudi Soc Agric Sci 19(5):332–338, ISSN 1658–077X. https://doi.org/10.1016/j.jssas.2020.03.003
Tully K, Sullivan C, Weil R, Sanchez P (2015) The state of soil degradation in Sub-Saharan Africa: baselines, trajectories, and solutions. Sustainability 7(6):6523–6552
UBOS (Uganda Bureau of Statistics) (2022) Statistical abstracts. Government of the Republic of Uganda, Kampala, p 336p
Vasu D, Srivastava R, Patil NG, Tiwary P, Chandran P, Singh SK (2018) A comparative assessment of land suitability evaluation methods for agricultural land use planning at village level. Land Use Policy 79:146–163
Wanyama D, Mighty M, Sim S, Koti F (2021) A spatial assessment of land suitability for maize farming in Kenya. Geocarto Int 36(12):1378–1395

Open Access This chapter is licensed under the terms of the Creative Commons Attribution 4.0 International License (http://creativecommons.org/licenses/by/4.0/), which permits use, sharing, adaptation, distribution and reproduction in any medium or format, as long as you give appropriate credit to the original author(s) and the source, provide a link to the Creative Commons license and indicate if changes were made.

The images or other third party material in this chapter are included in the chapter's Creative Commons license, unless indicated otherwise in a credit line to the material. If material is not included in the chapter's Creative Commons license and your intended use is not permitted by statutory regulation or exceeds the permitted use, you will need to obtain permission directly from the copyright holder.

Effects of Biochar Feedstock Type and Rates on Growth, Yield and Grain Quality in Soybean (*Glycine max* L.)

Mirriam Zeka, Keston Njira, Joseph Chimungu, Patson Nalivata, and Ivy S. Ligowe

Abstract Soil fertility decline pose a great threat to crop production in Malawi. Biochar application enhances crop yield and productivity; however, feedstock type and rates may influence its performance. The chapter illustrates how biochar application at different rates and feedstock types contribute to significant effects on plant height, stem diameter, leaf canopy and leaf area of soyabean causing a 13.8%, 20.6%, 50.6% and 19.1% increase, respectively. Biochar made from maize cores can cause an improvement in above ground biomass, below ground biomass, 100 seeds weight, and number of seeds per pod as compared to groundnut shell biochar. However, from this study reported in this chapter biochar did not affect grain weight and number of pods per plant. Data in the current chapter also reveals significant effects of biochar on Fe and Ca concentration in soybean grain, with no effects observed on Na, Mg, P, and Zn concentration. Nutrient uptake of P, Zn and Mg can also be affected by biochar, as compared to Na, K, Ca, and Fe. In this chapter, we have provided a synthesis of the impact of feedstock types and rates on growth, grain nutritional quality and yield of soybean. The chapter recommends a long-term field trial to conclusively ascertain the extent of biochar effects on soybean growth and yield, in different soil types and agro-ecological zones.

Keywords Biochar · Growth parameters · Nutrient uptake · Feedstock types · Yield

1 Introduction

Biochar is a carbon rich and highly stable, charcoal material produced by a thermochemical process called pyrolysis, where biomass is heated in the absence of oxygen (Baronti et al. 2010). It is produced from organic wastes such as agricultural residues, sewage sludge, or wood chips (Głodowska et al. 2017), grass, cow manure, wood chips, rice husk, wheat straw, and cassava rhizome. The extent of its effect on growth

M. Zeka · K. Njira (✉) · J. Chimungu · P. Nalivata · I. S. Ligowe
Lilongwe University of Agriculture and Natural Resources, Lilongwe, Malawi
e-mail: knjira@luanar.ac.mw

© The Author(s) 2026

B. R. Singh et al. (eds.), *Sustainable Food Systems in Sub-Saharan Africa*,
https://doi.org/10.1007/978-3-032-11288-0_6

and yield of food crops is determined by several factors including feedstock type, rate, as well as pyrolysis methods as these impact changes in soil features induced by the physical and chemical properties of biochar (Van Zwieten et al. 2010; Steiner 2009; Glaser et al. 2002; Zhu et al. 2015). On the other hand, Wang et al. (2016) in an on farm trial revealed that increasing biochar rate improved the plant growth parameters of soybean. Regarding feedstock types on biochar effects, Singh et al. (2010) observed that biochar had different properties and nutrient content depending on raw materials used. In addition, Filiberto and Gaunt (2013) found that biochar produced from wood, paper pulp, wood chips and poultry litter has positive impact on biomass yield. Zhang et al. (2012) also revealed that pyrolysis methods affected biochar properties such as biochar yield, soil pH, particle size, and surface area. Biochar is known to have several benefits including improved soil physical and chemical properties, improved biological nitrogen fixation (BNF), nodulation, growth and yield in different crops (Fig. 1).

Low yield is a major limiting factor in soybean production in Malawi as farmers obtain 800 kg ha^{-1} on average in comparison to its potential yield of over 4000 kg ha^{-1} (MoAFS 2012). This is due to a number of factors including poor soil fertility (Kananji et al. 2013), parasitic weeds, poor crop husbandry practices,

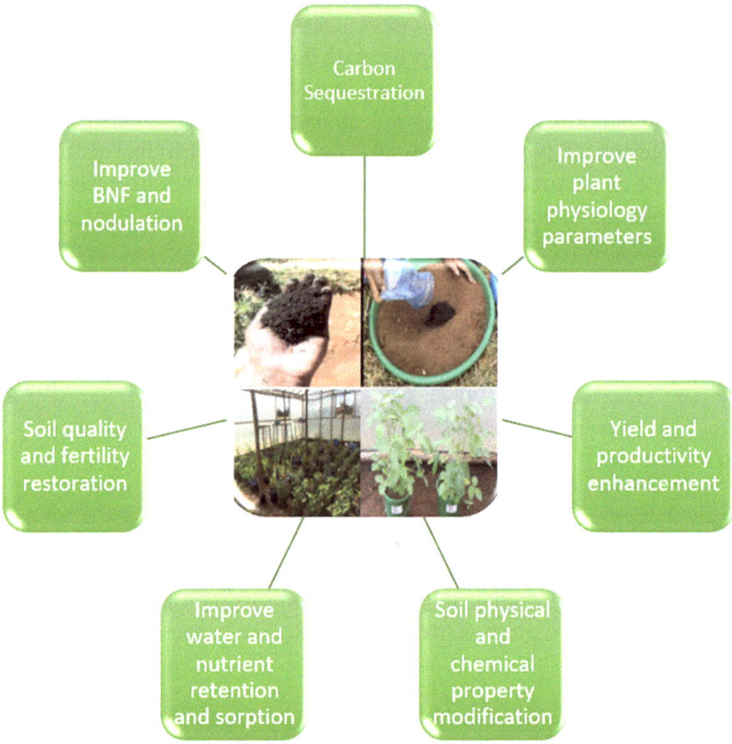

Fig. 1 Summary of the benefits of biochar to soil and plant health

drought, pests and diseases. Interestingly, biochar has the potential to improve plant growth (Ndor et al. 2016) and yield (Baronti et al. 2010) due to its ability in nutrient retention, encouraging nutrient availability for plant uptake which later can be utilized for plant growth and other metabolic processes hence improve yield. Various biochar feedstock types such as hydrochar and paper pulp at different rates, have been evaluated in other countries for their impact on yield and growth of various crops; however, data regarding use of maize cores and groundnut shell feedstock types at rates that could potentially improve growth and yield in soybean to improve food and nutrition security in Malawi is scanty. The objective of this chapter is to address specific biochar feedstock types and rates for locally available and economically feasible feedstock types for smallholder farmers in Malawi, with the aim of improving soybean growth, yield and grain nutritional quality at a minimal or no cost. The goal is to highlight the rates at which growth, yield and grain quality are enhanced in soybean in these feedstock types. Content in this chapter was derived from a study implemented to evaluate effects of biochar feedstock types and rates on growth, grain quality and yield of soybean in a green house at the crop and soil sciences students' research farm, LUANAR.

2 An Overview of Biochar

Biochar is amongst some of the many organic approaches in improving soil properties, plant yield and grain quality. Biochar is produced through a process known as pyrolysis from any organic material. It is a thermochemical process in which biomass is converted to biochar through heating with limited oxygen supply (Utomo 2014).

2.1 Effects of Pyrolysis Conditions and Biochar Feedstock Types on Soil Amendment Quality

Biochar has different properties depending on the raw materials and pyrolysis conditions used (Bonelli et al. 2010) which also affect the nutrients content of biochar (Singh et al. 2010). According to Nguyen et al. (2004), biochar produced from various organic materials under different conditions will have different effects on soils. The raw material and pyrolysis temperature used influence the quantity and dynamics of the release of nutrients and contaminants from the biochar (Tronto 2017). Furthermore, the biochar properties, such as elemental composition, structure, and chemical stability, depend on the raw material and the temperature of pyrolysis. Biochar from sewage sludge has the potential to improve the soil through release of P and micronutrients (Lu et al. 2013). It also reduces the volume of the solid residue, eliminates pathogens and organic compounds, and may also be used as a fertilizer or in remediation of contaminated soils. Demirbas (2004), Zhang et al. (2010), Hossain et al. (2011), observed that the charcoals produced from residues of grain hulls, husk, straw, and manure generally produce biochar with high ash content, which may be a result of the silica and the mineral content of the raw material. The result is a gradual loss of C, H, and O during processing. This calls for the evaluation of the impact of different biochar feedstock types and rates, and their amelioration effects on soil physical and chemical properties, growth, BNF, yield and nutrient uptake in soybean. Increasing pyrolysis temperature results in decreased biochar yield and nitrogen content due to volatilisation in biochar, and increased CEC, carbon content, surface area, pH, and ash content in biochar (Gaskin et al. 2008; Novak et al. 2009). Pyrolysis conditions that increase biochar yield include lower maximum temperature, slower heating rates, higher pressure, and a high concentration of lignin in the feedstock (Antal and Grønli 2003; Amonette and Joseph 2009). The most influential pyrolysis parameter on biochar yield is temperature, followed by heating rate (Amonette and Joseph 2009). Biochar possesses a diverse range of surface functionality including various amounts of acidic, basic, hydrophilic, and hydrophobic properties depending on the composition of the feedstock biomass and the pyrolysis conditions (Amonette and Joseph 2009). The CEC of biochar increases with increasing pyrolysis temperatures (Gaskin et al. 2008) due to the increased negative charges on its surface at high temperature as compared to low temperatures.

2.2 *Biochar Stability*

Biochar chemical and physical properties, as well as environmental factors such as temperature and rainfall, influence the decomposition rate of biochar (Major et al. 2010; Lehmann 2009). The stability of biochar is due to the transformation of the native carbon structure of the biomass to aromatic ring structures that takes place during the thermal treatment of organic matter (Tang and Bacon 1964). Glaser et al. (2001) reported that biochar is very stable due to its polycyclic aromatic carbon structures and can resist physical and microbial breakdown, allowing it to persist in soil. A study in Columbia demonstrated that biochar produced from mango trees via a simple earthen kiln mineralises very slowly in the soil (savanna Oxisol, sandy clay loam) with only 2.2% (when applied at 23.2 tonnes ha^{-1}) being lost by respiration over two years (Major et al. 2010). Surface area and particle size may also influence the decomposition rate by allowing more surface area for microbial and chemical reactions to occur (Lehmann et al. 2009).

2.3 *Biochar Effects on Soil Physical and Chemical Properties*

Malawi soils are dominated by Alfisols, Oxisols and Ultisols where their texture is generally sands and sandy loam (Snapp 2016). Application of biochar is known to have great impact on soil physical and chemical properties; however, this depends on the soil type, biochar rate, as well as biochar type. Recent studies have shown that the effects of biochar on the physical and chemical properties of soil are significantly influenced by both the type of soil and the specific characteristics of the biochar used. Application of biochar with a high acid neutralizing capacity (ANC) to acidic soils can effectively increase soil pH, while biochar with low ANC may not produce this effect. Furthermore, biochar can improve plant-available water in sandy soils by enhancing water retention; conversely, in heavy clay soils, it may reduce water content due to increased drainage. Biochar properties are also known to have been influenced by several factors including pyrolysis method among many. In another study, it was also observed that biochar from biomass pyrolysis has been proposed as an effective way of producing bioenergy, while at the same time improving some soil properties and sequestering carbon (Lehmann et al. 2006). Biochar addition to soil can produce changes in the soil's chemical and physical properties, including nutrient availability, CEC, pH, soil strength, and water holding capacity. Chan et al. (2008) reported that changes occur in the chemical properties of soil after biochar application. Studies have found that biochar addition to soil increases total Carbon (Van Zwieten et al. 2010), total N, pH, CEC, available P, and exchangeable cations (for example Ca, Mg, Na, and K) in soil (Chan et al. 2008). Similarly, Major et al. (2010) found that biochar addition increases available Ca, Mg, and pH in soil, but this may vary depending on the biochar rate and type used. Chan et al. (2007) also reported that addition of biochar produced from green waste (a mixture of grass clippings, cotton trash, and plant prunings) to soil resulted in increased organic carbon, available Na, K, and Ca, extractable P, and decreased available Al in the soil. Generally, these changes to soil characteristics are also proportional to the amount of biochar applied (Chan et al. 2007). One explanation of some biochar's ability to increase plant nutrient uptake is the formation of carboxylic groups on the edges of the aromatic carbon backbone that results from oxidation, leading to a greater ability to hold nutrients as the biochar weathers (Glaser et al. 2001). Among the many benefits associated with the use of biochar are the improvement of soil physical properties, including structure, density and pore size distribution, with implications for water holding capacity, aeration, soil workability as well as plant growth (Coumaravel et al. 2015).

2.4 Effects of Biochar on Nodulation and Biological Nitrogen Fixation

Nodule formation is a multi-step process involving movement of rhizobia by chemotaxis to the root hair. The attachment of rhizobia to the root hair induces deformation of growth in the root hair (Oke and Long 1999). This stimulates the division of the host cortical cells, culminating in infection and nodule development (Long 1990). Legumes are known to have a significant role in fixing nitrogen in the soil through a process called biological nitrogen fixation, estimated to contribute approximately

17.2 × 107 tons of nitrogen to soils globally each year (Azeem et al. 2019). In general, leguminous crops are estimated to contribute approximately half of the global symbiotic BNF at an estimated 21.5 × 106 tonnes (Azeem et al. 2019). Furthermore, Herridge et al. (2008) calculated soybean to fix 16.4 Tg N annually, representing 77% of the N fixed by the crop legumes. On the other hand, Güereña et al. (2015) discovered that addition of biochar could also improve the annual N fixed in the soil and improve nodulation and BNF. Regarding nodulation in soybean, Tagoe et al. (2008) and Ogawa and Okimori (2010) reported increased nodulation in response to short-term biochar addition due to increased population of soil microorganisms, including rhizobia bacteria which affect nodulation in legumes. The findings were contrary to a study by Quilliam et al. (2013) which showed no significant difference in the total number of root nodules of clover plants grown in soil having contained biochar for three growing seasons, regardless of the application rate. This raises a need to further explore various locally available feedstock types at rates that will enhance BNF in soybean.

3 Impact of Different Biochar Feedstock Type and Rates on Selected Soybean Growth Parameters

In assessing effects of biochar on yield and plant growth data on plant height (cm), leaf length (cm), leaf width (cm), leaf canopy (cm) and stem diameter (mm) was collected. These variables were measured using a 100 cm ruler. For leaf area, leaf length and width were measured consecutively which was then used to calculate the leaf area for the soybean leaves. Stem diameter was measured using a caliper, which was placed at the bottom of the stem and the diameter was recorded. Leaf area was estimated from non-destructive sample by using direct method to measure lamina length and width of all leaves in the pot of each treatment using a ruler.

3.1 Plant Height and Stem Diameter as Influenced by Different Biochar Rate and Feedstock Types

Results show significant effects of biochar rate on plant height and stem diameter. Biochar at the rate of 15 tonnes ha^{-1} increased stem diameter and plant height values significantly compared to the control treatment as more nutrients for growth and development were available. A similar trend was observed by Wang et al. (2016), where biochar increased stem diameter and plant height at the increasing rate of biochar applied to rice in tropical riparian wetlands, Indonesia. The study indicated that biochar incorporated in the soil improved growth parameters, including stem diameter and plant height. This could have been attributed to the fact that biochar improves soil properties, including CEC which have a nutrient retention ability that improves processes like photosynthesis, hence improving plant height and stem diameter. The findings corresponded with Tian et al. (2018) where increased biochar rate improved soil pH, reduced nutrients leaching especially of N, and improved CEC (Yooyen et al. 2015). Nitrogen is vital in forming part of chlorophyll component which plays a role in photosynthesis, thereby improving plant growth through photo-assimilates. This nutrient is usually available to plants in the form of nitrate and nitrate nitrogen. Ammonium nitrogen can be adsorbed on to soil colloids in the soil, and is readily exchangeable, as it easily dissolves in the soil solution, and can be directly absorbed by the plant (Wang et al. 2016). Increasing rates of biochar helps to improve availability of essential nutrients in the soil, which in turn aids in improving soybean growth parameters. The ability of biochar to improve nutrient retention and availability for plant use is evidenced by the significant differences in plant height and stem diameter of soybean growth parameter, which act as a morphological trait showing the availability of growth resource (Khan et al. 2008).

3.2 Leaf Canopy and Leaf Area in Soybean, as Affected by Biochar

Adding biochar shows significant increase in leaf canopy and leaf area. As the rate of biochar increases from 0 to 15 tonnes ha^{-1}, leaf canopy and leaf area showed the highest values in pots with 15 tonnes ha^{-1}(Table 1). A similar trend to this one is reflected in a study by Profile (2019) where rice husk biochar improved the leaf area of maize. Wang et al. (2016) also conducted a similar study of soybean in tropical riparian wetlands in Indonesia and found that every addition to the biochar rate increased leaf area in soybean. This might be attributed to the ability of biochar to retain nutrients. These nutrients play a big role in vital metabolic processes like cell division and photosynthesis, directly affecting plant growth in soybean. The result is improved growth and leaf expansion, which in turn enhances leaf canopy. Similarly, leaf area values were also positively influenced by biochar rate, where every addition of biochar depicted an increase in leaf area, resulting from improved photosynthesis and other vital processes in the plant. The increase in leaf area in soybean can also be attributed to the availability of nutrients and moisture, which allows sufficient N uptake and synthesis of nucleic acids, amides and amino acids (Mahboob et al. 2023), resulting in cell multiplication and increase of leaf area. Increased leaf area and canopy by biochar is a very vital development as it helps to improve growth and nutrient production in soybean, hence improved yield.

Table 1 Biochar feedstock types and rates in enhancing plant growth parameters in soybean

	Stem Diameter (mm)		Plant height (cm)			Leaf area (cm^2)			Leaf Canopy (cm)			
Biochar rate	G/nut Shell biochar	Maize cores biochar	Mean rate	G/nut Shell biochar	Maize cores biochar	Mean rate	G/nut Shell biochar	Maize cores biochar	Mean rate	G/nut Shell biochar	Maize cores biochar	Mean rate
0 tonnes ha^{-1}	2.65	2.51	2.57^a	38.35	36.29	37.32^a	40.40	36.60	38.47^a	30.80	29.67	30.24^a
5 tonnes ha^{-1}	2.85	2.71	2.78^b	39.68	37.55	87.62^a	41.90	38.70	40.31^b	32.09	31.53	31.35^b
10 tonnes ha^{-1}	2.91	2.86	2.88^c	41.34	39.80	40.57^b	44.70	40.70	42.70^c	31.56	31.15	31.81^b
15 tonnes ha^{-1}	3.23	3.09	3.16^d	42.80	43.13	42.97^c	49.00	45.50	47.22^d	32.46	31.31	31.89^b
Mean type	2.90	2.79		40.54	39.19		44.0	40.40				
F pr. type	0.59			0.367			0.07			0.082		
F pr. rate	< 0.01			< 0.001			< 0.001			0.015		
Fpr.(Type*Rate)	0.696			0.371			0.874			0.803		

Key**LSD* least significant difference, *CV* is coefficient of variation, *T* biochar type, *R* Biochar rate, *G/nut* groundnut. Within row means followed by the same letter are not significantly different

4 Biochar and Soybean Yield

Biochar rate showed significant effects on 100 seeds weight, which is a trait of yield components of soybean, and positively related with grain yield (Table 2). The increase in 100 seeds weight were attributed to the ability of biochar to improve nutrient retention, hence allowing for sufficient nutrients and water, which aid seed development. This in turn helps produce large and healthy seeds, hence increasing its weight. Results in this study stipulate that increasing rates of biochar will increase seeds weight significantly, which is in agreement with studies done by Yooyen et al. (2015) and Rab et al. (2016). Data regarding seeds/pod in the study also showed significant interactive effects of biochar type and rate ($p = 0.046$). This could be caused by the ability of biochar in improving C:N ratio in the soil (Rab et al. 2016), and hence improving the availability of nitrogen which in turn helps in seed pod formation. Biochar rate or feedstock type did not affect grain weight and pods/plant, but the increase in seeds per pod could be attributed to the ability of biochar to retain water as this improves soil porosity and helps in moisture retention (Li et al. 2021). Additionally, maize cores biochar contributed to higher yield as compared to groundnut shell biochar. Results also revealed higher variation in pods per plant as compared to seeds per pod. This was very likely because of the role of biochar in improving nutrients and water availability, causing an increase in vegetative growth, dry matter accumulation and longer linear growth rate, which directly influences higher number of pods per plant. This is in agreement with Berihun et al. (2017) who found that the number of pods was significantly affected by biochar. Above and below ground biomass of soybean was significantly affected by biochar rate and type, respectively as shown in Table 2. Plant biomass was measured at 50% flowering and harvesting, where destructive sampling was conducted and one plant in a pot was removed for biomass analysis. The total fresh plant weight was observed, which was composed of below and above ground biomass. After this, the stalks were oven dried at 60 °C for 56 h in order to observe the dry biomass on the selected plants. The biomass increase could be related to an early increase in leaf canopy which allowed enough area for photosynthesis, thereby improving accumulation of dry matter in the crop. This agrees with the results presented by Seadh and Abido (2013), where early leaf canopy closure enhanced maximum leaf area index, consequently increasing dry matter accumulation. In the current study, biochar rate affected leaf canopy from as early as week two, which allowed enough surface area for photosynthesis, and thus led to improved yield and growth. Similarly, Baronti et al. (2010) found that a minimum biochar rate of 10 tonnes ha^{-1} increased the above ground biomass by 23% compared to the control. Biochar caused significant effects for rate ($p = 0.07$) and interaction between biochar rate and type ($p = 0.037$) in total biomass of soybean crop. These findings are in agreement with the results by Barus (2017) where mixed rice husk biochar and compost manure significantly increased dry grain weight and biomass of soybean compared to the control. Biochar improves part of chlorophyll content in the leaf tissue, which enhances biomass production thereby increasing grain yield (Mete et al. 2015).

Table 2 Biochar rates and feedstock types in influencing soybean yield and seed parameters

Yield component	Biochar rate	Maize Cores biochar	G/nut shell biochar	Mean (Rate)	Yield component	Biochar Rate	Maize Cores Biochar	G/nutShell Biochar	Mean (Rate)
100 seeds dry(g)	0 tonnes ha^{-1}	12.08	11.16	11.62[a]	Pods/plant	0 tonnes ha^{-1}	30	11	20.5
	5 tonnes ha^{-1}	13.97	11.94	12.96[a]		5 tonnes ha^{-1}	17	22	19.5
	10 tonnes ha^{-1}	16.53	15.45	15.99[b]		10 tonnes ha^{-1}	19	17	18
	15 tonnes ha^{-1}	17.58	15.61	16.60[b]		15 tonnes ha^{-1}	19	28	23.5
	Mean (Type)	13.54	15.04			Mean (Type)	21.25	19.5	
	F pr. Rate		0.010			F pr. Rate		0.741	
	F pr. Type		0.533			F pr. Type		0.735	
	F pr.(T*R)		0.964			F pr.(T*R)		0.066	
	LSD (0.05)		6.616			LSD (0.05)		18.4	
	CV%		26.2			CV%		41.5	
Seedspod^{-1}	0 tonnes ha^{-1}	1.85[abc]	1.77[ab]	1.81	BG biomass(g/pot)	0 tonnes ha^{-1}	6.34	2.40	4.37
	5 tonnes ha^{-1}	1.70[a]	1.93[bc]	1.82		5 tonnes ha^{-1}	6.45	3.76	5.11
	10 tonnes ha^{-1}	1.85[abc]	1.85[abc]	1.85		10 tonnes ha^{-1}	4.41	4.13	4.27

(continued)

Table 2 (continued)

Yield component	Biochar rate	Maize Cores biochar	G/nut shell biochar	Mean (Rate)	Yield component	Biochar Rate	Maize Cores Biochar	G/nutShell Biochar	Mean (Rate)
	15 tonnes ha^{-1}	1.87^{abc}	2.02^{c}	1.95		15 tonnes ha^{-1}	6.18	4.31	5.24
	Mean (Type)	1.82	1.90			Mean (Type)	3.65	5.85	
	F pr. Rate		0.655			F pr. Rate		0.484	
	F pr. Type		0.230			F pr. Type		0.001	
	F pr.(T*R)		0.046			F pr.(T*R)		0.157	
	LSD (0.05)		0.197			LSD (0.05)		0.314	
	CV%		13.4			CV%		27.5	
Grain weight (g)	0 tonnes ha^{-1}	13.90	13.65	13.78	AG biomass(g/pot)	0 tonnes ha^{-1}	5.43	6.23	5.83^{a}
	5 tonnes ha^{-1}	12.52	13.90	13.21		5 tonnes ha^{-1}	6.80	7.10	6.95^{a}
	10 tonnes ha^{-1}	13.39	16.40	14.90		10 tonnes ha^{-1}	7.02	7.84	7.43^{a}
	15 tonnes ha^{-1}	17.18	17.03	17.11		15 tonnes ha^{-1}	11.72	9.96	10.84^{b}
	Mean (Type)	14.25	15.25			Mean (Type)	7.74	7.78	
	F pr. Rate		0.152			F pr. Rate		0.004	
	F pr. Type		0.630			F pr. Type		0.646	
	F pr.(T*R)		0.743			F pr.(T*R)		0.238	

(continued)

Table 2 (continued)

Yield component	Biochar rate	Maize Cores biochar	G/nut shell biochar	Mean (Rate)	Yield component	Biochar Rate	Maize Cores Biochar	G/nutShell Biochar	Mean (Rate)
	LSD (0.05)		6.200			LSD (0.05)		4.477	
	CV%		32.3			CV%		37.4	

5 Effects of Biochar on Grain Quality

Grain quality includes a range of properties that can be defined in terms of physical (moisture content, kernel size), sanitary (fungi and mycotoxin count), and intrinsic (fat content, protein content, hardness, starch content) quality characteristics (Ratnavathi et al. 2016). In this study we have focused on quality characteristics as a component in defining grain quality. Results from the grain analysis showed significant effects of biochar type, rate and their interactions on the nutrient uptake and concentration (Table 3). These effects were attributed to the nutrients present in biochar as well as their major role in improving CEC and nutrient retention (Sriburi 2016). Furthermore, Horel et al. (2019) found that biochar improves soil nutrients availability, making them available to the plant hence improving uptake, and creates a conducive environment for nutrient uptake. Significant interactive effects of biochar type and rate improved Mg uptake, with biochar type affecting Ca uptake. This was as a result of improved cation exchange capacity as caused by biochar, in line with Martinsen et al. (2017). The improved CEC results in enhanced cation availability at ion exchange sites, allowing availability of more cations including Ca and Mg to be made available for plant uptake, hence its availability in the grain. Additionally, biochar rate, type and interaction and their interactions significantly, ($p < 0.05$) affected phosphorus uptake in grain. This could be attributed to the ability of biochar to influence plant access to soil P (Zwetsloot et al. 2016) making it available for plant uptake. Fe concentration and Zn uptake were also significantly ($p < 0.05$) affected by biochar interactions between biochar rate and type as well as biochar rate, respectively.

6 Conclusion and Recommendations

The study found that biochar significantly enhanced soybean growth, yield, grain quality, and biomass accumulation, particularly at an application rate of 15 tonnes per hectare. The effectiveness of biochar varied based on the application rate and the type of feedstock used. Groundnut shell biochar notably improved growth parameters, while maize cores biochar influenced yield. Increased biochar application also positively affected seed weight, a key yield component. Grain analysis revealed that biochar type and rate significantly impacted nutrient uptake. As biochar rates increased from 0 to 15 tonnes ha^{-1}, leaf canopy and area improved the most at the highest rate. Although all biochar types at 5 tonnes ha^{-1} benefited soybean growth, the most significant improvements were seen at 15 tonnes ha^{-1}. The study recommends using groundnut shell and maize cores biochar at this rate, while suggesting further research on varying rates and conditions to maximize benefits across different soils and agro-ecological zones.

Table 3 Soybean grain nutrient concentrations and uptake as influenced by biochar

			Biochar rate (tonnes ha^{-1})								
Nutrient	Conc./uptake	Biochar type	0	5	10	15	Fpr. T	F pr. R	Fpr.T*R	LDS	CV%
Na	Conc. (mg/kg)	G/nut shell	254	253	255	259	0.543	0.78	0.62	243.2	38.7
		Maize cores	271	320	347	395					
	Uptake (g)	G/nut shell	3.79	3.85	4.96	5.88	0.801	0.843	0.278	0.0038	32.9
		Maize cores	3.55	3.72	3.03	3.86					
Mg	Conc. (mg/kg)	G/nut shell	4529	4776	5328	5533	0.788	0.443	0.238	938.1	10.6
		Maize cores	4645	4729	4795	5282					
	Uptake (g)	G/nut shell	57.3	63.9	94.3	95.5	0.617	0.063	**0.015**	0.03431	14.9
		Maize cores	73	63.9	62.5	73.2					
P	Conc. (mg/kg)	G/nut shell	4666	4827	4455	5231	0.476	0.417	0.390	734.3	7.6
		Maize cores	4517	4639	4737	4666					
	Uptake (g)	G/nut shell	41.5	63.5	60.9	91.6	**0.004**	**0.001**	**0.030**	0.016	15.5
		Maize cores	73.9	59.1	67	85.9					
K	Conc. (mg/kg)	G/nut shell	20,673	20,731	21,409	21,527	0.429	0.903	0.597	2871.1	7.6
		Maize cores	20,535	21,627	21,730	21,878					
	Uptake (g)	G/nut shell	259	283	349	367	0.731	0.093	0.096	0.122	15.4

(continued)

Table 3 (continued)

			Biochar rate (tonnes ha^{-1})								
		Maize cores	331	265	273	333					
Ca	Conc. (mg/kg)	G/nut shell	6687	7767	8035	8286	**0.036**	0.381	0.782	506.7	14.8
		Maize cores	6685	7375	7278	7021					
	Uptake (g)	G/nut shell	102	112	149	243	0.215	0.444	0.466	0.0604	27.7
		Maize cores	103	88	92	106					
Fe	Conc. (mg/kg)	G/nut shell	1633	2191	1074	1297	0.485	0.492	**0.019**	1312.2	32.5
		Maize cores	2169	1061	1845	2445					
	Uptake (g)	G/nut shell	28.1	32.7	20.5	21.9	0.931	0.666	0.173	0.023	47.6
		Maize cores	29	13.1	23.3	35.6					
Zn	Conc. (mg/kg)	G/nut shell	47.8	52.3	50.3	57.6	0.679	0.407	0.579	10.68	11.3
		Maize cores	48.9	50.3	52.7	51.1					
	Uptake (g)	G/nut shell	5.27	7.22	7.62	10.22	0.763	**0.003**	0.198	0.002	20.9
		Maize cores	7.69	63.6	7.21	8.7					

Key* *SD* standard deviation, *CV* coefficient of variation, *LSD* Least significant difference, *F pr.* F probability, *T* biochar type, *R* Biochar rate, *Conc.* concentration, Uptake values were converted to kg pot^{-1}, all nutrient concentration were rated at mg kg^{-1}. Bold F probability values represent points of Significance

References

Allen EJ (1989) Relationship between light interception, ground cover and Leaf area index in potatoes. J Agric Sci 113:355–359. https://doi.org/10.1017/S0021859600070040

Akmal M, Asim M, Akbar H (2010) Response of maize varieties to nitrogen application for leaf area profile, crop growth, yield and yield components. Pak J Bot. 42:1941–1947. https://www.researchgate.net/publication/236853659. Accessed 10 May 2017

Amonette JE, Joseph S (2009) Characteristics of biochar: microchemical properties. In: Lehmann J, Joseph S (ed) Biochar for environmental management: Science and technology. Earthscan, London, pp 33–52

Antal MJ Jr, Grønli M (2003) The art, science, and technology of charcoal production. Ind Eng Chem Res 42:1619–1640. https://doi.org/10.1021/ie0207919

Azeem M, Hayat R, Hussain Q, Ahmed M (2019) Biochar improves soil quality and wheat productivity under different cropping systems. Agron 9:1–18. https://doi.org/10.3390/agronomy9030138

Baronti S, Alberti G, National I, Vedove, GD (2010) The biochar option to improve plant yields : first results from some field and pot experiments in Italy. Ital J Agron 5:3. https://doi.org/10.4081/ija.2010.3

Barus J (2017) Soil Chemical properties and soybean yield due to application of biochar and compost of plant waste. J Trop Soils 21:1–7. https://doi.org/10.5400/jts.2016.v21i1.1-7

Bouyoucos GJ (2010) Hydrometer method improved for making particle size analyses of soils. Agron J 54:464. 34/agronj1962.00021962005400050028x

Chan KY, Van Zwieten L, Meszaros I, Downie A, Joseph S (2007) Agronomic values of greenwaste biochar as a soil amendment. Aust J Soil Res 45:629–634. https://doi.org/10.1071/SR07109

Chan KY, Van Zwieten L, Meszaros I, Downie A, Joseph S (2008) Using poultry litter biochars as soil amendments. Aust J Soil Res 46:437–444. https://doi.org/10.1071/SR08036

Coumaravel K, Santhi R, Maragatham S (2015) Effect of biochar on soil properties and nutrient uptake in maize. Madras Agri J 102:1–3

Demirbas A (2004) Effects of temperature and particle size on biochar yield from pyrolysis of agricultural residues. Energ Sources 26:245–252. https://doi.org/10.1080/00908310490256550

Filiberto D, Gaunt J (2013) Practicality of biochar additions to enhance soil and crop productivity. Agriculture 3:715–725. https://doi.org/10.3390/agriculture3040715

Gaskin JW, Steiner C, Harris K, Das KC, Bibens B (2008) Effect of low-temperature pyrolysis conditions on biochar for agricultural use. Trans ASABE 51:2061–2069. https://doi.org/10.13031/2013.25378

Glaser B, Lehmann J, Zech W (2001) Ameliorating physical and chemical properties of highly weathered soils in the tropics with charcoal: a review. Biol Fertil Soils 35:219–230. https://doi.org/10.1007/s003740100343

Glaser B, Lehmann J, Zech W (2002) Ameliorating physical and chemical properties of highly weathered soils in the tropics with charcoal—a review. Biol Fertil Soils 35:219–230. https://doi.org/10.1007/s00374-002-0466-4

Głodowska M, Schwinghamer T, Husk B, Smith D (2017) Biochar based inoculants improve soybean growth and nodulation. Agric Sci 8:1048–1064. https://doi.org/10.4236/as.2017.89076

Güereña DT, Lehmann J, Walter T (2015) Short-term effects of biochar on soil and plant productivity in Kenyan maize systems. Agri Ecosyst Environ 212:85–94. https://doi.org/10.1016/j.agee.2015.06.016

Horel Á, Gelyb G, Poty I (2019) Soil nutrient dynamics and nitrogen fixation rate changes over plant growth in temperate soil. Agronomy 9:1–14. https://doi.org/10.3390/agronomy9040179

Hossain MK, Strezov V, Chan KY, Ziolkowski A, Nelson PF (2011) Influence of pyrolysis temperature on production and nutrient properties of biochar. J Environ Manag 92:223–228. https://doi.org/10.1016/j.jenvman.2010.09.008

Ingestad T (1980) Growth, nutrition, and nitrogen fixation in grey alder at varied rate of nitrogen addition. Physiol Plant 50:353–364. https://doi.org/10.1111/j.1399-3054.1980.tb04113.x

Islam Md JS, Mannan MA, Khaliq QA, Rahman MM (2019) Growth and yield response of maize to rice husk biochar. Aust J Crop Sci 12:1813–1819. https://doi.org/10.21475/ajcs.18.12.12p944
Kananji GAD, Esnart Yohane E, Siyeni D, Kachulu L, Mtambo L, Chisama F B, Malaidza H, Tchuwa F (2013) A guide to soybean production in Malawi. Department of Agricultural Research Services, Lilongwe, Malawi. https://doi.org/10.13140/2.1.4982.2723
Khan H, Khan AZ, Khan R, Matsue N, Henmi T (2008) Soybean leaf area, plant height and reproductive development as influenced by zeolite application and allophanic soil. J Plant Sci 3:277–286. https://doi.org/10.3923/jps.2008.277.286
Lehmann J (2009) Biochar for environmental management: an introduction. In: Lehmann J, Joseph S (ed) Biochar for environmental management. Earthscan, London, pp 1–12
Lehmann J, Gaunt J, Rondon M (2006) Bio-char sequestration in terrestrial ecosystems: a review. Mitig Adapt Strat Glob Change 11:403–427. https://doi.org/10.1007/s11027-005-9006-5
Li L, Zhang Y, Novak J, Yang A, Wang Y, J(2021) Role of biochar in improving sandy soil water retention and resilience to drought. Water 13:407. https://doi.org/10.3390/w1304040
Long SP (1990) Primary productivity in grasslands and coniferous lforests in relation to climate change. Funct Ecol 4:523–527. https://doi.org/10.2307/2389316
Lu H, Zhang W, Yang Y, Huang X, Wang S, Qiu R (2013) Relative distribution of Pb^{2+} sorption mechanisms by sludge-derived biochar. Water Res 46:854–862. https://doi.org/10.1016/j.watres.2013.02.019
Mahboob W, Hussain Q, Fahad S, et al. (2023) Biochar for sustainable soil management and climate resilience: a global review. Sustain 15:1–25. https://doi.org/10.3390/su15010522
Major J, Rondon M, Molina D, Riha SJ, Lehmann J (2010) Maize yield and nutrition after 4 years of biochar application to a Colombian savanna oxisol. Plant and Soil 333:117–128. https://doi.org/10.1007/s11104-010-0327-0
Malawi Government, Ministry of Agriculture, Irrigation and Water Development (2012) Guide to agriculture and natural resource management. Agricultural Communication Branch, Ministry of Agriculture, Irrigation and Food Security. Lilongwe, Malawi
Martinsen V, Mulder J, Shitumbanuma V (2017) Farmer-led maize biochar trials: effect on crop yield and soil nutrients under conservation agriculture. Agri Ecosyst Environ 239:110–123. https://doi.org/10.1016/j.agee.2017.01.009
Mehlich A (1984) Mehlich 3 soil test extractant: a modificationof Mehlich 2 extractant. Communications in soil science and plant analysis 15:1409–1416. https://doi.org/10.1080/00103628409367568. Accessed 25 Sept 2019
Mete FZ, Mia S, Dijkstra FA, Abuyusuf M, Hossain ASMI (2015) Synergistic effects of biochar and NPK fertilizer on soybean yield in an alkaline soil. Pedosphere 25:713–719. https://doi.org/10.1016/S1002-0160(15)30052-7
Munera-Echeverri JL, Martinsen V, Strand LT, Zivanovic V, Cornelissen G, Mulder J (2018) Cation exchange capacity of biochar: an urgent method modification. Sci Total Environ 642:190–197. https://doi.org/10.1016/j.scitotenv.2018.06.017
Ndor E, Ogara JI, Bako DA, Osuagbalande JA (2016) Effect of biochar on macronutrients release and plant growth on degraded soil of Lafia, Nasarawa State, Nigeria. Asian Res J Agric 2:1–8. https://doi.org/10.9734/ARJA/2016/30926
Nguyen BT, Lehmann J, Kinyangi J, Saito M, Engelhard MH (2004) Long-term black carbon dynamics in cultivated soil. Biogeochemistry 89:295–318. https://doi.org/10.1007/s10533-008-9220-9
Novak JM, Lima I, Xing B (2009) Characterization of designer biochar produced at different temperatures and their effects on a loamy sand soil. Ann Environ Sci 3:195–206
Ogawa M, Okimori Y (2010) Pioneering works in biochar research, Japan. Aust J Soil Res 48:489–500. https://doi.org/10.1071/SR10006
Oke T, Long N (1999) The effect of biochar on soil temperature and water retention. Soil Sci 164:1–9
Profile (2019) Biochar production and utilization: an overview. Soil Use Manag 35:1–10

Quilliam RS, Glanville HC, Wade SC, Jones DL (2013) Life in the 'charosphere'—does biochar in agricultural soil provide a significant habitat for microorganisms? Soil Biol Biochem 65:287–293. https://doi.org/10.1016/j.soilbio.2013.06.004
Rab A, Asim M, Afridi MZ, Arif M, Pakhtunkhwa K (2016) Impact of biochar on mungbean yield and yield components. Pure Appl Biol 5:632–640. https://doi.org/10.19045/bspab.2016.50082
Ratnavathi CV, Komala VV, Chavan UD (2016) Mycotoxin contamination in sorghum. In sorghum biochemistry: an industrial perspective. Agric Biol Sci 358. https://doi.org/10.1016/B978-0-12-803157-5.00003-4
Seadh SE, Abido WAE (2013) How soybean cultivars canopy affect yield and quality. J Agron 12:46–52. https://doi.org/10.3923/ja.2013.46.52
Sigua GC, Stone KC, Hunt PG, Cantrell KB, Novak JM (2016) Increasing biomass of winter wheat using sorghum biochars. Agron Sustain Dev 35:739–748. https://doi.org/10.1007/s13593-014-0253-2
Singh BP, Hatton BJ, Singh B, Cowie AL, Kathuria A (2010) Influence of biochars on nitrous oxide emission and nitrogen leaching from two contrasting soils. J Environ Qual 39:1224–1235. https://doi.org/10.2134/jeq2009.0138
Sriburi SWT (2016) Increasing yield of soybean by adding biochar. J Environ Res Dev 9:1066–1074. Available at: https://www.researchgate.net/publication/296671721. Accessed December 12 December 2017
Snapp SS (2016) Soil nutrient management for sustainable intensification: managing trade-offs. J Integr Agri 15:2499–2510. https://doi.org/10.1016/S2095-3119(16)61391-1
Steiner C (2009) Soil charcoal amendments maintain soil fertility and establish a carbon sink—research and prospects. Soil Ecol Res Dev 125:438–452. http://www.biochar.org/joomla/images/stories/Expert%20Comment%20Steiner-PE.pdf. Accessed on 13 June 2018
Tagoe SO, Horiuchi T, Matsui T (2008) Effects of carbonized and dried chicken manure on the growth, yield, and N content of soybean. Plant and Soil 306:211–220. https://doi.org/10.1007/s11104-008-9571-x
Tang JC, Bacon PE (1964) Soil carbon dynamics and charcoal interactions. J Soil Sci 15:196–205
Tian F, Zhang RH, Wang X (2018) A coupled ocean physics-biology modeling study on tropical instability wave-induced chlorophyll impacts in the pacific. J Geophys Res: Ocean 123:5160–5179. https://doi.org/10.1029/2018JC013992
Tronto J (2017) Biochar systems and sustainable soil management. Earthscan, London
Utomo WH (2014) Biochar for soil improvement: characteristics and application. J Degraded Min Lands Manag 1:1–8. https://doi.org/10.15243/jdmlm.2014.011.001
Van Zwieten L, Kimber S, Morris S, Chan KY, Downie A, Rust J, Joseph S, Cowie A (2010) Effects of biochar from slow pyrolysis of papermill waste on agronomic performance and soil fertility. Plant Soil 327:235–246. https://doi.org/10.1007/s11104-009-0050-x
Wang Y, Wei Y, Sun J (2016) Biochar application promotes growth parameters of Soybean and reduces the growth difference. Commun Soil Sci Plant Anal 47:1493–1502. https://doi.org/10.1080/00103624.2016.1194988
Yooyen J, Wijitkosum S, Sriburi T (2015) Increasing yield of Soybean by adding biochar. J Environ Res Dev 9:1066–1074. https://www.researchgate.net/publication/296671721. Acccssed 2 Dec 2018
Zhang A, Liu Y, Pan G, et al. (2010) Effect of biochar amendment on maize yield and greenhouse gas emissions from a Chinese agricultural soil. J Environ Qual 39:1537–1545. https://doi.org/10.2134/jeq2009.0131
Zhang A, Pan G, Hussain Q, Zhang X (2012) Effect of biochar amendment on maize yield and greenhouse gas emissions from a soil organic carbon poor calcareous loamy soil from Central China Plain. Plant Soil 351:263–275. https://doi.org/10.1007/s11104-011-0957-x
Zhu Q, Peng X, Huang T (2015) Contrasted effects of biochar on maize growth and N use efficiency depending on soil conditions. Int Agro Phys 29:257–266. https://doi.org/10.1515/intag-2015-0023

Zwetsloot MJ et al (2016) Phosphorus availability from bone char in a P-fixing soil influenced by root-mycorrhizae-biochar interactions. Plant Soil 408:95–105

Open Access This chapter is licensed under the terms of the Creative Commons Attribution 4.0 International License (http://creativecommons.org/licenses/by/4.0/), which permits use, sharing, adaptation, distribution and reproduction in any medium or format, as long as you give appropriate credit to the original author(s) and the source, provide a link to the Creative Commons license and indicate if changes were made.

The images or other third party material in this chapter are included in the chapter's Creative Commons license, unless indicated otherwise in a credit line to the material. If material is not included in the chapter's Creative Commons license and your intended use is not permitted by statutory regulation or exceeds the permitted use, you will need to obtain permission directly from the copyright holder.

Impact of Biochar Application on Soil Health and Crop Productivity in Sub-Saharan Africa

Shimbahri Mesfin, Lars Olav Eik, Odd Evjen, Sveinung Sægrov, Eliminata Awet, and Bal Ram Singh

Abstract Biochar is the carbon-rich soil amendment produced by the thermal decomposition (pyrolysis) of biomass. It attracted much interest because it contributes to sustainable soil fertility management and climate change mitigation. This paper reviews the potential impact of biochar on sustainable soil and crop productivity, carbon (C) sequestration and reduction of greenhouse gas (GHG) emissions. The unique characteristics of biochar, such as high surface area, surface charges, low bulk density, alkaline pH being recalcitrant to decomposition, high stability, and nutrient content are factors that enable it to sustainably improve soil and crop productivity and mitigate climate change. However, the pyrolysis technique, pyrolysis temperature and feedstock affect the characteristics and functions of biochar. Several studies have confirmed that biochar application improves soil

S. Mesfin · L. O. Eik
Department of International Environment and Development Studies (Noragric), LANDSAM, Norwegian University of Life Sciences, As, Norway
e-mail: shimbahri.mesfin.gebreslase@nmbu.no; shimbahri.mesfin@mu.edu.et

L. O. Eik
e-mail: lars.eik@nmbu.no

O. Evjen
Norwegian Church Aid (NCA), Oslo, Norway
e-mail: odd.evjen@nca.no

S. Sægrov
Faculty of Engineering, Department of Civil and Environmental Engineering, Norwegian University of Science and Technology (NTNU), Trondheim, Norway
e-mail: sveinung.sagrov@ntnu.no

E. Awet
4CCP (Four Corners Cultural Program), Haydom, Mbulu, Manyara, Tanzania

B. R. Singh (✉)
Faculty of Environmental Sciences and Natural Resource Management, Norwegian University of Life Sciences, As, Norway
e-mail: balram.singh@nmbu.no

S. Mesfin
Land Resources Management and Environmental Protection, Mekelle University, Mekelle, Ethiopia

© The Author(s) 2026
B. R. Singh et al. (eds.), *Sustainable Food Systems in Sub-Saharan Africa*,
https://doi.org/10.1007/978-3-032-11288-0_7

properties such as soil porosity, structure and aggregation. It also enhances soil moisture holding capacity, pH, organic carbon (OC), CEC, total nitrogen (TN), available phosphorus (P), exchangeable potassium (K), and reduces phosphorus (P) fixation (immobilization). The above characteristics of biochar and volatile organic compounds provide a conducive habitat for microbes, enhance microbial population and enzymatic activities as well as nitrogen mineralization. They also help microbes to perform their functions, including decomposing organic matter, recycling nutrients, preserving soil structure, and releasing plant nutrients. However, biochar is more effective in sustainably improving crop production and contributing to food security and mitigating climate change when it is integrated with compost and manure application. Biochar provides a stable and inert form of C sequestration for long-term with a minimum risk of its return into the atmosphere. It also reduces GHG emissions through reducing nutrient losses, retaining, and cycling nutrients, and improving nutrient use efficiency. Though smallholder farmers in Sub-Saharan Africa (SSA) can produce biochar from unused crop residues and leftover fruits for home garden agriculture, this technology needs to be incorporated into soil management policy and transformed into large-scale schemes through using large feedstocks, including invasive species, municipality wastes and industry biproducts.

Keywords Carbon sequestration · Greenhouse gas emissions · Soil and crop productivity · SSA

1 Introduction

Biochar is a carbon-rich product derived from pyrolysis of biomass such as wood, fruit shells, residues of plants, manure, and waste under thermochemical reactions of organic materials in oxygen-limited conditions (Wang et al. 2018; Oni et al. 2019). Biochar production and use are helping to recycle wastes, retain nutrients, avoid pollutants, including heavy metals in the soil, and to offset some greenhouse gas (GHG) emissions (Pang 2018). It has also received attention for its contribution to two major global agendas; sustainable soil health and fertility management and climate change mitigation (Qambrania et al. 2017). This is because of its ability to amend soil physical, chemical, and biological properties on long term basis. Declining agricultural productivity and reduced soil health in most tropical soils in the sub-Saharan Africa (SSA) countries are associated with soil degradation such as soil erosion and compaction, fertility depletion, reduction of soil organic matter (SOM) and nutrient imbalances (Mesfin et al. 2018; Lema et al. 2019; Tadesse et al. 2016). Other soil degradation processes include deforestation, overgrazing, acidification, soil organic carbon (SOC) depletion and overexploitation (Mesfin et al. 2020a; Gwenzi et al. 2015). On the other hand, the smallholder farmers lack access to electricity as an energy source and hence rely on biomass and cow dung for cooking, which otherwise could be used to improve soil fertility. The soil nutrient depletion and nutrient mining are important concerns and have emerged as major constraints to the sustainability of crop production, and are directly linked to food insecurity (Mesfin et al. 2020b; Agegnehu et al. 2017). These challenges create an incentive to optimize use of biochar for soil amendment to improve soil health and enhance carbon (C) sequestration in SSA. Many research findings have proposed biochar as an option for improving soil fertility, restoring degraded land, and mitigating GHG emissions (Das et al. 2020; Brassard et al. 2019; Weber and Quicker 2018).

Application of biochar has several functions in soils (Fig. 1); it increases crop biomass and input of fresh organic carbon (OC), improves soil structure, promotes soil aggregate formation, and provides an ideal habitat for soil microorganisms (Obia et al. 2016). Biochar based soil amendment has numerous impacts on soil fertility through modifying the physical, chemical, and biological properties of the soil (Oni et al. 2019). For example, it increases OC content, water holding capacity, CEC, soil nutrient retention, available P, and microbial activity in soil, thereby improving soil quality and crop productivity (Cornelissen et al. 2013). The characteristics of biochar, such as high specific surface area, enable it to retain moisture and nutrient (Brassard et al. 2019). The surface functional group and specific surface area of biochar are the main properties related to its adsorption capacity for heavy metals (Qiu et al. 2021).

Although biochar application is recognized in improving soil and agricultural farming, its application at smallholder farmers in arid and semi-arid regions of SSA is lacking or quite limited. Several laboratory and field trials have confirmed that application of biochar improves soil and crop productivity, environmental resilience, and mitigates climate change (Das et al. 2020; Brassard et al. 2019; Oni et al. 2019; Weber and Quicker 2018). However, the resource is not well optimized and utilized by most smallholder farmers to sustainably manage their soil and enhance their crop productivity. Smallholder farmers face competition for biomass resource, and they

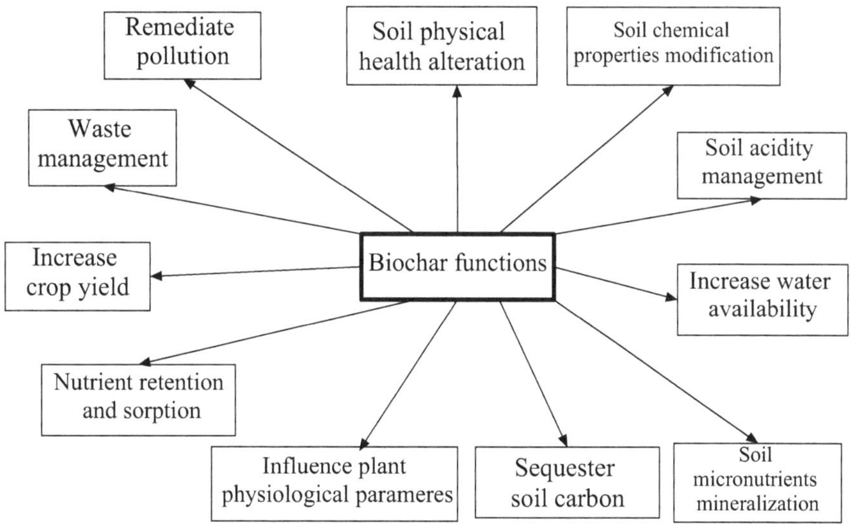

Fig. 1 The functions of biochar in amended soils

have both soil fertility and animal feed problems. However, farmers utilize crop biomass for animal feed and hence, they face supply challenges with resources for biochar production. Furthermore, use of wood for biochar production may not be feasible because it aggravates the deforestation problem in the SSA, as this region is vulnerable to deforestation and other degradations.

SSA also faces environmental degradation due to resources overexploitation, which challenges landscape productivity. The small-scale biochar production can work for small plots such as home garden and small fields of smallholder farmers, and its effect may last four or five years. However, other sources of biomass options for biochar production are needed to apply it on a large scale and in a sustainable way. In such situations, use of invasive species, residues of hardy crops such as pigeon pea, maize, sorghum etc. for biochar production is very crucial to maximize the use of biochar in a resource limited SSA countries. Production and utilization of biochar produced from invasive species such as *Prosopis juliflora* (Gebremedhin et al. 2015) may solve a thorny global problem, which seriously threatens the indigenous biodiversity, economy, and human well-being. This will help farmers to achieve two-fold objectives of controlling invasive spread and enhance fertility and water retention capacity of their soil through application of biochar (Xu et al. 2024). Thus, scaling up of invasive species-based biochar technology may be important in SSA, because invasive species like *Prosopis juliflora* has large coverage in Africa. The focus of this review is on biochar utilization and optimization to enhance sustainable soil nutrient management, crop productivity, C sequestration, and mitigate GHG emission as an emerging amendment in the present climate change scenario.

2 Characteristics of Biochar

Biochar is a porous organic product (Koné and Galiegue 2023) which is formed by the thermal decomposition of biomass under limited supply of oxygen. The process is done as slow or fast pyrolysis, gasification, and torrefaction (Brassard et al. 2016). Unlike most organic matter, biochar is a stable solid that is rich in pyrogenic C, which is highly resistant to microbial decomposition (Weber and Quicker 2018); and it is effective for long-term C sequestration. Biochar is composed of numerous aromatic compounds and other functional groups that are produced from lignin-based biomasses (Qambrania et al. 2017). The important physicochemical properties of biochar are a relatively high surface area, porosity, SOC, cation exchange capacity (CEC), neutral to high pH and low bulk density (Jindo et al. 2014). Biochar can be produced from various feedstocks such as agricultural plant byproducts, invasive woody and herbaceous species, forest industry byproducts, rural and urban wastes (Wang et al. 2020a, b). The aromatic porous structure of biochar is used as trenches to retain nutrients, moisture, and metallic ions (Weber and Quicker 2018; Qambrania et al. 2017). The characteristics of biochar such as porosity, specific surface area, surface charge, and nutrient and water retention capacity are very important for adsorption process (Nguyen et al. 2023). However, feedstock, pyrolysis temperature, method, and duration, influence the physicochemical properties of biochar (Tomczyk et al. 2020).

2.1 Physical and Morphological Properties

The nature of biomass and its cellulose, hemicellulose, and lignin contents influence the extent of devolatilization, product yield, and morphology of the biochar produced during the pyrolysis process (Ghysels et al. 2020; Tomczyk et al. 2020). These physical characteristics of biochar are important to amend soil physical properties including compaction, such as breaking of crusting and penetration resistance, mechanical, structural, hydraulic, and thermal properties of soils (Obia et al. 2017; Li et al. 2019). The morphological characteristics of biochar such as porosity and specific surface area are the most crucial properties for amending soil physical properties. The specific surface area rapidly increases with an increase in pyrolysis temperature greater than 400 °C. Specific surface area (ranging from 250 to 500 $m^2\ g^{-1}$) is one of the unique characteristics of biochar used for soil amendment (Tomczyk et al. 2020).

2.2 Proximate Properties

The proximate properties of biochar are its moisture, ash, volatile matter, and fixed C contents (Li and Chen 2018). During the pyrolysis, some volatile C can be lost while the recalcitrant C remains in the biochar. For example, Weber and Quicker noted that the C content was increased from 10 to 90% under 700 °C when volatile matter content was reduced from 90 to 10% (Weber and Quicker 2018). Ash content of biochar is undesirable because it might block the inner biochar pores and limits its sorption abilities. Ash content of biochar of some feedstocks such as pigeon pea increased with an increase in pyrolysis temperature; however, its fixed C content was reduced with an increase in temperature (Sahoo et al. 2021). This may be due to the escape of some C in the form of volatile substances at higher pyrolysis temperatures (Sun et al. 2017). However, according to Sahoo et al. (2021), the fixed C content of biochar produced from bamboo was increased from 82 to 85% with an increase in pyrolysis temperature. Moreover, the researcher reported that biochar produced at higher pyrolysis temperature (500 and 600 °C) have high ash recovery of 54.8% for pigeon pea and 78.5% for bamboo biomass. Tag et al. (2016) determined that biochar produced from fruit piles and vine pruning have low ash content and surface area (1.2 $m^2\ g^{-1}$ and 8.1 $m^2\ g^{-1}$) because the pores within the biochar are dead ended. The biochar obtained at lower pyrolysis temperature (400 °C) contained high volatile substances due to incomplete carbonization. The high content of volatile substances is not suitable for plant growth because the high microbial activity stimulated by the labile C decreases the nitrogen (N) availability due to immobilization (Yu et al. 2018). However, the high content of volatile substances can be beneficial for plants in soils with low C and high N.

2.3 Chemical Properties

The chemical characteristics of biochar such as high pH and C content, nutrient exchange, and higher potential for C sequestration in soil make it good for soil amendment (Ding et al. 2016). Subsequently, these chemical properties of biochar promote the overall enhancement of nutrient retention in the soil. The higher pH values of biochar, which is more sensitive to pyrolysis conditions (temperature and pyrolysis type) than feedstock type (Tomczyk et al. 2020), is one of the important chemical properties. Even though most biochars are alkaline, some biochars produced from wood feedstocks, biosolids and herbaceous feedstocks under pyrolysis temperature lower than 400 °C have neutral to slightly acidic pH values (Li et al. 2019).

3 Methods of Biochar Production

Following the broad biochar applicability and interests, its production techniques are being continuously simplified and modernized to improve the production rate and quality (Gabhane et al. 2020). The common thermochemical techniques used for biochar production include pyrolysis, hydrothermal carbonization, gasification, flash carbonization and torrefaction (Yaashikaa et al. 2020). Pyrolysis is the most common method for converting biomasses into biochar under limited oxygen supply and at high temperature (Panwar et al. 2019). Based on the temperature of pyrolysis, Yaashikaa et al. (2020) classified the process into slow pyrolysis (temperature < 300 °C), moderate pyrolysis (temperature from 300 to 500 °C) and quick pyrolysis (temperature > 500 °C). Thus, the average effective temperature range for pyrolysis was found to be 300–700 °C (Suárez-Abelenda et al. 2017).

3.1 Traditional Approaches

Smallholder farmers use traditional biochar production methods such as earthen pits and drum kilns (Sakhiya et al. 2020). Smallholder farmers use flame curtain kilns carbonize biomass in an open fire (Jayakumar et al. 2023; Schmidt and Taylor 2014) but have negative effects on the environment because these exhaust volatiles to the atmosphere and pollute the air (Sakhiya et al. 2020).

Theses traditional biochar production techniques are still in use in most SSA rural areas (Table 1). Even though the traditional methods of biochar production are low (10–35%) and have negative impacts on the environment, they are still preferred by smallholder farmers in the rural areas because of their simplicity, low construction cost and are easy to operate (Panwar et al. 2019). A study in Ghana by Duku et al. (2011) performed biochar production using an earthen mound kiln taking wood as a feedstock and found that the ground prevents entry of oxygen during the carbonization process. Similarly, Masek (2017) conducted an experiment using an earthen mound kiln, which gave yield of 10% biochar production. For their part, Kristofferson and Bokalders (2013) constructed a rectangular kiln using masonry blocks and poured concrete combined with steel reinforcements and produced up to 33% yield. In the traditional biochar production system, yield is low, and there are significant CH_4 and CO_2 emissions of incomplete combustion products, which adversely affect human health and the environment.

Although, the traditional approach of using batch processes is cheap and portable with simple technology, it is inefficient, slow, low biochar yield (10–35%) and releases harmful gasses to the atmosphere (Table 1). The other drawback of the traditional method of biochar production is that the slow pyrolysis takes several hours to complete the process of biochar production.

3.2 Modern Approaches

The modern pyrolysis has primarily been conducted via flash, vacuum, and microwave pyrolytic processes. Among these processes, the flash pyrolysis is fast and advanced pyrolysis (Cho et al. 2024). The advanced pyrolysis degrades the biomass in a short time (< 1 min) at high temperatures (>1000 °C) with the rate of heating > 1000 °C s^{-1} (Gabhane et al. 2020). These types of pyrolysis are important for large scale commercial sectors because of the maximum yield, energy efficiency and its quality (Pawar et al. 2019). However, they are less appropriate for smallholder farmers because they demand advanced technology and high financial resources. This advanced biochar production can be used in commercial agriculture or for those who can afford the financial cost and high skill. The modern biochar production technologies currently available include drum type pyrolizer, screw type pyrolizer and rotary kilns (Duku et al. 2011). The drum type pyrolizer is horizontally mounted to a drum kiln in which feedstock is pyrolyzed in continuous operation and heated externally at a temperature of 450 °C. The paddles help the drum for moving biomass and increased kiln efficiency, so that 90% of the recovered heat is used for drying feedstock (Pawar et al. 2019). The flash carbonization is the latest biochar production method- in which feedstock is quickly and efficiently converted into biochar with a yield of 40–50% and C content of 70–80% (Table 1).

Table 1 Biochar production methods under different pyrolysis processes

Process type	Biochar production technique	Temperature (°C)	Carbon content (%)	Biochar yield (%)	Conversion efficiency	Advantage	Disadvantage	References
Batch	Earthen and mound kiln	500	10–30	≥ 10	Inefficient	– Simple – Technology – Cheap and portable	– Inefficient – Low yield – No heat recovery – Release gas and vapors – Environmental pollution	Panwar et al. (2019) Masek (2017) Duku et al. (2011) Kristofferson and Bokalders (2013)
	Brick, concrete and metal kiln	400–600	20–30	20–35	Inefficient			
	Retorts	450–510	20–30	30–35	Inefficient			
Continuous	Drum type pyrolizer	400–500	50	25–30	Relatively efficient	– Higher yields – Has heat recovery – Easy to operate – relatively proven technology – Portable	Complex systems, expensive than batch processes. No usable by-products	Panwar et al. (2019) Yaashikaa et al. (2020)
	Screw type pyrolyzer	900	85	25–30	Relatively efficient			
	Rotary kiln (Lambiotte)	400–650	–	30–35	Relatively efficient			
Novel processes	Flash carbonization	> 1000	70–80	40–50	Efficient	– High yield – Produce more at shorter time	More complex systems; expensive than batch processes	Gabhane et al. (2020)

4 Biochar as a Soil Amendment

Biochar, one of the crucial materials for soil amendment, has attracted attention due to its ability to improve soil physical and chemical properties for longer time (Hossain et al. 2020). There is a growing interest in using biochar to improve soil quality and crop production, and to reduce environmental pollutions because application of inorganic fertilizer alone is not feasible due to high cost involved for poor farmers and it mya have negative impacts on the soil physical and chemical properties and the environment (Xie et al. 2023; Shifa et al. 2024). Thus, integration of biochar with organic manure, compost, and other organic fertilizers can be a good approach to amend soil physical and chemical properties, improve soil fertility, and enhance crop productivity (Zhang et al. 2024; Khan et al. 2024). Several studies conducted on the soil amending effects of biochar confirmed that the majority of soil physical, chemical, and biological properties are improved by the application of biochar (Hossain et al. 2020; Fig. 2). A study conducted in the highlands of Ethiopia by Bayabil et al. (2015) indicated that wood charcoal is a viable low-cost alternative for soil amendment in places where high-tech biochar is not available.

4.1 Soil Physical Properties

Biochar as a soil amendment significantly improves physical properties such as soil porosity, soil bulk density, soil moisture, aggregation, and structure due to the porous internal structure (Dayoub et al. 2023). However, the effect of biochar varies, depending on the nature and pyrolysis method of the biochar. Biochar produced at both high and low pyrolytic temperatures significantly decreased bulk density by 31% and 28%, respectively, over its control (Singh et al. 2022). Biochar with low density (300 kg m^{-3}) and high organic C content has a greater potential to reduce bulk density and penetration resistance than biochar with higher density and lower C content (Gwenzi et al. 2015). The reduced soil bulk density caused by biochar application showed the ability to retain soil water (Dayoub et al. 2023). The reduced bulk density is associated with increased soil porosity, better soil aggregation and surface functionality (Fig. 2). The high soil porosity also increases water and nutrient retention in the soil and facilitate easy water infiltration (Nyambo et al. 2021). The low density and high porosity characteristics of biochar enhance soil porosity, either through direct contribution of pores in the biochar particles or by the formation of new macropores and rearrangement of soil particles.

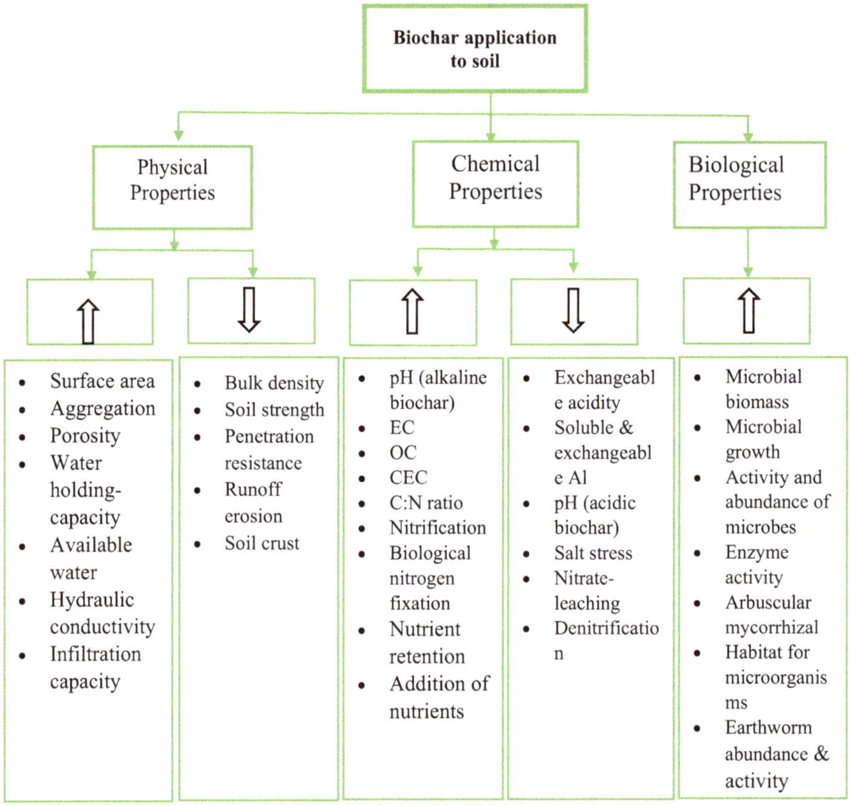

Fig. 2 Effect of biochar on soil properties. Adopted from Premalatha et al. (2023). *The upward arrows indicate that biochar application to soil increases the soil properties while downward arrows indicate that biochar decreases the soil properties*

4.2 Soil Chemical Properties

Application of biochar to soils increases soil pH, OC, CEC, TN, available P, Exchangeable K and reduces Al (Agegnehu et al. 2017; Fig. 2). Improving soil CEC is a significant effect, because it influences soil quality and ions exchange of cations calcium (Ca^{2+}), magnesium (Mg^{2+}), potassium (K^+), sodium (Na^+), and ammonium (NH_4^+) absorption. The increased CEC by biochar application improves nutrient availability and thereby increases microbial activity and soil fertility (Singh et al. 2022). Although the feedstock sources for biochar affects biochar's surface properties, biochar produced at higher pyrolysis temperature showed higher CEC as compared with the same feedstock generated at lower pyrolysis temperatures (Igalavithana et al. 2016). The properties of biochar such as high surface area and CEC, low bulk density, neutral to alkaline pH, high C content, high stability and nutrient content make it an ideal soil conditioner for poor fertility tropical clay and sandy soils in SSA (Gwenzi et al. 2015).

4.3 Soil Biological Properties

Biochar application has shown the ability to improve the soil biological properties through provision of conducive habitat for microorganisms (Premalatha et al. 2023). Biochar releases nutrients, creates conducive habitat for microbes, enhances microbial population and helps them to perform their functions such as decomposing organic matter, recycling nutrients, preserving soil structure, and releasing plant nutrients (Dangi et al. 2020). By modifying soil porosity, biochar improves aggregation, water holding capacity, surface area, pH, OC, CEC, nutrient release (Na^+, N, K^+, Mg^{+2} and P) and surface volatile organic compounds, and enhances microbial community makeup and activities. Thus, biochar plays a great role in enabling the functioning of microbial activities such as enhancing enzymatic activities, nitrogen mineralization, respiration of the soil and bacteria to fungi ratio (Premalatha et al. 2023; Agegnehua et al. 2017; Fig. 2) and thus leads to improved soil quality and crop productivity.

4.4 Soil Amelioration and Environmental Remediation

Biochar is an effective multi-beneficial product used for soil amelioration and crop productivity because it improves soil quality through improving its properties like porosity, bulk density, electrical conductivity (EC), OC content, CEC, moisture content, and pH level (Nguyen et al. 2023). The increased soil pH enabled by biochar application enhances activity of the enzymes to breakdown the organic matter in acidic soils. Biochar modifies the habitats and populations of soil microorganisms through improving essential soil properties for microbial growth and functions. It improves soil fertility by slow nutrient release and retention, and stimulates the microorganisms' functions through increasing soil porosity, water retention, better oxygen level and aggregate stability (Song et al. 2022).

Biochar plays a great role in soil remediation, caused by its electrostatic interaction and precipitation mechanism for inorganic contaminants such as heavy metals, and due to surface adsorption, partition, and sequestration of organic contaminants (Guo et al. 2020). Biochar's sorption, catalysis, and redox interaction abilities with other substances creates interface between biochar and pollutants. These surface biochar characteristics, including carboxyl, carbonyl, aromatic, hydroxyl, surface charges, and free radicals, are very important in creating interfaces with organic and inorganic pollutants. In addition, removal of some contaminants is also achieved by partitioning, electrostatic interaction, and pore-filling between biochar and pollutants (Mohan et al. 2014).

5 Biochar for Carbon Sequestration and Reduction of GHG Emission

Climate change, together with human interference, are the main causes for soil degradation and low agricultural production worldwide, including SSA (Koné and Galiegue 2023). Soil plays a crucial role in the C cycle affecting climate change. Mitigating climate change and food security can be addressed by enhancing SOC sequestration through environmentally sound land management practices. The '4 per 1000' aims to increase C storage in agricultural soils and therefore contributes to mitigating climate change and increasing food security (Rumpel et al. 2020). The aromatic structure of biochar enables the soil to sequester C and reduces GHG emissions (Yaashikaa et al. 2020). As C sequestration increases SOM that enhance nutrient retention, it increases water holding capacity, thus reducing leaching or surface runoff. Thus, N_2O and CO_2 emissions are reduced because of C sequestration in soils. Biochar application to soils is one of the strategies advocated to improve soil fertility and sequester C in soil and to reduce GHG (CO_2, CH_4 and N_2O) emissions, because biochar is rich in recalcitrant C content (Wang et al. 2020a, b; Wang and Wang 2019; Yaashikaa et al. 2020). Agricultural soils and grazing lands are the main sources of N_2O emissions, and application of biochar decreases this emission and contributes to the reduction of global warming (Guo 2020). The high surface area and porosity of biochar enables the soil to retain more water and nutrients (ammonium), and in turn reduces GHG emissions to atmosphere (Blanco-Canqui 2017).

6 Nutrient Dynamics and Crop Responses to Biochar Application

Many research reports have demonstrated that biochar has significantly enhanced soil fertility and crop productivity (Liu et al. 2013; Kätterer et al. 2022), because it reduces soil nutrient losses, increases nutrient retention and availability, and improve plant nutrient uptake. The unique properties of biochar such as high surface area, low bulk density, high CEC, neutrality to alkaline pH, high C content, high stability and nutrient content play a great role in amending tropical sandy soils of SSA (Gwenzi et al. 2015). A study conducted in Ghana and cited by Koné and Galiegue (2023) reported a 5% increase in nitrate recovery when biochar was applied to a maize farm on a sandy soil. This is because the aromatic structure of biochar has the ability to retain nutrients and moisture. Whereas Gwenzi et al. (2015) found decreased leaching of total nitrogen (TN) and dissolved P by 11% and 69%, respectively when amended with 20 g kg^{-1} biochar to a loamy soil in SSA, use of biochar modifies soil properties such as reducing soil penetration resistance like crusting and self-sealing and creates room for nutrient retention. Biochar produced from different feedstocks and pyrolyzed at different temperatures have shown varying yield levels

(Table 2). The smallholder farmers used different biochar application methods such as broadcasting, placement in furrows and ridges, and mixing manually with the soil (Table 2).

The increased C sequestration enabled by biochar plays a great role in boosting soil nutrients and minimizing nutrient leaching (El-Naggar et al. 2015). Biochar also enhanced the mycorrhizal colonization of crop roots and nutrient availability, resulting in increased P uptake and reduced P fixation. Mandal et al. (2018) suggested that the application of poultry manure biochar significantly enhanced concentrations of water-soluble nutrients P, K, Ca, Mg, Fe, Zn, Cu, Mn, and reacted positively to increased nutrient availability for improved plant growth. Positive effects of biochar application on soil properties including SOC, soil pH, CEC, porosity, water holding capacity, nutrient retention and soil aggregation were also observed by other researchers (Hussain et al. 2017; Gwenzi et al. 2015; Liu et al. 2014).

Nitrogen content of biochar is generally low because of its loss during pyrolysis (Li et al. 2019), and hence its plant availability is probably low. However, research findings have demonstrated that several potential mechanisms could increase N retention and availability to plants following biochar application such as improving N use efficiency by crops, enhanced root growth, reduced leaching, reduced gaseous N emissions, and suppression of nitrification (Plaimart et al.2021; He et al. 2017; Bruun et al. 2014). Application of biochar also enhances atmospheric N fixation, which has a great impact on increased N uptake by crops. Biochar also enhanced soil N through increasing decomposition rate of SOM (Kätterer et al. 2019). Other studies conducted in Ethiopia by Abrham and Tekilil (2019) also confirmed that biochar sustainably improved soil nutrient contents of exchangeable Na, K, Ca, and Mg (Agegnehua et al. 2016). Liu et al. (2014) conducted an experiment on maize and wheat with application of biochar for five cropping seasons and found that biochar sustainably increased SOC and available P and crop yield (Fig. 3). Similarly, Diatta et al. (2020) reported that soils amended with biochar produced from wheat crop residue applied a rate of 20 and 40 t ha^{-1} in paddy fields increased soil N contents by 5.4 and 18.8%, respectively.

Phosphorus (P) is one of the essential macronutrients for plants which is added as phosphate fertilizer to soils; however, only a portion of the fertilized P is available to plants, because a substantial amount of P associated with soil particles is washed out as surface runoff and pollute surface water (Gao et al. 2019). Thus, biochar application increases P nutrient use efficiency through retaining it by saving it from erosion and other loses; and this also helps to minimize environmental impacts and reduces the cost of inorganic fertilizer. Biochar directly enhances available P in the soil because it has a significant amount of soluble P salts produced during the carbonizing (Ghodszad et al. 2020). On the other hand, the liming effect of biochar on acidic soils through precipitating Al as Al $(OH)_3$ and Fe as Fe $(OH)_3$ increases P availability in the soil system (Dayoub et al. 2023). This was confirmed by Gao et al. (2019) who found that biochar application increased available P in the topsoil by 45%, and P in microbial biomass by 48%, caused primarily by the liming effect of biochar, and thus reducing P fixation by Al and Fe oxides (Dai et al. 2017).

Table 2 Effects of biochar application on crop yields in SSA

Biochar	Application method	Application rate t ha^{-1}	Crop type	Yield increase	Study Place	Reference
Coffee husks	Placement in furrow	10	Maize	4 t ha^{-1}	Kenya	Kätterer et al. (2022)
Coconut shells	Placement in furrow	10	Maize	3.2 t ha^{-1}	Kenya	Kätterer et al. (2022)
Maize cobs	Placement in furrow	10	Maize	2.8 t ha^{-1}	Kenya	Kätterer et al. (2022)
Prosopis juliflora	Broadcasting	4	Wheat	18.6%	Ethiopia	Gebremedhin et al. (2015)
Eucalyptu char	Broadcasting	12 t ha^{-1}	Teff	1.4–2.7 t ha^{-1} grain and 11.6–17.8 t ha^{-1} straw	Ethiopia	Abewa et al. (2014)
Corn cob	Broadcasting and mixed using hand hoes	20	– Amaranth – Lettuce – Carrot	39% (1st season) 17% (2nd season) 7% 11%	Burkina Faso	Manka'abusi et al. (2019)
Corn cob	Broadcasting	3	Maize	4–5%	Ghana	Yeboah et al. (2009)
Rice ball	Broadcasting	0.9–10.7	Lettuce	93%	Ghana	Steiner et al. (2018)
Variable	Broadcasting	Variable	Variable	1–3 t ha^{-1}	Tanzania	Rogers et al. (2017)
Rice straw and Corn stalk	Placement in the ridges	1–2	Corn, Peanut, and sweet potato	5–15%	–	Yang et al (2015)
Rice straw and Corn stalk	Placement in the ridges	4	Corn, peanut, &sweet potato	20%	–	Yang et al (2015)
Rice cob/ cow manure	Incorporated with rakes	5–10	Millet	0.1–0.45 t ha^{-1}	Senegal	Faye et al. (2021)
Wood	Broadcasting	10–30	Taro	Yield improved	Nigeria	Adekiya et al. (2020)
Rice ball	Broadcasting	20	Lettuce	15% (1st season) 9% (2nd season)	Ghana	Akoto-Danso et al. (2018)

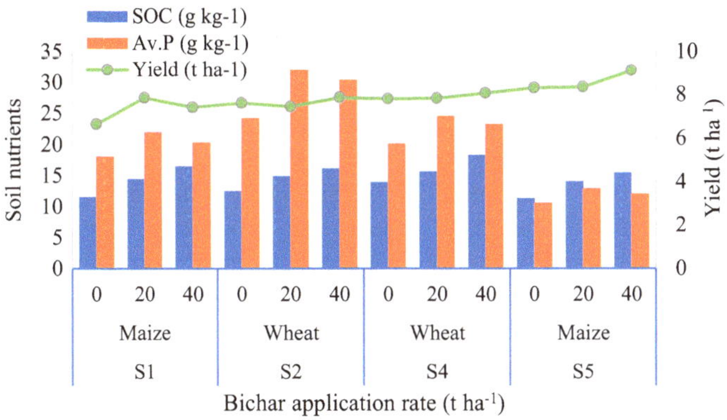

Fig. 3 Biochar sustainably improved soil nutrient and yield (Liu et al. 2014). Where, S1, S2, S4 and S5 are cropping seasons

The other important effect of biochar is its ability to enhance soil nutrient availability and reduce nutrient losses through leaching (Bilias et al. 2023). The nutrient availability of cations such as soil K availability was raised by biochar application (Agegnehua et al. 2016). A study by Berihun et al. (2017) in Ethiopia indicated that incorporation of 18 t ha^{-1} biochar produced from corn cobs, *Lantana camara* and *Eucalyptus globulus* increased exchangeable K of the soil from 32.7 cmol kg^{-1} to 38.16, 43.33 and 42.83 cmol kg^{-1}, respectively. The increased K content and plant uptake resulted in increased yield after biochar application to K-deficient soils improved soil K availability and reduced K leaching were reported by Bilias et al. (2023). This is because biochar by its properties has the ability to prevent leaching, and retains K nutrient in soil.

Like its functions on nutrient dynamics, biochar sustainably increases crop production. The meta-analysis by Singh et al. (2022) showed that biochar significantly increased crop yield of various crops by 34% in the greenhouse study and 30% in the field study. Similarly, application of 12 t ha^{-1} biochar increased teff grain yields from 1.4 to 2.7 t ha^{-1} and dry biomass from 11.6 to 17.8 t ha^{-1} in northwestern Ethiopia (Abewa et al. 2014). However, feedstock type, pyrolysis temperature, application rate, and soil texture influenced the effect of biochar on crop production (Table 2). For example, Li et al. (2019) reported that biochar produced at 400–500 °C was most effective in increasing crop productivity than biochar produced at high temperatures (> 600 °C). This is because biochar produced at high pyrolytic temperatures holds more water and dissolved minerals tightly, resulting in lower nutrient availability to plants. As shown in Table 2 Kätterer et al. (2022) reported different effects on maize yield of biochar produced at pyrolizer temperature (400–500 °C). Soil texture is one of the factors that influence impacts of biochar on crop productivity, because the effect of biochar is more visible in coarse textured soils than in fine textured soils. Herbaceous biochar has higher nutrient content than wood biochar because wood biochar is higher in lignin content and immobilize nutrients

(Singh et al. 2022). This indicated that smallholder farmers should prefer invasive herbaceous plants, like *parthenium* species, in Ethiopia than woody species.

Several studies provided different suggestions to justify the effect of biochar application on plant growth, nodulation, nitrogen fixation, and nutrient availability. Biochar improves legume crop growth and yield because it enhances biological nitrogen fixation (BNF) (Mia et al. 2014). Similarly, Asfaw et al. (2019) reported that biochar significantly increased legume crop production, caused by increased BNF and nutrient availability, especially when it was integrated with inoculation and P fertilizer. Application of 4-ton *Prosopis juliflora* biochar ha^{-1} combined with N and P (100 kg urea + 100 kg DAP) fertilizer significantly increased yield and yield components of wheat and post-harvest soil nutrients in Ethiopia (Gebremedhin et al. 2015). It increased grain yield and straw by 15.7 and 16.5%, respectively, compared to the NP applications alone (Gebremedhin et al. 2015). Similar effects of biochar on crop yield were also observed by other researchers (Manka'abusi et al. 2019; Akoto-Danso et al. (2018), as shown in Table 2.

7 Discussion and Outlooks

Biochar has been used for soil amendment in different regions for many years. However, the feedstock temperature of pyrolysis and type of pyrolysis determine biochar characteristics (Tomczyk et al. 2020). Feedstock and pyrolysis temperature are the most influential factors that modify the chemical nature and increases in the aromatic character of biochar. Biochar produced from different feedstocks, regardless of the pyrolysis temperature, increases C storage in soils. Thus, feedstock source followed by pyrolysis temperature are the key factors for biochar characteristics and its potential for soil amendment and environmental values.

The biochar characteristics and its potential to amend soil are highly influenced by the method of preparation, pyrolysis temperature and type of feedstock. Both traditional biochar production methods such as pits, earth mound, bricks, and metal kilns and modern ones such as drum type pyrolyzer, screw type pyrolyzer, rotary kilns and flash carbonization are shown in Table 1 (Panwar et al. 2019). The smallholder farmers use traditional biochar production because these are easy and can be done with minimum cost. Higher pyrolysis temperature increases biochar's surface area, porosity, carbonized fractions, pH and volatile matter but decreases CEC and content of surface functional groups (Tomczyk et al. 2020; Ding et al. 2014). Thus, biochar production methods should be adjusted based on the purpose of the biochar. Biochar's porosity and high surface area formed due to the water loss during the dehydration process are important for high sorption capacity. The chemical structure of C in biochar is transformed when heating to give an aromatic structure, which is resistant to microbial decomposition. Because of that, biochar is believed to be active for long-term C sequestration, because biochar with C compounds is very stable for a long period (Zhou et al. 2017). The increase in CEC due to high specific surface area is important for soil nutrient reaction, while pH is more suitable for the growth of microbes, especially for fungal hyphae. On the other hand, the porous structure of biochar increases the habitat for microbes' growth.

The application of biochar for soil amendment has several functions: it reduces soil bulk density, increases porosity, enhances soil aggregate stability and structure, improves water holding capacity and moisture retention, and breaks soil crusting and compaction as shown in Fig. 2 (Wang et al. 2020b, a). Biochar plays a great role in amending soil chemical properties by enhancing soil pH, increasing CEC, SOC and mineral N availability, dissolving OC, and improving microbial diversity (Singh et al. 2022). Therefore, use of biochar has several soil amending potentials such as increasing nutrient use efficiency, reducing nutrient leaching, increasing soil moisture and nutrient availability, reducing Al and Fe toxicity to plant roots and microorganisms, reducing heavy metal bioavailability and GHG emissions. For example, biochar application improves moisture retention in sandy soils with low water holding capacity, which are prone to drought periods during the crop growing season in arid and semi-arid regions.

Biochar also has the capacity to increase crop production in problematic soil, for example, in acidic sandy soils; it amends soil pH and retain more moisture (Martinsen et al. 2014). Biochar application increases soil K^+ and Ca/Al ratio and reduces exchangeable Al^{3+}, acid saturation and in turn increases crop production. The other main reason why biochar increases crop production is because it improves soil physical, chemical, and biological properties (Cornelissen et al. 2018). Biochar also improves soil nutrient availability such as N, P, K and Ca through improving soil CEC, pH and OC content and reducing nutrient losses (leaching, erosion, and gases volatilization). The strong effect of biochar is explained by a combination of nutrient retention (NH_4^+, Mg^{2+}, Ca^{2+}, K^+) (Cornelissen et al. 2018); and it improves plant-available water and moderates soil temperature, which is essential for crop productivity and climate smart agriculture (Obia et al. 2020). However, research findings by Gao et al. (2019) showed some negative effects of biochar on soil inorganic N. In particular, much of the C in the biochar's high C/N was found to be resistant to the decomposition of microorganisms, and thus incapable of stimulating microbial N immobilization. In this case, application of biochar together with organic fertilizer like compost and manure can offset the negative effects. Biochar reduces toxicity of Al and Fe to plant roots and microorganisms because of its liming effect, and thus reducing heavy metal bioavailability, increasing plant nutrient availability, increasing nutrient use efficiency, and decreasing GHG emissions from agricultural soils (Das et al. 2020).

Agricultural practices emit considerable amount of GHG, contributing to global warming. Some intensive agricultural farming practices also result in increased soil erosion and contamination of water resources. Hussain et al. (2017) reported that application of biochar to soils resulted in the reduction of GHG (CO_2, CH_4 and N_2O) emissions and increased SOC sequestration. Agricultural activities contribute to global warming through releasing GHGs (Laborde et al. 2021). However, biochar as soil amendment has the potential to mitigate the losses of different soil nutrients by retaining soil nutrients that could be emitted to the atmosphere and leached down or eroded as runoff, and thus leading to enhanced nutrient use efficiency (Fungo et al. 2014).

Therefore, the role of biochar be linked to climate change mitigation practices in SSA smallholder farming systems. Biochar has both direct impacts on GHG emissions, such as stabilization and C sequestration in the soil, and indirect impacts such as improving properties of the soil and reducing emissions. The C in biochar is highly recalcitrant to decomposition compared with the ordinary soil C, which is characterized by high turnover and release of CO_2. Thus, converting biomass into biochar is changing C from easily degradable into gradual degradable C or converting waste material into biochar (beneficial resources) that reduces CH_4 and CO_2 emissions in the landfill and waste dumps. Therefore, incorporation of biochar into soil as nutrient management in the smallholder farming system in SSA can sequester C in a soil, reduce GHG emissions and thus contribute to climate change mitigation. The traditional pyrolysis kilns emit significant amount of GHG such as CH4 and CO_2 to the atmosphere and affect human health; however, CH4 is considered more dangerous GHG than CO_2 (Adam 2009).

Biochar application to soils has both direct effects on soil fertility such as enhancing nutrient (K, P, Ca, Mg and S) availability, and indirect effects such as improved physical, chemical, and biological properties of the soil (Singh et al. 2022). However, the way it works to improve soil fertility depends on the pyrolysis method and feedstock. The rise in pH value of the soil with biochar application is attributed to the higher ash concentration of biochar and has a liming effect on the soil, and reduces the solubility and activity of Al and Fe in soils. This increases available P in the soil. The application of biochar to soil also increases soil CEC due to its high surface area, extremely porous structure and variable charge of the organic material, surface sorption capacity, and base saturation (Asfaw et al. 2019). The increase in soil OC and TN as a result of application of biochar to soil is due to its high C and organic matter concentration. Biochar application also raises the availability of most cations and P concentrations in the soil.

Studies in SSA confirmed that application of biochar has increased crop growth and performance (Rogers et al. 2022; Gebremedhin et al. 2015) because biochar application enhances moisture and nutrient retention by increasing soil pH and soil organic matter (Mendez et al. 2012). In addition to improving nutrient availability, biochar significantly increased biological nitrogen fixation by rhizobium and improved biomass production of plants (Asfaw et al. 2019; Rondonet al. 2007). This is because biochar adsorbs soluble organic matter and gases, and the nutrients are provided with a suitable habitat for rhizobia to grow and reproduce (Sun et al. 2021). Despite the fact that biochars are more stable and provide slower release of nutrients, application of biochar alone may not release enough nutrient (Marcinczyk and Oleszczuk 2022). As a result, biochar has less impact on crop yield in the first year, but sustainably increases yield over a longer time. This indicated that integrated application of biochar with compost, manure, and inorganic fertilizer in the first year is very important (Omara et al. 2020). Soil amendment using biochar produced from invasive species (*Prosopis juliflora*) in Ethiopia significantly increased wheat crop root biomass, grain, and straw yield (Gebremedhin et al. 2015) due to biochar characteristics which enable it to retain more water and nutrient, and releases nutrient to the soil. This indicated that producing biochar from invasive species which are available in their local area is best option for smallholder farmers who have feed shortage, due to crop residue being used as animal feed.

The unique characteristics of biochar such as high surface area, high porosity, its ash and sustainable release of plant nutrients, and its capacity to act as a medium for microorganisms are the main potentials of biochar to improve soil properties, enhance nutrient uptake and boost crop yield. The physical structure of biochar such as porosity provides refugia for beneficial soil organisms such as fungi (mycorrhiza) and bacteria, which are key components in the development of improved agricultural efficiency. As the effects of biochar on soil last for seasons or years, it increases soil and crop productivity, contributes to food security for an increasing global population and mitigates changes in climate. Therefore, application of biochar is beneficial to the improvement of soil fertility, increases nutrient uptake (it is more effective when it is integrated with compost and manure application), and ameliorates soil pollution. Biochar provides a stable and inert form of C sequestration which lasts for a long time, with a minimum risk of return into the atmosphere. Therefore, use of biochar as a source of soil C sequestration appears to provide economic benefits and mitigate climate change. Biochar has a long-term capacity for agricultural improvement compared to inorganic fertilizers. However, the economics of biochar production depends on the type of feedstock, conversion technology, C sequestration subsidies and C credits, reflecting the social value of GHG emission reductions (Li et al. 2023). Implementing smart farming practices could increase crop yields and improve the economic situation of smallholder farmers while reducing the adverse effects of climate change (Haeldermans et al. 2020).

8 Concluding Remarks

The literature data indicates that biochar-based soil amendment has a promising potential for the development of sustainable agricultural systems in SSA. The unique physical and chemical characteristics of biochar provide several benefits, such as enhancing crop production through improving soil fertility, contributing to reducing GHG emissions, and sequestering C within a soil. In addition, biochar improves the soil biochemical properties through increasing the bacterial and fungal populations, which is very important for organic matter mineralization. However, biochar functions depend on the type of feedstock, the pyrolytic conditions, the rate of application and soil type. Since biochar has low soil nutrients, its role is effective when it is combined with manure, compost, conservation tillage and conservation farming. Biochar provides a stable and inert form of C sequestration which lasts for a long time with a minimum risk of return into the atmosphere. Therefore, optimizing biochar's benefits to establish economic feasibility and mitigate climate change in the long run are interesting issues in the biochar technology development and application in SSA agriculture.

However, smallholder farmers in SSA face both biochar supply resource and production constraints. Because the farmers use crop residue for animal feed, there is tough competition for biochar feedstock; and therefore, alternative feedstocks for biochar production must be sought. In this case, local resource availability and use prioritization need to come into consideration. For example, using invasive species as feedstock for biochar production does not only save production costs, it also reduces negative impacts of invasive species, increases efficient resource utilization, and makes them widely used in the field of environment, agriculture, and energy. Although smallholder farmers can produce biochar from unused crop residues and leftover fruits for home garden agriculture, this technology needs to be incorporated into soil management policy and transformed into large-scale schemes through using large feedstocks including invasive species, municipality wastes and industry bi-products in the SSA countries.

References

Abrham B, Tekilil W (2019) Effect of biochar of grass, chat and inorganic fertilizer on the growth and yield of garden vegetable at Benishangu lgumze Region, Ethiopia. Sci J Energy Eng 7(4). https://doi.org/10.11648/j.sjee.20190704.15

Abewa A, Yitaferu B, Gebreselassie Y, Amare T (2014) The role of biochar on acid soil reclamation and yield of teff (Eragrostis tef [Zucc] Trotter) in Northwestern Ethiopia. J Agric Sci 6:1–12

Adam JC (2009) Improved and more environmentally friendly charcoal production system using a low-cost retort–kiln (Eco-charcoal). Renew Energy 34(2009):1923–1925. https://doi.org/10.1016/j.renene.2008.12.009

Adekiya AO, Agbede TM, Olayanju A, Ejue WS, Adekanye TA, Adenusi TT, Ayeni JF (2020) Effect of biochar on soil properties, soil loss, and cocoyam yield on a tropical sandy loam alfisol. Hindawi Sci World J 2020:1–9. https://doi.org/10.1155/2020/9391630

Agegnehu G, Srivastava AK, Bird MI (2017) The role of biochar and biochar-compost in improving soil quality and crop performance: a review. Appl Soil Ecol 119:156–170. https://doi.org/10.1016/j.apsoil.2017.06.008

Agegnehua G, Nelsona PN, Bird MI (2016) Crop yield, plant nutrient uptake and soil physico-chemical properties under organic soil amendments and nitrogen fertilization on Nitisols. Soil Tillage Res 160:1–13

Akoto-Danso EK, Manka'abusi D, Steiner C, Werner S, Ha¨ring V, Nyarko G, Marschner B, Drechsel P, Buerkert A (2018) Agronomic effects of biochar and wastewater irrigation in urban crop production of Tamale, northern Ghana. Nutr Cycl Agroecosyst 115:231–247. https://doi.org/10.1007/s10705-018-9926-6

Amalina F, Razak ASA, Krishnan S, Zularisam AW, Nasrullah M (2022) A comprehensive assessment of the method for producing biochar, its characterization, stability, and potential applications in regenerative economic sustainability—a review. Clean Mater 3:1–18. https://doi.org/10.1016/j.clema.2022.100045

Asfaw E, Nebiyu A, Bekele E, Ahmed M, Astatkie T (2019) Coffee-husk biochar application increased AMF root colonization, P accumulation, N2 fixation, and yield of soybean grown in a tropical Nitisol, southwest Ethiopia. J Plant Nutr Soil Sci 1–10.https://doi.org/10.1002/jpln.201800151

Bayabil HK, Stoof CR, Lehman JC, Yitaferu B, Steenhuis TS (2015) Assessing the potential of biochar and charcoal to improve soil hydraulic properties in the humid Ethiopian highlands: the Anjeni watershed. Geoderma 243–244:115–123. https://doi.org/10.1016/j.geoderma.2014.12.015

Berihun T, Tadele M, Kebede F (2017) The application of biochar on soil acidity and other physico-chemical properties of soils in southern Ethiopia. Plant Nutr Soil Sci 180:381–388

Bilias F, Kalderis D, Richardson C, Barbayiannis N, Gasparatos D (2023) Biochar application as a soil potassium management strategy: a review. Sci Total Environ 858:1–14. https://doi.org/10.1016/j.scitotenv.2022.159782

Blanco-Canqui H (2017) Biochar and soil physical properties. Soil Sci Soc Am J 81:687–711

Brassard P, Godbout S, Raghavan V (2016) Soil biochar amendment as a climate change mitigation tool: key parameters and mechanisms involved. J Environ Manag 181:484–497

Bruun S, lauson-Kaas S, Bobulska L, Thomsen IK (2014) Carbon dioxide emissions from biochar in soil: role of clay, microorganisms and carbonates. Eur J Soil Sci 65:52–59. https://doi.org/10.1111/ejss.12073

Cornelissen G, Martinsen V, Shitumbanuma V, Alling V, Breedveld GD, Rutherford DW, Sparrevik M, Hale SE, Obia A, Mulder J (2013) Biochar effect on maize yield and soil characteristics in five conservation farming sites in Zambia. Agronomy 3:256–274. https://doi.org/10.3390/agronomy3020256

Cornelissen G, Jubaedah, Nurida NL, Hale SE, Martinsen V, Silvani L, Mulder J (2018) Fading positive effect of biochar on crop yield and soil acidity during five growth seasons in an Indonesian Ultisol. Sci Total Environ 634:561–568. https://doi.org/10.1016/j.scitotenv.2018.03.380

Dai Z, Zhang X, Tang C, Muhammad N, Wu J, Brookes PC, Xu J (2017) Potential role of biochars in decreasing soil acidification—a critical review. Sci Total Environ 581–582:601–611. https://doi.org/10.1016/j.scitotenv.2016.12.169

Dangi S, Gao S, Duan Y, Wang D (2020) Soil microbial community structure affected by biochar and fertilizer sources. Appl Soil Ecol 150:1–10. https://doi.org/10.1016/j.apsoil.2019.103452

Das SK, Ghosh GK, Avasthe R (2020) Application of biochar in agriculture and environment, and its safety. Biomass Convers Biorefinery 13:1359–1369. https://doi.org/10.1007/s13399-020-01013-4

Dayoub EB, Tóth Z, Anda A (2023) Biochar and its effects on soil properties and evapotranspiration: a sustainable solution for plant growth. Cogent Food Agric 9:1–27. https://doi.org/10.1080/23311932.2023.2256136

Diatta AA, Fike JH, Battaglia ML, Galbraith JM, Baig MB (2020) Effects of biochar on soil fertility and crop productivity in arid regions: a review. Arab J Geosci 13:595–611. https://doi.org/10.1007/s12517-020-05586-2

Ding W, Dong X, Ime IM, Gao B, Ma LQ (2014) Pyrolytic temperatures impact lead sorption mechanisms by bagasse biochars. Chemosphere 105:68–74

Ding Y, Liu Y, Liu S, Li Z, Tan X, Huang X, Zeng G, Zhou L, Zheng B (2016) Biochar to improve soil fertility. A review. Agro Sustain Dev 36(36):1–18. https://doi.org/10.1007/s13593-016-0372-z

Duku MH, Gu S, Ben HE (2011) Biochar production potential in Ghana: a review. Renew Sustain Energy Rev 15:3539–3551. https://doi.org/10.1016/j.rser.2011.05.010

El-Naggar AH, Usman ARA, Al-Omran A, Ok YS, Ahmad M, Al-Wabel MI (2015) Carbon mineralization and nutrient availability in calcareous sandy soils amended with woody waste biochar. Chemosphere 138:67–73

Faye A, Stewart ZP, Diome K, Edward CT, Fall D, Ganyo DKK, Prasad PV (2021) Single application of biochar increases fertilizer efficiency, C sequestration, and pH over the long-term in sandy soils of Senegal. Sustainability 13(21):1–12. https://doi.org/10.3390/su132111817

Fungo B, Guerena D, Thiongo M, Lehmann J, Neufeldt H, Kalbitz K (2014) N_2O and CH_4 emission from soil amended with steamactivated biochar. J Plant Nutr Soil Sci 177:34–38

Gabhane JW, Bhange VP, Patil PD, Bankar ST, Kumar S (2020) Recent trends in biochar production methods and its application as a soil health conditioner: a review paper. SN Appl Sci 2:1307–1327. https://doi.org/10.1007/s42452-020-3121-5

Gao S, DeLuca TH, Cleveland CC (2019) Biochar additions alter phosphorus and nitrogen availability in agricultural ecosystems: a meta-analysis. Sci Total Environ 654:463–472. https://doi.org/10.1016/J.SCITOTENV.2018.11.124

Gebremedhin GH, Haileselassie B, Berhe D, Belay T (2015) Effect of biochar on yield and yield components of wheat and post-harvest soil properties in Tigray, Ethiopia. J Fertil Pestic 6:158. https://doi.org/10.4172/jbfbp.1000158

Ghodszad L, Reyhanitabar A, Maghsoodi MR, Lajayer BA, Chang SX (2020) Biochar affects the fate of phosphorus in soil and water: a critical review. Chemosphere 283.https://doi.org/10.1016/j.chemosphere.2021.131176

Ghysels S, Acosta N, Estrada A, Pala M, De Vrieze J, Ronsse F, Rabaey K (2020) Integrating anaerobic digestion and slow pyrolysis improves the product portfolio of a cocoa waste biorefinery. Sustain Energy Fuels 4:43712–43725. https://doi.org/10.1039/D0SE00689K

Guo M (2020) The 3R principles for applying biochar to improve soil health. Soil Syst 4(9):1–16. https://doi.org/10.3390/soilsystems4010009

Guo M, Song W, Tian J (2020) Biochar-facilitated soil remediation: mechanisms and efficacy variations. Front Environ Sci 8:1–23. https://doi.org/10.3389/fenvs.2020.521512

Gwenzi W, Chaukura N, Mukome FND, Machado S, Nyamasoka B (2015) Biochar production and applications in sub-Saharan Africa: opportunities, constraints, risks and uncertainties. J Environ Manage 150:250–261

Haeldermans T, Campion L, Kuppens T, Vanreppelen K, Cuypers A, Schreurs S (2020) A comparative techno-economic assessment of biochar production from different residue streams using conventional and microwave pyrolysis. Bioresour Technol 318:124083. https://doi.org/10.1016/j.biortech.2020.124083

He Y, Zhou X, Jiang L, Li M, Du Z, Zhou G, Shao J, Wang X, Xu Z, Hosseini Bai S, Wallace H, Xu C (2017) Effects of biochar application on soil greenhouse gas fluxes: a meta-analysis. GCB Bioenergy 9(4):743–755. https://doi.org/10.1111/gcbb.12376

Hossain MZ, Bahar MM, Sarkar B, Donne SW, Ok YS, Palansooriya KN, Kirkham MB, Chowdhury S, Bolan N (2020) Biochar and its importance on nutrient dynamics in soil and plant. Biochar 2:379–420. https://doi.org/10.1007/s42773-020-00065-z

Hussain M, Farooq M, Nawaz A, Al-Sadi AM, Solaiman ZM, Alghamdi SS, Ammara U, Ok YS, Siddique KHM (2017) Biochar for crop production: potential benefits and risks. J Soils Sediments 17:685–716. https://doi.org/10.1007/s11368-016-1360-2

Igalavithana AD, Ok, YS, Usman AR, Al-Wabel MI, Oleszczuk P, Lee SS (2016) The effects of biochar amendment on soil fertility. In: Agricultural and environmental applications of biochar: advances and barriers, pp 123–144

Jayakumar A, Morrisset D, Koutsomarkos V, Wurzer C, Hadden RM, Lawton L, Edwards C, Maˇsek O (2023) Systematic evaluation of pyrolysis processes and biochar quality in the operation of low-cost flame curtain pyrolysis kiln for sustainable biochar production. Crsust 5:100213. https://doi.org/10.1016/j.crsust.2023.100213

Jindo K, Mizumoto K, Sawada Y, Sanchez-Monedero MA, Sonoki T (2014) Physical and chemical characterization of biochars derived from different agricultural residues. Biogeosciences 11:6613–6621. https://doi.org/10.5194/bg-11-6613-2014

Kätterer T, Roobroeck D, Andrén O, Kimutai G, Karltun E, Kirchmann H, Nyberg G, Vanlauwe B, Röing de Nowina K (2019) Biochar addition persistently increased soil fertility and yields in maizesoybean rotations over 10 years in sub-humid regions of Kenya. Field Crops Res 235:18–26. https://doi.org/10.1016/j.fcr.2019.02.015

Kätterer T, Roobroeck D, Kimutai G, Karltun E, Nyberg G, Cecilia Sundberg C, de Nowina KR (2022) Maize grain yield responses to realistic biochar application rates on smallholder farms in Kenya. Agron Sustain Dev 235:18–26. https://doi.org/10.1007/s13593-022-00793-5

Khan S, Irshad S, Mehmood K, Hasnain Z, Nawaz M, Rais A, Gul S, Wahid MA, Hashem A, Allah EFA, Ibrar D (2024) Biochar production and characteristics, its impacts on soil health, crop production, and yield enhancement: a review. Plants 13:166–184. https://doi.org/10.3390/plants13020166
Koné S, Galiegue X (2023) Potential development of biochar in Africa as an adaptation strategy to climate change impact on agriculture. Environ Manage 72:1189–1203
Kristoferson LA, Bokalders V (2013) Renewable energy technologies: their applications in developing countries. Elsevier, Amsterdam
Laborde D, Mamun A, Martin W, Piñeiro V, Vos R (2021) Agricultural subsidies and global greenhouse gas emissions. Nat Commun 12:2601–2609 https://doi.org/10.1038/s41467-021-22703-1
Land Use land cove of Africa image—Search (bing.com)
Lema B, Mesfin S, Kebede F, Abraha Z, Fitiwy I, Haileselassie H (2019) Evaluation of soil physical properties of long-used cultivated lands as a deriving indicator of soil degradation, north Ethiopia. Phys Geogr 40(4):323–338. https://doi.org/10.1080/02723646.2019.1568148
Li S, Chen G (2018) Thermogravimetric, thermochemical, and infrared spectral characterization of feedstocks and biochar derived at different pyrolysis temperatures. Waste Manag 78:198–207. https://doi.org/10.1016/j.wasman.2018.05.048
Li S, Harris S, Anandhi A, Chen G (2019) Predicting biochar properties and functions based on feedstock and pyrolysis temperature: a review and data syntheses. J Clean Prod 215:890–902. https://doi.org/10.1016/j.jclepro.2019.01.106
Li Y, Gupta R, Zhang Q, You S (2023) Review of biochar production via crop residue pyrolysis: Development and perspectives. Bioresour Technol 369:1–14. https://doi.org/10.1016/j.biortech.2022.128423
Liu X, Ye Y, Liu Y, Zhang A, Zhang X, Li L, Pan G, Kibue GW, Zheng J, Zheng J (2014) Sustainable biochar effects for low carbon crop production: a 5-crop season field experiment on a low fertility soil from Central China. Agric Syst 129:22–29
Liu X, Zhang A, Ji C, Joseph S, Bian R, Li L, Pan G, Paz-Ferreiro J (2013) Biochar's effect on crop productivity and the dependence on experimental conditions-a meta-analysis of literature data. Plant Soil 373:583–594. https://doi.org/10.1007/s11104-013-1806-x
Mandal S, Donner E, Vasileiadis S, Skinner W, Smith E, Lombi E (2018) The effect of biochar feedstock, pyrolysis temperature, and application rate on the reduction of ammonia volatilization from biochar-amended soil. Sci Total Environ 627:942–950
Manka'abusi D, Steiner C, Akoto-Danso EK, Lompo DJP, Hearing V, Werner S, Marschner B, Buerkert A (2019) Biochar application and wastewater irrigation in urban vegetable production of Ouagadougou, Burkina Faso. *Nutr Cycl Agroecosyst*, 115:263–279. https://doi.org/10.1007/s10705-019-09969-0
Marcinczyk M, Oleszczuk P (2022) Biochar and engineered biochar as slow- and controlled-release fertilizers. J Clean Prod 339:1–14. https://doi.org/10.1016/j.jclepro.2022.130685
Martinsen V, Mulder J, Shitumbanuma V, Sparrevik M, Børresen T, Gerard Cornelissen1 G (2014) Farmer-led maize biochar trials: effect on crop yield and soil nutrients under conservation farming. J Plant Nutr Soil Sci 177:681–695. https://doi.org/10.1002/jpln.201300590
Masek O (2017) Biochar production technologies. http://www.geos.ed.ac.uk/sccs/biochar/documents/BiocharLaunch-OMase.pdf
Mendez A, Gomez A, Paz-Ferreiro J, Gasco G (2012) Effects of sewage sludge biochar on plant metal availability after application to a Mediterranean soil. Chemosphere 89:1354–1359. https://doi.org/10.1016/j.chemosphere.2012.05.092
Mesfin S, Gebresamuel G, Haile M, Zenebe A (2020a) Modelling spatial and temporal soil organic carbon dynamics under climate and land management change scenarios, northern Ethiopia. Eur J Soil Sci 72:1298–1311. https://doi.org/10.1111/ejss.13060
Mesfin S, Gebresamuel G, Zenebe A, Haile M (2020b) Nutrient balances in smallholder farms in northern Ethiopia. Soil Use Manage 37:468–478. https://doi.org/10.1111/sum.12635

Mesfin S, Taye G, Hailemariam M (2018) Effects of integrated soil and water conservation measures on soil aggregate stability, soil organic matter and soil organic carbon stock of smallholder farmlands in semi-arid Northern Ethiopia. Carbon Manag 9(2):155–164. https://doi.org/10.1080/17583004.2018.1443641

Mia S, van Groenigen JW, van de Voorde TFJ, Oram NJ, Bezemer TM, Mommer L, Jeffery S (2014) Biochar application rate affects biological nitrogen fixation in red clover conditional on potassium availability. Agric Ecosys Environ 191:83–91

Mohan D, Sarswat A, Ok YS, Pittman CU (2014) Organic and inorganic contaminants removal from water with biochar, a renewable, low cost and sustainable adsorbent—a critical review. Biores Technol 160:191–202. https://doi.org/10.1016/j.biortech.2014.01.20

Nguyen T, Sherpa K, Bui X, Nguyen V, Vo T, Ho H, Chen C, Dong C (2023) Biochar for soil remediation: a comprehensive review of current research on pollutant removal. Environ Pollut 337:1–11. https://doi.org/10.1016/j.envpol.2023.122571

Nyambo P, Mupambwa HA, Nciizah AD (2021) Biochar enhances the capacity of climate-smart agriculture to mitigate climate change. Handb Clim Chang Manag 1–19.https://doi.org/10.1007/978-3-030-22759-3-319-1

Obia A, Cornelissen G, Martinsen V, Smebye AB, Mulder J (2020) Conservation tillage and biochar improve soil water content and moderate soil temperature in a tropical Acrisol. Soil Tillage Res 197:1–12. https://doi.org/10.1016/j.still.2019.104521

Obia A, Børresen T, Martinsen V, Cornelissen V, Mulder J (2017) Biochar also improves in breaking soil crusting and penetration resistance to facilitate hydraulic properties of two coarse-textured tropical soils. Soil Tillage Res 170:114–121. https://doi.org/10.1016/j.still.2017.03.009

Obia A, Mulder J, Martinsen V, Cornelissen G, Børresen T (2016) In situ effects of biochar on aggregation, water retention and porosity in light-textured tropical soils. Soil Tillage Res 155:35–44. https://doi.org/10.1016/j.still.2015.08.002

Omara P, Aula L, Oyebiyi FB, Eickhoff EM, Carpenter J, Raun WR (2020) Biochar application in combination with inorganic nitrogen improves maize grain yield, nitrogen uptake, and use efficiency in temperate soils. Agronomy 10:1241–12453. https://doi.org/10.3390/agronomy10091241

Oni BA, Oziegbe O, Olawole OO (2019) Significance of biochar application to the environment and economy. Ann Agric Sci 64:222–236. https://doi.org/10.1016/j.aoas.2019.12.006

Pang S (2018) Advances in thermochemical conversion of woody biomass to energy, fuels and chemicals. Biotechnol Adv 1–30. https://doi.org/10.1016/j.biotechadv.2018.11.004

Panwar NL, Pawar A, Salvi BL (2019) Comprehensive review on production and utilization of biochar. Review paper. SN Appl Sci 1:168–186. https://doi.org/10.1007/s42452-019-0172-6

Plaimart J, Acharya K, Mrozik W, Davenport RJ, Vinitnantharat S, Werner W (2021) Coconut husk biochar amendment enhances nutrient retention by suppressing nitrification in agricultural soil following anaerobic digestate application. Environ Pollut 268:1–9. https://doi.org/10.1016/j.envpol.2020.115684

Premalatha RP, Bindu JP, Nivetha E, Malarvizhi P, Manorama K, Parameswari E, Davamani V (2023) A review on biochar's effect on soil properties and crop growth. Front Energy Res 1–19.https://doi.org/10.3389/fenrg.2023.1092637

Qambrania NA, Rahmana M, Wonc S, Shima S, Raa C (2017) Biochar properties and eco-friendly applications for climate change mitigation, waste management, and wastewater treatment: a review. Renew Sustain Energy Rev 79:255–273. https://doi.org/10.1016/j.rser.2017.05.057

Qiu B, Tao X, Wang H, Li W, Ding X, Chu H (2021) Biochar as a low-cost adsorbent for aqueous heavy metal removal: a review. J Anal Appl Pyrol 155:1–14. https://doi.org/10.1016/j.jaap.2021.105081

Rogers PM, Fridahl M, Yanda P, Hansson A, Pauline N, Haikola S (2017) Socio- economic determinants for biochar deployment in the Southern Highlands of Tanzania. Tanzan J Sci 48(2):256–267

Rogers PM, Yanda PZ, Pauline NM, Haikola S, Hansson A, Fridahl M (2022) Effects of biochar on soil fertility and crop yields: experience from the Southern Highlands of Tanzania. Tanzan J Sci 48(2):256–267. https://doi.org/10.4314/tjs.v48i2.3

Rondon MA, Lehmann J, Ramı´rez J, Hurtado M (2007) Bio-logical nitrogen fixation by common beans (*Phaseolus vulgaris* L.) increases with biochar additions. Biol Fertil Soils 43:699–708

Rumpel C, Amiraslani F, Chenu C, Cardenas MG, Kaonga M, Koutika L, Ladha J, Madari B, Shirato Y, Smith P, Soudi B, Soussana J, Whitehead D, Wollenberg E (2020) The 4p1000 initiative: opportunities, limitations, and challenges for implementing soil organic carbon sequestration as a sustainable development strategy. Ambio 49:350–360. https://doi.org/10.1007/s13280-019-01165-2

Sahoo SS, Vijay VK, Chandra R, Kumar H (2021) Production and characterization of biochar produced from slow pyrolysis of pigeon pea stalk and bamboo. Clean Eng Technol 3:1–10. https://doi.org/10.1016/j.clet.2021.100101

Sakhiya AK, Anand A, Kaushal P (2020) Production, activation, and applications of biochar in recent times. Biochar 2:253–285. https://doi.org/10.1007/s42773-020-00047-1

Schmidt H, Taylor P (2014) Kon-Tiki flame cap pyrolysis for the democratization of biochar production. Biochar J 14–24. www.biochar-journal.org/en/ct/39

Shifa S, Worku M, Beyene A (2024) Co-application of compost and biochar improves soil properties and Desho grass growth on acidic soils in a tropical environment of Southwestern Ethiopia. Cogent Food and Agriculture 10:1–18. https://doi.org/10.1080/23311932.2023.2290338

Singh H, Northup BK, Rice CW, Prasad PVV (2022) Biochar applications influence soil physical and chemical properties, microbial diversity, and crop productivity: a meta-analysis. Biochar 4(8):1–17. https://doi.org/10.1007/s42773-022-00138-1

Song PP, Ma WJ, Gao XY, Ai SY, Wang J, Liu WR (2022) Remediation mechanism of Cu, Zn, As, Cd, and Pb contaminated soil by biochar-supported nanoscale zero-valent iron and its impact on soil enzyme activity. J Clean Prod 378:1–11. https://doi.org/10.1016/j.jclepro.2022.134510

Steiner C, Bellwood-Howard I, Häring V, Tonkudor K, Addai F, Atiah K, Abubakari AH, Kranjac-Berisavljevic G, Marschner B, Buerkert A (2018) Participatory trials of on-farm biochar production and use in Tamale, Ghana. Agron Sustain Dev 38(12):1–11. https://doi.org/10.1007/s13593-017-0486-y

Suárez-Abelenda M, Kaal J, McBeath AV (2017) Translating analytical pyrolysis fingerprints to Thermal Stability Indices (TSI) to improve biochar characterization by pyrolysis-GC-MS. Biomass Bioenerg 98:306–320. https://doi.org/10.1016/J.BIOMBIOE.2017.01.021

Sun J, He F, Pan Y, Zhang Z (2017) Effects of pyrolysis temperature and residence time on physicochemical properties of different biochar types. *Acta Agric*. Scand Sect B Soil Plant Sci 67:12–22. https://doi.org/10.1080/09064710.2016.1214745

Sun Y, Xiong X, He M, Xu Z, Hou D, Zhang W, Ok YS, Rinklebe J, Wang L, Tsang DCW (2021) Roles of biochar-derived dissolved organic matter in soil amendment and environmental remediation: a critical review. Chem Eng J 424:1–17. https://doi.org/10.1016/j.cej.2021.130387

Tadesse B, Mesfin S, Tesfay G, Abay F (2016) Effect of integrated soil bunds on key soil properties and soil carbon stock in semi-arid areas of northern Ethiopia. S Afr J Plant Soil 33(4):297–302. https://doi.org/10.1080/02571862.2016.1148788

Tag AT, Duman G, Ucar S, Yanik J (2016) Effects of feedstock type and pyrolysis temperature on potential applications of biochar. J Anal Appl Pyrol 120:200–206

Tomczyk A, Sokołowska Z, Boguta P (2020) Biochar physicochemical properties: pyrolysis temperature and feedstock kind effects. Rev Environ Sci Biotechnol 19:191–215. https://doi.org/10.1007/s11157-020-09523-3.2345

Verma M (2019) Biochar for soil amendment. Char Carbon Mater Deriv Biomass 109–146.https://doi.org/10.1016/B978-0-12-814893-8.00004-3

Wang J, Wang S (2019) Preparation, modification and environmental application of biochar: a review. J Clean Prod 227:1002–1022. https://doi.org/10.1016/j.jclepro.2019.04.282

Wang D, Jiang P, Zhang H, Yuan W (2020a) Biochar production and applications in agro and forestry systems: a review. Sci Total Environ 723:137775–137788. https://doi.org/10.1016/j.scitotenv.2020.137775
Wang L, Xue C, Nie X, Liu Y, Chen F (2018) Effects of biochar application on soil potassium dynamics and crop uptake. J Plant Nutr Soil Sci 181:635–643. https://doi.org/10.1002/jpln.201700528
Wang X, Guo Z, Hu Z, Zhang J (2020b) Recent advances in biochar application for water and wastewater treatment: a review. Peer J 8:e9164
Weber K, Quicker Q (2018) Properties of biochar. Fuel 217:240–261. https://doi.org/10.1016/j.fuel.2017.12.054
Xie Y, Wang H, Guo Y, Wang C, Cui H, Xue J (2023) Effects of biochar-amended soils as intermediate covers on the physical, mechanical and biochemical behavior of municipal solid wastes. Waste Manag 171:512–521. https://doi.org/10.1016/j.wasman.2023.10.004
Xu W, Huang X, Yuan J, Wang Y, Wu M, Ni H, Dong L (2024) The potential for synthesized invasive plant biochar with hydroxyapatite to mitigate allelopathy of Solidago canadensis. Ecol Appl 1–15.https://doi.org/10.1002/eap.2833
Yaashikaa PR, Senthil-Kumar PS, Varjani S, Saravan A (2020) A critical review on the biochar production techniques, characterization, stability and applications for circular bioeconomy. Biotechnol Rep 28:1–15. https://doi.org/10.1016/j.btre.2020.e00570
Yang Y, Ma S, Zhao Y, Jing M, Xu Y, Chen J (2015) A field experiment on enhancement of crop yield by rice straw and corn stalk-derived biochar in Northern China. Sustainability 7:13713–13725. https://doi.org/10.3390/su71013713
Yeboah E, Ofori P, Quansah GW, Dugan E, Sohi SP (2009) Improving soil productivity through biochar amendments to soils. Afr J Environ Sci Technol 3(2):34–41
Yu Z, Chen L, Pan S, Li Y, Kuzyakov Y, Xu J, Brookes PC, Luo Y (2018) Feedstock determines biochar-induced soil priming effects by stimulating the activity of specific microorganisms. Eur J Soil Sci 69(3):521–534
Zhang K, Khan Z, Khan MN, Luo T, Luo L, Bi J, Hu L (2024) The application of biochar improves the nutrient supply efficiency of organic fertilizer, sustains soil quality, and promotes sustainable crop production. Food Energy Secur 13:520–536. https://doi.org/10.1002/fes3.520
Zhou Y, Liu X, Xiang Y, Wang P, Zhang J, Zhang F, Wei J, Luo L, Lei M, Tang L (2017) Modification of biochar derived from sawdust and its application in removal of tetracycline and copper from aqueous solution: adsorption mechanism and modelling. Bioresource Technol 245:266–273

Open Access This chapter is licensed under the terms of the Creative Commons Attribution 4.0 International License (http://creativecommons.org/licenses/by/4.0/), which permits use, sharing, adaptation, distribution and reproduction in any medium or format, as long as you give appropriate credit to the original author(s) and the source, provide a link to the Creative Commons license and indicate if changes were made.

The images or other third party material in this chapter are included in the chapter's Creative Commons license, unless indicated otherwise in a credit line to the material. If material is not included in the chapter's Creative Commons license and your intended use is not permitted by statutory regulation or exceeds the permitted use, you will need to obtain permission directly from the copyright holder.

Biochar Technology for Sustainable Agriculture at Smallholder Farms: A Case Study of Ethiopia, Malawi, and Tanzania

Shimbahri Mesfin, Lars Olav Eik, Sveinung Sægrov, Odd Evjen, Hamisi J. Tindwa, Nyambilila A. Amuri, J. Chimungu, and Bal Ram Singh

Abstract Smallholder farmers in Ethiopia, Malawi and Tanzania are under pressure to sustainably improve agricultural production and manage the environment due to limited access to agricultural inputs. Limited application of biochar among farmers

---

S. Mesfin · L. O. Eik
Department of International Environment and Development Studies (Noragric), LANDSAM, Norwegian University of Life Sciences, As, Norway
e-mail: shimbahri.mesfin.gebreslase@nmbu.no; shimbahri.mesfin@mu.edu.et

L. O. Eik
e-mail: lars.eik@nmbu.no

S. Sægrov
Faculty of Engineering, Department of Civil and Environmental Engineering, Norwegian University of Science and Technology (NTNU), Trondheim, Norway
e-mail: sveinung.sagrov@ntnu.no

O. Evjen
Norwegian Church Aid (NCA), Oslo, Norway
e-mail: odd.evjen@nca.no

H. J. Tindwa
Department of Soil and Geological Sciences, College of Agriculture, Sokoine University of Agriculture, Morgoro, Tanzania
e-mail: tindwa.hamis@sua.ac.tz

N. A. Amuri · J. Chimungu
Crop and Soil Science Department, Lilongwe University of Agriculture and Natural Resources, Lilongwe, Malawi
e-mail: namuri@sua.ac.tz

B. R. Singh (✉)
Faculty of Environmental Sciences and Natural Resource Management, Norwegian University of Life Sciences, As, Norway
e-mail: balram.singh@nmbu.no

S. Mesfin
Land Resources Management and Environmental Protection, Mekelle University, Mekelle, Ethiopia

© The Author(s) 2026

B. R. Singh et al. (eds.), *Sustainable Food Systems in Sub-Saharan Africa*,
https://doi.org/10.1007/978-3-032-11288-0_8

is caused by scattered research findings. Thus, the focus of this review is to critically analyze the current state of knowledge regarding biochar technology in the cases of Ethiopia, Malawi, and Tanzania. The smallholder farmers in these countries commonly use cheap and portable tools and techniques with low cost for biochar production such as open pile burns and open earth pit-based biochar production. Depending on the availability of labor and farming system, farmers apply biochar in the fields by either broadcasting, furrow application, band placement, spot placement, or ring application. Farmers also integrate biochar with co-composting with nutrient-rich manure, and animal and human urine for soil fertility management. Several studies have confirmed that biochar has a great role in sustaining agriculture through improving soil fertility, crop production, adsorbing pollutants, and mitigating climate change. Biochar is also an important feed additive when mixed with animal diets for cattle, poultry, pigs, and fish, and improves the quality and amount of animal production. However, feed stock resources competition of crop residue for biochar is less preferred than its use for animal feed, mulching under conservation agriculture. Use of wood for biochar production may cause deforestation and degradation. Use of alternative resources such as invasive plant species and organic wastes for biochar production is important. The smallholder farmers hold diverse perceptions about biochar, shaped by their experiences, knowledge, and socio-economic context. In general, biochar is an input for amendment of degraded soils, maintenance of long-term soil functions and carbon sequestration in a soil and reduction of GHG emissions in developing countries.

Keywords Biochar · Environmental management · Smallholder farmers · Sustainable agriculture

1 Introduction

Agriculture is the backbone of the African economy; however, weathered, low nutrient concentration and high acidity soils in Africa create major challenges to meet food demands of a growing population (Koné and Galiegue 2023). Soil quality reduction due to soil fertility depletion driven by limited addition and maintenance of nutrient and exacerbated by climate change is threatening the sustainability of agricultural production in the developing countries (Agegnehua et al. 2016). Lack of adequate resources for adding adequate nutrients to maintain soil quality among smallholder farmers negatively affects productivity (Nyambo et al. 2021). To tackle this problem, farmers in the developing countries including Ethiopia, Tanzania and Malawi have been using inorganic fertilizers. However, the consumption of inorganic fertilizer per hectare of arable land is still very low, and is estimated at 9.3 kg ha^{-1}in Tanzania, 42.1 kg ha^{-1} in Ethiopia and 96.7 kg ha^{-1} in Malawi (World Bank 2024). An increased soil and environmental degradation tendency, especially nutrient mining and deforestation to expand area under cultivation, has been observed (Tefera et al. 2024). One feasible measure to improve soil quality is integrated utilization of organic resources (including biochar and manure compost) and inorganic fertilizers. (Singh et al. 2020). The agricultural use of biochar, in combination with other amendments, can sustainably improve smallholder farming productivity (Schmidt et al. 2021). The potential benefits of biochar in agricultural systems includes soil nutrient enrichment, reduced input requirement, conserve moisture, enhanced yields, land remediation/reclamation and carbon sequestration. Inclusion of biochar application in agricultural soils for carbon trading are now being investigated as potential sources of sufficient added value.

In principle, biochar can be made from any type of biomass (Scholz et al. 2014). The critical elements to every biochar production system are (a) the source of biomass and (b) the methods of biochar production. Primary factors determining the quality of biochar are the type of organic material used as feedstock and process temperature (Gaskin et al. 2008). Other factors affecting the properties of biochar include heating rate, heat transfer inside the vessel and pressure during production (Lehmann and Joseph 2009). Furthermore, biochar characteristics (e.g. chemical composition, surface chemistry, particle and pore size distribution) determine the effects of biochar on soil functions. Relative contribution of biochar in improving soil quality has been assessed in different geographical regions. Several research reports from Ethiopia (Fenta 2024; Seitz et al. 2017), Malawi (Burrow 2023), and Tanzania (Eltigani et al. 2022; Rogers et al. 2022; Hansson et al. 2021) have shown the potential of biochar technology on sustainable agriculture such as soil fertility management, crop production, carbon sequestration and reducing GHG emissions. However, these are scattered over the countries and occur under specific conditions, which makes it difficult for the smallholder farmers and policy makers to practically use and draw a conclusive picture about the traditional biochar potentials. Therefore, the aim of this review is to bring these scattered experiences and available information on practical biochar activities of these countries into scientific report and draw conclusions for further wide use of biochar technology in Ethiopia, Malawi, and Tanzania as well as other Sub-Sahara African (SSA) countries.

In this review a thorough literature search was performed using keywords mainly from research conducted in Ethiopia, Malawi, and Tanzania. The review uses recently published articles; however, some earlier published studies have also been included for the overall discussion. The review includes the overall smallholders' views and experience in biochar production and application, potential of biochar uses for sustainable agriculture and environmental management, and the major challenges and opportunities of biochar use in smallholder farming systems.

2 Biochar Sources and Production Systems

The feedstocks for the preparation of biochar from different countries tend to be diversified, mainly including plant sources (straw and crop residues, wood chips and barks, twigs, leaf litter, forest debris, rice husk and bagasse, rice and wheat straw, corn stalk straw), animal sources (poultry manure, pig manure, cattle manure), and waste source (sewage sludge and municipal solid waste). However, the source of feedstock plays a key role in the physical and chemical properties of biochar (Cao et al. 2019). Numerous investigations have shown that the surface morphology, functional groups, hydrophobicity, stability and other properties of biochar are affected by type of feedstock. For example, biochar produced from poultry litter have higher pH and extractable phosphorus and sodium, while plant-based biochar has higher carbon content and surface area compared to animal sources.

Biochar has been used for several years and its methods of production have been modified (Jagnade et al. 2023). Based on the advancement of technology, there are both traditional (in developing countries) and modern (in developed countries) types of biochar production (Panuwar et al. 2019). Despite the potential benefit of biochar at the smallholder farms (in Ethiopia, Malawi, and Tanzania), lack of capital and poor infrastructure prevents smallholder farmers from getting access to modern biochar production methods (Bayabil et al. 2015). However, smallholder farmers have experience in the traditional biochar production systems because they have been producing charcoal using simple earth burn pits instead of high-tech kilns in most rural areas of Africa (Lehman et al. 2006).

The farmers in these countries use different biochar production techniques such as earth con-pits, earth-mound kiln, brick kiln and concrete, micro gasifier cooking stoves and retort kilns (Sakhiya et al. 2020). The traditional earth con-pits, earth mound kilns or brick kilns usually deliver good quality biochar though only high-value wood logs are used as feedstock. These methods operate without gas recovery. This means that the pyrolysis gases and condensable gases are released directly to the ambient air without combustion and thus creating environmental pollution. In addition, yields are relatively low (10–20%) (MacCarty et al. 2008), and the pyrolysis process is very slow, taking several days.

Top-lit up-draft gasifier stoves present a compelling pathway to transform waste biomass into biochar for rural communities in targeted countries, which are typically smallholdings. In these stoves, the pyrolysis gases are generated and combusted in the upper part of the stove, allowing the heat to be used for cooking purposes. The advantage of top-lit up-draft gasifier stoves are that they burn cleanly, avoiding negative health effects due to indoor air emissions, can use various waste biomasses as feedstock and are fuel-efficient. The main drawback of these stoves is that they generate too little biochar (0.5–1 kg per run for household devices and up to 10 kg for the bigger community stoves) to supply enough quantities for farming. For example, the biochar for sustainable soils (B4SS 2017) project in Ethiopia, stated that the adoption of the biochar-making stoves could also provide a significant economic benefit to poor women by reducing their need to either purchase or collect wood.

The introduction of the retort kilns is another option for biochar production for smallholder farmers in these countries. In these retort systems, pyrolytic gases are combusted and re-fed into the combustion chamber and combusted internally, consequently enhancing the process by delivering combustion heat. This re-circulation of gases reduces emissions substantially, but it is costly and consumes a considerable amount of valuable ignition material such as wood during start-up. A recent option has been the introduction of the conical soil pit flame curtain kilns, which is rapidly spreading in these countries. This technology combines the simplicity of the traditional kiln with the combustion of pyrolysis gases in the flame curtain (similar to retort kilns), also avoiding use of external fuel for start-up. These flame curtain kilns have emerged as affordable technology for smallholder farmers. The flame curtain kilns carbonize biomass utilizing an airless combustion process that catches rising pyrolysis gases, including methane, and burns them before they are released into the atmosphere. The additional benefits of this technology include their cost-effectiveness, ability to produce high quality biochar on-site and shorter production cycles compared to other traditional methods (Cornelissen et al. 2023).

The local and low-cost pyrolytic stoves, such as pit and drum kilns designed using locally available materials and technical skills, are most ideal for smallholder households. They use flame curtain kilns that carbonize biomass in an open fire which is positioned on the earth surface, in a pit or in an open trough (Jayakumar et al. 2023; Schmidt and Taylor 2014). But these flame curtain kiln-based methods of biochar production have negative effects due to emissions of methane from the kiln (Owsianiak et al. 2021). The kiln is lit from the top to generate enough heat to make the upper layer of fuel emit combustible gases, which form the flame curtain (Schmidt and Taylor 2014). Though these traditional biochar production techniques such as earthen, brick, and steel kilns are low cost, but they emit volatiles, thus creating air pollution (Sakhiya et al. 2020). Beneath the flame curtain, the fuel itself does not combust, but instead carbonizes because the flame gases consume all available oxygen, creating a pyrolysis zone where the flame protects the fuel from combustion (Cornelissen et al. 2016; Schmidt and Taylor 2014). The fuel will continue carbonizing at the bottom part of the kiln as heat from combusting gases and pyrolysis at the top is transferred to the bottom, mainly through heat conduction as convectional and radiative heat transfers are minimal in the oxygen-deprived environment (Cornelissen et al. 2016). Heat conduction occurs as hot materials near the combustion zone transfer heat to the cold materials that are beneath them through molecular contact (Levenspiel 2014). Because they are open, most of the radiated and convicted heat escapes to the environment, leading to minimal useful heat to assist pyrolysis.

The biochar production flame curtain kilns can be open-pile burn (Fig. 1a), open earth pit (Fig. 1b), kon-tiki cone kilns (Fig. 1c), or Oregon kilns (Fig. 1d).

Flame curtain kilns carbonize feedstocks in an open fire, either on the earth surface, in a pit or in an open trough (Jayakumar et al. 2023). Biochar production kiln can be assembled by screwing a steel sheet into different shapes through incorporation of a chimney, firing point and air inlets (Venkatesh et al. 2015). The pyrolysis heat is transferred through a combination of conduction, convectional and radiation mechanisms (Suzuki et al. 2008). Convectional heat transfer occurs as hot flue gases move within the kiln before exiting through the chimney whereas the kiln wall is more effective in convectional heat transfer (Levenspiel 2014). Like to the kiln assembled from steel, an open pit and kilns made from concrete or brick can be assembled by excavating a rectangular or conical open pit, filling it with the fuel and lighting it from the top (Smebye et al. 2017; Fig. 2a, b). Whereas the Kon-tiki cone kiln is made by folding a steel sheet into an open-topped conical structure with a wall inclination of 63° (Cornelissen et al. 2023; Fig. 2c). The open Oregon kiln can also be shaped

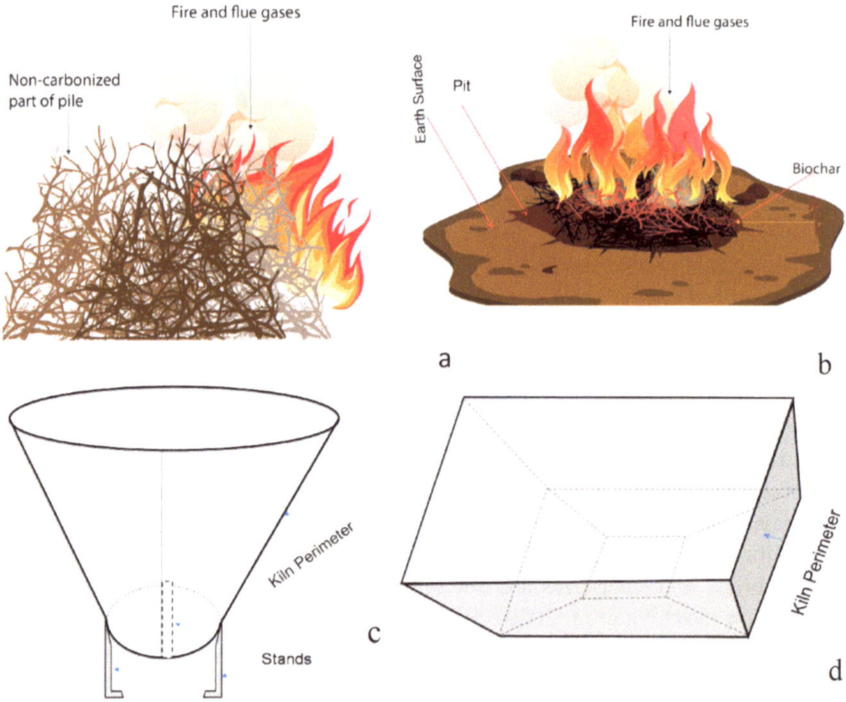

Fig. 1 Flame curtain kilns: **a** open pile burn, **b** open pit kiln, **c** Kon-tiki cone kiln and **d** Oregon kiln (Namaswa et al. 2023). *The authors give us permission to use this figure and cite in this chapter*

by folding and welding or screwing a steel to create an inverted, truncated pyramid (Puettmann et al. 2018).

In general, the traditional biochar productions are slow pyrolysis with low temperature, prolonged residence time and low heating rate compared to modern pyrolysis (Sakhiya et al. 2020). The operation and performance of the different curtain kilns are

Fig. 2 Schematic diagram of different kiln biochar productions. https://www.bing.com

summarized in Table 1. Although the open pile burns and open earth pit-based biochar production are low (below 15%) in their efficiency, they are cheap and portable with simple technology. However, the masonry kiln, Kon-Tiki Cone kiln and Oregon kiln require some installation cost and a skilled person, but they are better in efficiency. The top-lit cookstove is also used for biochar production because it is designed to generate biochar producing heat for cooking and heating instead of producing ash (Cornelissen et al. 2016). The cookstoves can be assembled from different used materials such as paint tins and tomato cans (Birzer et al. 2014). The open lit cookstoves can use small-sized traditional feedstocks such as withies, cuttings, husks, nuts, shavings, sawdust, and pellets, to produce useful heat and biochar, and they have relatively lower emissions compared to conventional biomass cookstoves (Hansson et al. 2021) which is important in reducing deforestation. The other important item for smallholder-based biochar production is masonry kilns. These are fixed kilns made from bricks, concrete and clay, with some additives sodium silicate to prevent cracking (Rodrigues and Junior 2019).

3 Biochar Uses for Various Purposes

The appropriate use of biochar can be achieved by selecting the right source of biochar, the place where to apply it, right application rate, and the right time for placement in the soil. Therefore, biochar application to soils requires the 4Rs (right source, right place, right rate, and right timing) strategies.

3.1 Right Source

As described under Sect. 2 (Biochar sources) selecting a right source is essential for improving its surface morphology, functional groups, hydrophobicity, stability and its efficiency in soil management and crop production.

3.2 Right Place

Soil texture, available nutrient level, and other properties influence the beneficial effects that could be attained through biochar amendment. Smallholder farmers apply biochar in planting basins, ripper lines, planting stations, and furrows. The most efficient method of applying biochar to improve the overall soil health is thorough incorporation into the soil. At occasions when incorporation is not feasible such as biochar amendment to perennial crops and grassland, surface application by broadcasting, banding, or side dressing (ring application) may be practiced. In case of broadcasting, biochar is uniformly spread manually on the surface of their field

Table 1 Types of pyrolysis kilns used or adopted for smallholder farmers-based biochar production

Parameters	Open pile burns	Open earth pit	Masonry kilns	Kon-Tiki Cone kiln	Oregon kiln	References
Feed stock	Branches and small wood	Solid organic residues	Solid organic residues	Solid organic residues	Solid organic residues	Jayakumar et al. (2023)
Construction material	No material	No material	Brick, cement, gravel	Mild steel	Mild steel	Namaswa et al. (2023)
Labor requirements	Little labor required	Labor intensive (digging pits)	Little labor required	Requires skilled labor	Requires little labor	Namaswa et al. (2023), Panwar et al. (2019)
Portability	Not permanent	Not permanent	Portable	Portable	Portable	Namaswa et al. (2023)
Source of heat	Partial oxidation	Partial oxidation	Partial oxidation	Partial oxidation	Partial oxidation	Namaswa et al. (2023)
Provider	Family and community member	Family and community member	Skilled family and community member	Skilled artisan	Skilled artisan	Jayakumar et al. (2023)
Initial investment cost	No	No	Little amount	Little amount	Little amount	Smebye et al. (2017)
Simplicity Needs	Indigenous knowledge	Indigenous knowledge	Require training	Require training	Require training	Panwar et al. (2019)
Pyrolysis temperature (∘C)	200–500	502	400–600	750–850	600	Panwar et al. (2019), Anika et al. (2022)

(continued)

Table 1 (continued)

Parameters	Open pile burns	Open earth pit	Masonry kilns	Kon-Tiki Cone kiln	Oregon kiln	References
Efficiency (%)	10–15	15	20–30	22–28	16	Page-Dumroese et al. (2017)
Biochar quality – Fixed Carbon (%) – Nitrogen (%)	28 0.22	46 1.0	20–35 –	75.5 0.69	65 0.6	Cornelissen et al. (2016), Page-Dumroese et al. (2017)

and incorporation, and it involves manually spreading and mixing it into the soil using a hoe or plowing it using an ox. In the case of spot method application, farmers dig a small hole and put the needed quantity of biochar per stand into the hole and covered it with soil. The ring method is also applied by making a ring of about 5 cm from where the crop or plant would be planted, followed by putting into the ring the required amount of biochar per stand, and covering it with soil. However, biochar application into the soil depends on many factors such as type of farming system, labor availability, and available power machinery (Duku et al. 2011).

3.3 Right Rate

Biochar application rates depend on soil types and crop. Also, biochar materials can differ widely in their characteristics, thus the nature of a specific biochar material (e.g. pH, ash content) also influences application rates. Published literature across the target countries has reported positive effects of biochar application on crop yields with rates of 5–50 tons of biochar per hectare, with appropriate nutrient management. However, excessively low and over application rates must be avoided to reduce their negative effects in soil management. For instance, Shifa et al. (2024) showed that compost and biochar (coffee-husks) co-applied at their full rates (100% compost and 100% biochar) to acidic soil in Ethiopia gave the highest improvements of soil properties and yield. Due to its recalcitrance to decomposition in soil, single applications of biochar can provide beneficial effects over several growing seasons in the field (Steiner et al. 2008). Therefore, biochar does not need to be applied with each crop, as is usually the case for manures, compost, and synthetic fertilizers. Depending on the target application rate, the availability of the biochar supply, and the soil management system, biochar amendments can be applied in increments. However, it is believed that beneficial effects of applying biochar to soil improve with time, and this may need to be taken into consideration when splitting applications over time. Biochar can also be applied directly or through mixing it with crop residue, compost, manure, and seed.

Smallholder farmers also incorporate biochar into soil by using simple power tiller, ox-plough and mixing biochar by hand depending on their economic status (Ndhlovu et al. 2017). These different biochar applications have been reported to increase crop production. For example, Cornelissen et al. (2013) reported an increased maize yield by 44% when biochar is mixed by hand in western Zambia. Yield for the same crop was also reported to have increased by 63% when the biochar was mixed by hand in Uganda (Deal et al. 2012). Similarly, Kimetu et al. (2008) reported 27% maize yield increase when biochar was incorporated to 10 cm depth by a hand hoe in Kenya. Many farmers also apply biochar by integrating it with organic fertilizers such as decomposed manure, compost, crop residues, and inorganic fertilizer to improve nutrient use efficiency over time, improve crop yield and mitigate climate change (Godlewska et al. 2017; Schmidt et al. 2017; Seitz et al. 2017; Gonzalez et al. 2015). Compost is a cheap and well-known source of organic soil nutrients (N, K, P, Ca,

Mg, or Na) and stimulates the activity of microorganisms and soil enzymes, which are very important for immobilization of nutrients and soil quality (Jiang et al. 2021; Bedada et al. 2014). On the other hand, biochar tends to be poor in nutrients and releases nutrient slowly but can immobilize pollutants, slow down and/or regulate the release of nutrients, improve soil properties, and sequester carbon (Ding et al. 2016). Because of these reasons, the combination of biochar and compost seems to be an interesting strategy to minimize the above limitations and meets the "win–win" principle (Rombel et al. 2022).

4 Biochar Use for Soil Fertility Management and Sustainable Agriculture

Smallholder crop production in Ethiopia, Malawi and Tanzania is practiced predominantly on infertile heavily weathered soils derived from granitic parent material. These soils have poor water holding capacity, low and declining soil fertility and are predominantly acidic in nature. Therefore, improving soil health plays a key role in improving agricultural production and yield levels in these countries. As stated earlier, use of biochar can reduce the problem through enhancing soil fertility (Hansson et al. 2021; Yeboah et al. 2020) and due to its recalcitrance to decomposition, it provides benefits to crops for long time (Fig. 3). Further, biochar application enhances the nutrient uptake, which reduces dependency on chemical fertilizer and is essentially important for smallholder farmers who cannot afford chemical fertilizers. A study in Tanzania by Rogers et al. (2022) also confirmed that biochar application has improved soil fertility and increased crop yields of coffee, maize, and beans from 1 to 3 tons per hectare.

The Biochar for Sustainable Soils (B4SS 2017) project formed to promote the adoption of sustainable land management practices in developing countries of Africa like Ethiopia has stated that biochar has great role in sustaining agriculture through improving soil fertility and crop yield (Diatta et al. 2020; Zhang et al. 2012) and increase soil C concentration in the long run (Alvum-Toll et al. 2011). Thus, biochar in combination with compost, manure and other traditional fertilizers is promising in sustainably increasing crop production in smallholder farmers (Burrow 2023; Hansson et al. 2021).

Many studies have documented that biochar-based soil amendment improves soil pH, CEC, SOC, reduces N leaching, improves nutrient availability and nutrient uptake, improves soil microbiological functions and mitigates climate change through storing carbon for several years and reducing N_2O and CH_4 emissions from the soil (Fenta 2024; Hansson et al. 2021; Semida et al. 2019; Premalatha et al. 2023) as shown in Fig. 3.

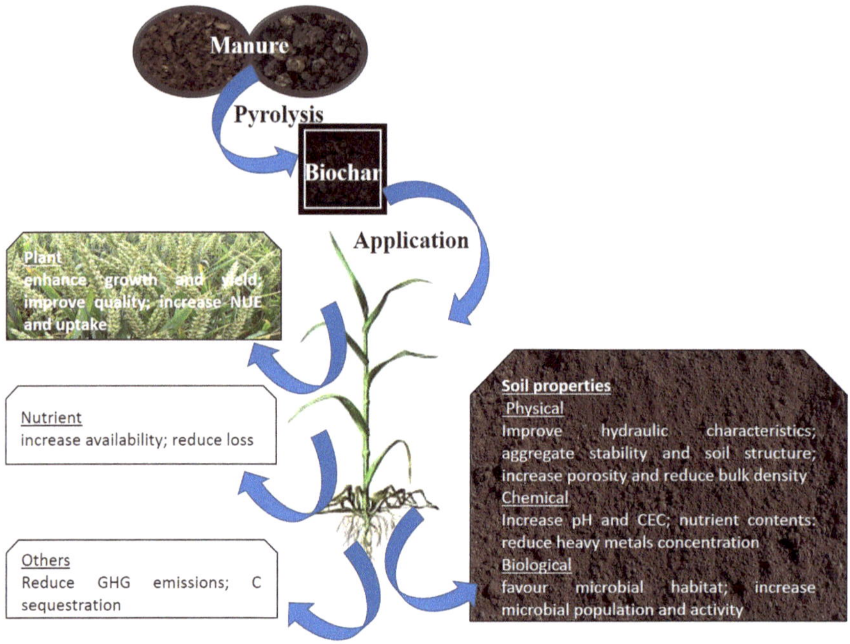

Fig. 3 Conceptual framework for impact of biochar on agriculture. Adopted from Hossain et al. (2020), *the authors give us permission to use this figure and cite in this chapter*

5 Biochar Use for Feed Supplement

Biochar is also an important feed supplement that improves animal health, increases nutrient intake efficiency and thus productivity (Parvathy et al. 2023; Schmidt et al. 2019). As biochar gets enriched with nitrogen-rich organic compounds during the digestion process, the excreted biochar-manure becomes a more valuable organic fertilizer with lower nutrient losses and GHG emissions during storage and soil application. Several studies have also investigated the usefulness of biochar in feed-stock and their effects as feed additives for livestock and poultry production. Biochar is usually mixed with animal diets for cattle, poultry, pigs, and fish (Schmidt et al. 2019) to improve the quality of feed and consequently their production potential, especially in the case of smallholder farming (Al-Azzawi et al. 2021; Schubert et al. 2021; Kalus et al. 2020). Mixing biochar with animal feeds, for example, has been reported to result in increased milk production by 3.43% and protein and fat content by 2.63 to 6.32%, respectively. Furthermore, it is reported to have reduced enteric methanogens of dairy cattle by 30%. This is quite important for smallholder farmers in Ethiopia, Malawi, and Tanzania cases (Al-Azzawi et al. 2021). Additionally, co-feeding with 2.5% biochar increased daily feed intake of laying hens, improved laying performance by 6%, and enhanced shell solidity and thickness by 10% and 6%, respectively (Kalus et al. 2020).

The most important benefits of biochar as feed supplement are: adsorption of mycotoxins, drugs, pesticides and environmental toxins; detoxification of plant toxins; mitigation of methane emission from ruminants; treatment of digestive disorders; and growth promotion (Usman et al. 2016; Devreese et al. 2014). Nair et al. (2023) also stated that the porous structure and large surface area of biochar has a high sorption potential for removing hazardous substances from the animal body and the environment. Other research by Galvano et al. (1996) cited by Nair et al. (2023) also confirmed that adding 2% biochar in animal feed lowered the concentration of extractable aflatoxin by up to 74% and reduced aflatoxin concentration in milk by up to 45%. Biochar also improves digestion, meat quality, egg production and quality; in addition, it increases feed intake and efficiency and prevents disease.

6 Resource Competition for Biochar Production

Biochar has a potential for sustainable agricultural production because it improves soil fertility and moisture supply to plants, which leads to improvement in crop productivity and environmental management (Fenta 2024; Burrow 2023; Eltigani et al. 2022; Rogers et al. 2022; Hansson et al. 2019). However, a strong competition exists for feedstock used for making biochar, because crop residues that are used for biochar production are also important sources of animal feeds. Consequently, smallholder farmers cannot produce biochar because they use the crop residues as animal feed (Seitz et al. 2017). In such cases, resource optimization for biochar production is quite important. Smallholder farmers face competition for biomass resource, and they have both soil fertility and animal feed problems. Customarily, smallholder farmers can produce biochar from feedstocks which are either regarded as waste or those that are in abundant supply. A case study in Malawi by Burrow (2023) confirmed that biochar can be produced by smallholder farmers using local resources which have less competition with animal feed, and this makes it a cost-effective option for smallholder farmers. Several feedstocks were recommended for production of biochar in the case of smallholder farmers in Tanzania and Ethiopia, including agricultural waste, residues from timber business and the organic portion of municipal solid waste, invasive species, and biproducts (Hansson et al. 20,219; Seitz et al. 2017). These are abundantly available and are sometimes a nuisance and cause problems with their disposal. For instance, converting invasive plant materials into biochar can be an option for the invasive plant management by solving their associated disposal problems (Gebremedhin et al. 2015). Table 2 shows resources that could be used to produce biochar, which have less competition among various end users and are relatively abundant in Ethiopia, Malawi, and Tanzania.

One strategy to abate resource competition in the biochar production industry is to maximize use of non-traditional resources, such as invasive species or crop biomass that is customarily left afield or burnt annually. Research shows that a total of 420,491 ha is cropped with sesame in Ethiopia mainly in Tigray, Oromia and Amhara

Table 2 Potential feedstocks for biochar production in Ethiopia, Malawi, and Tanzania

Potential feedstocks	Reported results and impacts	References
Ethiopia		
P. juliflora	It is tested at plot level and improved wheat yield	Gebremedhin et al. (2015)
Sugar cane wastes	Improve soil fertility	Bayrau et al. (2014)
Sesame stalks	The biochar produced from sesame residue improves soil fertility crop production	Gebresas et al. (2015)
Coffee husks	Converting the residue into important soil fertility management input	CSA (2015)
Animal bone	Income source for young unemployed through colleting and selling to local producers of bone char fertilizers	Gewin (2016)
Flower waste	Biochar produced from flower residues are rich in nutrients	Seitz et al. (2017)
Household waste	Widely practiced in smallholder farmers	Seitz et al. (2017)
Malawi		
Pig manure	Widely used by smallholder farmers, schools for fruits and vegetables	Burrow (2023)
Household waste	Traditionally used by many smallholder farmers	Jagger and Jumbe (2016)
Agriculture residues (corn stalks, cobs, and rice husks)	This is tested at plot level and improves soil fertility, improve crop production	Burrow (2023)
Factory biproducts	It is tested at plot level and improves oil fertility	Jagger and Jumbe (2016)
Left over animal feed	Improves soil fertility and crop production	Burrow (2023)
Tanzania		
Corn stalks and cobs, rice husks, coffee pulp	Tested in plots and increases crop yields	Hansson et al. (2021)
Urban biowaste	Results indicate a need to improve the recovery of heat	Rogers et al. (2022)
Residues from timber business	Unverified field trials report positive effects on yields	Hansson et al. (2021)
Agriculture residues	Soil improvements and increases yields	Eltigani et al. (2022)
Cashew nut shells	Unpublished assessments suggest that the system theoretically provides negative emissions	Hansson et al. (2021)

regions, and a straw yield of 2 Mg ha^{-1} (Gebresas et al. 2015) provides biochar feedstock potential of 235,475 Mg per year, which is otherwise left in the field or burned (Aregawi et al. 2013). On the other hand, the invasive species *P. juliflora*, spread to many areas of the lowlands (mainly Afar region), and can also be one potential feedstock for biochar production (Seitz et al. 2017). Research further shows the overall feedstock potential of sugar cane in Ethiopia ranges from 3,147,743 to 4,206,253 Mg yr^{-1}. The cane tops which have no competitive end use and are usually burned (Heuzé et al. 2016) afield are estimated to be 1,190,823 to 2,249,333 Mg yr^{-1}. Other potential feedstock sources for biochar production in Ethiopia with less competition are coffee residues (pulp, mucilage, and hull), flower residue and animal bones (Seitz et al. 2017). But many other biochar feedstock materials available at individual smallholder farm level are agricultural residues, household wastes, leftover fruits and husks, leftover animal feeds, crop biomass from irrigation fields that are not yet put to their full potential in terms of biochar production. Hansson et al. (2021) also documented the available feedstocks that can be used to produce biochar with relatively less competition in Tanzania. They include urban wastes, timber residues such as maize stalks, cashew nut shells and sesame stalks. Similarly, there are feedstocks with less competition in Malawi such as pig manure, urban wastes, agricultural residues and industrial biproducts (Burrow 2023; Table 2).

7 Biochar Use for Environmental and Climate (E. G. Reduction of GHG) Management

One of the major issues facing the modern world today is climate change (Patel and Panwar 2023). The release of greenhouse gases (GHGs) such as methane, carbon dioxide, and nitrous oxides produced by industries and factories, open burning of crop residue, breakdown of agricultural wastes lead to climate change and global warming (Mona et al. 2021). Collectively, the higher the concentration of greenhouse gases the higher is their potential to cause global warming and consequently climate change. Therefore, efforts to reduce GHGs emission from natural and anthropogenic activities on the earth's surface, for example by improving carbon sequestration processes-are paramount to limiting global warming. In recent years, biochar is declared as one of the most promising strategies to combat climate change, in addition to its crucial role in environmental protection and in agriculture such as removal of contaminants and improving soil health as a soil amendment, respectively (Tesfay and Elias 2023; Burrow 2023; Hansson et al. 2021; Sanka et al. 2020). Application of biochar made from agricultural waste in the soil is one of the strategies that lowers GHG emissions from crop residues (Gupta et al. 2020a, b). This is because, due its relatively high carbon content and recalcitrant nature, once applied to soil, biochar helps to hold large amounts of carbon in place, thereby curbing toxic gases emissions to the atmosphere.

On the other hand, biochar serves to ameliorate another environmental problem of heavy metal and aromatic compounds contamination of soils. Once incorporated into the soil, biochar adsorbs heavy metal ions and aromatic compounds on its surface because of its abundant surface area, porosity, and high affinity of functional groups and aromatic compounds on its surface and hence removes contaminants from wastewater and soil (Sakhiya et al. 2020; Sanka et al. 2020). The pollutant adsorption capacity of biochar depends on specific surface area and aromaticity of the biochar (Qiu et al. 2022). Biochar, due to its sorption behavior, is also used for water and wastewater treatment to remove common and emerging organic and inorganic contaminants (Gwenzi et al. 2017).

Soil pollution by inorganic and organic contaminants is one of the environmental challenges faced by the world (Qiu et al. 2022). The heavy metals pollutant in soils (Zn, Cu, Cr, Cd, and Pb) have severe health implications for humans and animals via the food chain and direct exposure (Yuan et al. 2019; Oliveria 2012). The aromatic structure, surface area, active functional groups on its surface, and cation exchange capacity of biochar has tremendous potential in removing heavy metal ions through adsorption (Gupta et al. 2020a, b; Qiu et al. 2022; Sanka et al. 2020). Desta et al. (2016) performed an experiment in Ethiopia on poultry litter biochar application into soil and found decreased Cd concentrations of soil by 71–83% compared to the control. Similarly, Tesfay and Elias (2023) in Ethiopia found that brewery sludge and farmyard biochar reduced the extractable Cd reflected by the decreased concentration of the metal in the shoots of leafy vegetables plants in heavy-metal polluted soils. Therefore, biochar effectively immobilizes metals in contaminated soils and has received increasing attention for use in soil remediation. It creates a recalcitrant soil carbon pool that is carbon-negative, serving as a net withdrawal of atmospheric carbon dioxide that is stored in highly recalcitrant soil carbon stocks.

8 Small-Holders Experiences and Training

The multi-faceted perceptions on the benefits of biochar that farmers witnessed biochar positively impacted soil health and crop yield (Burrow 2023; Rogers et al. 2022). In Tanzania, biochar use as soil amendment has been hindered by lack of a pyrolysis system (Hansson et al.2021), and by lack of integration into the government extension service (Rogers et al. 2022). These barriers call for a continued effort to create awareness and build capacity that are locally adapted in the farming system for wider adoption. The adoption and scaling up of biochar technology in smallholder farming systems can be an input to improve food security and livelihoods at low to no cost to farmers currently struggling to afford synthetic fertilizers (Burrow, 2023; Rogers et al. 2022). However, several studies confirm that lack of awareness on biochar production and utilization has been a challenge to engagement in biochar technology by smallholder farmers (Jha et al. 2020; Latawiec et al. 2017). The smallholder farmers in Ethiopia, Tanzania and Malawi have diverse understanding on the impacts of biochar on soil health, plant growth, and environmental remediation in the context of sustainable agriculture and environmental managements. These perceptions mainly depend on their experiences, knowledge, and socio-economic context (Fenta 2024; Burrow 2023; Seitz et al. 2017).

9 Challenges and Opportunities of Biochar Use for Economic Development

The overall understanding of the opportunities and challenges of biochar technology in Ethiopia, Malawi, Tanzania, and indeed across the entire African continent, is hindered by several complications ranging from policy and legal frameworks, institutional, socio-economic, ecological, health, and technical issues (Burrow 2023; Hansson et al. 2021; Gwenzi et al. 2015; Scholz et al. 2014). This is because of lack of workable policies and legal frameworks highlighting the rationale behind the real benefits of biochar.

9.1 Opportunities

When biochar is used as soil amendment, it reduces the release of GHG into the atmosphere (Koné and Galiegue 2023), strengthens the resilience of agricultural land to extreme climatic phenomena and sequestrates carbon in a soil. However, its potential depends on the biomass used, pyrolysis time, and temperature (Ding et al. 2016). One of the opportunities to produce biochar in Ethiopia, Malawi and Tanzania is the potential availability of feedstocks (Table 1), which can be enhanced through use of invasive species, household wastes, sugar cane wastes, coffee husks, fruit peels, and animal bones. These biochar feedstocks have less competition in all the three countries (Fenta 2024; Burrow 2023; Eltigani et al. 2022; Seitz and Solomon 2017).

9.2 Challenges

The bottlenecks to wide use and upscaling of biochar technology mainly in the smallholder farmers are the negative attitudes and perceptions surrounding it, related to the risk-averse and resources competition (Omulo 2020). In such situations, use of feedstocks which have less importance to the farmers' economy and the environments, such as invasive species, may reduce the economic and associated environmental problems (Gebremedhin et al. 2015). Though biochar has a wide range of benefits, there are still several constraints associated with production and application. Some of which are listed here:

a. The smallholder farmers have less awareness in using safety tools, especially in the face of gaseous particles of biochar that are harmful to human health.
b. The production methods of biochar in smallholder families do not follow the optimization techniques for specific application.
c. The costs and operational variables in the performance of small-scale biochar production kilns used in low-income countries is also a challenge (Namaswa et al. 2023).
d. Poor farmers expect to get increased crop production with sole application of biochar. However, it insignificantly improves soil nutrient (Marcinczyk and Oleszczuk 2022). This calls for an integrated application of biochar with other organic fertilizers (Omara et al. 2020).
e. Obtaining resources for biochar production competition is also one of the challenges. In this case, local resource availability and use prioritization such as invasive species, household wastes, agricultural wastes and others need to come into consideration.
f. Lack of the necessary policy and legal framework that develop and promote biochar production and utilization at smallholder farmers' level.

10 Outlooks and Policies

This review revealed that biochar technology has potential to reduce agriculture and environmental problems. However, appropriate biochar feedstock and pyrolysis techniques that are suitable to the existing situation of smallholder farmers are necessary for them to acquire the benefits. Moreover, the best way of biochar application is through integrating with other organic fertilizers (compost and/or manure) to improve soil nutrient and soil health (Agegnehu et al. 2016; Shifa et al. 2024). This biochar, together with manure and compost, is an important alternative, especially for those poor smallholder farmers, because they cannot afford the cost of chemical fertilizer to improve their soil nutrient and crop production (Nepal et al. 2023). This is because biochar improves crop use efficiency by increasing nutrient uptake and decreasing nutrient losses, and hence minimizing environmental pollution (Hossain et al. 2020). On the other hand, biochar as a feed additive has the potential to improve animal health, feed efficiency and livestock productivity (Nair et al. 2023). However, there are challenges for the small holder farmers, associated with lack of awareness and skill on biochar production, resources competition, policy and legal framework gaps, and lack of finance. Therefore, addressing these challenges, for example by using invasive plants and wastes for biochar production, reduces resource competition (Seitz et al. 2017; Gebremedhin et al. 2015). This also has other benefits because it solves ecological problems caused by invasive plants. The other important reason for the smallholder farmers to use biochar is that they are heavily exposed to the environmental risks and lack adequate resources for adaptation (Nyambo et al. 2021). Therefore, use of biochar is one of the climate change adaptation mechanisms (Patel and Panwar 2023; Gupta et al. 2020a, b) that lead to reduction in GHG emissions.

However, farmers in the developing countries are sometimes reluctant to accept new technologies brought to them. As a result, it is challenging to practice biochar technology, even when it is clearly beneficial (Rogers et al. 2022; Hasson et al. 2021). The adoption of biochar technology by smallholder farmers is hindered by several complicated factors that hinder them from engaging in its production and application (Gwenzi et al. 2015). One of the necessary interventions is to raise farmers' awareness and optimal utilization of available feedstock resources with less competition in these areas.

11 Conclusions, Future Perspectives and Recommendations

Low soil fertility is a common problem across Ethiopia, Malawi, and Tanzania, which needs to be overcome. Use of organic fertilizers like biochar can help to improve the soil health of degraded lands. This review emphasizes the local potential of biochar for agriculture and environmental management, locally available feed stocks by adopted pyrolysis techniques. Studies carried out in these countries confirmed that the application of biochar to agricultural soils improved soil functioning and soil carbon sequestration, and reduced GHG emissions. Thus, giving more attention to biochar use and bringing it into policies has the potential to sustainably improve agricultural production. Moreover, biochar technology is adapted to the local conditions, backed up with current research evidence and proper capacity building. Thus, use of biochar made from agricultural wastes represents a promising approach for sustainably agricultural production and climate change mitigation. Though biochar use in these countries faces several constraints, it is possible to minimize the challenges via:

- Raising farmers' awareness and attitudes, field schools and on-farm demonstrations of biochar technologies practices can increase biochar production and utilization to sustainably improve food security, social, and to address the concerns of environmental degradation.
- Though biochar is a good soil ameliorant, it has less nutrient content; and thus, the synergistic effect of biochar and organic fertilizer like manure, compost and crop rotation can sustainably improve agricultural productivity and mitigate climate change.
- Collaborative efforts are quite important among stakeholders, policymakers, and researchers to promote the adoption of biochar and sustainable agricultural practices for a resilient and productive agricultural sector in smallholder farmers in these countries.
- The recommendations provided offer a roadmap for future endeavors in harnessing the potential of biochar to transform smallholder agriculture in the countries.

References

Agegnehua G, Nelsona PN, Bird MI (2016) Crop yield, plant nutrient uptake and soil physicochemical properties under organic soil amendments and nitrogen fertilization on Nitisols. Soil Tillage Res 160:1–13

Al-Azzawi M et al (2021) Addition of activated carbon into a cattle diet to mitigate GHG emissions and improve production. Sustainability 13:8254. https://doi.org/10.3390/su13158254

Alvum-Toll K, Karlsson T, Ström H (2011) Biochar as soil amendment—a comparison between plant materials for biochar production from three regions in Kenya. Institutionen För Mark Och Miljö, SLU, Examensarbeten 12:1–74

Anika OC, Nnabuife SG, Bello A, Okoroafor ER, Kuang B, Villa R (2022) Prospects of low and zero-carbon renewable fuels in 1.5-degree net zero emission actualisation by 2050: a critical review. Carbon Capture Sci Technol 5:1–17. https://doi.org/10.1016/j.ccst.2022.100072

Aregawi T, Animut G, Kassa H (2013) Utilization and nutritive value of sesame (*Sesamum indicum* L.) straw as feed for livestock in the Northwestern lowlands of Ethiopia. Livest Res Rural Dev 25, Article 124. http://www.lrrd.org/lrrd25/7/areg25124.htm

Bayabil HK, Stoof CR, Lehmann JC, Yitaferu B, Steenhuis TS (2015) Assessing the potential of biochar and charcoal to improve soil hydraulic properties in the humid Ethiopian Highlands: the Anjeni watershed. Geoderma 243–244:115–123. https://doi.org/10.1016/j.geoderma.2014.12.015

Bayrau A, Bekele F, Assefa B, Mihiretu M (2014) An institutional assessment of the CotMg and sugarcane commodities in Ethiopia: The climate change perspective. Ethiopian Development Research Institute (EDRI). Research Report 17. 36 pp

Bedada W, Karltun E, Lemenih M, Tolera M (2014) Long-term addition of compost and NP fertilizer increases crop yield and improves soil quality in experiments on smallholder farms. Agric Ecosyst Environ 195:193–201. https://doi.org/10.1016/j.agee.2014.06.017

B4SS (2017) The biochar for sustainable soils (B4SS) project. Available online: http://biochar.international/

Birzer C, Medwell P, MacFarlane G, Read M, Wilkey J, Higgins M, West T (2014) A biochar-producing, dung-burning cookstove for humanitarian purposes. Procedia Eng 78:243–249

Burrow BJ (2023) Biochar impacts on soil health, and plant growth. A case study of smallholder farms in Malawi. Master's Thesis, pp 1–55

Cao Y, Jing Y, Hao H, Wang X (2019) Changes in the physicochemical characteristics of peanut straw biochar after freeze-thaw and dry-wet aging treatments of the biomass. BioResources 14 (2), 4329–4343.https://doi.org/10.15376/biores.14.2.4329-4343

Cornelissen G, Pandit NR, Taylor P, Pandit BH, Sparrevik M, Schmidt HP (2016) Emissions and char quality of flame-curtain "kon tiki" kilns for farmer-scale charcoal/biochar production. PLoS ONE 11:e0154617. https://doi.org/10.1371/journal.pone.0154617

Cornelissen G, Sørmo E, Rosa RKA, Ladd B (2023) Flame curtain kilns produce biochar from dry biomass with minimal methane emissions. Sci Total Environ 166547.https://doi.org/10.1016/j.scitotenv.2023.166547

Cornelissen G, Martinsen V, Shitumbanuma V, Alling V, Breedveld G, Rutherford D, Sparrevik M, Hale S, Obia A, Mulder J (2013) Biochar effect on maize yield and soil characteristics in five conservation farming sites in Zambia. Agronomy 3:256–274. https://doi.org/10.3390/agronomy3020256

Central Statistical Agency (CSA) (2015) Large and medium scale commercial farms sample survey 2014/15 (2007EC) results at country and regional level. Report 3:1–41

Deal C, Brewer CE, Brown RC, Okure MAE, Amoding A (2012) Comparison of kiln-derived and gasifier-derived biochars as soil amendments in the humid tropics. Biomass Bioenerg 37:161–168. https://doi.org/10.1016/j.biombioe.2011.12.017

Desta W, Pay D, Bernard K, Marschner B, Fisseha I, Heluf G (2016) Effects of biochar and alkaline amendments on cadmium immobilization, selected nutrient and cadmium concentrations of lettuce (*Lactuca sativa*) in two contrasting soils. Springerplus 5:397

Devreese M, AnMgissen G, DeBacker P, Croubels S (2014) Efficacy of active carbon towards the absorption of deoxynivalenol in Pigs. Toxins 6(10): 2998–3004. https://doi.org/10.3390/toxins 6102998
Diatta AA, Fike JH, Battaglia ML, Galbraith JM, Baig MB (2020) Effects of biochar on soil fertility and crop productivity in arid regions: a review. Arab J Geosci 13:595–611. https://doi.org/10.1007/s12517-020-05586-2
Ding Y, Liu Y, Liu S, Li Z, Tan X, Huang X, Zeng G, Zhou L, Zheng B (2016) Biochar toimprove soil fertility. A review. Agron Sustain Dev. https://doi.org/10.1007/s13593-016-0372-z
Duku MH, Gu S, Hagan EB (2011) Biochar production potential in Ghana—a review. Renew Sustain Energy Rev 15(8):3539–3551. https://doi.org/10.1016/j.rser.2011.05.010
Eltigani A, Olsson A, Krause A, Ernest B, Fridahl M, Yanda P, Hansson A (2022) Exploring lessons from five years of biochar-producing cookstoves in the Kagera region, Tanzania. Energy Sustain Dev 71:141–150. https://doi.org/10.1016/j.esd.2022.09.015
Fenta AA (2024) State of the art of biochar in Ethiopia. A review. Heliyon 10:1–9. https://doi.org/10.1016/j.heliyon.2024.e24934
Galvano F, Pietri A, Fallico B, Bertuzzi T, Scirè S, Galvano M, Maggiore R (1996) Activated carbons: in vitro affinity for aflatoxin B1 and relation of adsorption ability to physicochemical parameters. J Food Prot 59(5):545–550. https://doi.org/10.4315/0362-028X-59.5.545
Gaskin JW, Steiner C, Harris K, Das KC, Bibens B (2008) Effect of low temperature pyrolysis conditions on biochar for agricultural use. T. Asabe 51:2061e2069
Gebremedhin GH, Haileselassie B, Berhe D, Belay T (2015) Effect of biochar on yield and yield components of wheat and post-harvest soil properties in Tigray, Ethiopia. J Fertil Pestic 6:158. https://doi.org/10.4172/jbfbp.1000158
Gebresas A, Asmelash H, Berhe H, Tesfay T (2015) Briquetting of charcoal from sesame stalk. J Energy 2015:6 Article ID 757284. https://doi.org/10.1155/2015/757284
Gewin V (2016) Can mountains of animal bones boost food security in Ethiopia? Ensia Magazine. Retrieved 23 Nov 2016, from: http://ensia.com/articles/can-mountains-ofanimal-bones-boost-food-security-in-ethiopia
Gwenzi W, Chaukura N, Noubactep C, Mukome FND (2017) Biochar-based water treatment systems as a potential low-cost and sustainable technology for clean water provision. J Environ Manage 197:732–749. https://doi.org/10.1016/j.jenvman.2017.03.087
Godlewska P, Schmidt HP, Ok, YS, Oleszczuk P (2017) Biochar for composting improvementand contaminants reduction. A review. Bioresour Technol. https://doi.org/10.1016/j.biortech.2017.07.095
Gonzalez ME, Cea M, Medina J, González A, Diez MC, Cartes P (2015) Evaluation of biodegradable polymers as encapsulating agents for the development of a urea controlled-release fertilizer using biochar as support material. Sci Total Environ 505:446–453
Gupta DK, Gupta CK, Dubey R, Fagodiya RK, Sharma G, Keerthika A, Noor Mohamed MB, Dev R, Shukla AK (2020) Role of biochar in carbon sequestration and greenhouse gas mitigation. In: Singh J, Singh C (eds) Biochar applications in agriculture and environment management. Springer, Cham, pp 141–165. https://doi.org/10.1007/978-3-030-40997-5_7
Gupta S, Sireesha S, Sreedhar I, Patel CM, Anitha K (2020b) Latest trends in heavy metal removal from wastewater by biochar-based sorbents. J. Water Process Eng. 38:101561. https://doi.org/10.1016/j.jwpe.2020.101561
Gwenzi W, Chaukura N, Mukome FND, Machado S, Nyamasoka B (2015) Biochar production and applications in sub-Saharan Africa: opportunities, constraints, risks and uncertainties. J Environ Manage 150:250–261
Hansson A, Haikola S, Fridahl M, Yanda P, Mabhuye E, Pauline N (2021) Biochar as multi-purpose sustainable technology: experiences from projects in Tanzania. Environ Dev Sustain 23:5182–5214. https://doi.org/10.1007/s10668-020-00809-8
Heuzé V, Tran G, Archimède H, Lebas F (2016) Sugarcane tops. Feedipedia, a programme by INRA, CIRAD, AFZ and FAO. 14:52. http://www.feedipedia.org/node/558

Hossain M, Bahar M, Sarkar B, Donne SW, Ok YS, Palansooriya KN, Kirkham MB, Chowdhury S, Bolan N (2020) Biochar and its importance on nutrient dynamics in soil and plant. Biochar 2:379–420. https://doi.org/10.1007/s42773-020-00065-z

Jagger P, Jumbe C (2016) Stoves or sugar? Willingness to adopt improved cookstoves in Malawi. Energy Policy 92:409–419. https://doi.org/10.1016/j.enpol.2016.02.034

Jagnade P, Panwar NL, Gupta T, Agrawal C (2023) Role of biochar in agriculture to enhance crop productivity: an overview. Biointerference Res Appl Chem 13(5):429–449. https://doi.org/10.33263/BRIAC135.429

Jayakumar A, Morrisset D, Koutsomarkos V, Wurzer C, Hadden RM, LawMg L, Edwards C, Maˇsek O (2023) Systematic evaluation of pyrolysis processes and biochar quality in the operation of low-cost flame curtain pyrolysis kiln for sustainable biochar production. Crsust 5:100213. https://doi.org/10.1016/j.crsust.2023.100213

Jha S, Kaechele H, Lana M, Amjath-Babu TS, Sieber S (2020) Exploring farmers' perceptions of agricultural technologies: a case study from Tanzania. Sustainability 12:998. https://doi.org/10.3390/su12030998

Jiang Z, Zheng H, Xing B (2021). Environmental life cycle assessment of wheat production using chemical fertilizer, manure compost, and biochar-amended manure compost strategies. Sci Total Environ 760.https://doi.org/10.1016/j.scitotenv.2020.143342

Kalus K et al (2020) Laying hens biochar diet supplementation effect on performance, excreta N content, NH_3 and VOCs emissions, egg traits and egg consumers acceptance. Agriculture 10:237. https://doi.org/10.3390/agriculture10060237

Kimetu JM, Lehmann J, Ngoze SO, Mugendi DN, Kinyangi JM, Riha S, Verchot L, Recha JW, Pell AN (2008) Reversibility of soil productivity decline with organic matter of differing quality along a degradation gradient. Ecosystems 11:726–739. https://doi.org/10.1007/s10021-008-9154-z

Kone S, Galiegue X (2023) Potential development of biochar in Africa as an adaptation strategy to climate change impact on agriculture. Environ Manage 72:1189–1203. https://doi.org/10.1007/s00267-023-01821-0

Latawiec AE, Królczyk JB, Kubo´n M, Szwedziak K, Drosik A, Polanczyk E, Grotkiewicz K, Strassburg BBN (2017) Willingness to adopt biochar in agriculture: the producer's perspective. Sustainability 9:655. https://doi.org/10.3390/su9040655

Lehmann J, Joseph S (2009) Biochar Systems. In: Lehmann J, Joseph S (eds) Biochar for environmental management: science and technology, London, Earthscan 146–68

Lehman J, Gaunt J, Rondon M (2006) Bio-charhar sequestration in terrestrial ecosystems a review. Mitig Adapt Strateg Glob Chang 11:403–427. https://doi.org/10.1007/s11027-005-9006-5

Levenspiel O (2014) The three mechanisms of heat transfer: Conduction, convection, and radiation. In: Levenspiel O (ed) Engineering flow and heat exchange. The plenum chemical engineering series. Springer, BosMg, MA. https://doi.org/10.1007/978-1-4615-6907-7-9

MacCarty N, Ogle D, Still D, Bond T, Roden C (2008) A laboratory comparison of the global warming impact of five major types of biomass cooking stoves. Energy Sustain Dev 12(2):56–65

Marcinczyk M, Oleszczuk P (2022) Biochar and engineered biochar as slow- and controlled-release fertilizers. J Clean Prod 339:1–14. https://doi.org/10.1016/j.jclepro.2022.130685

Mona S, Malyan SK, Saini N, Deepak B, Pugazhendhi A, Kumar SS (2021) Towards sustainable agriculture with carbon sequestration, and greenhouse gas mitigation using algal biochar. Chemosphere 275:129856. https://doi.org/10.1016/J.CHEMOSPHERE.2021.129856

Nair PS, Menon PSS, Suresh S, Sreekanth AJ, Sivasabari K, Krishna SA, Anuranj, PR, Krishnan N, Parvathy S, Chakraborty S, Chopra H, Akash S, Amin H, Dey A, Alagawany M, Chandran D, Dhama K (2023) Beneficial impacts of biochar as a potential feed additive in animal husbandry. J Exp Biol Agric Sci 11(3):479–499. https://doi.org/10.18006/2023.11(3).479.499

Namaswa T, David FRP, Burslem DFRP, Smith J (2023) Emerging trends in appropriate kiln designs for small-scale biochar production in low to middle income countries. Bioresour Technol Rep 24:101641. https://doi.org/10.1016/j.biteb.2023.101641

Ndhlovu M, Kiggundu N, Wanyama J, Banadda N (2017) Effects of incorporating biochar into the soil using power tiller and ox-plough. Sustain Agric Res 6(4):93–103. https://doi.org/10.5539/sar.v6n4p93

Nepal J, Ahmad W, Munsif F, Khan A, Zou Z (2023) Advances and prospects of biochar in improving soil fertility, biochemical quality, and environmental applications. Front Environ Sci 11:1114752. https://doi.org/10.3389/fenvs.2023.1114752

Nyambo P, Mupambwa HA, Nciizah AD (2021) Biochar enhances the capacity of climate-smart agriculture to mitigate climate change. Springer Nature Switzerland AG.https://doi.org/10.1007/978-3-030-22759-3_319-1

Omara P, Aula L, Oyebiyi FB, Eickhoff EM, Carpenter J, Raun WR (2020) Biochar application in combination with inorganic nitrogen improves maize grain yield, nitrogen uptake, and use efficiency in temperate soils. Agronomy 10:1241–12453. https://doi.org/10.3390/agronomy10091241

Omulo G (2020) Biochar potential in improving agricultural production in East Africa. Appl Biochar Environ Saf.https://doi.org/10.5772/intechopen.92195

Owsianiak M, Lindhjem H, Cornelissen G, Hale SE, Sørmo E, Sparrevik M (2021) Environmental and economic impacts of biochar production and agricultural use in six developing and middle-income countries. Sci Total Environ 755:142455–142464. https://doi.org/10.1016/j.scitotenv.2020.142455

Page-Dumroese DS, Busse MD, Archuleta JG, McAvoy D, Rousse E (2017) Methods to reduce forest residue volume after timber harvesting and produce black carbon. Hindawi, 1–8. https://doi.org/10.1155/2017/2745764

Panwar NL, Pawar A, Salvi BL (2019) Comprehensive review on production and utilization of biochar. Review paper. SN Appl Sci 1:168–186. https://doi.org/10.1007/s42452-019-0172-6

Parvathy SN, Sivani MPS, Shreya S, Sreekanth AJ, Sivasabari K, Adithya KS, Anuranj PR, Nayana K, Parvathy S, Sandip C, Hitesh C, Shopnil A, Ruhul A, Abhijit D, Mahmoud A, Deepak C, Kuldeep D (2023) Beneficial impacts of biochar as a potential feed additive in animal husbandry. J Exp Biol Agric Sci 11(3):479–499

Patel MR, Panwar NL (2023) Biochar from agricultural crop residues: environmental, production, and life cycle assessment overview. Resour, Conserv Recycl Adv 19:1–16. https://doi.org/10.1016/j.rcradv.2023.200173

Premalatha RP, Bindu JP, Nivetha E, Malarvizhi P, Manorama K, Parameswari E, Davamani V (2023) A review on biochar's effect on soil properties and crop growth. Front Energy Res 1–19.https://doi.org/10.3389/fenrg.1092637

Puettmann M, Wilson K, Oneil E (2018) Life cycle assessment of biochar from postharvest forest residues: final report. Comm Res Renew Inst Mater. https://greenyourhead.typepad.com/files/w2w-biochar-lca-final-report.pdf

Qiu M, Liu L, Ling Q, Cai Y, Yu S, Wang S, Fu D, Hu B, Wang X (2022) Biochar for the removal of contaminants from soil and water: a review. Biochar 4:19. https://doi.org/10.1007/s42773-022-00146-1

Rodrigues T, Junior B (2019) Charcoal: a discussion on carbonization kilns. J Anal Appl Pyrolysis 143:104670. https://doi.org/10.1016/j.jaap.2019.104670

Rogers PM, Yanda PZ, Pauline NM, Haikola S, Hansson A, Fridahl M (2022) Effects of biochar on soil fertility and crop yields: experience from the Southern Highlands of Tanzania. Tanzan J Sci 48(2):256–267. https://doi.org/10.4314/tjs.v48i2.3

Rombel A, Krasucka P, Oleszczuk P (2022) Sustainable biochar-based soil fertilizers and amendments as a new trend in biochar research. Sci Total Environ 816.https://doi.org/10.1016/j.scitotenv.2021.151588

Sakhiya AK, Anand A, Kaushal P (2020) Production, activation, and applications of biochar in recent times. Biochar 2:253–285. https://doi.org/10.1007/s42773-020-00047-1

Sanka PM, Rwiza MJ, Mtei KM (2020) Removal of selected heavy metal ions from industrial wastewater using rice and corn husk biochar. Water Air Soil Pollut 231:244.https://doi.org/10.1007/s11270-020-04624-9

Schmidt H, Kammann C, Hagemann N, Leifeld J, Bucheli TD, Monedero MAS, Cayuela ML (2021) Biochar in agriculture—a systematic review of 26 global meta-analyses. GCB Bioenergy 13:1708–1730. https://doi.org/10.1111/gcbb.12889

Scholz SM, Sembres T, Roberts K, Whitman T, Wilson K, Lehmann J (2014) Biochar systems for smallholders in developing countries. WashingMg, DC 20433 https://doi.org/10.1596/978-0-8213-9525-7

Schubert DC et al (2021) Effect of two different biochars as a component of compound feed on nutrient digestibility and performance parameters in growing pigs. Front Anim Sci. https://doi.org/10.3389/fanim.2021.633958

Schmidt H-P, Hagemann N, Draper K, Kammann C (2019) The use of biochar in animal feeding. PeerJ 7:1–54. https://doi.org/10.7717/peerj.7373

Schmidt H-P, Pandit BH, Cornelissen G, Kammann CI (2017) Biochar-based fertilization with liquid nutrient enrichment: 21 field trials covering 13 crop species in Nepal. Land Degrad Dev 28(3):2324–2342. https://doi.org/10.1002/ldr.2652

Schmidt H, Taylor P (2014) Kon-Tiki flame cap pyrolysis for the democratization of biochar production. Biochar J 14–24. www.biochar-journal.org/en/ct/39

Seitz T, Solomon D (2017) Synopsis and evaluation of biochar activities in Ethiopia-schemes for potential biochar systems. Report 2

Seitz T, Meyer S, Glaser B (2017) Potential analysis of biochar system for improved soil and nutrient management in Ethiopian agriculture. Report 2: synopsis and evaluation of biochar activities in Ethiopia—schemes for potential biochar systems

Semida WM, Beheiry HR, Setamou M, Simpson CH, Abd El-Mageed TA, Rady MM, Nelson SD (2019) Biochar implications for sustainable agriculture and environment: a review. S Afr J Bot 127:333–347. https://doi.org/10.1016/j.sajb.2019.11.015

Shifa S, Worku M, Beyene A (2024) Co-application of compost and biochar improves soil properties and Desho grass growth on acidic soils in a tropical environment of Southwestern Ethiopia. Cogent Food Agric 10:1–18. https://doi.org/10.1080/23311932.2023.2290338

Singh R, Srivastava P, Bhadouria R, Yadav A, Singh H, Raghubanshi AS (2020) Combined application of biochar and farmyard manure reduces wheat crop eco-physiological performance in a tropical dryland agro-ecosystem. Energ Ecol Environ 5(3):171–183. https://doi.org/10.1007/s40974-020-00159-1

Smebye AB, Sparrevik M, Schmidt HP, Cornelissen G (2017) Life-cycle assessment of biochar production systems in tropical rural areas: comparing flame curtain kilns to other production methods. Biomass Bioenerg 101:35–43. https://doi.org/10.1016/j.biombioe.2017.04.001

Steinera C, Dasb KC, Garciac M, Fo¨rsterd B, Zecha W (2008) Charcoal and smoke extract stimulate the soil microbial community in a highly weathered xanthic Ferralsol. Pedobiologia, 51:359–366

Suzuki T, Okazaki T, Yamamoto K, Nakata H, Fujita O (2008) Improvement in pyrolysis of waste in an externally heated rotary kiln: measurement of the overall heat transfer coefficient from the wall to wastes. J Therm Sci Technol 3:523–531. https://doi.org/10.1299/jtst.3.523

Tefera ML, Carletti A, Altea L, Rizzu M, Migheli Q, Seddaiu G (2024) Land degradation and the upper hand of sustainable agricultural intensification in sub-Saharan Africa - a systematic review. J Agric Rural Dev Trop Subtrop 125: 63–83

Tesfay W, Elias E (2023) A comprehensive review on: response of soil properties and crop performance to biochar application in Ethiopia. https://www.researchgate.net/publication/376834826

Usman AR, Ahmad M, El-Mahrouky M, Al-Omran A, Ok YS, Sallam A, El-Naggar AH, Al-Wabel MI (2016) Chemically modified biochar produced from conocarpus waste increases NO_3 removal from aqueous solutions. Environ Geochem Health 38(2):511–521. https://doi.org/10.1007/s10653-015-9736-6

Venkatesh G, Rao CS, Gopinath KA, Reddy S (2015) A low-cost kiln for biochar production from on-farm crop residues. Indian Farming 64:9–12. https://www.researchgate.net/publication/344350437

World bank (2024) Fertilizer Price Shocks in Smallholder Agriculture Cross-Country Evidence from High-Frequency Phone Surveys in Sub-Saharan Africa. Policy Research Working Paper 10842

Yeboah E, Asamoah G, Ofori P, Amoah B, KOA (2020) Method of biochar application affects growth, yield and nutrient uptake of cowpea. Agriculture 5:352–360. https://doi.org/10.1515/opag-2020-0040

Yuan P, Wang J, Pan Y, Shen B, Wu C (2019) Review of biochar for the management of contaminated soil: preparation, application, and prospect. Sci Total Environ 659:473–490

Zhang AF, Bian R, Pan G, Cui L, Hussain Q, Li L, Zheng J, Zheng J, Zhang X, Han X, Yu X (2012) Effects of biochar amendment on soil quality, crop yield and greenhouse gas emission in a Chinese rice paddy: a field study of two consecutive rice growing cycles. Field Crops Res 127:153–160. https://doi.org/10.1016/j.fcr.2011.11.020

Open Access This chapter is licensed under the terms of the Creative Commons Attribution 4.0 International License (http://creativecommons.org/licenses/by/4.0/), which permits use, sharing, adaptation, distribution and reproduction in any medium or format, as long as you give appropriate credit to the original author(s) and the source, provide a link to the Creative Commons license and indicate if changes were made.

The images or other third party material in this chapter are included in the chapter's Creative Commons license, unless indicated otherwise in a credit line to the material. If material is not included in the chapter's Creative Commons license and your intended use is not permitted by statutory regulation or exceeds the permitted use, you will need to obtain permission directly from the copyright holder.

The State of Agroecology in Malawi: A Review

Keston O. W. Njira, Frank Tchuwa, Trust K. Donga, Simon Mng'omba, Hector Malaidza, Edna Chinseu, Mayamiko Kakwera, Carolyne Phiri, Esnart Kanyenda, Wezi Mhango, Amos Ngwira, Kefasi Kamoyo, Maxwell Mkondiwa, and Vernon H. Kabambe

Abstract Agricultural productivity is facing various challenges in many parts of the world including Malawi. These challenges are mostly associated with soil fertility decline and the increase in crop pests and diseases. The effects of the challenges are exacerbated by increasing human population that puts pressure on the land resources, which results in climate change that manifests in the form of droughts and extreme

---

K. O. W. Njira (✉) · T. K. Donga · S. Mng'omba · E. Kanyenda · W. Mhango · V. H. Kabambe
Crop and Soil Sciences Department, Lilongwe University of Agriculture and Natural Resources, Lilongwe, Malawi
e-mail: knjira@luanar.ac.mw

T. K. Donga
e-mail: tdonga@luanar.ac.mw

S. Mng'omba
e-mail: smngomba@luanar.ac.mw

W. Mhango
e-mail: wmhango@luanar.ac.mw

F. Tchuwa · M. Kakwera · C. Phiri
Extension Department, Lilongwe University of Agriculture and Natural Resources, Lilongwe, Malawi
e-mail: ttchuwa@luanar.ac.mw

M. Kakwera
e-mail: mkakwera@luanar.ac.mw

H. Malaidza · E. Chinseu · A. Ngwira
Chitedze Agricultural Research Station, Department of Agricultural Research Services, Ministry of Agriculture, Irrigation and Water Development, Lilongwe, Malawi

K. Kamoyo
Department of Land Resources Conservation, Ministry of Agriculture, Irrigation and Water Development, Lilongwe, Malawi

M. Mkondiwa
Centre for Agricultural Research and Development, Lilongwe University of Agriculture and Natural Resources, Lilongwe, Malawi

CIMMYT, G2, B Block, National Agricultural Science Centre Complex (NASC), New Delhi, India

© The Author(s) 2026

B. R. Singh et al. (eds.), *Sustainable Food Systems in Sub-Saharan Africa*,
https://doi.org/10.1007/978-3-032-11288-0_9

rainfall events. Furthermore, there is evidence of negative effects on food systems and environment, as a result of the influence of chemicals such as pesticides and fertilizers that are applied to agroecosystems. In recent years, countries have been promoting various agroecological practices that consider ecological and social aspects, with the aim of sustaining productivity of agroecosystems. The aim of this paper is to review the use of agroecological practices for increased agricultural productivity in Malawi. Efforts are made to check which of the FAO agroecological elements are commonly promoted in the practices. Published literature on Malawi complemented by results of a scoping study indicate that while some agroecological practices such as diversified legume cropping systems are becoming popular among smallholder farmers, uptake of a good number of technologies such as conservation agriculture (CA), agroforestry and compost manure still faces challenges. On the other hand, technologies such as biofertilizers and biopesticides need more research, infrastructure, and human capacity development in Malawi. Furthermore, diversity and recycling are FAO agroecology elements that are more prominently addressed by papers on Malawi's agricultural and food systems. From this review it is noted that there is still need for more effort in the promotion of agroecological interventions with efforts on ensuring its sustained uptake by farmers.

1 Introduction

FAO (2018a, b) defined agroecology as an integrated approach that simultaneously applies ecological and social concepts and principles to the design and management of food and agricultural systems. Agroecology as an approach is designed to achieve sustainable agriculture. Sustainable agriculture generally reflects the agenda of ensuring that there is continued increase in agricultural productivity while safeguarding the environmental quality. The achievement of sustained production is hampered by many factors including soil degradation, pest and diseases, failure to implement good agricultural practices, market instability and climate variability and change effects. Soil degradation, with highly pronounced fertility decline, is considered as one of the major challenges facing sub-Saharan Africa that leaves the soils deficient of various essential nutrients (Zingore et al. 2015; Lal. 2015). This is aggravated by increased population pressures on the limited land resources (Ricker-Gilbert et al. 2014), poor farming methods and challenges in finding sustainable ways of replacing nutrients.

The proposed and tested solutions face various criticisms on both biophysical and socioeconomic aspects. For instance, retention of crop residues on the fields faces competing uses with livestock (Giller et al. 2009; Rusinamodzi et al. 2015), while use of composts is hampered by not only quality attributes in terms of the nutrients contained, but also the bulkiness with an implication on transportation costs. Use of livestock-based manure faces the challenge of dwindling numbers of livestock in the farming communities (Ajayi et al. 2007). Inorganic fertilizers face challenges of rising costs that do not match most of the sub-Saharan smallholder farmer economic situation. Furthermore, consideration of the long-term effects of increased use of inorganic fertilizers shows a contradiction with the agenda of ensuring environmental quality and sustainable agriculture. The use of hazardous chemicals in the control of plant diseases, pests, and weeds also raises environmental concerns. Exacerbating the challenges in the quest for achieving sustainable agriculture is the reality of the effects of climate variability and change, which lead to heavy precipitation, frequent dry spells and shortening of rainy seasons (Gbangou et al. 2018).

The prominent sustainable agricultural production practices being advocated by agricultural agencies include conservation agriculture, permaculture, organic farming, ecological agriculture, and agroecology. Definitions from various authors show similarities or overlaps and differences among the mentioned farming systems, but one key issue about these farming systems is the aim of achieving sustainable agriculture (Pretty 2007). For the clear conceptual understanding and definitions of the farming systems mentioned, readers are referred to a number of already published articles (Sharma 2008; Watts and Williamson 2015). There are various arguments about the science, mechanisms and substantial evidence as to how each of the systems can ably lead to safeguarding of the environment while achieving continued high productivity, thus sustainability of agriculture (Pretty 2007; Mockshell and Villarino 2019; and Smith et al. (2017). It should be noted that arguments about the dependability of these systems in the achievement of agricultural sustainability are beyond the scope and essence of this paper.

This paper uses available evidence on the benefits accrued by farmers and the environment when these farming systems are followed. Of particular importance in this paper is agroecology. Besides being an old approach, agroecology is recently becoming popular through promotions by several organisations in many parts of the world including sub-Saharan Africa, in an effort to guide countries to achieve Sustainable Development Goals (SDGs). The Food and Agricultural Organisations of the United Nations (FAO) is one of the organisations that are at the fore front of promoting agroecology. In ensuring that agroecology is mainstreamed in agriculture, FAO outlined the ten elements that need to be followed which are: diversity, synergy, efficiency, resilience, recycling, co-creation and sharing of knowledge, human and social values, culture and food traditions, circular and solidarity economy, and responsible governance (FAO 2018a, b). The ten elements mentioned here for agroecology are well defined in the FAO document and by Barrios et al. (2020), but in this paper their elaborations are covered as part of the discussion. Readers are also referred to a publication by the High-Level Panel of Experts on Food Security and Nutrition for more insights on agroecology (HLPE, 2019).

As noted at the beginning of the paper, the other terms to note before going further with the paper are sustainable agricultural intensification (SAI) and agroecological intensification (AEI). Pretty et al. (2003) considered sustainable intensification as a practice that depends on natural, social and human capital assets, combined with the use of best available technologies and inputs such as best genotypes and best ecological management that minimise or eliminate harm to the environment. On the other hand, Wezel et al. (2015) observe that there are overlaps and a blurriness in the definitions of agricultural, ecological and agroecological intensification. They note however, that in agroecological intensification there is a notion of systems approach that integrates more cultural and social perspectives in the concept. A careful analysis of them however shows that they are both intended to achieve agricultural and environmental sustainability. Hence, in this paper agroecology and agroecological intensification are used interchangeably and form the centre point.

Considering the efforts that researchers, governments and other advocates of good agricultural practices have put in so far, it is important to take stock of what has been done in terms of intensification of agroecological practices. Therefore, this paper is aimed at presenting a review of research findings on agroecological technologies or practices, paying attention to the type of practices, and the benefits, challenges and opportunities noted mostly in Malawi. Efforts are made to refer to the 10 elements of agroecology according to FAO (2018a, b) in discussing the agricultural practices mostly reported on Malawi.

2 Methodology

We conducted a critical review of literature according to procedures suggested by Aboujaoude et al. (2018) and synthesized the evidence from over 80 peer reviewed papers obtained from the internet. Papers containing general and background information, as well as information that is specific to Malawi, were easily retrieved through Google scholar and research gate. An effort was made to ensure that a large proportion of the papers were those indexed in Web of Science and Scopus. Masters and PhD thesis work were also considered. The papers were appraised for element(s) of agroecology being addressed in Malawi by critically looking at the key results in reference to FAO (2018a, b). Different from most review articles that are based on only published work, this paper is complemented by a scoping study in which information was also obtained from various stakeholders through focus group discussions (FGD) with farmers and key informant interviews with extension officers (Table 1).

3 Soil Fertility Enhancing Technologies

Soil fertility decline has, over the years, been considered as the leading limiting factor for crop productivity in Malawi, as it leads to low yields and poor returns to agricultural investments (FAO, 2018b). To address this challenge, a number of farm management practices that include elements of agroecology are advocated by government, non-governmental organisations (NGOs), and public and private research institutions. Use of compost, livestock manure, green manure, legume-based cropping systems, agroforestry, organic and biofertilizers/inoculants, inorganic fertilizers and an integration of inorganic and organic fertilizers are among the practices being promoted in Malawi.

Table 1 Number of people participating in focus group discussions and key informants

EPA	Number of farmers participating			Key informants (Extension officers)
	FGD with males	FGD with females	All FGDs	
Bembeke (Dedza district)*	16	19	35	2
Kandeu (Ntcheu)**	6	9	15	1
Mitundu (Lilongwe district)***	11	18	29	2
Mkanakhoti (Kasungu)****	9	13	22	2
Total	42	51	101	7

* Members of Mwaiwathu cooperative (lead farmers included)
** FRN members of Ganya village
*** Members of Ching'oma and Jeremaya farmer field schools (lead farmers included)
*** * FRN members of Falasi village

3.1 Legume-Based Cropping Systems

Over the years, legumes have played key roles in Malawian agricultural and food systems. A wide range of legumes, including common bean (*Phaseolus vulgaris* L.), groundnuts (*Arachis hypoaea* L.), soybean (*Glycine max* (L.) Merr), cowpea (*Vigna unguiculata* [L.] Walp), pigeon pea (*Cajanus cajan* [L.] Millsp) and Bambara nuts (*Vigna subterranea*) are grown in Malawi (Kamanga et al. 2014). There have been changes in crop production systems as influenced by government policies and increasing population pressure. Snapp et al. (2014) reported that 37.9% of the fields surveyed in Malawi contained maize-legume intercrop. In recent times, legume-based systems have increased in popularity as they are now grown in diversified cropping systems in different parts of the country. Legumes are commonly grown as sole crops and in cereal-legume intercrops. However, in recent years, some farmers are growing legumes in intercrops with other legumes, a practice popularly referred to as "doubled up." The compatible combinations in this practice include a tall and a short legume that ensures reduced interspecific competition for nutrients, water and radiation (Chikowo et al. 2020). For instance, pigeon pea is intercropped with groundnuts or pigeon pea or with soybean. These diversified legume systems provide a number of benefits to the farmers including monetary advantages (Njira et al. 2021), improved soil fertility mostly through biological nitrogen fixation (BNF) (Table 2), increased nitrogen use efficiency (NUE) and increased yields of the cereal that follow in rotation (Mhango 2011; Smith et al. 2016; Njira et al. 2020; Phiri and Njira 2022). Legume-cereal short rotations improve yields of subsequent crops through their positive influences on soil health, more especially when legume residues are retained in the soil. Increases in NUE in form of partial productivity of applied N (PFP_N) of greater than 70 kg grain per kg N applied, indicating well managed systems, in the maize crop following both maize-legume intercrop and doubled-up legumes have been reported (Njira et al. 2020). Additionally, food grain legumes provide a major source of proteins, crop and dietary diversity and other ecosystem services such as contributions to soil carbon sequestration (Kerr et al. 2019; Snapp et al. 2019; Katengeza et al. 2019).

3.2 Organic Manures and Composts

Different types of manures including composts, green manure and livestock manure are used by smallholder farmers in Malawi. Compost making has been promoted by government as a matter of policy (Kabambe et al. 2012), and farmers have been making different types of composts including pit compost, liquid compost, bokash, windrow and changu (Nalivat 2007; Matsui et al. 2016; Ngwira et al. 2013a, b). Composts are used in soil management to serve a number of functions, mostly to add plant nutrients and soil organic matter that has the effect of improving soil structure. Contrasting results about use of composts have been noted in which on one hand improvement of some soil health indicators such as raising of soil pH of acid soils, and increase in N, available phosphorus (P) and potassium (K) have been reported (Ngwira et al. 2013a, b). On the other hand, the quality of the composts and compost making knowledge of farmers are known challenges associated with use of composts by smallholder farmers in Malawi (Mustafa-Msukwa et al. 2011; Ngwira et al. 2013a, b).

Not many studies have been done to evaluate the use of composts by smallholder farmers in Malawi. The number of farmers who use compost manure in their fields vary among different sites in Malawi. For instance, in one study based on a survey from 432 households conducted in the northern region of Malawi in the districts of Mzimba, Rumphi and Nkhata Bay it was noted that use of composts is not very common. Only 30% of farmers were reported to have been able to apply composts to half of their fields, while 21% of farmers experienced increased yields due to application of composts (Matsui and Cornelius 2017). Some of the major challenges to compost use noted in the aforementioned study was collection and transportation (Matsui and Cornelius 2017). This was corroborated by Place et al. (2003), in reference to the bulkiness of manure and its associated transport challenges. Chinangwa (2006) also reported that labour demand in the production and transportation, and availability of materials, were perceived by most farmers interviewed in Zomba and Machinga districts in southern Malawi to be the most important factors affecting use of compost manure. On the other hand, livestock manure use is also reported to be limited by its holding sizes (Chinangwa 2006). However, in instances where livestock manure is used, its influence can be remarkable. In a study in which a wide range of maize yields (0.4 to 12 t ha^{-1}) was reported from farmers' fields in Dedza district of Central Malawi, it was noted that higher yields were obtained from where farmers integrated livestock manure, inorganic fertilizer and improved varieties (Tamene et al. 2016).

Table 2 Biological nitrogen fixation from diversified legume cropping systems in different sites across Malawi

Crop	Cropping system and farm description	Method of estimating nitrogen fixation	Amount of N fixed (kg N ha^{-1})	References
Pigeon pea	Intercropped with maize; farmers field in various parts of Malawi	^{15}N isotope dilution method	37–117.2	Adu-Gyamfi et al. (2007)
Soybean (local inoculated)	Sole cropped and grown in varied environments in Mchinji, Dowa and Salima districts	^{15}N Natural abundance	75.7	van Vugt et al. (2018)
Soybean (Nasoko inoculated)	Sole cropped and grown in varied environments in Mchinji, Dowa and Salima districts	^{15}N Natural abundance	57.8	van Vugt et al. (2018)
Cowpea	Sole cropped on a sandy clay loam soil with medium soil organic matter and N in Dowa district	N Difference	62.5	Njira et al. (2017)
Pigeon pea	Sole cropped on a sandy clay loam soil with medium soil organic matter and N in Dowa district	N Difference	92.9	Njira et al. (2017)
Groundnut	Sole cropped; On-farm; researcher designed farmer managed trials on the sandy soils in Mzimba district	^{15}N Natural abundance	21–102	Mhango et al. (2017)
Soybean	Soybean intercropped with pigeon pea; researcher designed farmer managed on-farm trial on the sandy soils in Kasungu district	N Difference	27.3	Njira et al. (2012)
Pigeon pea + cowpea doubled-up	Pigeon sole intercropped with cowpea; On-farm; researcher designed farmer managed trials on a sandy clay loam soil with medium soil organic matter and N in Dowa district	N Difference	82.9	Njira et al. (2017)
Pigeon pea	Sole cropped; On-farm; researcher designed farmer managed trials on the sandy soils in Mzimba district	^{15}N Natural abundance	11–64	Mhango et al. (2017)

(continued)

Table 2 (continued)

Crop	Cropping system and farm description	Method of estimating nitrogen fixation	Amount of N fixed (kg N ha^{-1})	References
Pigeon pea + soybean doubled-up	Pigeon intercropped with soybean; researcher designed farmer managed on-farm trial on the sandy soils in Kasungu district	N Difference	53.6	Njira et al. (2012)
Pigeon pea + groundnut doubled-up	Pigeon pigeon pea intercropped with groundnut; researcher designed farmer managed trial on the sandy soils in Kasungu district	N Difference	82.8	Njira et al. (2012)

Note No inorganic fertilizers were applied to the legumes presented in this Table 2; Amount of N fixed (kg N ha^{-1}) indicated as doubled up in the column of crops is the contribution (sum) from the two legumes

3.3 *Biofertilizers*

Addressing the decline in soil fertility is a challenge that raises a big dilemma on the amount of external inputs, especially chemical fertilizer usage, into agroecosystems. While it is known that chemical fertilizers can immediately alleviate low soil fertility and ensure food security to some extent, the negative effects that chemical fertilizers bring to the environment are well documented (Nadarajan and Sukumaran, 2021; Pahalvi et al, 2021. Therefore, one of the options that is more environment friendly is the use of biofertilizers, also referred to as microbial inoculants, which are organic materials or substances that contain living microorganisms and are packaged for use in soil fertility or plant nutrition improvement (Kaewchai et al. 2009). They can be applied to the seed, soil or roots of seedlings. They work through a number of mechanisms including fixation of nitrogen in rhizobium-legume symbiosis, solubilisation of unavailable nutrients such as phosphorus (P) and zinc (Zn) into available forms, increasing root surface area and enhancing nutrient uptake for the case of mycorrhizae and production of growth hormones with the possibility of improving nutrient acquisition (Khan et al. 2009).

Microbial inoculants have been in use in Malawi for many years, mostly coordinated at Chitedze Research station where rhizobial inoculants are produced and tested. Earlier N_2 fixing microbial inoculation studies by Chibwana and Wood (1995) and Lupayi and Mkandawire (1996) resulted in improved yields in pigeon pea and common bean, respectively. However, over the years most of the studies conducted were on N_2 fixing microbial inoculants in soybean. Njira et al. (2013) reported a range of 26 to 56 kg N ha^{-1} biologically fixed nitrogen from soybean fields inoculated with *Bradyrhizobium japomicum* compared with 5 to 9 kg N ha^{-1} in non-inoculated soybean on the sandy soils in Kasungu District. Furthermore, van Vugt et al. (2018) reported averages of 45, 57.8 and 76.7 kg N ha^{-1} biologically fixed nitrogen from non-inoculated, inoculated with *Bradyrhizobium japonicum*, and inoculated with *Bradyrhizobium japonicum* plus inorganic fertilizer fields, respectively, from studies conducted in Mchinji, Dowa and Salima Districts in Malawi. Inoculation and increased plant population in soybean led to significant increases in yields and higher profitability with a value cost ratio (VCR) > 2 (Van Vugt et al. 2016). Inoculated legumes also offered better residual soil fertility effects to subsequent crops.

Some progress has been made in inoculant production in the sense that now Chitedze Research Station checks the quality of the products, but production has been scaled out to the private sector. For instance, a company by name of AISL produces inoculants for soybean and groundnuts. Nitrogen fixing inoculants are the more common ones, but not much research has been done on inoculants that can ensure balanced plant nutrition like phosphorus, sulphur and zinc solubilizers. Studies show that in many African countries, including Malawi, production and adoption of different biofertilizer types are derailed by inadequate research, low technology development and ineffective frameworks to regulate the processes (Raimi et al., 2021).

3.4 Agroforestry

Several agroforestry technologies have been tried and tested in Malawi. Agroforestry, an important initiative in the improvement of land management, has a great potential to end or reduce hunger and poverty, and rebuild resilient environments (Nair and Garrity 2012). Agroforestry systems are diverse with a key feature of including trees in combination with crops, and/or pastures and livestock (Mosquera-Losada et al. 2009). It mostly involves integration of leguminous trees (referred to as fertilizer trees) into cropping systems to mitigate land degradation and consequently improve crop yields. In Africa, agroforestry has been promoted mainly to address soil fertility challenges (Glover et al. 2012), but other benefits include soil erosion control, and tree products and services (Garrity 2012). In southern Africa, several research studies undertaken for more than two decades have shown that the use of agroforestry tree species in cropping systems increased soil stability, plant nutrients, soil organic matter, resistance to rainfall impact, infiltration rates and microbial activities (Beedy et al. 2010; Sileshi et al. 2012), hence improved crop yields. According to Sileshi et al. (2008), fertilizer trees have doubled maize yields compared with continuous maize production without fertilizer. A meta-analysis of 94 studies showed that fertilizer trees could double or triple maize yields (Sileshi et al. 2008). Also, an economic assessment of the fertilizer trees shows that they are profitable compared with farmers' traditional practices (Franzel 2004; Ajayi et al. 2009).

Some fertilizer tree species which are being used in southern Africa including Malawi include *Gliricidia sepium* (Jacq.) Walp (gliricidia), *Sesbania sesban* (L.) Merr. (sesbania), *Tephrosia vogelii* Hook, f. (tephrosia), and *Leucaena leucocephala*

(Lam.) (leucaena); *Faidherbia albida* (Kwesiga et al. 2003). However, research has been limited to a few agroforestry tree species to establish their efficiency to improve crop yields. Agroforestry tree biomass, when incorporated into crop fields, improves nutrient cycling and increases soil organic carbon (Barrios et al. 2012). Increased cotton and maize yields, according to studies conducted in Malawi, have been achieved with half dosage of inorganic fertilizer and gliricidia biomass combination (Mng'omba and Akinnifesi 2019). The use of agroforestry tree species is said to be a sustainable land management as application of inorganic fertilizer without the addition of organic matter poses a danger of increasing soil acidity where there has been continuous cultivation, prolonged use of inorganic fertilizers and burning of crop residues (Munthali 2007). Despite the merits derived from agroforestry technologies, wide promotion and adoption in Malawi have been slow. Smallholder farmers do not realize the long-term benefits due to a myriad of constraints such as the lack of technical support and expertise at the local level (especially from extension workers), the absence of multi-locational trials across the country to test the promising tree species and technologies developed, poor project visibility, and lack of consistent national policy on agroforestry. There has been limited awareness about the long-term benefits and economic returns of agroforestry technologies considering that many farmers prefer to use inorganic fertilizers for a quick fix, unlike fertilizer trees which would ensure a sustainable crop production and better land use (Toth et al. 2017b, a).

Although many agroforestry tree species have proved beneficial for crop production, there have been variations in their performance across soil types, climates and fertilizer tree biomass application (Sileshi et al. 2010). This presents a major challenge in scaling up and adoption of fertilizer trees. This requires an investigation into the right fertilizer tree for each place. The use of the right agroforestry tree species at the right place (ecological area) would play a vital role in improving crop yields.

4 Crop Livestock Integration

Crop-livestock integration in Malawi and other countries in sub-Saharan Africa forms an important part of sustainable agricultural strategies. This integration entails a symbiotic interaction between the crop and livestock components in which crop residues are used as animal feed while animals provide manure, draught power and income for procurement of other inputs into the system. Synergy(ies) as one of the elements of agroecology is being exploited in this integration. For instance, integration of livestock in Malawi's farming systems brings positive contributions, including provision of organic manure, income for acquiring various agricultural inputs (Kaumbata et al. 2020), and ensures smallholder farmer's dietary diversity (Jones et al. 2014). Makate et al (2017) noted that farmers in the Chinyanja triangle (Southern and Central Malawi, Eastern Zambia and Northern Mozambique) tend to keep livestock units to diversify their livelihoods as mitigation factor to climate change or variability effects.

The contribution of livestock to integrated systems, and the challenges associated with it, vary from one area to another as influenced by the density of both human and livestock populations. A comparative study, conducted in Mzimba district in Malawi, Nyaki in southern Zimbabwe and Changara in the Tete province of Mozambique, revealed that competing uses for residues is still a big challenge even in Mzimba (northern Malawi district), with lower livestock numbers than the other sites studied (Homann-Kee et al. 2013). In the aforementioned study, although farmers considered the declining soil fertility as a big challenge, they had limited residues to spare for mulching (Homann-Kee et al. 2013). It was noted that the trade-off of reallocating crop residue from feed to mulch was high (Homann-Kee et al. 2013). Low livestock ownership and small herd sizes, implying small amounts of manure and lack of community bye-laws to curb livestock movements that lead to depletion of residues in farmers' fields, are some of the challenges negatively affecting the contribution of livestock in integrated systems (Ajayi et al. 2007; Zulu et al. 2018).

5 Conservation Agriculture

Conservation agriculture (CA) is one of the technologies that have been studied and promoted fairly extensively in Malawi. This is a farming system that has been evolving from what used to be popularly referred to as zero tillage in North Africa, and it became very popular in 1930s in the USA after the dust bowl disaster. It was developed as a measure to deal with the degradation of soils due to over cultivation that was leading to excessive water and wind erosion. It incorporated other aspects from different farming systems in other parts of the world; and today, it is commonly recognized by the implementation of three major principles, which are: minimum soil disturbance, soil cover or mulching and crop rotations or intercropping. Realising the importance of this farming system, the Ministry of Agriculture, Irrigation and Water Development (MoAIWD), through the National Conservation Agriculture Task Force (NCATF) (2016), produced national guidelines for implementing conservation agriculture in Malawi in which the three major principles were complemented with a number of other good agricultural practices (Fig. 1).

CA is also noted to work well where agroforestry tree species have been used; this is referred to as Conservation Agriculture with trees (CAWT). As crop residues are dwindling due to the said competition with livestock for feed, agroforestry tree species are used to provide adequate biomass to cover the soil. While conservation agriculture (CA) and agroforestry have evolved separately as research disciplines, there are also opportunities to integrate them. Individually, each holds a tremendous potential for climate change adaptation and mitigation as well as achieving sustainable land management (SLM). It is the integration of the two practices that is referred to as Conservation Agriculture with Trees (CAWT). With trees in CAWT, the extra benefits over and above those drawn from conservation agriculture include further improvement in crop yields, carbon sequestration, and pest control and weed suppression (Garrity 2010).

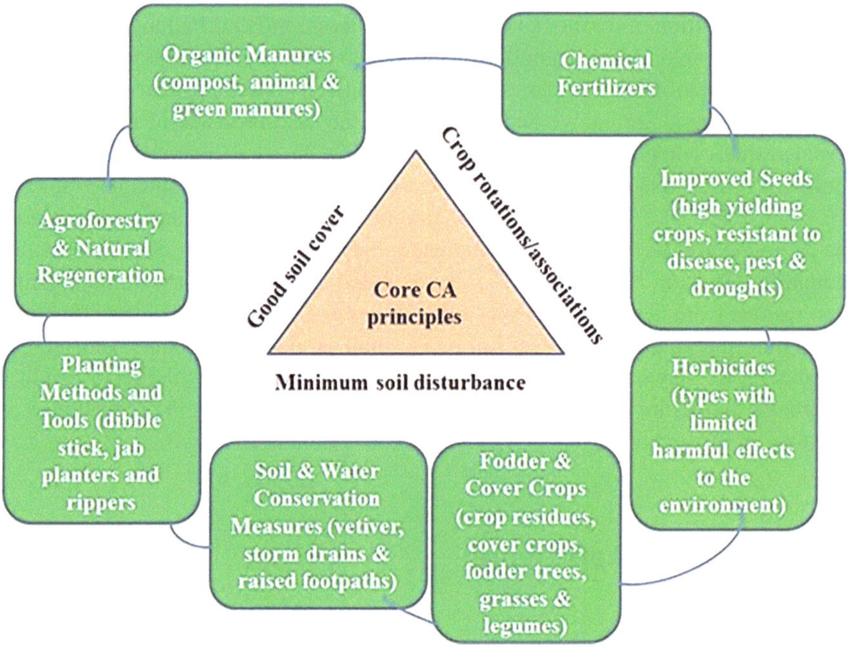

Fig. 1 Malawi's system of CA with complementary practices available. *Source* NCATF (2016)

The Government of Malawi, together with development partners, NGOs such as Total Land Care, Concern Worldwide, Care Malawi, Concern Universal and World Vision International has been at the forefront promoting CA (Dougill et al. 2017). The long term benefits of CA have been reported as improvement of soil health and increase in crop yields from Malawi soils and elsewhere (Thierfelder and Wall 2009; Thierfelder et al. 2013, 2015). However, the most notable shortfall for CA in Malawi is low adoption and substantial dis-adoption in which smallholder farmers do not sustain the uptake of the technology (Chinseu et al. 2019). Conservation agriculture by country as a percentage of arable land uptake was reported as 1.7% (FAOSTAT, 2014). Ngwira et al. (2014) reported 18% adoption from a survey conducted in 8 Extension Planning Areas (EPAs) in which 15,854 was the number of smallholder farmers' households. On the other hand, Bouwman et al. (2021), from a study in Central Malawi, noted that most of those considered practising CA do not practise all the three major principles simultaneously; rather, they do one or two of the principles which they are able to adapt. Factors cited for low adoption and dis- adoption of CA in Malawi include total cultivated land, duration of practising CA and lack of sufficient technical support (Ngwira et al. 2014; Chinseu et al. 2019). Dougill et al. (2017) noted that knowledge gaps and institutional barriers influence land management and affect CA uptake.

6 Permaculture and Organic Farming

Permaculture is one of the agroecological approaches. Holmgren (2002) defined it as consciously designed landscapes which mimic the patterns and relationships found in nature, while yielding an abundance of food, fiber and energy for delivery of local needs; and he expounded it by outlining its 12 elements. Generally, it highlights the positive role of humans as ecosystem managers (Morel et al. 2018). Permaculture literature on Malawi is very scanty. Rivett et al. (2018) reported on permaculture based on data from the borehole-garden project (2009–2017), which spilt groundwater for water-food-health security across rural Chikwawa District, in Southern Malawi in 123 sites (6% of the District water points). Early data from the borehole-garden study affirmed garden profitability, but it was noted that the uptake and performance were dependent on the diversity of technical, economic, social and governance. On the other hand, organic farming is similar to other agroecological approaches in promoting the use of organic resource except that organic farming strictly prohibits use of chemicals or synthetics, which includes pesticides, herbicides and inorganic fertilizers (Watts and Williamson 2015). For the case of Malawi, information about the extent to which organic farming is practised is scanty. However, it should be noted that there are NGOs, like the Malawi Organic Growers Association (MOGA),which are responsible for coordinating activities of organic farmers and Kusamala Institute for Agriculture and Ecology that promotes and offers short training courses in agroecology.

7 Pest and Disease Management

Agroecological pest management puts emphasis on biological processes to reduce the use of external inputs (Deguine et al. 2008). It is based on preventive measures rather than reactive response to manage pests. Therefore, to build and enhance biological system, cultural practices are supposed to be given priority. For Malawi, it is evident that several hundred pests affect agricultural production (Donga and Eklo 2018). This poses a big challenge to crop yields and storage. Similar to what is being done in many countries, responses of controlling pests include use of synthetic pesticides and other integrated approaches. However, the high costs and ineffectiveness of synthetic pesticides against new and emerging pests is forcing farmers to find alternatives. The need for safer and effective alternatives to synthetics is also necessitated by recent restrictions on some pesticides such as methyl bromide and international markets such as Fairtrade, which are acting as change agents.

It should be noted that there are several innovative ways of increasing yield sustainably with which researchers and some farmers in Malawi are familiar. In this section, the focus is on agroecological practices in the broad sense, which are becoming popular such as botanicals based on plant material and biocontrol method, which involve living organisms, including microorganisms and other natural enemies.

Isman (2017) argued that there are already a good number of plant species occurring in Africa that are either endemic or introduced, which have been verified to possess bioactive ingredients for insecticide production. However, it was noted in Isman (2017) that what is needed is to develop simple methods of propagation and cultivation, extraction, validation, and demonstration of application strategies.

One important crop protection challenge to note is that there is a higher aflatoxin contamination risk among the Malawian populace. Magamba et al. (2017) reported that 72% of branded groundnut products found in the country's supermarkets had aflatoxins above the regulatory limit for Malawi (3 μg/kg), European Union (4 μg/kg) and United States of America (20 μg/kg). The same authors also found that the total aflatoxin levels ranged from 1.5 to 1200 μg/kg in raw groundnuts and 83–820 μg/kg in groundnut flour from vendors. A survey by Matumba et al. (2014) reported that aflatoxin contamination is also high in locally processed maize-based baby food sold in popular markets. With funding from United States Agency for International Development (USAID) through the Feed the Future Malawi Improved Seed Systems and Technologies (MISST) Project, and in collaboration with IITA, government facilitated the development of biological control products for aflatoxin management in Malawi. Two products Aflasafe MW-MZ01 and Aflasafe MW02 were launched in April 2019. An aflatoxin testing laboratory at Chitedze Research Station in Lilongwe was also handed over to the Malawi Government by USAID MISST Project.

8 Summarized Appraisal of Elements of Agroecology Covered in Papers on Malawi's Agricultural and Food Systems

This section presents a summary of the appraisal of practices in reference to the ten elements of agroecology according to FAO (2018a, b) that are being addressed in Malawi. This is based on literature review; therefore, what is covered is what has been reported in published papers. Key aspects addressed in the papers, matched against the ten elements of agroecology, show that some elements have been addressed in research more than others. For instance, diversity in terms of cropping systems, varieties and diets (Kerr et al. 2019; Snapp et al. 2019; Mhango et al. 2012; Njira et al. 2012; Njira et al. 2017; Mhango et al. 2017; Njira et al. 2020; Njira et al. 2021), efficiency in terms of nutrient use (Makumba 2003; Mhango 2011; Snapp et al. 2014; Njira et al. 2020; Khataza et al. 2017; Franke et al. 2014; Ngwira et al. 2012b; Ngwira et al 2013a, b) and nutrient cycling (Nyagumbo et al. 2016; Mhango 2011; Njira et al. 2013; Njira et al. 2020; Mupangwa et al. 2017; Kalasa et al. 2018) are more conspicuous in the papers reviewed. An aspect of responsible governance is noted on the roles of the Malawi government on demand driven extension policy and its weaknesses, promotion of CA guidelines and land reform bills (Álvarez-Mingote et al. 2018a, b; Toth et al. 2017b, a; Dougill et al. 2017; Zulu et al. 2018; NCATF,

2016). It should be noted that high reporting in these aspects such as diversification implies that these areas are reasonably more studied than other elements. On the other hand, published work or information on circular and solidarity economy pertaining to Malawi's agricultural and food systems is scanty. Some aspects of circular economy are covered under recycling, but studies on economic implications of these are scanty.

9 AEI Innovations and Indigenous Knowledge Existing and Practiced by Smallholder Farmers: Evidence from the Scoping Study

On existing and practiced innovations, including those based on indigenous knowledge, the findings of this study reveal that several innovations related to AEI were promoted and tested by farmers in the study areas. Some innovations were being showcased in demonstration plots and others tested in farmer-managed experimental plots. The different innovations belonged to two categories: The science-based (e.g., hybrid varieties and planting patterns) and local knowledge (e.g., organic manure, sand, fish soup and plant extracts for controlling pests). In other instances, a combination (adaptation) of both scientific and local knowledge was observed (e.g., mixing inorganic fertilizer with organic manure in maize-legume rotation systems). Whilst some AEI innovations were observed in all the study sites (e.g., manure and legume-maize rotation for soil health improvement, and application of sand to control fall-army worms), there was also some AEI innovations which were not observed in some areas. For example, double-up legume-maize rotation was not observed in Mitundu and Bembeke EPAs. Similarly, application of fish soup to control army worms was only observed in Mitundu EPA. The findings of the study also show that there were variations in the focus of different approaches used to engage farmers in AEI related demonstrations and experimentations. For instance, improving soil health through different rotation systems, including double-up legume practice, was the focus of farmers participating in the farmer research networks (FRNs). Rotation of maize with a single legume combined with conservation agriculture and modern planting patterns was the focus of lead farmers (farmers with experience and used as extension agents).

The key informants reported that in terms of application of the different AEI innovations in main fields, farmers were adopting new planting patterns, especially one–one planting in maize and double rows in soya beans; maize-legume intercropping and rotation; manure application as well as some elements of conservation agriculture (e.g., minimum tillage). During the discussions with the farmers, it was learned that they were willing to apply soil health options such as crop diversification through integration of different legumes, which potentially contributed to improvement of their soils and household food security and nutrition, as well as income (e.g., groundnuts, soya beans, and pigeon peas). The decreasing farm land size (due to population), as well as the limited farm labour, also prompted farmers to get interested in innovations which promoted land use intensification (e.g., intercropping legumes or legumes with cereals; ridge-realignment from 90 to 75 cm; planting patterns which were considered as increasing plant population per area (e.g., one maize seed per station planted at 25 cm apart, one soya bean seed seeded at 5 cm apart in double rows per ridge, one groundnut seed planted at 15 cm apart). Variations and unpredictability of the rainfall pattern were other reasons for farmers' willingness to apply soil health related innovations such as manure, agroforestry, incorporating crop residues, mulching and minimum tillage. Farmers explained that the organic matter (decomposed leaves, stems, and roots) helped to retain soil moisture during dry spells and increase water infiltration. Therefore, it reduced run-off and soil erosion.

10 Challenges and Opportunities

10.1 Challenges

Production inefficiencies: It is noted that one of the major challenges in some agroecological practices is the capacity of the smallholder farmers to implement the agricultural practices efficiently. For instance, using a directional input distance function (DIDF) in an appraisal of the value of symbiotic nitrogen fixation in legume-based systems in Malawi, Khataza et al. (2017) noted legume productivity inefficiency by smallholder farmers. The productivity inefficiency affects yields of both grain and biomass, leading to reduction in legume influences on soil fertility improvement. Similarly, Franke et al. (2014) noted that low resource endowed farmers face challenges of lack of needed inputs including inoculants to legumes.

Lack of capacity or knowledge: Capacity for handling some of the technologies is also a challenge. Mustafa-Msukwa et al. (2011) reported lack of composting knowledge by smallholder farmers as one of the challenges affecting adoption of compost application in Malawi. On the other hand, while some strides have been taken in biofertilizer production in Malawi, with an example of AISL company which is producing inoculants for soybean, infrastructure and human capacity in the research institutions is still a challenge as most of the laboratories are not well equipped.

Gender relations: One of the significant factors affecting agroecological intensification is the intra-household gender relations that influence decisions on time and labour allocation for given technologies. Me-Nsope and Larkins (2016) reporting from a study on pigeon pea value chain noted that men tend to dominate at the market stage because of cultural restrictions on women's mobility and gender disparities in transportation assets. This discourages women who are mostly providing labour at a production stage. Kerr et al. (2019) argued that agroecology being labour and knowledge intensive, there is need to address the factor of gender power relations both within and beyond households, in order for it to contribute effectively in achieving food sovereignty for smallholder farmers.

Marketing: One of the factors that encourage farmers to take up a technology is the issue of marketing of their products. For instance, legume production is very important in agroecology as it ensures food and nutrition security of smallholder farmers and improves soil health through biological nitrogen fixation; and the residue biomass also contributes to increasing soil organic matter. However, it is reported that poor market prospects and low farm gate prices discourage farmers from intensifying legume production (ICRISAT 2013). From a study conducted in Central and Southern Malawi, Ortega et al. (2016) reported market access as one of the major barriers affecting farmers' adoption of legume technologies. Dzanja et al. (2019) noted that market demand is more likely to influence uptake of multipurpose grain legume cropping systems intensification.

Other challenges: Some challenges are technology specific. For instance, it is noted that there are some agroecological technologies that have some specific challenges that complicate farmers' decisions in adopting them. For instance, the bulkiness of various types of manure, and the challenges associated with hand weeding under CA. Other challenges are to do with awareness and knowledge intensiveness. For instance, not many farmers are knowledgeable about use of botanicals, organic farming and permaculture.

10.2 Opportunities

While challenges that retard or slow down the promotion of agroecology in Malawi are noted, it is also noted that opportunities that can resuscitate agroecology are available. It is generally noted that the Government of Malawi accommodates agroecology in the quest for achieving sustainable agriculture. For instance, the Malawi National Agricultural Policy priority area 1 "Sustainable Agricultural Production and Productivity" includes clauses that encourage agroecological practices (MoAIWD 2016). A good example of government initiatives is the establishment of the National Conservation Agriculture Taskforce, among whose achievements was the development of the guidelines for implementing CA in Malawi (NCATF 2016). Furthermore, over the past 20 years, the Government of Malawi has paid special attention to the agriculture sector through implementation of farm input subsidies that includes fertilizers and seeds (Denning et al. 2009; Chirwa and Dorward 2013). Scientists and stakeholders that advocate for agroecological principles can take the huge interest of the government in agriculture as an opportunity to influence policies that enhance technologies aimed at reducing use of external inputs, as opposed to subsidizing huge amounts of chemical fertilizers.

Another opportunity is the presence of many organisations (government and non-governmental) that are involved in promoting agroecological approaches. Department of Agricultural Research Services, Department of Land Resources Conservation, FAO Malawi, Forestry Research Institute of Malawi, Malawi Organic Growers Association (MOGA), Kusamala Agriculture and Ecology Research Institute, Total Land Care, Care-Malawi and World Agroforestry Centre are among organisations operating in Malawi, promoting practices that are agro-ecological (Dougill et al. 2017). This wide range of organisations is an opportunity that can be exploited in increasing awareness on agroecology.

11 Conclusion and Recommendations

From this review, it is noted that a few agroecological practices are being promoted by various institutions. Practices including legume-based cropping systems, conservation agriculture, use of various types of manure, biofertilizers, botanicals or pesticidal plants, agroforestry, organic farming, and permaculture are some of the agroecological practices being promoted by various organizations in Malawi. It is further noted that more research work has been conducted and reported for some practices such as legume-based systems, CA, and use of manures. On the other hand, not much has been reported on other practices such as biofertilizers, organic farming, and permaculture. Adoption of these practices vary from one technology/practice to another. For instance, it is noted that grain legume-based cropping systems are becoming popular while literature shows very little progress has been achieved in adoption of CA. On top of the scantiness in literature on organic farming and permaculture, for technologies that strictly ban use of external inputs, the scoping study for the sites as reported in this paper never revealed anything about these technologies. An appraisal of coverage of FAO elements of agroecology on the work reported in papers show variation in that diversity, efficiency and recycling are among the most addressed aspects for papers on Malawi agricultural and food systems. Common challenges retarding the progress in agroecology include production inefficiencies in some technologies such as legume cropping systems by smallholder farmers, lack of capacity or knowledge in implementing some technologies, unbalanced gender relations that affects execution of some agroecological activities, and lack of markets, and this frustrates efforts by smallholder farmers. On the other hand, it is noted that opportunities to promote agroecology are available as evidenced by National Agricultural Policy that accommodates some aspects of agroecology and previous initiatives such as the NCATF 2016 that was instituted to produce guidelines for CA which is an example of an agroecological practice. The presence of a good number of governmental and non-governmental organisations that are promoting agroecological practices is another opportunity that can be embraced in the quest for promoting agroecology.

From this review, the following are the recommendations:

- There is a need for comprehensive studies on evaluating the efficiency being achieved by smallholder farmers when they are implementing various agroecological intensification strategies such as diversified cropping systems that include legumes, use of different organic manures, use of biofertilizers, use of botanicals, and biopesticides.
- There is a need for studies to assess the extent and efficiency of crop-livestock integration in Malawi.
- There is a need for more studies on use of the right agroforestry tree species at the right place (ecological area) and implications on improving crop yields.
- There is a need for more research on organic farming and permaculture in Malawi, including assessing their state in the institutional curricula.
- There is a need for a situation analysis of organisations that are promoting agroecological practices, including knowing the extent the agroecological practices are being implemented.

Acknowledgements This study was conducted as part of the contribution of the Malawi Agroecology (AE) Hub with funding from the McKnight Foundation. The review work benefited from comments by stakeholders who attended the Malawi AE Hub 2021 annual review meeting.

References

Aboujaoude MK, Feghali K, Kfouri C (2018) Planning the research. In: Marx Gómez J, Mouselli S (eds) Modernizing the academic teaching and research environment. Progress in IS. Springer, Cham. https://doi.org/10.1007/978-3-319-74173-4_2

Adu-Gyamfi JJ, Myaka FA, Sakala WD, Odgaard R, Vesterager JM, Høgh-Jensen H (2007) Biological nitrogen fixation and nitrogen and phosphorus budgets in farmer-managed intercrops of maize–pigeonpea in semi-arid southern and eastern Africa. Plant Soil 295:127–136. https://doi.org/10.1007/s11104-007-9270-0

Ajayi OC, Akinifesi FK, Sileshi G, Chakeredza S (2007) Adoption of soil fertility replenishment technologies in southern Africa: lessons learnt and the way forward. Nat Res Forum 31:306–317. https://doi.org/10.1111/j.1477-8947.2007.00163.x

Ajayi OC, Akinnifesi FK, Sileshi G, Kanjipite W (2009) Labour inputs and financial profitability of conventional and agroforestry-based soil fertility management practices in Zambia. Agrekon 48(3):276–292. https://doi.org/10.1080/03031853.2009.9523827

Álvarez-Mingote C, McNamara PE (2018a) Demands, responsibility, and influence in Malawi's participatory agricultural extension services. Dev Pract 28(1):81–94. https://doi.org/10.1080/09614524.2018.1397104

Álvarez-Mingote C, McNamara PE (2018b) Evaluating agricultural extension and advisory services through a Governance Lens. J Int Agric Ext Educ 25:71–86. https://doi.org/10.5191/jiaee.2018.25206

Barrios E, Sileshi GW, Shepherd K, Sinclair F (2012) Agroforestry and soil health: linking trees, soil biota and ecosystem services. In: Wall DH (ed) Soil ecology and ecosystem services. Oxford University Press, Oxford, UK, pp 315–330. https://doi.org/10.1093/acprof:oso/9780199575923.003.0028

Barrios E, Gemmill-Herren B, Bicksler A, Siliprandi E, Brathwaite R, Moller S, Batello C, Tittonell P (2020) The 10 Elements of Agroecology: enabling transitions towards sustainable agriculture

and food systems through visual narratives. Ecosystems and People 16(1):230–247. https://doi.org/10.1080/26395916.2020.1808705
Bayala J, Sileshi G, Coe R, Kalinganire A, Tchoundjeu Z, Sinclair F, Garrity D (2012) Cereal yield response to conservation agriculture practices in drylands of West Africa: a quantitative synthesis. J Arid Environ 78:13–25. https://doi.org/10.1016/j.jaridenv.2011.10.011
Beedy TL, Snapp SS, Akinnifesi FK, Sileshi GW (2010) Impact of Gliricidia sepium intercropping on soil organic matter fractions in a maize-based cropping system. Agric Ecosyst Environ 138:139–146. https://doi.org/10.1016/j.agee.2010.04.008
Chibwana PAD, Wood M (1995) Estimation of nitrogen fixation in pigeon peas (*Cajanus cajan* L Millsp) grown in pure or mixed stand. In: Mloza-Banda HR, Kanyama-Phiri GY, Sambo E, Saka VW (eds) African crop science conference proceedings, Blantyre, Malawi 19–24 Feb 1995, pp 211–21
Chikaya-Banda J, Chilonga D (2020) Key challenges to advancing land tenure security through land governance in Malawi: impact of land reform processes on implementation efforts. Land Use Policy 104994.https://doi.org/10.1016/j.landusepol.2020.104994
Chikowo R, Chimonyo V, Gwenambira C, Snapp S (2020) Ecosystem services in doubled-up legume systems. In: The role of ecosystem services in sustainable food systems. Academic Press, pp 171–180. https://doi.org/10.1016/B978-0-12-816436-5.00009-3
Chinangwa LLR (2006) Adoption of soil fertility improvement technologies among smallholder farmers in southern Malawi. MSc Thesis, Norwegian University of Life Sciences, Oslo, Norway
Chinseu E, Dougill A, Stringer L (2019) Why do smallholder farmers dis-adopt conservation agriculture? Insights from Malawi. Land Degrad Dev 30(5):533–543. https://doi.org/10.1002/ldr.3190
Chirwa E, Dorward A (2013) Agricultural input subsidies: the recent Malawi experience. Oxford University Press, p 320. https://doi.org/10.1093/acprof.oso/9780199683529.001.0001
Deguine J-P, Ferron P, Russell D (2008) Sustainable pest management for cotton production: a review. Agron Sustain Dev 28:113–137. https://doi.org/10.1051/agro:2007042
Denning G, Kabambe P, Sanchez P, Malik A, Flor R, Harawa R, Nkhoma P, Zamba C, Banda C, Magombo C, Keating M, Wangila J, Sachs J (2009) Input subsidies to improve smallholder maize productivity in Malawi: Toward an African green revolution. PLoS Biol 7(1) https://doi.org/10.1371/journal.pbio.1000023
Donga TK, Eklo OM (2018) Environmental load of pesticides used in conventional sugarcane production in Malawi. Crop Prot 108:71–77. https://doi.org/10.1016/j.cropro.2018.02.012
Dougill AJ, Whitfield S, Stringer LC, Vincent K, Wood BT, Chinseu EL, Steward P, Mkwambisi DD (2017) Mainstreaming conservation agriculture in Malawi: Knowledge gaps and institutional barriers. J Environ Manage 195:25–34. https://doi.org/10.1016/j.jenvman.2016.09.076
Dzanja JK, Kankwamba H, Matita M (2019) Assessing market prospects for grain legumes in Malawi. Afr J Agri Res Econ 12:204–216
FAO (2018) The 10 elements of agroecology. Guiding the transition to sustainable food and agricultural systems. FAO, Rome, Italy
FAO (2018) Soil and nutrient loss in Malawi. An economic assessment. FAO, http://www.fao.org/3/CA2663EN/ca2663en.pdf Accessed Oct 2021
FAOSTAT (2014) FAOSTAT. http://faostat.fao.org. Accessed Oct 2020
Franke AC, Van Den Brand GJ, Giller KE (2014) Which farmers benefit most from sustainable intensification? An ex-ante impact assessment of expanding grain legume production in Malawi. Eur J Agron 58:28–38. https://doi.org/10.1016/j.fcr.2011.12.014
Franzel S (2004) Financial analysis of agroforestry practices. In: Alavalapati JRR, Mercer DE (eds) Valuing agroforestry systems, Chap. 2. Dordrecht: Kluwer Academic Publishers. https://doi.org/10.1007/1-4020-2413-4_2
Garrity D (2012) Agroforestry and the Future of global land use. In: Nair PKR, Garrity D (eds) Agroforestry—the future of global land use, advances in agroforestry, vol 9. Springer. Science + Business Media Dordrecht, pp 21–27. https://doi.org/10.1007/978-94-007-4676-3

Garrity DP, Akinnifesi FK, Ajayi OC, Weldesemayat SG, Mowo JG, Kalinganire A, Larwanou M, Bayala J (2010) Evergreen agriculture: a robust approach to sustainable food security in Africa. Food Sec 2(3):197–214. https://doi.org/10.1007/s12571-010-0070-7

Gbangou T, Sylla MB, Jimoh OD, Okhimamhe AA (2018) Assessment of projected agro-climatic indices over Awun river basin, Nigeria for the late twenty-first century. Clim Change 151:445–462. https://doi.org/10.1007/s10584-018-2295-y

Giller KE, Witter E, Corbeels M, Tittonell P (2009) Conservation agriculture and smallholder farming in Africa: the heretics view. Field Crop Res 114:23–24. https://doi.org/10.1016/j.fcr.2009.06.017

HLPE (2019) Agroecological and other innovative approaches for sustainable agriculture and food systems that enhance food security and nutrition. A report by the high level panel of experts on food security and nutrition of the committee on world food security, Rome. file:///C:/Users/HP/Documents/HLPE%20agroecology.pdf

Holmgren D (2002) Permaculture: principles & pathways beyond sustainability. Holmgren Design Services, Hepburn. https://www.amazon.com/Permaculture-Principles-Pathways-beyond-Sustainability/dp/0646418440#

Homann-Kee Tui S, Adekunle A, Lundy M, Tucker J, Birachi E, Schut M, Klerkx L, Ballantyne PG, Duncan AJ, Cadilhon JJ, Mundy P (2013) What are innovation platforms? Innovation platforms practice brief 1. Innov Platforms Pract Brief 1, 1–7

ICRISAT (2013) Bulletin of tropical legumes. http://www.icrisat.org/TropicallegumesII/Bulletin-of-Tropical-Legumes.htm

Isman MB (2017) Bridging the gap: moving botanical insecticides from the laboratory to the farm. Ind Crops Prod 110:10–14. https://doi.org/10.1016/j.indcrop.2017.07.012

Jones AD, Shrinivas A, Bezner-Kerr R (2014) Farm production diversity is associated with greater household dietary diversity in Malawi: findings from nationally representative data. Food Policy 46:1–12. https://doi.org/10.1016/j.foodpol.2014.02.001

Kabambe VH, Chilimba ADC, Ngwira A, Mbawe M, Kambauwa G, Mapfumo P (2012) Using innovation platforms to scale out soil acidity-ameliorating technologies in Dedza district in central Malawi. Afr J Biotech 11(3):561–569. https://doi.org/10.5897/AJB10.2227

Kaechai S, Soyton K, Hyde KD (2009) Mycofungicides and fungal biofertilizers. Fungal Diversity 38:25–50

Kalasa JB, Mhango WG, Kanyama-Phiri GY, Lowole MW (2018) Maize response to doubled-up legumes, compost manure and inorganic fertiliser on smallholder farms in Ntcheu district of Malawi. Afr J Agri Res 13(21):1101–1110

Kamanga BC, Kanyama-Phiri GY, Waddington SR, Almekinders CJ, Giller KE (2014) The evaluation and adoption of annual legumes by smallholder maize farmers for soil fertility maintenance and food diversity in central Malawi. Food Security 6(1):45–59. https://doi.org/10.1007/s12571-013-0315-3

Katengeza SP, Holden ST, Fisher M (2019) Use of integrated soil fertility management technologies in Malawi: impact of dry spells exposure. Ecol Econ 156:134–152. https://doi.org/10.1016/j.ecolecon.2018.09.018

Kaumbata W, Banda L, Mészáros G, Gondwe T, Woodward-Greene MJ, Rosen BD, Wurzinger M (2020) Tangible and intangible benefits of local goats rearing in smallholder farms in Malawi. Small Ruminant Res 187:106095. https://doi.org/10.1016/j.smallrumres.2020.106095

Kerr RB, Hickey C, Lupafya E, Dakishoni L (2019) Repairing rifts or reproducing inequalities? Agroecology, food sovereignty, and gender justice in Malawi. J Peasant Stud 46(7):1499–1518. https://doi.org/10.1080/03066150.2018.1547897

Kerr RB, Rahmanian M, Owoputi I, Batello C (2019) Agroecology and nutrition: transformative possibilities and challenges. In: Sustainable diets: linking nutrition and food systems. CABI, Wallingford UK, pp 53–63

Khan AA, Jilani G, Akhtar MS, Naqvi SMS, Rasheed M (2009) Phosphorus solubilization: Occurence, mechanisms and their role in crop production. J Agr Biol Sci 1(1):48–58

Khataza RR, Hailu A, Kragt ME, Doole GJ (2017) Estimating shadow price for symbiotic nitrogen and technical efficiency for legume-based conservation agriculture in Malawi. Aust J Agri Res Econ 61(3):462–480. https://doi.org/10.1111/1467-8489.12212

Kwesiga FR, Akinnifesi FK, Mafongoya PL, McDermott MH, Agumya A (2003) Agroforestry research and development in Southern Africa during 1990s: review and challenges ahead. Agrofor Syst 53:173–186. https://doi.org/10.1023/B:AGFO.0000005222.68054.38

Lal R (2015) Restoring soil quality to mitigate soil degradation. Sustainability 7(5):587–5895. https://doi.org/10.3390/su7055875

Lupwayi NZ, Mkandawire ABC (1996) Phosphorus, sulphur and molybedenum effects pn response of Phaseolus beans to inoculation with Rhizobium in Malawi. Afr Crop Sci J 4(3):359–364

Magamba K, Matumba L, Matita G, Gama AP, Singano L, Monjerezi M, Njoroge S (2017) Aflatoxin risk management in commercial groundnut products in Malawi (Sub-Saharan Africa): a call for a more socially responsible industry. J Consum Prot Food Saf 12:09–316. https://doi.org/10.1007/s00003-017-1129-6

Makate C, Makate, M, Mango N (2017) Smallholder farmers' perceptions on climate change and the use of sustainable agricultural practices in the Chinyanja Triangle, Southern Africa. Soc Sci 6(1):30. https://doi.org/10.3390/socsci6010030

Makumba WIH (2003) Nitrogen use efficiency and carbon sequestration in legume tree-based agroforestry systems. A case study in Malawi. PhD Thesis, Wagenigen University and Research Centre, Wagenigen, The Netherlands

Matsui N, Nakata K, Cornelius C, Macdonald M (2016) Chemical properties and fermentation behavior of the composts prepared by three composting methods in Malawi. Afr J Agric Res 11(25):2213–2223. https://doi.org/10.5897/AJAR2016.10919

Matumba L, Van Poucke C, Biswick T, Monjerezi M, Mwatseteza J, De Saeger S (2014) A limited survey of mycotoxins in traditional maize based opaque beers in Malawi. Food Control 36(1):253–256. https://doi.org/10.1016/j.foodcont.2013.08.032

Me-Nsope N, Larkins M (2016) Beyond crop production: Gender relations along the pigeon pea value chain and implications for income and food security in Malawi. J Gend, Agric Food Secur (Agri-Gend) 1(302-2016-4760). https://doi.org/10.19268/JGAFS.132016.1

Mhango WG (2011) Nitrogen budgets in legume based cropping systems in Northern Malawi. PhD dissertation, Michigan State University, East Lancing, USA

Mhango WG, Snapp S, Kanyama-Phiri GY (2017) Biological nitrogen fixation and yield of pigeonpea and groundnut: Quantifying response on smallholder farms in northern Malawi. Afr J Agric Res 12(16):1385–1394. https://doi.org/10.5897/AJAR2017.12232

Mhango WG, Snapp SS, Phiri GY (2012) Opportunities and constraints to legume diversification for sustainable maize production on smallholder farms in Malawi. Renew Agri Food Syst 28(3): 234–244. https://doi.org/10.1017/S1742170512000178

Ministry of Agriculture, Irrigation and Water Development (MoAIWD) (2016) National agricultural policy. MoAIWD. https://cepa.rmportal.net/Library/government-publications/national-agriculture-policy-2016/view. Accessed on 15th Nov 2020

Mng'omba SA, Akinnifesi FK (2019) Effect of soil amendment with Gliricidia sepium and Tephrosia vogelii biomass on maize yield at Makoka in Malawi. Agroforest Syst https://doi.org/10.1007/s10457-019-00405-4

Mockshell J, Villarino ME (2019) Agroecological intensification: Potential and limitations to achieving food security and sustainability, In: Ferranti P, Berry EM, Anderson JR (eds) Encyclopedia of food security and sustainability, vol 3. Elsevier, pp 64–70 https://doi.org/10.1016/B978-0-08-100596-5.22211-7

Morel K, Léger F, Ferguson RS (2018) Permaculture. Encyclopedia of ecology, 2nd edn. Elsevier Inc.https://doi.org/10.1016/B978-0-12-409548-9.10598-6

Mosquera-Losada MR, McAdam JH, Romero-Franco R, Santiago-Freijanes JJ, Rigueiro-Rodróguez A (2009) Definitions and components of agroforestry practices in Europe. In: Agroforestry in Europe: current status and future prospects, pp 3–19.https://doi.org/10.1007/978-1-4020-8272-6_1

Mupangwa W, Thierfelder C, Ngwira A (2017) Fertilization strategies in conservation agriculture systems with maize-legume cover crop rotations in Southern Africa. Exp Agric 53(2):288. https://doi.org/10.1017/S0014479716000387

Munthali WW (2007) Integrated soil fertility management technologies: a counteract to existing milestone in obtaining achievable economical crop yields in cultivated lands of poor smallholder farmers in Malawi. In: Bationo A et al (eds) Advances in integrated soil fertility management in sub-Saharan Africa: challenges and opportunities. Springer, Dordrecht, the Netherlands, pp 531–536. https://doi.org/10.1007/978-1-4020-5760-1_50

Mustafa-Msukwa AK, Mutimba JK, Masangano C, Edriss AK (2011) An assessment of the adoption of compost manure by smallholder farmers in Balaka District, Malawi. S Afr J Agric Ext 39(1):17–25

Nadarajan S, Sukumaran S (2021) Chemistry and toxicology behind chemical fertilizers. In: Controlled release fertilizers for sustainable agriculture. Academic Press, pp 195–229

Nair PKR, Garrity D (2012) Agroforestry—the future of global land use. Advances in agroforestry 9. Springer, Springer. Science + Business Media Dordrecht. https://doi.org/10.1007/978-94-007-4676-3

Nalivata PC (2007) Evaluation of factors affecting the quality of compost made by smallholder farmers in Malawi. PhD thesis, Cranfield University, UK

National Conservation Agriculture Task Force (2016) National guidelines for implementing conservation agriculture in Malawi. NCATF of Malawi

Ngwira AR, Aune JB, Mkwinda S (2012) On-farm evaluation of yield and economic benefit of short term maize legume intercropping systems under conservation agriculture in Malawi. Field Crop Res 132:149–157. https://doi.org/10.1016/j.fcr.2011.12.014

Ngwira AR, Nyirenda M, Taylor D (2013a) Toward sustainable agriculture: an evaluation of compost and inorganic fertilizer on soil nutrient status and productivity of three maize varieties across multiple sites in Malawi. Agroecol Sustain Food Syst 37(8):859–881. https://doi.org/10.1080/21683565.2013.763889

Ngwira AR, Thierfelder C, Lambert DM (2013b) Conservation agriculture systems for Malawian smallholder farmers: long-term effects on crop productivity, profitability and soil quality. Renew Agric Food Syst 28(4):350–363. https://doi.org/10.1017/S1742170512000257

Ngwira A, Johnsen FH, Aune JB, Mekuria M, Thierfelder C (2014) Adoption and extent of conservation agriculture practices among smallholder farmers in Malawi. J Soil Water Conserv 69(2):107–119. https://doi.org/10.2489/jswc.69.2.107

Njira KOW, Nalivata PC, Kanyama-Phiri GY, Lowole MW (2013) An assessment for the need of soybean inoculation with *Bradyrhizobium japonicum* in sites of Kasungu district, Central Malawi. Int J Curr Microbiol App Sci 2(8):60–72

Njira KOW, Nalivata PC, Kanyama-Phiri GY, Lowole MW (2012) Biological nitrogen fixation in sole and doubled-up legume cropping systems on the sandy soils of Kasungu. Central Malawi J Soil Sci Environ Manage 3(9):224–230. https://doi.org/10.5897/JSSEM12.056

Njira KOW, Semu E, Mrema JP, Nalivata PC (2017) Biological nitrogen fixation by pigeon pea and cowpea in the doubled-up and other cropping systems on the Luvisols of Central Malawi. AfrJAgric Res 12(15):1341–1352. https://doi.org/10.5897/AJAR2017.12167

Njira KOW, Semu E, Mrema JP, Nalivata PC (2020) integrated soil fertility management based on Pigeon Pea and Cowpea cropping systems influences Nitrogen use efficiency, yields and quality of subsequent maize on alfisols in Central Malawi. In: Singh B, Safalaoh A, Amuri N, Eik L, Sitaula B, Lal R (eds) Climate impacts on agricultural and natural resource sustainability in Africa. Springer. https://doi.org/10.1007/978-3-030-37537-9_5

Njira KOW, Semu E, Mrema JP, Nalivata PC (2021) Productivity of pigeon pea, cowpea and maize under sole cropping, legume-legume and legume-cereal intercrops on Alfisols in Central Malawi. Agrofor Syst 95(2):279–291. https://doi.org/10.1007/s10457-020-00589-0

Nyagumbo I, Mkuhlani S, Pisa C, Kamalongo D, Domingos D, Mekuria M (2016) Maize yield effects of conservation agriculture based maize–legume cropping systems in contrasting agro-ecologies of Malawi and Mozambique. Nutr Cycl Agroecosyst 105:275–290. https://doi.org/10.1007/s10705-015-9733-2

Ortega DL, Waldman KB, Richardson RB, Clay DC, Snapp S (2016) Sustainable intensification and farmer preferences for crop system attributes: evidence from Malawi's central and southern regions. World Dev 87:139–151. https://doi.org/10.1016/j.worlddev.2016.06.007

Pahalvi HN, Rafiya L, Rashid S, Nisar B, Kamili AN (2021) Chemical fertilizers and their impact on soil health. In: Microbiota and biofertilizers, vol 2: ecofriendly tools for reclamation of degraded soil environs, pp 1–20

Phiri AT, Njira KOW (2022) Grain legume-based cropping systems' effects on soil organic carbon and nutrient dynamics. Agric Res 1–8.https://doi.org/10.1007/s40003-022-00619-6

Place F, Barrett CB, Freeman HA, Ramisch JJ, Vanlauwe B (2003) Prospects for integrated soil fertility management using organic and inorganic inputs: evidence from smallholder African agricultural systems. Food Policy 28(4):365–378. https://doi.org/10.1016/j.foodpol.2003.08.009

Pretty J (2007) Agricultural sustainability: concepts, principles and evidence. Philos Trans R Soc B: Biol Sci 363(1491):447–465. https://doi.org/10.1098/rstb.2007.2163

Pretty JN, Morison JI, Hine RE (2003) Reducing food poverty by increasing agricultural sustainability in developing countries. Agr Ecosyst Environ 95(1):217–234. https://doi.org/10.1016/S0167-8809(02)00087-7

Raimi A, Roopnarain A, Adeleke R (2021) Biofertilizer production in Africa: current status, factors impeding adoption and strategies for success. Sci Afr 11:e00694. https://doi.org/10.1016/j.sciaf.2021.e00694

Ricker-Gilbert J, Jumbe C, Chamberlin J (2014) How does population density influence agricultural intensification? Evidence from Malawi. Food Policy 48:114–128. https://doi.org/10.1016/j.foodpol.2014.02.006

Rivett MO, Halcrow AW, Schmalfuss J, Stark JA, Truslove JP, Kumwenda S, Miller AV (2018) Local scale water-food nexus: Use of borehole-garden permaculture to realise the full potential of rural water supplies in Malawi. J Environ Manage 209:354–370. https://doi.org/10.1016/j.jenvman.2017.12.029

Rusinamodzi L, Van Wijk MT, Corbeels M, Rufino MC, Giller K (2015) Maize crop residue uses and trade-offs on smallholder crop-livestock farms in Zimbabwe: economic implications of intensification. Agr Ecosyst Environ 214:31–45. https://doi.org/10.1016/j.agee.2015.08.012

Sharma AK (2008) A handbook of organic farming. Agrobios. Jodhpur, India

Silcshi G, Akinnifesi FK, Ajayi OC, Place F (2008) Meta-analysis of maize yield response to planted fallow and green manure legumes in Sub-Saharan Africa. Plant Soil 307:1–19. https://doi.org/10.1007/s11104-008-9547-y

Sileshi G, Akinnifesi FK, Debusho LK, Beedy T, Ajayi OC, Mng'omba S (2010) Variation in maize yield gaps with plant nutrient inputs, soil type and climate across sub-Saharan Africa. Field Crop Res 116:1–13. https://doi.org/10.1016/j.fcr.2009.11.014

Sileshi GW, Debusho LK, Akinnifesi FK (2012) Can integration of tree legumes Increase yield stability in rainfed maize cropping systems in southern Africa. Agron J 104:1–7. https://doi.org/10.2134/agronj2012.0063

Smith A, Snapp S, Chikowo R, Thorne P, Bekunda M, Glover J (2017) Measuring sustainable intensification in smallholder agroecosystems: a review. Glob Food Sec 12:127–138. https://doi.org/10.1016/j.gfs.2016.11.002

Smith A, Snapp S, Dimes J, Gwenambira C, Chikowo R (2016) Doubled-up legume rotations improve soil fertility and maintain productivity under variable conditions in maize-based cropping systems in Malawi. Agric Syst 145:139–149. https://doi.org/10.1016/j.agsy.2016.03.008

Snapp S, Bezner Kerr R, Ota V, Kane D, Shumba L, Dakishoni L (2019) Unpacking a crop diversity hotspot: farmer practice and preferences in Northern Malawi. Int J Agric Sustain 17(2):172–188. https://doi.org/10.1080/14735903.2019.1591657

Snapp S, Jayne TS, Mhango W, Benson T, Ricker-Gilbert J (2014) Maize-nitrogen response in Malawi's smallholder production systems. In National symposium on eight years of FISP—impact and what next, pp 14–15

Tamene L, Mponela P, Ndengu G, Kihara J (2016) Assessment of maize yield gap and major determinant factors between smallholder farmers in the Dedza district of Malawi. Nutr Cycl Agroecosyst 105(3):291–308. https://doi.org/10.1007/s10705-015-9692-7

Thierfelder C, Wall PC (2009) Effects of conservation agriculture techniques on infiltration and soil water content in Zambia and Zimbabwe. Soil and Tillage Research 105(2):217–227. https://doi.org/10.1016/j.still.2009.07.007

Thierfelder C, Bunderson WT, Mupangwa W (2015) Evidence and lessons learned from long-term on-farm research on conservation agriculture systems in communities in Malawi and Zimbabwe. Environments 2(3):317–337. https://doi.org/10.3390/environments2030317

Thierfelder C, Chisui JL, Gama M, Cheesman S, Jere ZD, Bunderson WT, Rusinamhodzi L (2013) Maize-based conservation agriculture systems in Malawi: long-term trends in productivity. Field Crop Res 142:47–57. https://doi.org/10.1016/j.fcr.2012.11.010

Toth GG, Nair PR, Jacobson M, Widyaningsih Y, Duffy CP (2017a) Malawi's energy needs and agroforestry: adoption potential of woodlots. Hum Ecol 45(6):735–746. https://doi.org/10.1007/s10745-017-9944-z

Toth GG, Nair PR, Duffy CP, Franzel SC (2017b) Constraints to the adoption of fodder tree technology in Malawi. Sustain Sci 12(5):641–656. https://doi.org/10.1007/s11625-017-0460-2

Van Vugt D, Franke AC, Giller KE (2016) Participatory research to close the soybean yield gap on smallholder farms in Malawi. Exp Agri 53(3):396–415

Van Vugt D, Franke AC, Giller KE (2018) Understanding variability in the benefits of N2-fixation in soybean-maize rotations on smallholder farmers' fields in Malawi. Agr Ecosyst Environ 261:241–250. https://doi.org/10.1016/j.agee.2017.05.008

Watts M, Williamson S (2015) Replacing chemicals with biology: phasing out highly hazardous pesticides with agroecology. Pesticide Action Network (PAN) International

Wezel A, Soboksa G, McClelland S, Delespesse F, Boissau A (2015) The blurred boundaries of ecological, sustainable, and agroecological intensification: a review. Agron Sustain Dev 35(4):1283–1295. https://doi.org/10.1007/s13593-015-0333-y

Zingore S, Mutegi J, Agesa B, Tamene L, Kihara J (2015) Soil degradation in Sub-Saharan Africa and crop production options. Better Crops 99(1):24–26. https://cgspace.cgiar.org/handle/10568/68702

Zulu LC, Adams EA, Chikowo R, Snapp S (2018) The role of community-based livestock management institutions in the adoption and scaling up of pigeon peas in Malawi. Food Policy 79:141–155. https://doi.org/10.1016/j.foodpol.2018.06.007

Open Access This chapter is licensed under the terms of the Creative Commons Attribution 4.0 International License (http://creativecommons.org/licenses/by/4.0/), which permits use, sharing, adaptation, distribution and reproduction in any medium or format, as long as you give appropriate credit to the original author(s) and the source, provide a link to the Creative Commons license and indicate if changes were made.

The images or other third party material in this chapter are included in the chapter's Creative Commons license, unless indicated otherwise in a credit line to the material. If material is not included in the chapter's Creative Commons license and your intended use is not permitted by statutory regulation or exceeds the permitted use, you will need to obtain permission directly from the copyright holder.

Nutrient-Dense and Biofortified Food

Interventions in Soil–Plant-Animal-Human Being Nexus for Food Security, Nutritional Security and Prevention of Malnutrition

Amit Anil Shahane and Yashbir Singh Shivay

Abstract The production activities of agriculture and allied sectors are increasingly focusing on nutritional security after achieving food security, with the capacity to export food grains (referred to elsewhere as food resilience). The increase in agricultural production is an indication of the prospects of the agriculture sector to achieve nutritional security. In addition to that, policy support through the release of biofortified varieties and different schemes and programmes for health care and alleviating malnutrition has additive and synergistic effects on agricultural activities intended to achieve nutritional security. These attempts were conditioned and constrained by the widespread problem of malnutrition, hidden hunger, the disparity in distribution, non-availability of nutritious food in the diet and low purchasing power. Presently two billion people are affected by hidden hunger; while nearly half of the world's population affected by micronutrient deficiency lives in India. Children and women are the groups most affected by malnutrition. The trade-off between prospects and constraints for nutritional security will be achieved through nexus-based attempts (soil–plant-animal-human being). This chapter aims to discuss the status of malnutrition and different biases of nutritional security (resource, development and knowledge bias) to highlight their potential to contribute to the alleviation of the problem of food security. The nexus of soil–plant-animal-human beings is elaborated in this review, with their condition and potential towards achieving nutritional security being highlighted. Being the ultimate source, the soil needs to be conserved, utilized as per capabilities, and restored for the benefit of future generations, with the adoption of interventions that contribute to both its conservation and prevention of degradation. The implementation and evaluation of nutrient-sensitive agricultural practices (input provision and services), interventions in post-harvest processing (value addition), attention to the nutritional quality of produce from farm to consumer plate, and the delivery of food to targeted population, need to be attended. Besides that, interventions in animal-based enterprises (upgrading the quality of breeds, feed and

A. A. Shahane
College of Agriculture (CAU, Kyrdemkulai), Ri-Bhoi District, Meghalaya, India

Y. S. Shivay (✉)
Division of Agronomy, ICAR-Indian Agricultural Research Institute, New Delhi, India
e-mail: ysshivay@gmail.com

© The Author(s) 2026
B. R. Singh et al. (eds.), *Sustainable Food Systems in Sub-Saharan Africa*,
https://doi.org/10.1007/978-3-032-11288-0_10

supplementation, product diversity, product preservation, and value addition) are reported to have significant impacts on catering to nutritional needs and preventing malnutrition. The capacity building for household production of nutritional foods (nutritional and kitchen gardens, backyard poultry, vertical farming, terrace farming), awareness and access to quality foods, identification of malnourished sections and effective implementation of diet diversification, supplementation and medication, are all needed.

Keywords Africa · Asia · Bio-fortification · Millets · Nutrient-dense crops · Soil degradation · Undernutrition

1 Introduction to Nutritional Security

The importance of agriculture and allied fields in the generation of different provisional sources (Nelson et al. 2011; Montoya et al. 2019) is being highlighted due to significant changes in challenges in the present world (Pretty et al. 2010; Nchuchuwe and Adejuwon, 2012; Calicioglu et al. 2019). The importance can be realized from seventeen sustainable development goals (SDGs). Among the SDGs, no poverty, zero hunger, responsible consumption and production, and life on land, are directly related to agriculture (Agro-ecosystem); while affordable and clean energy, climate action and life below water are indirectly related to agriculture and natural resources. Among the SDGs, zero hunger and no poverty deal with nutritional security; while others (listed above) are one or multiple ways related to food and nutritional security of human beings. Besides that, good health and well-being, as well as clean water and sanitation, are related to the food and nutritional security of human beings. Food security, food sufficiency, food resilience and nutritional security, recommended daily allowance and recommended dietary intake, and calories/different food commodity availability per capita, are the terminologies commonly used to indicate the status of food and nutritional security. Food security is defined in various ways due to its dynamic, multidimensional, conceptual dependence, time dimension changes, and variation in emphasis on indicators and parameters used (Venugopal 1999; Pinstrup-Andersen 2009; Pangaribowo et al. 2013; Cafiero et al. 2014; Ansah et al. 2019; Peng and Berry 2019; Pawlak and Kołodziejczak 2020). Food security is defined as physical, social, and economic access to sufficient, safe, and nutritious food for all people and all the time, to meet their dietary needs and food preferences for healthy and active life (Peng and Berry 2019; Clapp et al. 2022). In addition, adequate status of nutrients in food, a good sanitary environment, adequate health services, proper care and feeding practices are needed to ensure nutritional security along with food security (Ingram 2020; Bilali et al. 2019; Ge et al. 2021). Tendall et al. (2015) define food system resilience as the capacity of the food system and its units at multiple levels over time to provide sufficient, appropriate and accessible food to all in the face of various disturbances.

The recommended daily allowance is the level of intake of all essential nutrients that are adequate to meet the known nutrient needs of healthy persons (based on scientific knowledge approved by the Food and Nutrition Board) (Anonymous 1989). Agencies such as the Food and Agricultural Organization, World Health Organization, World Food Programme, International Fund for Agriculture Development, International Food Policy Research Institute, United Nations International Children's Emergency Fund, High-Level Panel of Experts on Food Security and Nutrition, United Nation's Committee on World Food Security, United State Department on Agriculture, United State Agency for International Development are the major international agencies dealing with different aspects of food and nutritional securities, along with other national and international agencies.

Food and nutritional security has a resource bias, a development bias, and a knowledge and awareness bias. The resource bias is expressed based on the distribution of natural resources such as cultivable land, fresh water resource and plant genetic diversity. The negative and indirect indicator of resource bias includes degradation or desertification of land, monotony or monopoly of single or few crops, reducing per capacity availability of natural resources due to increasing population of human beings and domestic animals. Land degradation is expressed as physical degradation (wind erosion, water erosion and water logging), chemical degradation (predominance of saline soil, alkaline soil and acid soil) and biological degradation (Agrochemical residues build up and changes in microbial diversity). The over and under-application of resources and their faulty management modify the land resources beyond their resistance and resilience capacity, leading to degradation of natural resources which remain subtle and unidentified. The use of these lands without investment in reducing and preventing degradation will intensify the level of degradation and make this land unsuitable for crop production (Olsson et al. 2019). Olsson et al. (2019) reported that, < 10 million km^2 to 60 million km^2 of land is degraded globally; while Thiombiano and Tourino-Soto (2007) reported that, Africa accounts for 65% of the total extensive cropland degradation of the world. In India, the degraded land varies from 53.3 to 187.7 million ha, as reported by various agencies (Bhattacharyya et al. 2015). This difference indicates the existence of land with subtle degradation, and is still used for cultivation or productive purposes. The degradation of water resources is expressed based on the increasing footprint of human activities and climate change-induced variations in rainfall and changes in groundwater depth (Arto et al. 2016). There is a problem of disposal of black and grey water (collectively called wastewater), as well as that of seawater intrusion along the sea coast due to excessive pumping of groundwater, which are indications of the increasing footprint of human activities on water resources (Chenoweth et al. 2014). The increasing variation in the distribution of rainfall in time and space dimensions, as reported by Konapala et al. (2020), reduces the utilizable flow of river water, which in turn reduces the water available for agricultural and allied activities due to intensive competition from other sectors. These were additional indicators of water resource degradation (Baggio et al. 2021; D'Odorico et al. 2020). The degradation of the plant resource base is explained by the drop in the utilizable crop diversity for food and nutritional purposes, increased need for in-situ and ex-situ conservation of plant genetic resources, and the need for policy, research and development initiatives focusing on non-or under-utilized crops (Li et al. 2020; Hossain et al. 2021; Srivastava and Arya 2021). Natural resources such as soil, plants and water are primary and ultimate sources of nutrition to animals and human beings. Hence, meeting the nutrient requirement, with resource degradation, constraints and disparity in distribution, will be a challenging task.

The development bias is expressed by the availability of modern technology for input endowment and access to infrastructural facilities and institutional capacity to cater to the needs of growing populations. Besides that, the policies, fund availabilities and political will for emphasizing the development activities for catering to food and nutritional security needs are also a part of the developmental bias. The development and release of biofortified varieties (Yadava et al. 2017; 2020), policy promotion for processing and value addition in food items (Anonymous 2019; Olson et al. 2021), firm and robust quality control systems (Kamboj et al. 2020; Aworh 2021) are the indication of commitments of Government to enhancing the developmental agenda for achieving the food and nutritional security. Apart from these attempts, the reach of these policies to each individual, and their sustenance in the long term, is fund-demanding. As a result, they may sometimes get less priority. The population below the poverty line and people with low purchasing power or those just above the poverty line, are the first and major beneficiaries of such development activities. This section of the population has less access to nutritious foods and reported malnutrition problems (Discussed in the next section of this chapter). Besides these activities, medical and health care facilities were also significant development activities in reducing malnutrition arising out of deficient and imbalanced diet.

The knowledge bias of food and nutritional security is explained by awareness of changes in cultivation practices and input endowment at the soil and plant levels and knowledge about processing and value addition at the plant and animal levels. At the human being level/component of nexus, nutritional facts about the food consumed, as well as consumption patterns and awareness about self-health care and medication facilities/supplementation facilities available to meet the nutrition deficit, are the major areas of work for achieving food and nutritional security. These biases should deal with the soil–plant-animal-human being nexus, and should be dynamic (Fig. 1). The dynamism is influenced by the varied needs of each component, the suitability of different techniques for each component and the status of malnutrition at the human being level.

2 Status of Food and Nutritional Security

The status of food and nutritional security is a balance between sources of food and nutrition and demand, as modulated by interventions at all the four components of nexus. The interventions include the attempts to produce sufficient and nutritious food, attempts to ensure the supply to each individual and the status of different

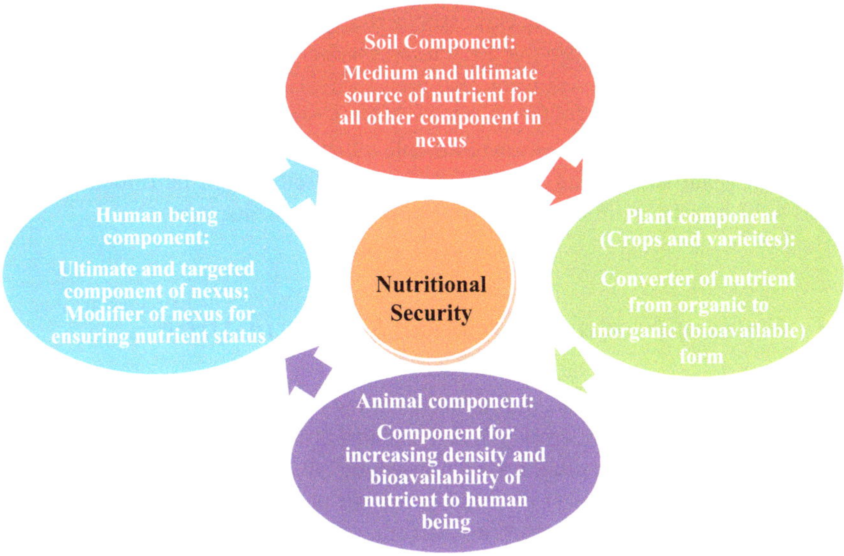

Fig. 1 Nexus of nutritional security

forms of malnutrition. The status of food and nutritious products is explained based on the food production and import–export of food and related items for the country. Besides edible/food crops, non-food crops also contribute indirectly to food and nutritional security through their role in the income security (purchasing power) of farms and farming families. Increase in production of rice (598.66–769.23 million tonnes from 2000 to 2020), wheat (582.0 to 774.55 million tonnes 2000 to 2020), maize (592.01 to 1210.23 million tonnes from 2000 to 2021), vegetables (682.43–1138.74 million tonnes from 2000 to 2020), fruits (576.65–899.56 million tonnes from 2000 to 2020), eggs (51,133–87,074 thousand metric tonnes from 2000 to 2020), milk (489–746 million tonnes from 2000 to 2020) and meat (233–337 million tonnes from 2000 to 2020) is indication of the prospects of the agriculture sector to achieve the nutritional security. The soil nutritional status, level of fertilizer consumption, status of nutrients in staple foods commodities and bioavailability of nutrients and presence of anti-nutritional factors in food commodities are other criteria of significant importance in determining nutritional security. Harper et al. (2022) mentioned adult and child hunger in households, self-reported household food sufficiency, food expenditure > 60% of monthly expenditure and household dietary diversity as food security indicators. Nutritional security is also explained based on the composition of diets consumed by the majority of the population. The attempt to supply food to all individuals is explained by different government schemes, programmes, projects and incentives that are also considered in accessing food and nutritional security. Another important criterion used for indicating the status of nutritional security is the status of malnutrition at the human being level of nexus. Malnutrition (as defined by WHO) is deficiencies or excesses or imbalances in a person's intake of energy and/or nutrients. Malnutrition is expressed in three major types viz., under-nutrition and over-nutrition and hidden hunger. The terms *double burden of malnutrition* and *triple burden of malnutrition* were also used to indicate one or multiple forms of malnutrition. (Peng et al. 2020; Were et al. 2020; Cristian and Dake 2021). Considering the status of malnutrition and equitable attention needed for each component of the soil–plant-animal-human being nexus, the discussion of nutritional security needs to be holistic with the incorporation of all components.

2.1 Status of Food and Nutritional Security in Africa

The threat of both food and nutritional security for the African population was reported widely (Moseley et al. 2020; Ogunniyi et al. 2020; Tanumihardjo et al. 2020). The major reason cited for such insecurity was poverty and lack of access to nutritious food (Militao et al. 2022; Adeyeye et al. 2023), market failure and constraints related to food production, marketing and consumption (Azomahou et al. 2022; Mohajan 2022), consumption of ultra-processed food (Reardon et al. 2021), commodity price fluctuation (Mekhasa et al. 2022), gender inequality and subsistence farming (Ashagidigbi et al. 2022), mother and child malnutrition during early growth stages (Ahinkorah et al. 2021) and Mycotoxin contamination in food and feed (Kebede et al. 2020). According to Owolade et al. (2022), poverty, overpopulation, unsuccessful small-scale agriculture, low educational status, climate change, corruption, wars and conflicts, fluctuation of food prices, etc. are the factors responsible for malnutrition in Sub-Saharan African countries. This malnutrition results in illness and chronic diseases. In another study, Ashagidigbi et al. (2022) reported that gender and occupation are considered major determinants of malnutrition in Nigeria. They found that the most malnourished children were male in female-headed households living in rural areas in north western zone of the country, and households following subsistence farming with the least returns were the more affected by malnutrition in Nigeria. Guevara-Romero et al. (2021) found that the presence of a mother, age, height, educational level, occupation, food intake, breast feeding, family income, family size and urbanization type are the factors associated with the double burden of malnutrition based on the review of 70 published articles from two major databases viz. *PubMed* and *Web of Science*. Alamu et al. (2020) conducted a study involving 660 mother–child pairs in Akwa Ibom state from the southern region of Nigeria. They found that stunting and wasting were problems in children, and underweight and overweight/obesity were observed in mothers. Among the studied population, 37.4% of the children were found stunted and 13% found wasted. Among the mothers, the underweight and overweight were reported to be 9% and 23.2%, respectively. Popkin et al. (2020) reported that low and middle-income countries (most African and Asian countries) changed the nutritional reality with more consumption of ultra-processed food products, increased control of the entire food chain by the food service provider (manufacturers, trailers and agribusinesses), and increase in the sale of non-essential foods and beverages, all of which is playing significant role in increasing the *double burden of malnutrition.* At the same time, increasing urbanization, low per capita income/GDP, working woman and their opportunity cost of time leads to increased demand of ready to eat food. This further facilitates the entry of less nutritious food as well as control of food service providers in the food chain, which further increases the double burden of malnutrition.

The meta-analysis of data from demographic and health surveys conducted in 32 sub-Saharan African countries from 2006 to 2016 was done to determine the prevalence of each malnutrition indicator such as wasting, stunting and underweight (Akombi et al. 2017). The results showed that in countries such as Burundi, Malawi, Niger, Mali, Sierra Leone, Nigeria Democratic Republic of Congo and Chad, severe stunting varies from 57.7% in Burundi to 36.8% in Nigeria. Wasting was highest in Niger (18.0%), with other countries like Burkina Faso, Mali Comoros, Ethiopia, Namibia, Chad and Sao Tome and Principle also reporting wasting. Underweight was found in the countries of Burundi, Ethiopia, Niger, Nigeria, Burkina Faso, Mali and Chad. Another study was conducted in 23 Sub-Saharan countries based on data from demographic and health surveys collected from 2008 to 2017 (Cristian and Dake 2021). It involves 145,020 households with the availability of valid data on the nutrition of women or reproductive age and children. They found that the double and triple burden of malnutrition was found in 8% and 5% of the households, with anaemia being the most common form of malnutrition, the prevalence of it being in 70% of households. Among the studied households, 63.4% of children and 41.3% of women were found anaemic; on the other hand, stunting in children (37.2%) and overweight among mothers (15.7%) were found to be important indicators of malnutrition. The severe concern of malnutrition in Sub-Saharan African countries was reported by Were et al. (2020), with 5 countries facing significant undernutrition, 11 countries with double burden of malnutrition and 18 countries with significant overnutrition in 34 countries from Sub-Saharan Africa. They found that region, wealth quintile, age group, marital status, residential setting, ethnicity, parity and occupation are the important predictors of nutritional status in studied countries. They categorize Niger, Democratic Republic of Congo, Mali, Uganda, Guinea, Gambia, Senegal, Namibia, Zambia, Nigeria and Congo with both underweight and overweight affecting $\geq$ 10% population; and the countries of Mozambique, Sierra Leone, Rwanda, Malawi, Ivory Coast, Benin and Liberia are categorized as countries with overweight, affecting $\geq$ 10% population. Wells et al. (2020) reported adverse effects of malnutrition in early life, which include the imbalance of the gut microbiome, inflammation, metabolic dysregulation and impaired insulin signalling. Besides that, early malnutrition followed by being overweight increases the risk of non-communicable disease due to high metabolic load on a depleted capacity for homeostasis and increased risk of childbirth complications in women. The analysis of data from demographic and health surveys in 52 low and middle-income countries was conducted to know about the occurrence of overweight, underweight, and obesity among women of reproductive age (1,099,187) (Alem et al. 2023). Among the studied population, 15.2% were found underweight, while overweight and obesity occurred in 19.0% and 9.1%, respectively. Factors such as educational level, household wealth, use of modern contraceptives, exposure to media and parity were associated with overweight and obesity; on the other hand, the risk of being underweight was higher among rural women in comparison to urban women. Harper et al. (2022) reported that, stunting, wasting and over-weight were found in 18.4%, 6.8% and 10.4% of the population, respectively. Diet diversity has a significant role in determining nutritional status. The medium diet diversity children (> 5 years of age) and adolescents had more stunting problems than the one with high diet diversity. In contrast to that, an increase in diet diversity increases the risk of overweight and obesity in adults. These reports of malnutrition indicate the multifaceted, complex and critical situation of malnutrition in African countries, which needs to be addressed at the soil, plant, animal, and human beings levels.

2.2 *Status of Food and Nutritional Security in Asia*

Asia is the most populous continent, with China and India accounting for around 35% of world's population. Along with this large human population, Asia is also home to a large number of domestic animals. Considering this vast population, the concern of food and nutritional security of this population is a global concern. The increasing population of Asia is also considered as an important cause of malnutrition in Asia. A meta-analysis of 33 original studies published from January 2008 to December 2018 on the double burden of malnutrition in the Asia–Pacific region (Peng et al. 2020) showed that the prevalence of under-nutrition was 8.8%, and over-nutrition was 23.3% in the same population. It was also concluded that the double burden of malnutrition was at an alarming rate and tilted toward over-nutrition. In the review of literature published from 1990 to 2017 (48 studies from 8 different countries) (Rachmi et al. 2018), it was found that a double burden of malnutrition was observed in 5% of households in Vietnam to 30.6% in Indonesia. They also reported that, a double burden of malnutrition occurred at the community, household and individual levels, with different types of undernutrition (stunting, underweight, wasting and thinness) as well as overweight and obesity. The demographic and health surveys were conducted in five South East Asian countries (Bangladesh, India, Pakistan, Maldives, and Nepal) for children between the age group of 24–59 months (146,996 children) from 2009 to 2016 to study the impact of socio-economic status (Hossain et al. 2020). The result showed that socio-economic status, education and maternal nutrition and household wealth index have an impact on the occurrence of under-nutrition or over-nutrition. The prevalence of underweight was 37% in Bangladesh, 38% in India, 19% in Maldives, 29% in Nepal and 28% in Pakistan. On the other hand, the prevalence of overweight was 2–4% in India, Bangladesh and Nepal, while it was 6% for Maldives. Geo et al. (2020) reported that the level of economic development influences over and under-nutrition across the 4 selected countries (India, China, Nepal and Pakistan); while the prevalence of overweight/obesity was found higher in an urban areas than in rural areas, irrespective of countries (with the exception of the childbearing woman in China). Farrell et al. (2021) reported the concern of triple burden of malnutrition which includes under-nutrition, diet-related chronic diseases and micronutrient deficiencies. They enumerated reasons for such burden of malnutrition which included changing food availability, with the predominance of less healthy food with high sugar, salt and fat, increasing cost of nutritious food and dependency on trade and import of healthy food, limited availability of nutritious food and promotion of unhealthy food items with some incentives (free gift, etc.). Harding et al. (2018) analyzed the data from India, Maldives, Bangladesh, Nepal, Pakistan and Afghanistan obtained from a national survey for children aged 0 to 59 months (N = 252,797) for the study of wasting. They reported that young children (0–5 months of age) had low body mass index (< 18.5 kg/m^2), and those months were associated with wasting in various countries. Besides these factors, birth order, being male, maternal illiteracy, short maternal stature, lack of improved water resources and household poverty are associated with waste in some countries, but not systematically in all countries. Adepoju and Allen (2019) reported that stunting affects almost 40% of children in South Asia and the number of stunted children in Africa is rising; 45% of all deaths under 5 age were attributable to undernutrition. In the entire world, 22.2% of children (0–59 months of age) are stunted, and the proportion of wasted and overweight children is 7.5% and 5.6%, respectively. Harding et al. (2017) reported that progress toward alleviating iodine deficiency is encouraging in South Asia, while attempts to alleviate Zn deficiency, anaemia and vitamin A deficiency is slow and need immediate policy action. Aziz et al. (2021) reported that, economic and physical access to food for poor populations is essential to reduce undernutrition in South Asia. In their cross-country analysis, they found that agricultural food production reduces the prevalence of undernourishment in India, Nepal, Sri Lanka and Pakistan. At the same time, an increase in the food price index reduces undernourishment in Afghanistan, Bangladesh, India and Nepal. Beal et al. (2021) reported that, iron deficiency, Zn deficiency, vitamin A deficiency, folate deficiency and vitamin B12 deficiency were the most commonly reported deficiencies among 6–23 month age children in the eight South Asian countries of India, Pakistan, and Bangladesh, Nepal, Afghanistan, Sri Lanka, Bhutan and the Maldives. Fagbamigbe et al. (2020) analyse the data from the Demographic and Health Survey of 51 low and middle-income countries consisting of 532,680 under 5 years of age children. They reported that factors such as children's gender (males), age (infants), low birth weight, a mother with no formal education, and poor household access to any media are strong indicators of the likelihood of suffering from severe acute malnutrition. Besides that, factors such as children from rural areas, neighbourhoods with high illiteracy and unemployment rates are also likely to have severe acute malnutrition. In the literature review for adolescent undernutrition in South Asia involving 131 studies (out of 6972 publications), it was observed that thinness (Low body mass index for age), stunting (low height for age) and micronutrient deficiencies were indicators used for expressing undernutrition in adolescence (Querol et al. 2022). Wali et al. (2021) reported that the prevalence of stunting among 0–23 months age children was 30% in the five South Asian countries of India, Bangladesh, Maldives, Nepal and Pakistan; and stunting rates among 24–59 months age and 0–59 months age children were 38% and 35%, respectively.

2.3 Status of Food and Nutritional Security in India

India is the most populous country with 1.4 billion population and large diversity with respect to socio-economic, cultural and food habits and the concerns of malnutrition exist with different cause-effect relationships. The significant role of gender, age, wealth inequalities and residence (rural and urban) on under and over-nutrition was reported in the analysis of data from national family health survey data (Nguyen et al. 2021). India has the world's largest number of stunted and wasted children. At the same time 23% and 20% of women and men, respectively were underweight; 21% of women and 19% man were overweight or obese (Kapur and Suri 2020). Based on the comparison of data for 2006 and 2016 for children and adults, it was found that decrease in child stunting (48% in 2006 to 38% in 2016), a high prevalence of stunting in rural and urban slum households and a large impact of wealth inequalities for stunting were major changes occurred in said period for children. For adults, major changes in malnutrition were a decrease in underweight prevalence, an increase in overweight and obesity and a large impact of wealth inequalities on both underweight (among poor households) and overweight (wealthier households). Chopra et al. (2023) conducted a review of literature published from 1976 to 2006 to identify the factors associated with child malnutrition. They found that inadequate dietary intake, illness, lifestyle, nutritional practices, unsafe water, food insecurity, lack of sanitation and basic hygiene, unhealthy feeding and caring practices, inadequate health infrastructure and suboptimal implementation of government nutrition schemes are the factors associated with child malnutrition. These factors act intensely and create a threat of the triple burden of malnutrition (malnutrition, overnutrition and micronutrient deficiency). The data from the national family health survey-IV conducted in 2015–16 was used to determine the prevalence and factors associated with the triple burden of malnutrition (sample size 168,784) (Kumar et al. 2021a, b). Factors such as the age of the mother, educational status of the mother, caesarean section delivery, birth size of the baby, wealth status of the household and place of residency affect the occurrence of the triple burden of malnutrition. Among the studied population, 5.7% of mother–child pairs were suffering from a triple burden of malnutrition and 57.8% of children were found anaemic; and the proportion of stunted and underweight children was 38.9% and 35.8%, respectively. The risk factors and determinants of malnutrition were established by studying the severe acute malnourished children aged 6 months to 59 months in Vellore, Tamil Nadu, India (David et al. 2020). Severe acute malnutrition is associated with birth weight, failure to consistently breastfeed for 6 months, inadequate calorie intake and mother being underweight. Gosh (2020) reviewed articles published from 1990 to 2019 from online databases (Scopus, web of Science and Elsevier) and found that a mother's knowledge about her children's health, her education level and her place of residence are the major factors affecting the prevalence of child malnutrition. Besides these factors, other factors such as income and ethnicity of households, household infrastructure, birth order, sex of child, maternal characteristics, maternal employment status and domestic violence also have significant impacts on the prevalence of child malnutrition.

Mahapatra et al. (2021) found a positive association between the vulnerability of agriculture to climate change and the prevalence of child malnutrition based on the national family health survey and data on climate-related parameters from the Indian Meteorological Department. They reported that the odds of children suffering from stunting, wasting, being underweight and anaemia increased by 32%, 42%, 45% and 63%, respectively if the child belonged to a district categorised as very highly vulnerable and compared to those districts with very low vulnerability. Khandpan et al. (2023) studied 28,004 adults (60 + age) in India and found that 26.7% and 22.2% of adults were underweight and overweight, respectively. Besides that, socio-economic factors such as years of schooling, living in Scheduled Castes (SC) or Scheduled Tribes (ST) household and living in rural areas are more related with underweight. The most vulnerable categories for food insecurity and malnutrition were individuals who were in old age, widowed, divorced/ separated/deserted, SC, ST, economically weaker and persons without or with only a few years of formal education. Panda et al. (2020) reported that half of the children from really poor households (economically poor and have welfare cards for the public distribution system) and excluded poor (economically poor and do not have welfare cards for the public distribution system).

3 Soil–Plant-Animal-Human Being Nexus for Nutritional Security

3.1 Soil Components

The soil is a whole and basic medium for crop nutrition, and the major sources of nutrients in soil are mineral matter, organic matter and immobilized nutrients in microbial biomass (Brevik et al. 2020; Lal et al. 2021). The soil microbial population turnover and plant living and dead roots are the recurring nutrient resources (Hirt 2020). Interventions such as input addition and management practices (such as tillage, land configuration and seedbed preparation) influence the nutrient availability to the plant. These interventions are needed in agro-ecosystems considering the nutrient removal by growing crops and the amount of nutrients removed in the form of crop output taken from the system (open nutrient cycling) and the level of land degradation (Bhattacharyya et al. 2015). The attempts to reduce soil degradation and build soil health (Muchane et al. 2020; Coban et al. 2022) are forerunners in retaining the productive/ nutrient-supplying capacity of the soil. The various positive effects of soil health-building practices on soil properties were widely reported (Shahane and Shivay 2021). The nutrient status of the soil is expressed as the amount of plant-available nutrients and different soil properties affecting the nutrients' plant-availability. The plant-available pool for nutrients in soil is dynamic in different dimensions. These include: length, width, depth, time, space, forms in which mineral nutrients are present, variable nutrient acquisition capacity of crops plants and differences in the manipulation of soil, nutrient forms, and cultivation methods. Therefore, in order to get generalized values of plant-availability nutrients, extensive and frequent sampling and analysis for nutrient determination is needed. The nutrient status of the soil is represented based on the fertility map for different regions (Chen et al. 2020; Hengl et al. 2021; Sharma et al. 2021). Among the major primary nutrients, nitrogen has been highlighted for a long time to its significant role in protein formation, universal deficiency and lack of presence in the mineral composition of the soil and the highest plant/crop response among all nutrients. In the case of phosphorus, the major concerns and considerations were: inequitable distribution of phosphorus minerals over the earth's surface, phosphorus deficiency in arable land, its low use efficiency, fixation in soil, and high cost per unit nutrient application through fertilizer (Alewell et al. 2020; Doydora et al. 2020; Hanyabui et al. 2020; Kisinyo and Opala 2020). Potassium deficiency is a rising concern in the human population and plants (Sinha and Tandon 2020; Sun and Weaver 2020; Soumare et al. 2023) with its higher uptake due to its definite mineral reserve (Roberts and Stewart 2002; Rawashdeh 2020). The rising concerns over multi-nutrient deficiencies (Shukla et al. 2021), micronutrient deficiency (Shukla and Behera 2011; 2012; 2019) and negative nutrient balance (NAAS 2018) were cited in multiple works of literature. This indicates the need for attention to soil plant-available nutrient pool enrichment through different sources and the conservation of soil nutrient pools from different losses. The two nutrients viz., zinc and iron are very important for humans considering their wide-scale deficiencies in humans leading to hidden hunger (Lowe 2021; Weffort and Lamounier 2023). Hence, micronutrient fertilizations take a lead at all components of nexus including soil fertilization (Koc and Karayigit 2022; Teklu et al. 2023). The need for Zn fertilization due to a significant increase in grain yield, and the increase in Zn concentration in milled rice grain by 0.7–0.8 mg/kg and wheat grain by 2.7–3.4 mg/kg, respectively due to zinc fertilization (5 kg Zn ha^{-1} through zinc sulphate heptahydrate soil application) was reported (Shahane et al. 2018; 2019). In a study conducted in 7 different countries, the mean increase in Zn concentration in wheat grain was 3.1 mg kg^{-1} with Zn fertilization (soil application @ 50 kg $ZnSO_4.7H_2O$) (Zou et al. 2012). Besides fertilization, other practices such as green manuring have recorded significant increases in rice and wheat grain Zn and Fe concentration (Mendi et al. 2022). Puli et al. (2019) reported a 7.67 and 3.54 mg kg^{-1} increase in grain Zn content in the first and second year, respectively due to farm yard manure application. They also reported a positive effect of green manuring on Zn content in rice grain. In another study, an increase in pearl millet grain Zn concentration by 2.14 mg kg^{-1} with the application of Zn 5 kg ha^{-1} through zinc oxide was reported (Kumar et al. 2021a, b). The high-yield varieties and hybrids of cereal crops are the backbone of food security and have been promoted to a great extent over the last six decades. The nutrient requirements of these varieties are higher hence soil fertilization, manuring and amendment application are needed and it will play a great role in meeting the nutritional requirements of plants, animals and ultimately human beings (Fig. 2).

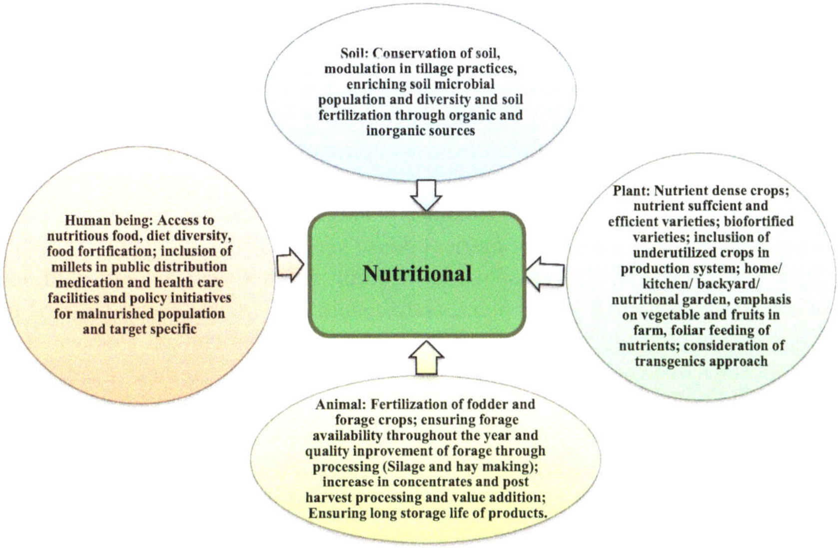

Fig. 2 Factors affecting food and nutritional security at each component level

3.2 Plant Component

The crop plants are intermediate between soil and animal or human beings and act as carriers of nutrients from soil to human beings. Direct use of soil as a source of nutrition is very rare and also quite limited in its extent (Geography in Africa and Asia) (Davies 2022; Bonglaisin et al. 2022). This component plays another and most important role in converting the nutrient from inorganic to organic form (required by animals and human beings). This component is discussed at the production, distribution and consumption levels. At the production level, the major aspects of the nutritional point of view are conservation and utilization of natural crop/plant diversity (Siddique et al. 2021), development of new cultivars with nutrient-dense varieties (also called genetic bio-fortification) (Yadava et al. 2020), utilizing the under-utilized plants/crops and testing of varieties and landraces for their response to different soil-based modification (amendments application and management practices). Genetic erosion is one of the most commonly reported problems in the present conventional modern agrochemical-based agriculture. The narrowing of the genetic base for getting desired characteristics (high yield) are a most widely reported problem in several food crops as well as non-food crops. Institutions such as the National Bureau of Plant Genetic Resources, International Bureau of Plant Genetic Resources, different CGIAR institutions as well as different ICAR institutions are actively involved in these activities. The conservation of species and varietal diversity of millets (Joshi et al. 2021; Seetharam and Bhat 2021) were reported. Besides these attempts, integrated farming systems (Das et al. 2021), vertical farming and nutritional gardens in backyards (Lal 2020) are the other interventions at the plant level.

The development of new and nutrient-dense varieties is gathering momentum in India, with increasing involvement by the Government (Ramdasa et al. 2020; Sheoran et al. 2022). The crops such as rice, wheat, maize, pearl millet (Satyavathi et al. 2021), finger millet, little millet, linseed, lentil, soybean, cauliflower, groundnut, mustard, green gram, cowpea, soybean, groundnut, potato, sweet potato, greater yam, and pomegranate have availability of nutrient-dense varieties; and among the nutrients, protein, iron, zinc, lysine, tryptophan, provitamin A, oleic acid, linoleic acid, low erucic acid and glucosinolate content, anthocyanin and vitamin C are targeted for biofortification, and are mainly designated for nutrient enrichment (Yadava et al. 2020). Besides these attempts, the development of genetically engineered crops with higher nutrient content was explored with the development of golden rice (Amna et al. 2020; Glover et al. 2020). The attempts at the production level are conditional, and are constrained by different factors such as the bioavailability of nutrients from the crops produced and the presence of anti-nutritional factors (Samtiya et al. 2020) as well as availability and affordability. Besides that, high investment in research and development, socio-economic acceptability, affordability to the majority of the population and environmental safety concerns are among the major hurdles reported in developing nutrient-dense crops through a genetic engineering approach (Gaharwar et al. 2021).

The staple food has a major role in the nutrition of the majority of the population. Mughal and Sers (2020) reported that a 1% increase in cereal production and yield results in a decrease of 0.84% undernourishment in South Asia. At the same time, a deficiency of major cereals (rice, wheat, maize) in micronutrients, leading to its deficiency in humans (hidden hunger), was reported (White and Broadly 2005; Bouis et al. 2011). The research on biofortification for enhancing the concentration of Zn and Fe in rice, wheat, maize, sorghum and finger millet was widely reported (Shivay et al. 2008a, b, c). The increase in Zn concentration in rice grain by 2–2.2 mg kg^{-1} to 19–22 mg kg^{-1} with foliar application of Zn at different growth stages was reported by Shivay et al. (2015) and Cakmak et al. (2010). Similarly, an increase in iron concentration in grain by 0.2–0.4 to 1.0–5.0 mg kg^{-1} due to foliar fertilization was reported in the same study, indicating the potential of micronutrient enrichment for major cereals. The increase in Zn concentration in milled rice grain by 0.9 and 1.0 mg kg^{-1} in the first and second year, respectively due to the application of microbial inoculations (*Anabaena-Pseudomonas* biofilmed formulations) was reported (Shahane et al. 2019). Similarly, the application of the same microbial consortia in wheat increased wheat grain Zn concentration by 3.2 and 1.9 mg kg^{-1}, respectively in the first and second years (Shahane et al. 2018).

The role of underutilized and neglected crops in alleviating malnutrition was recognized by Li et al. (2020) and Srivastava and Arya (2021). They reported that, underutilized crops such as finger millet, proso millet, pearl millet and foxtail millet are superior in nutrient content over the staple foods (rice and wheat), and hence need to be incorporated into the diet to increase diet diversity. Besides that, lentils, buckwheat and taro can be used to increase the nutrient content in the diet. Hossain et al. (2021) reported that around 1,000 underutilized and/or neglected crop species have the potential to serve as nutrient-dense food along with their wider adaptability to different stress conditions in climate change conditions. Ali and Bhattacharjee (2023) reported that, out of 4,523 research studies, 99 studies noted the nutrient superiority of underutilized crops and neglected crops, and also found the presence of health-promoting bioactive compounds. Siddique et al. (2021) also reported the role of underutilized and neglected crops in alleviating hidden hunger by citing the example of Nepal and relating it to the zero hunger initiative in Asia. This shows the potential of underutilized crops and neglected crops in alleviating malnutrition-related problems.

The scientific information on accountable relation/contribution of agricultural assets and inputs used on nutrition and malnutrition-related aspects is very limited (Gupte and Longhurst 2019; Shankar et al. 2019); both were dealing with two different levels of nexus viz. soil–plant and human being and will be very heterogeneous and complex considering diversity. One attempt was noted by Shankar et al. (2019). They evaluated 10,618 items from different online databases and screened 44 full texts, and finally selected 14 published materials. They studied the linkage between agricultural assets and inputs used and their nutritional outcomes for humans in South Asia. They did not observe any clear association of land ownership and size with nutritional outcomes of farm households; they noted however that land productivity (which is increased/modulated by inputs-seeds, fertilizers and irrigation water) was more clearly associated with nutrition outcomes. Besides that, there was a strong linkage between livestock keeping and dietary and nutrition outcomes. This relation is more particular for animals reared for milk where improvement in household milk intake had a strong influence on the growth of children. Besides that, model-based estimates were made available by using different indicators or data analysis techniques. Gupte and Longhurst (2019) reported that an integrated approach to addressing food and nutrition security at the nexus level involves addressing the nutritional values in underutilized crops, maintaining the genetic resource bases, incorporating women's concerns in the local rural economy, approaching problems in a participatory manner, and setting up (effective) inter-ministerial connections at national and region (province/ state level). Bird et al. (2019) reported that there is a potential impact of agricultural interventions on nutritional outcomes for humans from their review of nutrient-sensitive intervention studies published in India (2), Bangladesh (2) and Nepal (2). The nutrient-sensitive interventions they studied include livestock, home gardens and poultry and aquaculture. At the same time, nutrient-sensitive interventions in agriculture promoted in the same study area by different organizations include homestead vegetable gardens for female adolescents, household duck rearing, biochar fertilizer, nutrient sensitive farming systems in tribal communities, kitchen gardens, farming systems for nutrition, farming systems for improved nutrition, agricultural technology and woman access to agricultural assets. The interventions in agriculture and agri-food systems for addressing food and nutritional security by Glover and Poole (2019) from the literature review in South Asia include home/kitchen gardens, collective vegetable gardening by young women, small livestock (poultry, duck, etc.) rearing, nutrient-dense and biofortified crops, food product fortification, biofortified transgenic crops, nutrient cycling using biochar, a farming system for nutrition, integrating nutrient advice into agricultural extension platforms, promotion of nutritious traditional crops and foods and inclusion of millets in the public distribution system of food items. Dizon et al. (2021) reported that the cultivation of home gardens, animal agriculture, and the use of biofortification and post-harvest fortification are promising pathways to improve nutrition; on the other hand, public food transfer alone (making food available and affordable to poor populations) is not sufficient to cater for the needs of nutrition in South Asia.

3.3 Animal Component

The role of animal components in nutritional security is discussed at two levels viz., animal management level and animal product post-harvest technology level. The animal management component consists of diet management through the incorporation of more concentrate (Anonymous 2006) and providing medical facilities at an affordable cost (Khan et al. 2020; Smulski et al. 2020). Balehegn et al. (2020) reported the constraints in meeting the need for quality feed demand in low and middle-income countries. The major constraints were grazing management, fertilization of degraded pastureland, introducing new forage species, and multipurpose forage trees, forage seed production, supplementation with concentrates and availability of multi-mineral blocks. By overviewing the published literature, they also showed that, overcoming these constraints leads to a positive impact on animal productivity. Windsor et al. (2021) reported an improvement in buffalo daily milk production by 24–31% with supplementation with high-quality cow-calf molasses nutrient block containing 10% urea. The medical facilities increase the survival rate and number of calving and milking periods. The vaccination against New Castle disease in poultry birds leads to an increase in egg productivity by 80% and chicken production by 36% (Bessell et al. 2020). At the same time, the use of new technology—such as artificial insemination—significantly increases the yielding potential of animal components (Consentini et al. 2021). Practices such as the replacement of low-yielding and/or unproductive animals help in increasing productivity at the national level. In post-harvest technology, preservation plays a significant role due to the perishable nature of animal products, followed by processing (Getabalew et al. 2020). Animal products are more nutrient-dense products than plant-origin products; they hence contribute more efficiently to the nutritional security of the animal population. Beal and Ortenzi (2022) reported that different animal-origin products such as small fish, goat, beef, egg, milk, canned fish with bones, mutton, cheese, goat milk and pork rank top as sources of micronutrients. Beal et al. (2021) reported that small fish, ruminant meat, chicken liver, and eggs are nutrient-dense and whole food sources of Fe, Zn, Vitamin A, folate and vitamin B12. The milk is amenable to processing and large arrays of processed products were reported. In the case of meat and eggs, it is their availability and consumption patterns that determine the extent which they enhance nutrition.

3.4 Human Components

The human component in nexus is the ultimate and most highlighted component in terms of attempts made to achieve nutritional security. The attempts at this stage are classified as food-based interventions and medication approaches. Food-based interventions are getting prime attention due to their policy interventions at the national and international levels. The harvest plus programme (Bouis and Saltzman 2017; Sanjeeva Rao et al. 2020), food fortification programme (Ohanenye et al. 2021) and diet diversification through legumes, oilseed and underutilized crops (Desire et al. 2021; Teye et al. 2020) are some of such interventions. The food fortification may be a mass fortification (iodized salt in India), target fortification (targeting children and women), or market-driven due to an increase in demand and awareness (Tripathi and Mishra 2020). The fortification is also classified as a classical fortification (for mass fortification), industrial fortification (market-driven), biofortification (incorporating nutrients during the production process through soil fertilization and plant foliar application), food-to-food fortification (replacing less nutritious food with nutrient-dense food) and point-of-use/home fortification (adding minerals and vitamin in food when ready cook and/or consume) (Thakur et al. 2023). Bromage et al. (2019) reported that awareness and understanding (mandatory or voluntary) and designing health promotion and social marketing for creating an enabling environment for the supply and demand of fortified foods are needed for industrial fortification. They found that 55% of the studied population (311 healthy men and women of age $\geq$ 18 years old) favoured mandatory fortification of food, 14% disapproved of it and 31% were uncertain about fortification. Chadare et al. (2018) reported that iron, zinc, vitamin A, vitamin D, calcium, iodine and folic acid are the major fortificants in fish sauce, rice meal, wheat flour, maize meal, soy sauce, porridge cereals, rice, cereal flour, margarine, milk and dairy products, soybean oil, condiments and processed foods for classical fortification. On the other hand, protein, fat, ash, phosphorus, iron, manganese, copper, zinc, potassium, vitamin C, sugar, crude fibre, carbohydrates, lipids, vitamins of B group (B1, B2), vitamin E, vitamin A, silicon and calcium are used as a fortifier in food-to-food fortification. They also reported that socio-economic factors, technical limits and communication factors greatly influence the extent to which food fortification is practiced; and along with food fortification, all possible avenues need to be stressed rather than depending only on food fortification. Mkambula et al. (2020) reported that low coverage at the population level and quality of fortified food (wheat, rice, maize, oil/ghee and salt) are important concerns in large-scale food fortification. They found 84 countries with new large-scale food fortification programmes. They outlined a four-point strategy for food fortification, which includes: (a) establishment of new large-scale food fortification programmes wherever necessary, (b) systems innovations by the implementation of research to address coverage and quality gaps, (c) advocacy to form new partnerships and resources, particularly with the private sector, and d) exploration of new fortificants and vehicles and other innovations that can address existing challenges. In another study, Osendarp et al. (2018) reported that food fortification programmes are highly efficient and cost-effective; however, these programmes require continued interest and investment by the government in order to ensure sustained impact. The major challenges in food fortification reported include: ensuring effective coverage, accessibility and equality of fortified foods and lack of determination of evidence-based gaps. The major components for a successful large-scale food fortification programme include an appropriate choice of food vehicle which is reaching a large section of the population in affordable and adequate quantity, coordination with other programmes and making the food fortification programme part of national micronutrient deficiency control strategy. Other factors are effective quality control, strengthening of capacity and resources, governance and policy commitments and ensuring safe and impactful programme with frequent review of the assumptions of the programmes. Another study (Begin et al. 2001) reported that appropriate advocacy and communication, collaboration among several sectors, food regulation and standards and quality assurance and monitoring are needed to ensure effective implementation of vitamin A fortification programmes. They also reported that proven vehicles for vitamin A fortification relevant for Africa include sugars, oils, fats and cereal flours.

Ohanenye et al. (2021) overviewed different compounds used for food fortification in content of African countries for zinc, iron and vitamin A. They observed that ferrous sulphate is the most used water-soluble fortificant in food fortification, and ferrous fumarate is the most popular, and also the cheapest iron source. Other compounds used for food fortification were ferrous gluconate, ferric ammonium citrate, ferrous bis-glycinate, ferrous lactate and sodium iron ethylene diamine tetraacetic acid (Na Fe EDTA). In the case of zinc, commonly used compounds in fortification are sulphate, gluconate, oxide, chloride and stearate with zinc oxide being the cheapest fortificant. In India fortified food products and fortificants include iodized salt, vitamin A fortified oil, vitamin D fortified oil, vitamin A fortified milk, vitamin D fortified milk, Atta and Maida fortified with iron, folic acid and vitamin B12 and fortified rice with iron and vitamin B12 (Sirohi et al. 2018; Thakur et al. 2023). A survey was conducted to study the effectiveness of fortified wheat flour (fortified with vitamins, A, B1, B2, niacin, folic acid, zinc and iron) involving 300 adult females (aged 20–60 years) in China (Huo et al. 2012). They found a significant improvement in serum retinol content, serum iron and free erythrocyte protoporphyrin (for iron status) in the population feeding on fortified food. Besides that, an increase in haemoglobin levels also occurred due to the consumption of fortified wheat flour.

Another study found that the inclusion of groundnut, cowpea and sesame along with cereal crops in the diet helps in increasing protein, vitamins, zinc, iron, sodium, magnesium, calcium and potassium content in the diet (Desire et al. 2021). Teye et al. (2020) reported a large number of under-utilized plant species (Grain amaranths, groundnut, soybean, moringa, cowpeas, green peas, pearl millet and Plantin (Banana used as vegetables)) which can be used to enrich nutrient-poor cereal and carbohydrate-rich crops (maize, sweet potato, cassava, sorghum, yam and rice) in the diet of Sub-Saharan African populations. These crops help increase the nutrient density of magnesium, potassium, zinc, iron and phosphorus in the diet.

Madzorera and Fawzi (2020) reported that women's empowerment is an integral part of nutrient-sensitive interventions and optimizing food and nutrient supply to families. Besides that, the major concern of child malnutrition can be addressed effectively with a nutritious and diverse diet for women. Considering its central role, emphasis should be given to women's empowerment, access to resources and increasing women's control of production resources. Gaincola et al. (2022) reported that, the double burden of malnutrition needs to be addressed through the different call-of-action avenues in Sub-Saharan Africa. These include locally led programmes supporting diverse food systems and health education programmes, and awareness creation among school children and young adults about the cause and effect of non-communicable diseases. They also reported that, increased attention towards children and adolescents, protecting citizens from big food industries and implementation of WHO best buys (involving replacing trans-fat with unsaturated fatty acids and reduction of salt in diet) are needed to address the double burden of malnutrition in Sub-Saharan African countries. In India, an alternative nutrient-dense food for the management of acute malnutrition in India viz., 'Balamrutham+' was developed and tested on severe acute malnourished and moderate acute malnourished children (Kumar et al. 2020). After two weeks of feeding, improvement in health of 22.3% of children from moderate acute malnutrition was observed; while 17.7% of severe acute malnourished children achieved the status of moderate acute malnutrition. They also reported that, there were 42 locally available food products suitable for the development of alternative nutrient-dense food. Panda et al. (2020) reported the role of the public distribution system (PDS) in alleviating malnutrition in India by showing higher children malnutrition among households without welfare cards of PDS. Therefore, improving the quality of food grain and widening the food basket of PDS will play an important role in reducing malnutrition. In addition to the direct efforts to provide nutritious food, the proper assessment of the situation in terms of major problems faced, solutions available and strengthening the capacity to act at regional and global levels play significant roles in addressing the problem. Gillespie et al. (2019) reported that commitment, power, accountability, coherence, data, leadership and capacity play a great role in strengthening agri-food systems for caring for nutritional needs.

Diet diversification and increasing the animal-origin food products in diet is another food-based intervention for achieving the nutritional security constrained by the higher price of these nutrient-dense products (Moura et al. 2023). It is also important to increase the supply of nutrient-rich food items such as millets, pulses and oil at affordable (subsidized) rates. Food fortification in processing, and increasing the availability of nutrient-dense ready-to-eat products and the other variant practices at the human being level to ensure nutrient-rich food availability, should also be emphasized. Besides these approaches, medication plays an important role, because of the extent of malnutrition in India, and the constraints experienced in using the food-based approach to reach the neediest population. The medication approach consists of a food supplementation approach and providing medical assistance at an affordable cost. Supplementation is the target-oriented approach that involves providing higher doses of deficient nutrients, especially to women and children. This approach has age, location and policy support bias. The efforts made in this approach have a direct and significant impact on human nutrients, and they require continuous policy support to sustain them over a long period. The nexus-based interventions are helpful in ensuring accountability for different programme and activities undertaken at different levels, sustaining these attempts and avoiding the overlapping of different interventions. This also makes nutritional security a completely mapped activity with transparent governance and minimum chances of malpractices.

4 Conclusion

Agricultural production activities complement health care policies to alleviate malnutrition thereby helping in achieving nutritional security. The investment for achieving nutritional security should be equitable at the nexus level rather than over-emphasizing the interventions at the human level alone. At the level of soil and plant health, fertilization to increase nutrient accumulation in plants, identification and utilization of nutrient-dense crops/landraces, and development of nutrient-dense (bio-fortified) varieties need to be sustained as a long-term target. At the animal component level, improvement in animal health and value addition during processing needs significant attention. At the human being level, medication and supplementation are short-run and need-based interventions. An increase in the share of animal components in diet and supply of fortified food at affordable cost is the sustainable and long-lasting affordable avenue for achieving nutritional security.

References

Adepoju AA, Allen S (2019) Malnutrition in developing countries: nutrition disorders, a leading cause of ill health in the world today. Paediatr Child Health 29(9):394–400. https://doi.org/10.1016/j.paed.2019.06.005

Adeyeye, Samuel Ayofemi O, Tolulope J, Ashaolu, Olusola T, Bolaji, Titilope A, Abegunde, Adetola O, Omoyajowo (2023) Africa and the Nexus of poverty, malnutrition and diseases. Critcal Rev Food Sci Nutr 63(5):641–656. https://doi.org/10.1080/10408398.2021.1952160

Ahinkorah BO, Amadu I, Seidu A-A, Okyere J, Duku E, Hagan JE Jr, Budu E, Archer AG, Yaya S (2021) Prevalence and factors associated with the triple burden of malnutrition among mother-child pairs in Sub-Saharan Africa. Nutrients 13(6):2050. https://doi.org/10.3390/nu13062050

Akombi BJ, Agho KE, Merom D, Renzaho AM, Hall JJ (2017) Child malnutrition in sub-Saharan Africa: A meta-analysis of demographic and health surveys (2006–2016). PLoS ONE 12(5):e0177338. https://doi.org/10.1371/journal.pone.0177338

Alam AZ, Yeshaw Y, Liyew AM, Tessema ZT, Worku MG, Tasema GA, Alamneh TS, Teshale AB, Chilot D, Ayalew HG (2023) Double burden of malnutrition and its associated factors among women in low and middle income countries: findings from 52 nationally representative data. BMC Public Health 23:1479. https://doi.org/10.1186/s12889-023-16045-4

Alamu AO, Eyinla TE, Sanusi RA, Maziya-Dixon B (2020) Double burden of malnutrition: Evidence from a selected Nigerian population. J Nutr Metabolism. https://doi.org/10.1155/2020/5674279

Alewell C, Ringeval B, Ballabio C, Robinson DA, Panagos P, Borrelli P (2020) Global phosphorus shortage will be aggravated by soil erosion. Nat Commun 11(1):4546. https://doi.org/10.1038/s41467-020-18326-7

Ali A, Bhattacharjee B (2023) Nutrition security, constraints, and agro-diversification strategies of neglected and underutilized crops to fight global hidden hunger. Front Nutr 10:1144439. https://doi.org/10.3389/fnut.2023.1144439

Amna QS, Tantray AY, Bashir SS, Zaid A, Wani SH (2020) Golden rice: genetic engineering, promises, present status and future prospects. In: Roychoudhury A (eds) Rice research for quality improvement: genomics and genetic engineering. Springer, Singapore. https://doi.org/10.1007/978-981-15-5337-0_25

Anonymous (1989) Recommended dietary allowances, National Research Council (US) subcommittee on the tenth edition of the recommended dietary allowances, 10th edn. National Academies Press Washington (DC) (US)

Anonymous (2006) Improving animal productivity by supplementary feeding of multi-nutrient blocks, controlling internal parasites and enhancing utilization of alternate feed resources. International Atomic energy Agency (IAEA-TECDOC-1495), p 278; Available online at: https://www-pub.iaea.org/MTCD/Publications/PDF/te_1495_web.pdf. Accessed on 29th Jan 2024

Anonymous (2019) National Food Processing Policy 2019; Ministry of Food processing Industries, Government of India, New Delhi, India p 34. Available on: https://www.mofpi.gov.in/sites/default/files/draft-nfpp_0.pdf. Access on 25th Jan 2023

Ansah IGK, Gardebroek C, Ihle R (2019) Resilience and household food security: a review of concepts, methodological approaches and empirical evidence. Food Secur 11(6):1187–1203. https://doi.org/10.1007/s12571-019-00968-1

Arto I, Andreoni V, Rueda-Cantuche JM (2016) Global use of water resources: a multiregional analysis of water use, water footprint and water trade balance. Water Resour Econ 15:1–14. https://doi.org/10.1016/j.wre.2016.04.002

Ashagidigbi WM, Ishola TM, Omotayo AO (2022) Gender and occupation of household head as major determinants of malnutrition among children in Nigeria. Sci Afr 16:e01159. https://doi.org/10.1016/j.sciaf.2022.e01159

Aworh OC (2021) Food safety issues in fresh produce supply chain with particular reference to sub-Saharan Africa. Food Control 123:107737. https://doi.org/10.1016/j.foodcont.2020.107737

Aziz N, He J, Raza A, Sui H, Yue W (2021) Elucidating the macroeconomic determinants of undernourishment in south Asian countries: building the framework for action. Front Public Health 9:696789. https://doi.org/10.3389/fpubh.2021.696789

Azomahou TT, Boucekkine R, Kazianga H, Korir M, Ndung'u N (2022) Guest editors' introduction: the role of policy in reducing malnutrition in sub-Saharan Africa. Food Policy 113:102378 https://doi.org/10.1016/j.foodpol.2022.102378

Baggio G, Qadir M, Smakhtin V (2021) Freshwater availability status across countries for human and ecosystem needs. Sci Total Environ 792:148230. https://doi.org/10.1016/j.scitotenv.2021.148230

Balehegn M, Duncan A, Tolera A, Ayantunde AA, Issa S, Karimou M, Zampaligré N, André K, Gnanda I, Varijakshapanicker P, Kebreab E (2020) Improving adoption of technologies and interventions for increasing supply of quality livestock feed in low-and middle-income countries. Glob Food Secur 26:100372 https://doi.org/10.1016/j.gfs.2020.100372

Beal T, Ortenzi F (2022) Priority micronutrient density in foods. Front Nutr 7(9):806566. https://doi.org/10.3389/fnut.2022.80656

Beal T, White JM, Arsenault JE, Okronipa H, Hinnouho GM, Murira Z., Torlesse H, Garg A (2021) Micronutrient gaps during the complementary feeding period in South Asia: a comprehensive nutrient gap assessment. Nutrition Rev 79(S1):26–34.https://doi.org/10.1093/nutrit/nuaa144

Begin F, Cervinskas J, Mannar V (2001) Food fortification with vitamin A: the potential for contributing to the elimination of vitamin A deficiency in Africa. Food Nutr Bull 22(4):408–415. https://doi.org/10.1177/156482650102200411

Bessell PR, Woolley R, Stevenson S, Al-Riyami L, Opondo P, Lai L, Gammon N (2020) An analysis of the impact of Newcastle disease vaccination and husbandry practice on smallholder chicken productivity in Uganda. Prev Veterinary Med 177:104975. https://doi.org/10.1016/j.prevetmed.2020.104975

Bhattacharyya R, Ghosh BN, Mishra PK, Mandal B, Rao CS, Sarkar D, Das K, Anil KS, Lalitha M, Hati KM, Franzluebbers AJ (2015) Soil degradation in India: challenges and potential solutions. Sustainability 7(4):3528–3570. https://doi.org/10.3390/su7043528

Bird FA, Pradhan A, Bhavani RV, Dangour AD (2019) Interventions in agriculture for nutrition outcomes: a systematic review focused on South Asia. Food Policy 82:39–49. https://doi.org/10.1016/j.foodpol.2018.10.015

Bonglaisin JN, Kunsoan NB, Bonny P, Matchawe C, Tata BN, Nkeunen G, Mbofung CM (2022) Geophagia: benefits and potential toxicity to human—a review. Front Public Health 10:893831. https://doi.org/10.3389/fpubh.2022.893831

Bouis HE, Hotz C, McClafferty B, Meenakshi JV, Pfeiffer WH (2011) Biofortification: a new tool to reduce micronutrient. Biofortification: a new tool to reduce micronutrient. Food Nutr Bull 32(1):S31–S40. https://doi.org/10.1177/15648265110321S105

Bouis HE, Saltzman A (2017) Improving nutrition through biofortification: a review of evidence from HarvestPlus, 2003 through 2016. Glob Food Secur 12:49–58. https://doi.org/10.1016/j.gfs.2017.01.009

Brevik EC, Slaughter L, Singh BR, Steffan JJ, Collier D, Barnhart P, Pereira P (2020) Soil and human health: current status and future needs. Air Soil Water Res 13:1178622120934441. https://doi.org/10.1177/1178622120934441

Bromage S, Gonchigsumlaa E, Traeger M, Magsar B, Wang Q, Bater J, Li H, Ganmaa D (2019) Awareness and attitudes regarding industrial food fortification in Mongolia and Harbin. Nutrients 11(1):201. https://doi.org/10.3390/nu11010201

Cafiero C, Melgar-Quiñonez HR, Ballard TJ, Kepple AW (2014) Validity and reliability of food security measures. Ann N Y Acad Sci 1331(1):230–248. https://doi.org/10.1111/nyas.12594

Cakmak I, Pfeiffer WH, McClafferty B (2010) Biofortification of durum wheat with zinc and iron. Cereal Chem 87(1):10–20. https://doi.org/10.1094/CCHEM-87-1-0010

Calicioglu O, Flammini A, Bracco S, Bellù L, Sims R (2019) The future challenges of food and agriculture: an integrated analysis of trends and solutions. Sustainability 11(1):222. https://doi.org/10.3390/su11010222

Chen S, Lin B, Li Y, Zhou S (2020) Spatial and temporal changes of soil properties and soil fertility evaluation in a large grain-production area of subtropical plain, China. Geoderma 357:113937. https://doi.org/10.1016/j.geoderma.2019.113937

Chenoweth J, Hadjikakou M, Zoumides C (2014) Quantifying the human impact on water resources: a critical review of the water footprint concept. Hydrol Earth Syst Sci 18(6):2325–2342 https://doi.org/10.5194/hess-18-2325-2014

Chopra H, Paul B, Virk A, Pandey G, Lahariya C (2023) Triple burden of malnutrition among children in India: current scenario and the way forward. Indian J Pediatr. https://doi.org/10.1007/s12098-023-04739-x

Clapp J, Moseley WG, Burlingame B, Termine P (2022) The case for a six-dimensional food security framework. Food Policy 106:102164. https://doi.org/10.1016/j.foodpol.2021.102164

Coban O, De Deyn GB, van der Ploeg M (2022) Soil microbiota as game-changers in restoration of degraded lands. Science 375(6584):abe0725. https://doi.org/10.1126/science.abe0725

Consentini Consentini CE, Wiltbank MC, Sartori R (2021) Factors that optimize reproductive efficiency in dairy herds with an emphasis on timed artificial insemination programs. Animals 11(2):301. https://doi.org/10.3390/ani11020301

Ctistian AK, Dake FAA (2021) Profiling household double and triple burden of malnutrition in sub-Saharan Africa: prevalence and influencing household. Public Health Nutr 25(6):1563–1576. https://doi.org/10.1017/S1368980021001750

D'Odorico P, Chiarelli DD, Rosa L, Bini A, Zilberman D, Rulli MC (2020) The global value of water in agriculture. Proc National Acad Sci 117(36):21985–21993. https://doi.org/10.1073/pnas.2005835117

Das A, Datta D, Samajdar T, Idapuganti RG, Islam M, Choudhury BU, Mohapatra KP, Layek J, Babu S, Yadav GS (2021) Livelihood security of small holder farmers in eastern Himalayas, India: pond based integrated farming system a sustainable approach. Curr Res Environ Sustain 3:100076. https://doi.org/10.1016/j.crsust.2021.100076

David SM, Pricilla RA, Paul SS, George K, Bose A, Prasad JH (2020) Risk factors for severe acute malnutrition among children aged 6–59 months: a community-based case-control study from Vellore, Southern India. J Fam Med Prim 9:2237–2243. https://doi.org/10.4103/jfmpc.jfmpc_211_20

Davies TC (2022) Current status of research and gaps in knowledge of geophagic practices in Africa. Front Nutr 9:1084589. https://doi.org/10.3389/fnut.2022.1084589

Desire MF, Blessing M, Elijah N, Ronald M, Agather K, Tapiwa Z, Florence MR, George N (2021) Exploring food fortification potential of neglected legume and oil seed crops for improving food and nutrition security among smallholder farming communities: a systematic review. J Agric Food Res 3:100117 https://doi.org/10.1016/j.jafr.2021.100117

Dizon F, Josephson A, Raju D (2021) Pathways to better nutrition in South Asia: evidence on the effects of food and agricultural interventions. Glob Food Secur 28:100467. https://doi.org/10.1016/j.gfs.2020.100467

Doydora S, Gatiboni L, Grieger K, Hesterberg D, Jones JL, McLamore ES, Peters R, Sozzani R, Van den Broeck L, Duckworth OW (2020) Accessing legacy phosphorus in soils. Soil Syst 4(4):74. https://doi.org/10.3390/soilsystems4040074

El Bilali H, Callenius C, Strassner, Probst (2019) Food and nutrition security and sustainability transitions in food systems. Food Energy Secur 8(2):e00154

Fagbamigbe AF, Kandala NB, Uthman OA (2020) Severe acute malnutrition among under-5 children in low-and middle-income countries: a hierarchical analysis of associated risk factors. Nutrition 75:110768. https://doi.org/10.1016/j.nut.2020.110768

Farrell P, Rachmi CN, Mulcahy G, Helble M, Thow AM (2021) Food environment research is needed to improve nutrition and well-being in Asia and the Pacific. Public Health Nutr 24(14):4706–4710. https://doi.org/10.1017/S136898002100241X

Gaharwar US, Verma A, Singh R, Prasad T, Rajamani P (2021). Health risks and environmental concerns of GM crop adoption. In: Policy issues in genetically modified crops. Academic, pp 371–400. https://doi.org/10.1016/B978-0-12-820780-2.00017-0

Gao L, Bhurtyal A, Wei J, Akhtar P, Wang L, Wang Y (2020) Double burden of malnutrition and nutrition transition in Asia: a case study of 4 selected countries with different socioeconomic development. Adv Nutr 11(6):1663–1670

Ge J, Polhill JG, Macdiarmid JI, Fitton N, Smith P, Clark H, Dawson T, Aphale M (2021) Food and nutrition security under global trade: a relation-driven agent-based global trade model. R Soc Open Sci 8(1):201587. https://doi.org/10.1098/rsos.201587

Getabalew M, Alemneh T, Zewdie D (2020) The milk processing: status, challenges and opportunities in Ethiopia. Int J Veterinary Sci Res 6(1):052–057. https://doi.org/10.17352/ijvsr.000054

Giancola A, Hauerslev M, Ndinda C, Farmer M, Musah L, Msengi G, Bhutta Z (2022) Addressing the double-burden of malnutrition among children and adolescents in sub-Saharan Africa. J Glob Health Rep 6:e2022033. https://doi.org/10.29392/001c.36904

Gillespie S, Bold MVD, Hodge J (2019) Nutrition and the governance of agri-food systems in South Asia: a systematic review. Food Policy 82:13–27. https://doi.org/10.1016/j.foodpol.2018.10.013

Glover D, Kim SK, Stone GD (2020) Golden Rice and technology adoption theory: a study of seed choice dynamics among rice growers in the Philippines. Technol Soc 60:101227. https://doi.org/10.1016/j.techsoc.2019.101227

Glover D, Poole N (2019) Principles of innovation to build nutrition-sensitive food systems in South Asia. Food Policy 82:63–73. https://doi.org/10.1016/j.foodpol.2018.10.010

Gosh P (2020) Factors responsible for childhood malnutrition: a review of the literature. Curr Res Nutr Food Sci 8(2):360–370. https://doi.org/10.12944/CRNFSJ.8.2.01

Guevara-Romero E, Flórez-García V, Egede LE, Yan A (2021) Factors associated with the double burden of malnutrition at the household level: a scoping review. Crit Rev Food Sci Nutr. https://doi.org/10.1080/10408398.2021.1908954

Gupte J, Longhurst R (2019) How do the state's organisational capacities at the micro- and macro-levels influence agriculture-nutrition linkages in fragile contexts? Food Policy 82:74–83. https://doi.org/10.1016/j.foodpol.2018.10.016

Hanyabui E, Obeng Apori S, Agyei Frimpong K, Atiah K, Abindaw T, Ali M, Yeboah Asiamah J, Byalebeka J (2020) Phosphorus sorption in tropical soils. AIMS Agric Food 5(4):599–616. https://doi.org/10.3934/agrfood.2020.4.599

Harding KL, Aguayo VM, Webb P (2017) Hidden hunger in South Asia: a review of recent trends and persistent challenges. Public Health Nutr 21(4):785–795. https://doi.org/10.1017/S1368980017003202

Harding KL, Aguayo VM, Webb P (2018) Factors associated with wasting among children under five years old in South Asia: implications for action. PLoS ONE 13(7):e0198749. https://doi.org/10.1371/journal.pone.0198749

Harper A, Goudge J, Chirwa E, Rothberg A, Sambu W, Mall S (2022) Dietary diversity, food insecurity and the double burden of malnutrition among children, adolescents and adults in South Africa: findings from a national survey. Front Public Health 10:948090. https://doi.org/10.3389/fpubh.2022.948090

Hengl T, Miller MA, Križan J, Shepherd KD, Sila A, Kilibarda M, Antonijević O, Glušica L, Dobermann A, Haefele SM, McGrath SP (2021) African soil properties and nutrients mapped at 30 m spatial resolution using two-scale ensemble machine learning. Sci Rep 11(1):6130. https://doi.org/10.1038/s41598-021-85639-y

Hirt H (2020) Healthy soils for healthy plants for healthy humans: How beneficial microbes in the soil, food and gut are interconnected and how agriculture can contribute to human health. EMBO Rep 21(8):e51069. https://doi.org/10.15252/embr.202051069

Hossain A, Islam MT, Maitra S, Majumder D, Garai S, Mondal M, Ahmed A, Roy A, Skalicky M, Brestic M, Islam T (2021) Neglected and underutilized crop species: are they future smart crops in fighting poverty, hunger and malnutrition under changing climate? In: Zargar SM, Masi A, Salgotra RK (eds) Neglected and underutilized crops—towards nutritional security and sustainability. Springer, Singapore. https://doi.org/10.1007/978-981-16-3876-3_1

Hossain FB, Shawon MSR, Al-Abid MSU, Mahmood S, Adhikary G, Bulbul MdSI (2020) Double burden of malnutrition in children aged 24 to 59 months by socioeconomic status in five South Asian countries: evidence from demographic and health surveys. BMJ Open 10:e032866. https://doi.org/10.1136/bmjopen-2019-032866

Huo J, Sun J, Huang J, Li W, Wang L, Selenje L, Gleason GR, Yu X (2012) Effectiveness of fortified flour for enhancement of vitamin and mineral intakes and nutrition status in northwest Chinese villages. Food Nutr Bull 33(2):161–168. https://doi.org/10.1177/156482651203300210

Ingram J (2020) Nutrition security is more than food security. Nature Food 1(1):2–2. https://doi.org/10.1038/s43016-019-0002-4

Joshi DC, Meena RP, Chandora R (2021) Genetic resources: collection, characterization, conservation, and documentation. In: Millets and pseudo cereals. Woodhead Publishing, pp 19–31. https://doi.org/10.1016/B978-0-12-820089-6.00003-3

Kamboj S, Gupta N, Bandral JD, Gandotra G, Anjum N (2020) Food safety and hygiene: a review. Int J Chem Stud 8(2):358–368

Kapur K, Suri S (2020) Towards a malnutrition-free India: best practices and innovations from POSHAN Abhiyaan, ORF special report No. 103, observer research foundation. Available online at https://www.orfonline.org/research/towards-a-malnutrition-free-india-63290. Accessed on 15th Jan 2024

Kebede H, Liu X, Jin J, Xing F (2020) Current status of major mycotoxins contamination in food and feed in Africa. Food Control 110:106975. https://doi.org/10.1016/j.foodcont.2019.106975

Khan MZ, Liu L, Zhang Z, Khan A, Wang D, Mi S, Usman T, Liu G, Guo G, Li X, Wang Y (2020) Folic acid supplementation regulates milk production variables, metabolic associated genes and pathways in perinatal Holsteins. J Anim Physiol Anim Nutr 104(2):483–492. https://doi.org/10.1111/jpn.13313

Khandpan B, Pradhan I, Pradhan J (2023) Food insecurity and malnutrition among Indian older adults: findings from longitudinal ageing study in India, 2017–18. J Popul Aging 16:639–669. https://doi.org/10.1007/s12062-022-09378-2

Kisinyo P, Opala P (2020) Depletion of phosphate rock reserves and world food crisis: reality or hoax? Afr J Agric Res 16(9):1223–1227. https://doi.org/10.5897/AJAR2020.14892

Koc E, Karayiğit B (2022) Assessment of biofortification approaches used to improve micronutrient-dense plants that are a sustainable solution to combat hidden hunger. J Soil Sci Plant Nutr 22(1):475–500. https://doi.org/10.1007/s42729-021-00663-1

Konapala G, Mishra AK, Wada Y, Mann ME (2020) Climate change will affect global water availability through compounding changes in seasonal precipitation and evaporation. Nat Commun 11(1):3044. https://doi.org/10.1038/s41467-020-16757-w

Kumar M, Singh U, Raiger PR, Singh LN, Parewa HP, Chouhan R (2021) Micronutrient Fortification Micronutrient Fortification of Pearl Millet [*Pennisetum glaucum* (L.) R. Br.] Hybrids using customized fertilizer formulation. Int J Bio-Resour Stress Manag 12(5):416–425 https://doi.org/10.23910/1.2021.2438

Kumar P, Chauhan S, Patel R, Srivastava S, Bansod DW (2021) Prevalence and factors associated with triple burden of malnutrition among mother-child pairs in India: a study based on National Family Health Survey 2015–16. BMC Public Health 21:391.https://doi.org/10.1186/s12889-021-10411-w

Kumar P, Sriswan S, Arlappa N, Tiwari K, Rohatgi S, Sarika G, Roy D, Geddam JJB, Hemalata R (2020) Development and use of alternative nutrient-dense foods for management of acute malnutrition in India. Field Exchange 63:53–56. www.ennonline.net/fex

Lal R (2020) Home gardening and urban agriculture for advancing food and nutritional security in response to the COVID-19 pandemic. Food Secur 12(4):871–876. https://doi.org/10.1007/s12571-020-01058-3

Lal R, Bouma J, Brevik E, Dawson L, Field DJ, Glaser B, Hatano R, Hartemink AE, Kosaki T, Lascelles B, Monger C (2021) Soils and sustainable development goals of the United Nations: an international union of soil sciences perspective. Geoderma Regional 25:e00398. https://doi.org/10.1016/j.geodrs.2021.e00398

Li X, Yadav R, Siddique KHM (2020) Neglected and underutilized crop species: the key to improving dietary diversity and fighting hunger and malnutrition in Asia and the Pacific. Front Nutr 7:593711. https://doi.org/10.3389/fnut.2020.593711

Lowe NM (2021) The global challenge of hidden hunger: perspectives from the field. Proc Nutr Soc 80(3):283–289. https://doi.org/10.1017/S0029665121000902

Mahapatra B, Walia M, Rao CAR, Raju BMK, Saggurti N (2021) Vulnerability of agriculture to climate change increases the risk of child malnutrition: Evidence from a large-scale observational study in India. PLoS ONE 16(6):e0253637. https://doi.org/10.1371/journal.pone.0253637

Mandi S, Shivay YS, Prasanna R, Kumar D, Purakayastha TJ, Pooniya V, Nayak S, Raihan O, Baral K, Pal M, Hussain S (2022) Improving micronutrient density in basmati rice and durum wheat through summer green manuring and elemental sulfur fertilization. Crop Pasture Sci 73:804–816. https://doi.org/10.1071/CP21313

Mekhasa TJ, Molla KG, Tarp F, Aikaeli J (2022) Commodity price fluctuations and child malnutrition. World Dev 158:105927. https://doi.org/10.1016/j.worlddev.2022.105927

Militao EM, Salvador EM, Uthman OA, Vinberg S, Macassa G (2022) Food insecurity and health outcomes other than malnutrition in Southern Africa: a descriptive systematic review. Int J Environ Res Public Health 19(9):5082. https://doi.org/10.3390/ijerph19095082

Mkambula P, Mbuya MN, Rowe LA, Sablah M, Friesen VM, Chadha M, Osei AK, Ringholz C, Vasta FC, Gorstein J (2020) The unfinished agenda for food fortification in low-and middle-income countries: quantifying progress, gaps and potential opportunities. Nutrients 12(2):354. https://doi.org/10.3390/nu12020354

Mohajan HK (2022) Food insecurity and malnutrition of Africa: a combined attempt can reduce them. J Econ Dev, Environ People 11(1):24–34

Montoya D, Haegeman B, Gaba S, De Mazancourt C, Bretagnolle V, Loreau M (2019) Trade-offs in the provisioning and stability of ecosystem services in agroecosystems. Ecol Appl 29(2):e01853. https://doi.org/10.1002/eap.1853

Moseley WG, Battersby J (2020) The vulnerability and resilience of African food systems, food security, and nutrition in the context of the COVID-19 pandemic. Afr Stud Rev 63(3):449–461. https://doi.org/10.1017/asr.2020.72

Moura MAFE, Martins BDA, Oliveira GPD, Takahashi JA (2023) Alternative protein sources of plant, algal, fungal and insect origins for dietary diversification in search of nutrition and health. Crit Rev Food Sci Nutr 63(31):10691–10708. https://doi.org/10.1080/10408398.2022.2085657

Muchane MN, Sileshi GW, Gripenberg S, Jonsson M, Pumariño L, Barrios E (2020) Agroforestry boosts soil health in the humid and sub-humid tropics: a meta-analysis. Agr Ecosyst Environ 295:106899. https://doi.org/10.1016/j.agee.2020.106899

Mughal M, Sers CF (2020) Cereal production, undernourishment and food insecurity in South Asia. https://hal.science/hal-02089616v2

NAAS (2018) Soil health: new policy initiatives for Farmers welfare. Policy Brief No. 3. National Academy of Agricultural Sciences, New Delhi, p 19

Nchuchuwe FF, Adejuwon KD (2012) The challenges of agriculture and rural development in Africa: the case of Nigeria. Int J Acad Res ProgIve Educ Dev 1(3):45–61

Nelson E, Wood S, Koo J, Polasky S (2011) Provisioning and regulatory ecosystem service values in agriculture. Natural Capital: theory and practice of mapping ecosystem services. Oxford University Press, Oxford, pp 150–167

Nguyen PH, Scott S, Headey D, Singh N, Tran LM, Menon P et al (2021) The double burden of malnutrition in India: trends and inequalities (2006–2016). PLoS ONE 16(2):e0247856. https://doi.org/10.1371/journal.pone.0247856

Ogunniyi AI, Mavrotas G, Olagunju KO, Fadare O, Adedoyin R (2020) Governance quality, remittances and their implications for food and nutrition security in Sub-Saharan Africa. World Dev 127:104752. https://doi.org/10.1016/j.worlddev.2019.104752

Ohanenye IC, Emenike CU, Mensi A, Medina-Godoy S, Jin J, Ahmed T, Sun X, Udenigwe CC (2021) Food fortification technologies: influence on iron, zinc and vitamin A bioavailability and potential implications on micronutrient deficiency in sub-Saharan Africa. Sci Afr 11:e00667. https://doi.org/10.1016/j.sciaf.2020.e00667

Olson R, Gavin-Smith B, Ferraboschi C, Kraemer K (2021) Food fortification: the advantages, disadvantages and lessons from sight and life programs. Nutrients 13(4):1118. https://doi.org/10.3390/nu13041118

Olsson L, Barbosa H, Bhadwal S, Cowie A, Delusca K, Flores-Renteria D, Hermans K, Jobbagy E, Kurz W, Li D, Sonwa DJ, Stringer L (2019) Land degradation. In: Shukla PR, Skea E, Calvo Buendia V, Masson-Delmotte HO, Pörtner DC, Roberts P, Zhai R, Slade S, Connors J, van Diemen R, Ferrat M, Haughey E, Luz S, Neogi S, Pathak M, Petzold J, Portugal Pereira J, Vyas P, Huntley E, Kissick K, Belkacemi M, Malley J (eds) Climate change and land: an IPCC special report on climate change, desertification, land degradation, sustainable land management, food security, and greenhouse gas fluxes in terrestrial ecosystems. https://doi.org/10.1017/9781009157988.006

Osendarp SJ, Martinez H, Garrett GS, Neufeld LM, De-Regil LM, Vossenaar M, Darnton-Hill I (2018) Large-scale food fortification and biofortification in low-and middle-income countries: a review of programs, trends, challenges, and evidence gaps. Food Nutr Bull 39(2):315–331. https://doi.org/10.1177/0379572118774229

Owolade AJJ, Abdullateef RO, Adesola RO, Olaloye ED (2022) Malnutrition: an underlying health condition faced in sub Saharan Africa: challenges and recommendations. Ann Med Surg 82:104769. https://doi.org/10.1016/j.amsu.2022.104769

Panda BK, Mohanty SK, Nayak I, Shastri VD, Subramanian SV (2020) Malnutrition and poverty in India: does the use of public distribution system matter? BMC Nutr 6:41. https://doi.org/10.1186/s40795-020-00369-0

Pangaribowo EH, Gerber N, Torero M (2013) Food and nutrition security indicators: a review. ZEF Working Paper No. 108, Available at https://doi.org/10.2139/ssrn.2237992

Pawlak K, Kołodziejczak M (2020) The role of agriculture in ensuring food security in developing countries: considerations in the context of the problem of sustainable food production. Sustainability 12(13):5488. https://doi.org/10.3390/su12135488

Peng W, Berry EM (2019) The concept of food security. In: Ferranti P, Berry EM, Anderson JR (eds) Encyclopedia of food security and sustainability, vol 2, no 1. Elsevier, pp 1–7. https://doi.org/10.1016/B978-0-08-100596-5.22314-7

Peng W, Mu Y, Hu Y, Li B, Raman J, Sui Z (2020) Double burden of malnutrition in the Asia-Pacific Region—a systematic review and meta-analysis. J Epidemiol Glob Health 10(1):16–27https://doi.org/10.2991/jegh.k.191117.002

Pinstrup-Andersen P (2009) Food security: definition and measurement. Food Security 1(1):5–7. https://doi.org/10.1007/s12571-008-0002-y

Popkin BM, Corvalan C, Grummer-Strawn L (2020) Dynamics of the double burden of malnutrition and the changing nutrition reality. Lancet 395(10217):65–74. https://doi.org/10.1016/S0140-6736(19)32497-3

Pretty J, Sutherland WJ, Ashby J, Auburn J, Baulcombe D, Bell M, Bentley J, Bickersteth S, Brown K, Burke J, Campbell H (2010) The top 100 questions of importance to the future of global agriculture. Int J Agric Sustain 8(4):219–236. https://doi.org/10.3763/ijas.2010.0534

Puli MR, Prasad PRK, Babu PR, Rao KL, Subbaiah G (2019) Effect of organic and inorganic sources of nutrients on secondary and micronutrient content in rice at various growth periods. Oryza-An Int J Rice 56(3):312–317. https://doi.org/10.35709/ory.2019.56.3.7

Querol SE, Gill P, Eqbal R, Kletter M, Ozdemir N, Al-Khudairy (2022) Adolescent undernutrition in South Asia: a scoping review. Nutr Res Rev 35:39–49. https://doi.org/10.1017/S0954422421000068

Rachmi CN, Li M, Baur LA (2018) The double burden of malnutrition in Association of South East Asian Nations (ASEAN) countries: a comprehensive review of the literature. Asia Pac J Clin Nutr 27(4):736–755. https://doi.org/10.3316/ielapa.761708622605017

Ramadas S, Vellaichamy S, Ramasundaram P, Kumar A, Singh S (2020) Biofortification for enhancing nutritional outcomes and policy imperatives. In: Wheat and barley grain biofortification. Woodhead Publishing, pp 309–327. https://doi.org/10.1016/B978-0-12-818444-8.00013-4

Rawashdeh RA (2020) World peak potash: an analytical study. Resources Policy 69.https://doi.org/10.1016/j.resourpol.2020.101834

Reardon T, Tschirley D, Liverpool-Tasie LSO, Awokuse T, Fanzo J, Minten B, Vos R, Dolislager M, Sauer C, Dhar R, Vargas C (2021) The processed food revolution in African food systems and the double burden of malnutrition. Glob Food Sec 28:100466. https://doi.org/10.1016/j.gfs.2020.100466)

Roberts TL, Stewarts WM (2002) Inorganic phosphorus and potassium production and reserves. Better Crops 86(2):6–7

Samtiya M, Aluko RE, Dhewa T (2020) Plant food anti-nutritional factors and their reduction strategies: an overview. Food Prod, Process Nutr 2:6. https://doi.org/10.1186/s43014-020-0020-5

Sanjeeva Rao D, Neeraja CN, Madhu Babu P, Nirmala B, Suman K, Rao LS, Surekha K, Raghu P, Longvah T, Surendra P, Kumar R (2020) Zinc biofortified rice varieties: challenges, possibilities, and progress in India. Front Nutr 7:26. https://doi.org/10.3389/fnut.2020.00026

Satyavathi CT, Ambawat S, Khandelwal V, Govindaraj M, Neerja CN (2021) Micronutrient rich pearl millet for nutritionally secure India. ICAR-All India co-ordinated research project on pearl millet, Jodhpur, Rajasthan, India, p 18

Seetharam A, Bhat P (2021) Diversity and small millets in India. In: Padulosi S, Israel ED, King O, Hunter D, Swaminathan MS (eds) Orphan crops for sustainable food and nutrition security: promoting neglected and underutilized species, pp 68–77. https://doi.org/10.4324/9781003044802-6

Shahane AA, Shivay YS (2021) Soil health and its improvement through novel agronomic and innovative approaches. Frontiers in Agronomy 3:680456. https://doi.org/10.3389/fagro.2021.680456

Shahane AA, Shivay YS, Kumar D, Prasanna R (2018) Role of zinc, nitrogen and phosphorus fertilization, and use of microbial inoculation in zinc nutrition of wheat (*Triticum aestivum*) under different crop establishment methods. Indian J Agric Sci 88(9):1434–1441

Shahane AA, Shivay YS, Kumar D, Prasanna R (2019) Zinc nutrition of rice as influenced by crop establishment methods, rates of nitrogen and phosphorus fertilization and inoculation with microbial consortia. J Plant Nutr 42(16):1967–1981. https://doi.org/10.1080/01904167.2019.1636066

Shankar B, Poole N, Bird FA (2019) Agricultural inputs and nutrition in South Asia. Food Policy 82(2019):28–38. https://doi.org/10.1016/j.foodpol.2018.10.011

Sharma RP, Chattaraj S, Vasu D, Karthikeyan K, Tiwary P, Naitam RK, Dash B, Tiwari G, Jangir A, Daripa A, Singh SK (2021) Spatial variability assessment of soil fertility in black soils of central India using geostatistical modeling. Arch Agron Soil Sci 67(7):876–888. https://doi.org/10.1080/03650340.2020.1766678

Sheoran S, Kumar S, Ramtekey V, Kar P, Meena RS, Jangir CK (2022) Current status and potential of biofortification to enhance crop nutritional quality: an overview. Sustainability 14(6):3301. https://doi.org/10.3390/su14063301

Shivay YS, Kumar D, Prasad R (2008a) Effect of zinc-enriched urea on productivity, zinc uptake and efficiency of an aromatic rice-wheat cropping system. Nutr Cycl Agroecosyst 81(3):229–243. https://doi.org/10.1007/s10705-007-9159-6

Shivay YS, Kumar D, Prasad R, Ahlawat IPS (2008b) Relative yield and zinc uptake by rice from zinc sulphate and zinc oxide coatings onto urea. Nutr Cycl Agroecosyst 80(2):181–188. https://doi.org/10.1007/s10705-007-9131-5

Shivay YS, Prasad R, Rahal A (2008c) Relative efficiency of zinc oxide and zinc sulphate-enriched urea for spring wheat. Nutr Cycl Agroecosyst 82:259–264. https://doi.org/10.1007/s10705-008-9186-y

Shivay YS, Prasad R, Singh RK, Pal M (2015) Relative efficiency of zinc-coated urea and soil and foliar application of zinc sulphate on yield, nitrogen, phosphorus, potassium, zinc and iron biofortification in grains and uptake by Basmati rice (*Oryza sativa* L.). J Agric Sci 7(2):161–173 https://doi.org/10.5539/jas.v7n2p161

Shukla AK, Behera SK (2011) Zinc management in Indian soils: past, present and future. Indian J Fertil 7(10):14–33

Shukla AK, Behera SK (2012) Micronutrient fertilizer and higher productivity. Indian J Fertil 8(4):100–117

Shukla AK, Behera SK, Prakash C, Patra AK, Rao CS, Chaudhari SK, Das S, Singh AK, Green A (2021) Assessing multi-micronutrients deficiency in agricultural soils of India. Sustainability 13(16):9136. https://doi.org/10.3390/su13169136

Shukla AK, Behera SK, Satyanarayana T, Majumdar K (2019) Importance of micronutrients in Indian agriculture. Better Crops-South Asia, pp 8–10

Siddique KHM, Li X, Gruber K (2021) Rediscovering Asia's forgotten crops to fight chronic and hidden hunger. Nature Plants 7:116–122. https://doi.org/10.1038/s41477-021-00850-z

Sinha D, Tandon PK (2020) An overview of nitrogen, phosphorus and potassium: key players of nutrition process in plants. In: Mishra et al (eds) Sustainable solutions for elemental deficiency and excess in crop plants, pp 85–117. https://doi.org/10.1007/978-981-15-8636-1_5

Sirohi A, Pundhir A, Ghosh S (2018) Food fortification: a nutritional management strategy in India. Innovare J Food Sci 6(2):1–8

Smulski S, Gehrke M, Libera K, Cieslak A, Huang H, Patra AK, Szumacher-Strabel M (2020) Effects of various mastitis treatments on the reproductive performance of cows. BMC Vet Res 16:1–10. https://doi.org/10.1186/s12917-020-02305-7

Soumare A, Djibril SARR, Diedhiou AG (2023) Potassium sources, microorganisms and plant nutrition: challenges and future research directions. Pedosphere 33(1):105–115. https://doi.org/10.1016/j.pedsph.2022.06.025

Srivastava S, Arya C (2021) Millets: malnutrition and nutrition security. In: Kumar A, Tripathi MK, Joshi D, Kumar V (ed) Millet and millet technology. Springer publication, pp 81–100. https://doi.org/10.1007/978-981-16-0676-2_4

Sun H, Weaver CM (2020) Rise in potassium deficiency in the US population linked to agriculture practices and dietary potassium deficits. J Agric Food Chem 68(40):11121–11127. https://doi.org/10.1021/acs.jafc.0c05139

Tanumihardjo SA, McCulley L, Roh R, Lopez-Ridaura S, Palacios-Rojas N, Gunaratna NS (2020) Maize agro-food systems to ensure food and nutrition security in reference to the sustainable development goals. Glob Food Sec 25:100327. https://doi.org/10.1016/j.gfs.2019.100327

Teklu D, Gashu D, Joy EJ, Amede T, Broadley MR (2023) Effectiveness of agronomic biofortification strategy in fighting against hidden hunger. Agronomy 13(8):2173. https://doi.org/10.3390/agronomy13082173

Tendall DM, Joerin J, Kopainsky B, Edwards P, Shreck A, Le QB, Krütli P, Grant M, Six J (2015) Food system resilience: defining the concept. Glob Food Sec 6:17–23. https://doi.org/10.1016/j.gfs.2015.08.001

Teye E, Deha CI, Dadzie R, MacArthur RL (2020) Delivering the nutritional needs by food to food fortification of staples using underutilized plant species in Africa. Int J Food Sci. https://doi.org/10.1155/2020/8826693

Thakur S, Singh A, Insa B, Sharma S (2023) Food fortification in India as malnutrition concern: a global approach. Sustain Food Technol 1:681–695. https://doi.org/10.1039/D3FB00079F

Thiombiano L, Tourino-Soto I (2007) Status and trends in land degradation in Africa. In: Sivakumar MVK, Ndiang'ui N (eds) Climate and land degradation, environmental science and engineering. Springer, Berlin, Heidelberg. https://doi.org/10.1007/978-3-540-72438-4_2

Tripathi A, Mishra S (2020) An estimation of price effects of making food fortification mandatory in India. Food Nutr Bull 41(3):355–366. https://doi.org/10.1177/03795721209385

Venugopal KR (1999) Food security vs. nutrition security. Health Millions 25(2):18–19. PMID: 12295422

Wali N, Agho KE, Renzaho MN (2021) Factors associated with stunting among children under 5 years in five south Asian countries (2014–2018): analysis of demographic health surveys. Nutrients 12:3875. https://doi.org/10.3390/nu12123875

Weffort VRS, Lamounier JA (2023) Hidden hunger–a narrative review. Jornal de Pediatriahttps://doi.org/10.1016/j.jped.2023.08.009

Wells JC, Sawaya AL, Wibaek R, Mwangome M, Poullas MS, Yajnik CS, Demaio A (2020) The double burden of malnutrition: aetiological pathways and consequences for health. Lancet 395(10217):75–88. https://doi.org/10.1016/S0140-6736(19)32472-9

Were JM, Stranges S, Creed IF (2020) Fertility is a key predictor of the double burden of malnutrition among women of child-bearing age in sub-Saharan Africa. J Glob Health 10(2):020408. https://doi.org/10.7189/jogh.10.020408

White PJ, Broadly M (2005) Biofortifying crops with essential mineral elements. Trends Plant Sci 10(12):586–593. https://doi.org/10.1016/j.tplants.2005.10.001

Windsor P, Martin S, Khounsy S, Young J, Thomson P, Bush R (2021) Improved milk production from supplementation of swamp buffalo with molasses nutrient blocks containing 10% urea. Dairy 2(1):90–103. https://doi.org/10.3390/dairy2010009

Yadava DK, Choudhury PR, Hossain F, Kumar D (2017) Biofortified varieties: sustainable way to alleviate malnutrition. Indian Council of Agricultural Research, New Delhi, India, p 19 + v

Yadava DK, Choudhury PR, Hossain F, Kumar D, Mohapatra T (2020) Biofortified varieties: sustainable way to alleviate malnutrition, 3rd edn. Indian Council of Agricultural Research, New Delhi, 86p

Zou CQ, Zhang YQ, Rashid A, Ram H, Savasli E, Arisoy RZ, Ortiz-Monasterio I, Simunji S, Wang ZH, Sohu V, Hassan M (2012) Biofortification of wheat with zinc through zinc fertilization in seven countries. Plant Soil 361(1):119–130. https://doi.org/10.1007/s11104-012-1369-2

Open Access This chapter is licensed under the terms of the Creative Commons Attribution 4.0 International License (http://creativecommons.org/licenses/by/4.0/), which permits use, sharing, adaptation, distribution and reproduction in any medium or format, as long as you give appropriate credit to the original author(s) and the source, provide a link to the Creative Commons license and indicate if changes were made.

The images or other third party material in this chapter are included in the chapter's Creative Commons license, unless indicated otherwise in a credit line to the material. If material is not included in the chapter's Creative Commons license and your intended use is not permitted by statutory regulation or exceeds the permitted use, you will need to obtain permission directly from the copyright holder.

Promoting Agroecology to Improve Soil Health and Human Nutrition in the Context of Sub-Saharan Africa

Anna Brazier

Abstract This chapter outlines the connection between human nutrition and soil health, and describes practical technologies that can regenerate the soil while producing nutritious plant and animal-based food in Sub-Saharan Africa. It explains how improving soil management can lead to improved production of nutritious food as well as improvement of the human gut microbiome, both of which are vital for our health and well-being. The chapter begins with an overview of the Sub-Sahara African food system context, looking at soil conditions and land degradation as well as the food security and nutrition situation. The chapter continues with a discussion of the services provided by the soil ecosystem. It explores the connections between the microbiome of the soil, plants and the human gut and changes that have occurred in the human gut because of diet changes and environmental factors. This is followed by a description of the impacts of agriculture on soils and human health and a discussion of the benefits of cultivating soil microbial communities. The final section explores some principles and practices of the agroecology approach that can be applied in Sub-Saharan Africa to improve soil health, food security and nutrition. It describes various crop and livestock production methods that have been successfully implemented in African contexts. To develop more sustainable food systems and address the malnutrition problems facing the continent, it is vital that development agents embark on a continent-wide effort to improve soils using agroecological methods.

1 Introduction

In recent years, through advances in science, the critical link between soil health, human health and nutrition has been acknowledged and analysed. This work has reinforced the idea that the way food is produced can have a major impact on human nutrition and well-being. It was well understood that soil containing sufficient nutrients for plant growth leads to the production of nutritious crops. However, the connection between microbial communities in the soil and those in the human gut is a new

A. Brazier (✉)
Independent Researcher, Harare, Zimbabwe
e-mail: brazier.anna@gmail.com

© The Author(s) 2026

B. R. Singh et al. (eds.), *Sustainable Food Systems in Sub-Saharan Africa*,
https://doi.org/10.1007/978-3-032-11288-0_11

field that has major implications for agriculture and food system transformation. The human-soil microbiome connection underlines the need to phase out practices that damage the soil. These practices also have negative impacts on water, atmospheric systems and biodiversity, which are interlinked to soil and human health. This chapter discusses the connection between human nutrition and soil health and describes practical technologies that can regenerate the soil while producing nutritious plant and animal-based food in Sub-Saharan Africa.

The context of sub-Saharan African soils and nutrition

The food system in sub-Saharan Africa is diverse and fundamentally dependent on the existing climate and environmental conditions in the continent. Most African food producers are smallholder farmers cultivating land under 2 ha with minimal inputs, and in most areas relying on rainfall rather than irrigation. Apart from Africa's diverse and often challenging climatic conditions, most of Africa's soils are ancient, extremely weathered and have been farmed for millennia. Consequently, they tend to be highly acidic, lack essential nutrients and organic matter, and are vulnerable to erosion (Jones et al. 2013). This makes crop production difficult. Inappropriate land management techniques, including traditional shifting cultivation and introduced technologies such as the plough (Steiner and Bwalya 2003), have led to land degradation and soil fertility declines. This has been exacerbated leading to low levels of food production and food insecurity. These factors are now being exacerbated by climate change impacts. Today, Africa is the continent most severely affected by land degradation, with 45% of its land area undergoing desertification and the other 55% at high or very high risk of further degradation (ELD Initiative and UNEP 2015). Land degradation in Africa is leading to widespread hunger, food insecurity, poverty, and vulnerability to climate hazards.

Nearly 20% of the population in Africa faces hunger and 78% are unable to afford a healthy diet (FAO, IFAD, UNICEF, WFP and WHO 2023). The Global Nutrition Report[1] states that only 6 countries in Africa are on course to achieve their targets for stunting of children under 5 years, 38 countries have made only modest progress on stunting and 10 countries have made no progress or are worsening with respect to stunting prevalence. In addition, not a single country in Africa is on course to achieve the global targets for anaemia in women of reproductive age, low birth weight, diabetes among men, diabetes among women, obesity among men, and obesity among women. The report notes that only 23.8% of children under 5 years of age in Africa receive minimum dietary diversity, and only 13.2% receive a minimum acceptable diet. The prevalence of undernutrition in Africa has been rising since 2014. It is against this challenging backdrop that we will explore the relationship between soil health and nutrition in the Sub-Saharan African context, and explore ways to augment this relationship to improve food security and nutrition.

[1] https://globalnutritionreport.org/resources/nutrition-profiles/africa/

Definition of key terms

Microbes or microorganisms—microscopic organisms such as bacteria, protozoa, and some fungi and algae.

Microbiome—the community of microorganisms that inhabit a particular environment.

Soil health—"the continued capacity of soil to function as a vital living ecosystem that sustains plants, animals, and humans." Soil Health Institute 2018.

2 The Connection Between Soil Health and Human Health

This section will look at the services provided by the soil ecosystem. It explores the connections between the microbiome of the soil, plants and the human gut and changes that have occurred in the human gut because of diet changes and environmental factors.

2.1 The Many Services Provided by the Soil Ecosystem

Called "dirt" by some, the soil is better described as the fundamental medium necessary for most life on the planet. The composition of soil varies from place to place but essentially it is a complex ecosystem made up of rock particles, organic matter, water, air, living plant roots, fungi, invertebrates, and trillions of microorganisms. This diverse community supports the plants and animals that make up our food system and most of the planet's biodiversity (Brevik et al. 2020). Soil regulates the Earth's hydrological systems; cleaning, storing, and supplying us with water. It regulates the climate by nurturing forests, grasslands, and other plant communities and through its fundamental role in the water cycle. By storing organic carbon, nitrogen and methane, the soil reduces the dangerous greenhouse gases in the atmosphere. Soil also provides us with medicine in the form of antibiotics and the raw material for buildings, roads, and the manufacture of utensils (Brevik et al. 2020), without which human settlements would not exist as they do today.

2.2 Connections Between the Microbiome in the Gut, Soil, and Plants

Microbes (including bacteria, fungi, viruses, archaea, and protozoa) have been studied for centuries, mainly with respect to those which cause disease, or which can be harnessed commercially for making alcohol and other useful products. Most of us have been brought up to view microbes as disease-causing organisms that should be eradicated from our bodies and environment. In fact, most microbes are beneficial, and we cannot survive without them. Our bodies contain billions of microorganisms. There are around ten times more microbe cells than human cells in our body (Hirt 2020), most of which live in harmony with us. They clean the inside and outside of our bodies, help us to digest food and to access vital nutrients, boost our immune system, and even support our mental health (Brevik et al. 2020). In particular, the microbial communities in the human digestive system (the gut microbiome) have major impacts on people's metabolism health and well-being. Disturbing the gut microbiome can lead to obesity, cardiovascular conditions, and mental health problems (Blum et al. 2019). For optimal health, humans require a range of important substances that our bodies are unable to manufacture. Most of these are supplied through the digestion of food, but a number (including vitamin B12) are produced by gut microbes (Hirt 2020). Gut bacteria also produce immunosuppressant, anti-inflammatory and anticancer substances as well as antibiotics (Hirt 2020). As noted by Hirt (2020, p 5), both root and gut microbes "synthesise essential amino acids, vitamins and many other secondary metabolites that modulate their host immune system: as such the plant and gut microbiomes can be considered as meta-organs with paramount importance for the health of their hosts."

Influence of the soil on the gut microbiome

The number of active microorganisms that inhabit the human gut are almost equal to the number of active microorganisms in the soil, although species diversity is lower in the gut (Blum et al. 2019). These microorganisms most likely enter the gut through our continuous interaction with the soil, through touching it, inhaling soil particles, drinking water that has passed through the soil and eating plants grown in the soil (Blum et al. 2019). Soil has also been deliberately ingested as medicine in various cultures for millennia. It is evident that humans evolved in parallel with our internal microbial communities, and most of the relationships forged were symbiotic and commensal rather than pathogenic (Blum et al. 2019).

Each person's gut microbiome is unique. The types of microbes that live inside you vary according to your genetics, where you live, your lifestyle and your diet. A person's initial source of gut microbes is through transmission from your mother during and after birth. Scientists have discovered that the main factor influencing your microbiome is your environment. Due to less interaction with the soil, people who grow up and live in urban areas have less diverse gut microbiomes than those from rural areas (Blum et al. 2019). There is evidence that increased exposure to soil microbes during childhood is beneficial for the immune system. This might explain why people raised in urban areas and in more developed countries tend to suffer from more autoimmune diseases compared to those raised in rural areas in less developed countries with lower levels of sanitation (Blum et al. 2019). It is highly probable that the immune system is weakened by the low levels of exposure to both pathogenic and non-pathogenic microbes.

2.3 *Changes in the Human Gut Microbiome*

Numerous factors have caused major disruptions to the health of communities in our gut. These include the modern diet (with high levels of processed carbohydrates, fats and oils and animal products), the widespread and often unnecessary use of antibiotics, the industrialization of agriculture and improvements in sanitation and hygiene (Brevik et al. 2020). This is not to negate the fact that many of these factors have also led to radical improvements in human life expectancy, health, and well-being.

The human diet has changed radically throughout history. The first shift occurred when humans discovered how to cook food about 1.8 million years ago. Cooking probably led to changes in the shape of our teeth and skull, enabling our brain size to increase (Wrangham 2017); but this necessarily reduced exposure to soil microbes. Later, increasingly complex food preservation techniques began to address seasonal and spatial food insecurity. Another major shift occurred when many human populations moved from a hunter-gatherer lifestyle to an agriculture-based lifestyle (Crittenden and Schnorr 2017). This would have also reduced exposure to a diverse range of microbes, particularly exacerbated as time progressed, using agricultural implements and machinery as well as pesticides and fertilizers. The most recent change has come because of industrialization and urbanization.

As noted by Blum et al. (2019) the modern, western diet is characterized by a lack of diversity, abundance of highly processed food, reduced fibre content, excess fats and sugars and meat consumption. Apart from the usual health problems associated with this diet (including vitamin and mineral deficiencies, overweight and obesity, cardiovascular diseases, and cancers), the disruptions to the microbiome can lead to a vicious cycle whereby microbes favoured by the urban diet "manipulate host eating behaviour and may generate cravings for unhealthy foods that suppress their microbial competitors". Research has shown that our dietary habits influence our gut bacteria. For example, vegans and vegetarians have different species of bacteria from people who eat meat and dairy products. Based on recent research into the health benefits of the gut microbiome, it is recommended that a diet of reduced animal products and consumption of a wider range of diverse fibre-rich foods, grown using organic production methods, can be beneficial to our health.

3 The Impacts of Modern Agricultural Methods on Soil and Human Health

This section describes the impacts of agriculture on soils and human health, and discusses the benefits of cultivating soil microbial communities.

Scientists estimate that a typical gram of soil contains 90–100 million bacteria, and about 200,000 fungi (Glick 2018). Just as there is a diverse range of soil types across the globe, the composition of soil microbial communities is also highly varied

(Blum et al. 2019). We can determine whether the soil is healthy by the size and diversity of microorganisms, plants, and animals that it supports and the levels of organic matter. Healthy soil has a crumbly texture, about half of which comprises spaces containing water and air. Between these spaces, clumps of soil (aggregates) are formed by sticky substances produced by fungi and earthworms. These bind the soil, improving its structure, reducing erosion, and facilitating the infiltration of water into the soil. Organic matter provides food for microbes and improves the soil's ability to store nutrients and water, particularly when it is broken down by soil organisms to form the stable substance, *humus*. Brevik et al. (2020) note that the fertility and functioning of soils depend heavily on "the interaction between the soil mineral matrix, plants and microbes … [which] are responsible for the preservation and availability of nutrients".

Communities of microorganisms differ in different parts of the soil. This is mainly due to the abundance or dearth of carbon (a major source of energy for microbes) in different soil zones. The area closest to plant roots (the rhizosphere) is typically abundant in microbes because plant roots tend to produce carbon-rich exudates which feed soil microbes (Glick 2018). Some of these microbes are harmful to plants, but many are plant growth-promoting bacteria and are increasingly encouraged through sustainable agriculture techniques as well as direct inoculation of crops (Brevik et al. 2020). Plant roots also contain communities of internal microbes. Blum et al. (2019) have noted many similarities in terms of structure, function and microbial species between the human gut microbiome and the rhizosphere.

3.1 Impacts of Conventional Agricultural Techniques on Soil and Human Health

While making agriculture more efficient, tillage (the mechanical preparation of land for crops, and cultivation of land during crop management) has been shown to have extremely negative impacts on soil microbial life, soil structure, organic matter content and water-holding capacity. Ploughing destroys soil aggregates, reducing the air and water spaces where microbes can live. This compacts the soil, thus reducing the ability of rainwater to flow into it. This leads to reduced water being stored in soils, flooding, and soil erosion. Cultivation of monocultures, particularly of cereals, results in chronic depletion of minerals in the soil that are essential for human and animal nutrition. Crops grown in depleted soils have lower levels of essential nutrients (particularly zinc). More than 2 billion people worldwide suffer from micronutrient deficiencies, and this contributes significantly to the global disease burden. This disease burden is felt most in parts of the world where consumption of cereals produced in monoculture dominates (Brevik et al. 2020). Pesticides, herbicides, and synthetic fertilizers disrupt not only the organisms that they target; they affect the entire agroecosystem. Chlorphyriflos, phosphamidon, malathion, fenthion, methyl phosphorothioate, parathion, chlorfluazuron and cypermethrin all damage soil microbial communities. The widely used herbicide glyphosate is antimicrobial and antifungal, and has been classified by the WHO as a potential human carcinogen (Hirt 2020). Glyphosate may also inhibit several beneficial microbes in plants and the human gut. According to Hirt (2020), livestock production uses about four times more antibiotics than human medicine, mainly to increase the growth and vigour of the animals. This threatens soil microbes, and has also led to the emergence of antibiotic-resistant bacteria, which in turn reduces the potency of human antibiotics that are vital for fighting disease. There is also concern about the impacts of other veterinary chemicals on the soil microbiome, including those used to fight internal and external parasites that contaminate the soil through manure applications.

Agrochemicals negatively impact soil health by killing beneficial soil organisms and also present direct risks to human health (Hirt 2020). Brevik et al. (2020) cite numerous studies which show that agricultural pesticides and herbicides have been implicated in incidences of long-term diseases such as diabetes, cancer, and asthma.

3.2 Benefits of Cultivating Soil Microbe Communities

Cultivation of healthy, beneficial soil microbial communities can improve crop yields, the nutritional quality of crop and livestock products, soil nutrient cycling (reducing the need for synthetic fertilisers), soil texture, moisture-holding capacity; at the same time, it helps to inhibit pests and diseases (Brevik et al. 2020). Brevik et al. (2020, p 11) note that "[E]effectively managing soil microbial communities for increased long-term sustainability and ecosystem resilience will be critical to ensure secure food systems and maintain or improve human health and well-being under future global change conditions". However modern agricultural practices, including tillage, crop cultivation in monocultures, and the use of agrochemicals have had a severely deleterious impact on soil health. This includes the disruption of soil microbial communities and damaging beneficial soil macrofauna (such as earthworms). Soil erosion and climate change have also had a detrimental effect on soil microbial populations (Hirt 2020). Nurturing the soil microbiome could produce higher and better-quality yields, more nutritious food, as well as reducing the need for fertilizer, pesticides, and herbicides. This could also help to improve the diversity of the human gut microbiome through the consumption of nutritious organic produce.

4 Agroecology to Improve Food Security and Nutrition

This section explores some principles and practices of the agroecology approach that can be applied in different contexts in Sub-Saharan Africa. This approach promotes numerous methods which improve the structure and functioning of the soil to improve food security and nutrition. The section describes various crop and livestock production methods that have been successfully implemented in different African contexts.

4.1 Overview of Principles and Practices

In response to the health and environmental problems associated with the use of synthetic agrochemicals and other issues associated with conventional agriculture, multiple alternative agricultural approaches have arisen. These include organic agriculture, biodynamics, permaculture, conservation farming, climate-smart agriculture, regenerative agriculture, and agroecology. While the many different names can be confusing, one commonality is the desire to develop low-input systems that move away from synthetic fertilizers, pesticides, and other agrochemicals. The motivations for this include improving human health, animal welfare and environmental health, as well as reducing the dependency of farmers on corporations which manufacture, market, and promote expensive inputs. Most of these approaches are based on underlying principles, but many prescribe specific techniques for the management of soil, water, pests and diseases, crops, and livestock.

Agroecology holds particular promise for improving African food systems. It has many definitions, but essentially it involves the application of ecological principles to farming systems. Agroecology has evolved to encompass a scientific discipline, a set of practices and a social movement. Rather than focusing on promoting techniques, agroecology emphasizes socio-cultural, economic, and political dimensions. In a recent publication by the Alliance for Food Sovereignty in Africa (AFSA), agroecology is defined as "a system which promotes agricultural practices based on the science of ecology to enhance the sustainability of land-use systems and the livelihoods of the communities that depend on them. Agroecology is also a movement of like-minded people who aim to transform food systems and societies towards increased sustainability and food sovereignty" (AFSA/SKI 2021).

Agroecology deals with different scales, treating farms, communities, and landscapes as interconnected systems that have interrelated and interacting social, ecological, and economic elements. Many of the practices that are now promoted through agroecology have been in use for a long time. Some are based on indigenous agricultural systems; others have developed through organic agriculture and other approaches; and many more are based on new technological and scientific advances. There is a great advantage to being flexible about bringing these different approaches under one umbrella without being pedantic about terminology. Another important strength is that agroecology is a transitional approach and recognizes that adoption of the various principles and practices can only practically come about in stages rather than being all or nothing.

Agroecological movements have arisen across Africa and globally. In the last decade the Food and Agriculture Organization of the United Nations (FAO) has begun the promotion of agroecology and has set up an information-sharing hub https://www.fao.org/agroecology/overview/en/). On the agroecology information hub, practitioners can learn about the ten elements of agroecology developed through a global multistakeholder process. In Africa, agroecological movements have developed in different ways across the continent, and AFSA aims to link these different movements together to build the largest social movement in Africa (for more information visit https://afsafrica.org/).

4.2 *Crop Production Methods for Better Soil and Human Health*

We have seen that soil is the foundation of the food system, and it is an important direct source of nutrients from food and via the gut microbiome. We have also seen that poor management of the soil has severely impaired the services that it provides us. The better ways to manage the soil and the agroecological soil management practices have the following underlying principles (adapted from AFSA/SKI 2021).

- Disturb the soil as little as possible, avoiding excessive turning, digging, and ploughing
- Encourage a diverse range of plants to be grown
- Keep living roots in the soil as much as possible
- Keep the soil covered with living plants or their residue at all times
- Avoid synthetic inputs
- Integrate grazing animals and their urine and dung into soil management systems.

Soil and water management

The first step is to appreciate the interconnection between the soil and water systems, and to acknowledge that to conserve and manage soil, it is vital to manage the flow of water across and within the soil. Conversely, to increase the amount of water in the soil, rivers, and streams, it is important to reduce surface run-off of water, increase infiltration and improve the soil's water storage capacity. Water-harvesting and soil erosion reduction measures are vital; but rather than focussing on individual farmers, these methods are best introduced at the landscape level using participatory processes with whole communities.

An example of this is the Whole Landscapes, Whole Communities approach (promoted by the Seeds and Knowledge Initiative in Southern Africa),[2] a multi-stakeholder participatory process to assess and map problems along water catchments. Community representatives develop plans to improve catchment management through water harvesting, soil protection and improved crop and grazing measures. These are agreed by the whole community before implementation; they are then monitored and reviewed through regular discussions during community gatherings. Typical soil protection measures include revegetation of barren areas with grasses and trees, constructing gabions and stone bunds to reduce run-off of water, and to sink water into the soil. Typical water-harvesting measures include construction of

[2] https://www.seedandknowledge.org/wp-content/uploads/2021/08/BFG-Landscape-Cover-and-intro-pages-lores.pdf.

dead-level contours (swales) on slopes, half-moon ridges around trees and shrubs, and diversion drains and pits to harvest water off roads, rocks, and other hard surfaces. Construction of sand dams further increases the concentration of usable water along catchments by creating barriers across waterways to trap sand and silt. These areas become saturated with water, and protect it from pollution and evaporation. The water can be accessed using simple pumps for both agriculture and domestic use.

These landscape level measures can be complimented with in-field measures such as tied ridging, development of crop planting basins, infiltration pits and trenches, as well as application of mulch to cover the bare soil around crops. A whole landscape approach to soil and water conservation has been used successfully for decades in semi-arid areas of Zimbabwe by Dabane Trust.[3]

Planting drought-resistant crops is a crucial but much-neglected practice in many parts of dryland Africa (places that have short rainy seasons and low annual rainfall). For example, maize is widely grown in Southern and East Africa, often in places where rainfall and soil type are inadequate to support the crop. This leads to low crop yields or crop failure and food insecurity. The choice to plant maize over drought-resistant indigenous grains is driven by changing food preferences, lack of markets for indigenous grains (including millets and sorghum) and the fact that they are labour-intensive to produce and process. In Niger farmers are saving the seed and promoting a drought-tolerant, traditional variety of sorghum.[4] Meanwhile in Mutomo district in eastern Kenya farmers, are replacing maize with drought tolerant sorghum, millet, cowpeas and cassava.[5]

Biodiverse cropping

Growing diverse communities of crops was an indigenous practice in Africa, but it was discouraged by colonial and subsequent agricultural extension services. However, it has many benefits for African farmers. In many parts of pre-colonial Africa, it was common to clear only certain trees from land intended for cultivation. A proportion of trees was left to provide fruit, leaf litter, livestock fodder and shade. Fields were then sown with a variety of cereals, legumes, cucurbits, and vegetables. This system has been replaced with monocultures of cereals and other cash crops in many parts of the continent, leading to environmental degradation, erosion of diverse traditional diets and associated nutrition and health problems.

Biodiverse cropping (growing many different crops together) benefits the soil by boosting microbial communities through root exudates produced by plants. Leguminous crops, for example, benefit the soil and other crops grown with them through their associations with nitrogen-fixing bacteria. Growing many different crops together reduces the multiplication of pests and diseases as some crops have pest repellent properties. Others can be grown for livestock feed or slashed and left

---

[3] Impact of Integrated Water Resources Management (IWRM) in Gwanda District. Dabane Trust. https://www.youtube.com/watch?v=vJ3usSy-1z0.

[4] https://afsafrica.org/wp-content/uploads/2023/12/niger-sorghum-en.pdf.

[5] Finding a way out of the maize. In Agroecology the Bold Future of Farming in Africa (p 70) https://www.fao.org/family-farming/detail/en/c/853551/.

to decay on the soil surface as green manure. Growing crops of different heights saves space and low-growing cover crops help to smother weeds and keep the soil moist. Having many crops in fields provides farmer households with a diverse range of food and improves the ability of farming systems to cope with shocks such as drought, floods or pest and disease attacks. Diverse crops also provide an insurance policy against price fluctuations in local, national, and international markets that threaten cash crop farmers. In Emuhaya, Western Kenya, Bio Gardening Innovations (BIOGI) is promoting food forests, multilayered agroecosystems incorporate trees, shrubs, and annual crops.[6] Food forests have helped local communities diversify their diets and incomes while improving soil and water resources and pest and disease management.

Agroforestry involves incorporating perennial trees and shrubs into annual cropping systems and livestock grazing systems. Tree roots stabilize soil, while the trees provide shade and shelter to crops and livestock. Leguminous trees improve the nitrogen content of the soil. Trees also provide leaf material for compost, mulch and livestock fodder, as well as wood for fuel and timber. Trees also help address climate change by removing carbon dioxide from the atmosphere. Farmer Managed Natural Regeneration is an agroforestry approach that is gaining attraction in the Sahel, East and Southern Africa where land degradation and climate change have caused massive declines in crop yields. In this method, as in the indigenous farming system, certain trees are allowed by farmers to grow in crop fields and grazing areas. The trees are often pruned to prevent competition with crops but with their deep roots, the trees improve the water content of the soil by bringing water closer to the surface allowing surrounding crops to benefit from the extra moisture. Studies (Westerberg et al. 2019 and Bogie et al. 2018) have shown that crops grown with trees in this way in arid environments have better yields, and are able to withstand dry conditions better than crops grown without trees. In Ahferom district of Tigray State, Northern Ethiopia, Mekelle University College of Dryland Agriculture and Natural Resources is working with farmers to develop silvopastures, incorporating *Shibaka* trees (*Ficus thonningii-ahferom*) into livestock grazing areas. The leaves and branches of this drought tolerant tree can be harvested and fed to the livestock while the presence of the trees reduces soil erosion and improves water infiltration into the soil.[7]

Alternatives to synthetic inputs

As we have already discussed, synthetic fertilisers and pesticides cause major damage to microbial communities and other important fauna in the soil. Phasing out synthetic inputs improves soil texture and fertility, farmers' incomes and the natural resource base as well as reducing risks to human health.

Organic fertilisers: A wide range of organic methods, using locally available materials, can be promoted to help farmers improve long-term soil fertility. These include

---

[6] https://afsafrica.org/wp-content/uploads/2021/02/biogi_compressed.pdf.

[7] https://afsafrica.org/wp-content/uploads/2019/07/ficus-thonningi-eng-online.pdf.

applying livestock manure applications, compost and biofertilisers (fermented preparations that create a beneficial environment for soil microbes to improve soil fertility and plant growth). Compost has been widely promoted with successful results by women's associations in Burkina Faso.[8] Applying mulch in the form of crop residues, leaves, grass and other biodegradable materials is another organic method that improves soil fertility. It helps to suppress weeds and conserves soil moisture and has been promoted with success in Zimbabwe.[9]

Factors that inhibit their uptake include the fact that farmers find them labour-intensive, and some lack the materials for mulch or organic fertiliser preparation. Promoters are trying to address these issues by encouraging farmers to form groups for performing labour-intensive tasks, growing crops and trees specifically for the production of mulch material and organic fertilisers, and harvesting weeds and invasive plants to turn them into organic fertilisers. Some biofertilisers require ingredients that need to be purchased, such as yeast and molasses. To address the cost of these items, farmers in some countries have pooled resources to purchase the ingredients in bulk. Others have used a portion of the income from increased crop sales resulting from the use of biofertilisers. In some countries, farmers have begun to manufacture and sell biofertilisers as a business. Biofertilisers have been promoted with smallholder farmer groups in Zambia, Malawi, Zimbabwe, South Africa, Uganda, Kenya and Togo.[10]

Integrated pest and disease management: In terms of phasing out pesticides, the integrated pest and disease management approach has been widely promoted in agroecology. This involves the use of biological, mechanical, and physical methods, as well as improved crop management and the use of biopesticides made from local plant extracts. Some of these methods are based on traditional pest management methods, such as the use of ash and indigenous plant extracts as pesticides. The effectiveness of some biopesticides has been investigated by international and African academic institutions in Tanzania and Malawi partnership with farmers through farmer-led research (Belmain et al. 2022).

4.3 Livestock Production Methods for Better Soil and Human Health

Livestock are a crucial part of the farming system in all African countries and play a particularly important role in areas that are too arid for crop production. Hardy indigenous breeds of cattle, goats, sheep, pigs, poultry, and other livestock have developed in Africa over the millennia, able to survive on natural vegetation, and agricultural

[8] https://www.oaklandinstitute.org/sites/oaklandinstitute.org/files/compost_burkina_faso.pdf.

[9] https://www.oaklandinstitute.org/sites/oaklandinstitute.org/files/conservation_agriculture_zimbabwe.pdf.

[10] https://www.seedandknowledge.org/the-biofertilizer-story-2/; https://afsafrica.org/are-biofertilizers-a-part-of-the-solution-to-africas-concerns-with-soil-health-and-the-environment/.

and kitchen waste, to produce food and income for many rural African families. Livestock also represent an important means of storing wealth and an economic safety net that can be tapped into (slaughtered or sold) during times of hardship. In the past, most large livestock were grazed communally with traditional grazing systems involving rotating and resting areas to allow for recovery of pasture and veld. Encroaching settlements and cropping areas, land degradation and climate change have put considerable pressure on these traditional systems; and in many places, they no longer exist. This has led to increased decline of soil water and vegetation resources, reduced livestock productivity, increased pest and disease problems, conflicts between different land users and threats to livestock from wildlife.

One of the most promising agroecological methods promoted to address this is Holistic Land and Livestock Management. The framework was developed by Zimbabwean-born ecologist Allan Savory in the 1980s, and has expanded into a global movement including successful promotion in in Zimbabwe, Namibia, Botswana, South Africa, Zambia, Mozambique and Kenya (AFSA/SKI 2021). The foundational idea behind the approach is that livestock, if managed correctly, can play a vital role in regenerating degraded land. Through a participatory process, communities are brought together to develop agreed plans on how and where livestock will be grazed. Ideally, the communities' livestock are combined into one large herd that is directed to graze certain areas for a specific length of time before being moved to the next area. The impact of the animals' hooves and grazing breaks up the surface of the soil, thus encouraging water infiltration and plant growth. Also, the application of dung and urine encourages beneficial soil microbes to improve soil health. This practice regenerates grazing areas and encourages the growth of diverse nutritious vegetation.

In crop areas, movable livestock enclosures can be used to prepare the land during the dry season. This system is being tried with small poultry, pigs, sheep, and goats as well as cattle in various countries. The enclosed animals eat weeds and crop residues, apply manure and urine, and prepare the soil for crop planting, making ploughing unnecessary. Pigs and poultry have an added advantage as they eat insect pests during their time in an area. Once the land in an area has been cleared and prepared, the enclosure is moved to a new area and the prepared area is planted with crops. Uptake of this method has been hampered by resistance from many communities to herd their livestock communally. Lack of material for movable fencing is yet another barrier to the uptake of the movable enclosures system. Despite these barriers, the technique has significantly transformed landscapes in areas where it has been adopted and has increased the productivity of crops and livestock.

5 Conclusions and Recommendations

This chapter looked at the links between soil health, human health, and nutrition. It explained how improving soil management can lead to improved production of nutritious food while nurturing the human gut microbiome, both of which are vital for our health and well-being. While it has been long understood that soil health is important to the production of nutritious food, the critical role of microorganisms in the maintenance of soil fertility, structure and plant health has only recently been realized. Poor management of Africa's vulnerable soils has led to widespread land degradation, food insecurity, poverty, and vulnerability across the continent. To develop more sustainable food systems and address the malnutrition problems facing the continent, it is vital that development agents embark on a continent-wide effort to improve soils using agroecological methods. These methods can be used in both crop and livestock production systems. Although they have been tried in many African contexts, the uptake of these promising methods remains slow. More focus needs to be put into identifying and addressing the barriers and finding ways to motivate farmers and communities to adopt agroecological approaches.

References

AFSA/SKI (2021) Healthy soil for healthy communities: an introduction to soil health practices for Africa. https://www.seedandknowledge.org/publication-announcement-healthy-soil-for-healthy-communities-an-introduction-to-soil-health-practices-for-africa/

Belmain SR, Tembo Y, Mkindi A, Arnold SEJ, Stevenson P (2022) Elements of agroecological pest and disease management. Elementa Sci anthropocene 10:1. https://doi.org/10.1525/elementa.2021.00099. Elements of agroecological pest and disease management | Elementa: Science of the Anthropocene | University of California Press (ucpress.edu)

Blum WEH, Zechmeister-Boltenstern S, Keiblinger KM (2019) Does soil contribute to the human gut microbiome? Microorganisms 7(9):287. PMID: 31450753; PMCID: PMC6780873. https://doi.org/10.3390/microorganisms7090287

Bogie NA, Bayala R, Ibrahima, D. Conklin MH, Fogel ML, Dick RP, Ghezzehei TA (2018) Hydraulic redistribution by native Sahelian shrubs: bioirrigation to resist in-season drought. Front Environ Sci 6:18. https://doi.org/10.3389/fenvs.2018.000

Brevik EC, Slaughter L, Singh BR, Steffan JJ, Collier D, Barnhart P, Pereira P (2020) Soil and human health: current status and future needs. Air Soil Water Res 13:1–23

Crittenden A, Schnorr SL (2017) Current views on hunter-gatherer nutrition and the evolution of the human diet. Am J Phys Anthropol 162:84–109

ELD Initiative and UNEP (2015) The economics of land degradation in Africa: benefits of action outweigh the costs. Available from www.eld-initiative.org

FAO, IFAD, UNICEF, WFP and WHO (2023) The state of food security and nutrition in the world 2023. In: Urbanization, agrifood systems transformation and healthy diets across the rural–urban continuum. Rome, FAO. https://doi.org/10.4060/cc3017en

Glick BR (2018) Soil microbes and sustainable agriculture. Pedosphere 28(2):1–201. https://doi.org/10.1016/S1002-0160(15)60020-7. ISSN 1002-0160/CN 32-1315/P

Hirt H (2020) Healthy soils for healthy plants for healthy people: How beneficial microbes in the soil, food and gut are interconnected and how agriculture can contribute to human health. Science and Society. EMBO Rep 21:e51069. https://doi.org/10.15252/embr.202051069

Jones A, Breuning-Madsen H, Brossard M, Dampha A, Deckers J, Dewitte O, Gallali T, Hallett S, Jones R, Kilasara M, Le Roux P, Micheli E, Montanarella L, Spaargaren O, Thiombiano L, van Ranst E, Yemefack M, Zougmoré R (eds) (2013) Soil Atlas of Africa. European Commission, Publications Office of the European Union, Luxembourg, pp 176. https://esdac.jrc.ec.europa.eu/content/soil-map-soil-atlas-africa

Steiner K, Bwalya M (2003) The African conservation tillage network: enhancing conservation tillage in Africa. In: García-Torres L, Benites J, Martínez-Vilela A, Holgado-Cabrera A (eds) Conservation agriculture. Springer, Dordrecht. https://doi.org/10.1007/978-94-017-1143-2_14. https://link.springer.com/chapter/10.1007/978-94-017-1143-2_14#citeas

Westerberg V, Doku A, Damnyag L, Kranjac-Berisavljevic G, Owusu S, Jasaw G, Di Falco S (2019) Reversing land degradation in dry lands: the case for farmer managed natural regeneration (FMNR) in the upper west region of Ghana. https://www.eld-initiative.org/fileadmin/user_upload/ELD-Ghana-Report-final-240120.pdf

Wrangham R (2017) Control of fire in the Paleolithic. Evaluating the cooking hypothesis. Curr Anthropol 58(S16). https://doi.org/10.1086/692113

Open Access This chapter is licensed under the terms of the Creative Commons Attribution 4.0 International License (http://creativecommons.org/licenses/by/4.0/), which permits use, sharing, adaptation, distribution and reproduction in any medium or format, as long as you give appropriate credit to the original author(s) and the source, provide a link to the Creative Commons license and indicate if changes were made.

The images or other third party material in this chapter are included in the chapter's Creative Commons license, unless indicated otherwise in a credit line to the material. If material is not included in the chapter's Creative Commons license and your intended use is not permitted by statutory regulation or exceeds the permitted use, you will need to obtain permission directly from the copyright holder.

Soils, Climate Change and Carbon Sequestration

Climate Change Adaptability and Soil Carbon Sequestration

Hamisi Juma Tindwa and Ernest William Semu

Abstract There exists a continuous cross-talk between climate change and soil carbon sequestration processes, and how the latter affects communities' capability to adapt to the effects of a changing climate. While technology-assisted anthropogenic activities help to improve the production of material goods and the general quality of life, they have been linked to elevated production of Green House Gases (GHGs) and their ultimate emission into the atmosphere. Scientific reports show, for example, that before the climate changed, temperatures were relatively lower and optimal for plant growth. Rainfall levels and distribution were also favourable, and consequently the forces that replenish soil nutrients and soil carbon from atmospheric stocks were in check. As climate change progressively intensified, temperatures and rainfall patterns changed for the worse, the quality of soil deteriorated and agricultural productivity decreased. Climate change has disrupted the balance of the carbon concentrations between the soil and the atmosphere as the two largest carbon pools in the universe. Atmospheric CO_2 concentrations have risen from approximately 280 parts per million (ppm) prior to 1850, to 414.72 ppm in 2021. Communities, therefore, need to adapt to the effects of the changing climate by embracing efforts that lead to a reversal of the deteriorating trend in soil degradation and carbon balance between the soil and the atmosphere. In this chapter, we highlight the factors that enhance communities' and agricultural systems' adaptability to climate by promoting efforts to make use of soil carbon sequestration. We suggest that to be able to replenish SOC stocks in agro-ecosystems, there is a need to adopt a combination of practices that help to capture and store C in plants and soils, and those that enhance carbon stabilization in soils, hence effectively reducing the GHG emissions into the atmosphere.

Keywords Soil carbon sequestration · Green house gases · Climate change · Food security · Adaptability to climate change

H. J. Tindwa (✉) · E. W. Semu
Department of Soil and Geological Sciences, College of Agriculture, Sokoine University of Agriculture, Morogoro, Tanzania
e-mail: tindwa.hamis@sua.ac.tz

© The Author(s) 2026

B. R. Singh et al. (eds.), *Sustainable Food Systems in Sub-Saharan Africa*,
https://doi.org/10.1007/978-3-032-11288-0_12

1 Introduction

Charles Dickens, in the novel "*A Tale of Two Cities*", suggests that there are times when human-kind may be viewed as living in "the best of times" but also "in the worst of times" at the same time. The best of times in this particular case, are due to what may be called the scientific-technological-economic boom. Following the onset of the industrial revolution, the world has witnessed an unprecedented expansion in the scope and depth of science that led to the creation of a myriad of technologies which, in turn, resulted in a high level of economic growth and improved human welfare (Kniivilä 2007). The downside of this boom was the onset and intensification of climate change, along with its many ensuing palpable and negative effects on the environment and human livelihoods. The negative effects include higher temperatures, increases in precipitation patterns, rising sea levels, typhoons, increased drought and decreased food security. Thus, the downside of the boom, experienced together with the benefits, may also be viewed as the worst of times.

As climate change evolved, agriculture also advanced from that based on stone tools, to animal-drawn implements and later onto highly mechanized and computerized, precision agricultural methods. Consequently, economies transformed from agrarian to technologically advanced, attended by mass-production of goods and services. This transformation happened in congruency with the increasing intensity of land degradation and rising levels of Green House Gases (GHGs) emissions that affected and changed climate negatively. As early as 1912, for example, the world burned about 2 billion tons of coal annually (Anonymous 1912) emitting, in turn, about 7 billion tons of CO_2 gas into the atmosphere. Over a period of about 1.1 centuries, the global average temperature has increased by nearly 1.5 °C, mainly attributable to GHGs emissions. Estimates show that if present trends of GHGs emissions continue, the global temperature will rise by as much as 2.5–3 °C by the end of this century (Mauritsen and Pincus 2017; Raftery et al. 2017). The rising temperatures have already resulted in accelerated melting of glaciers in the Polar Regions (Bamber et al. 2019; Holland et al. 2020; Braddock et al. 2022) and increased frequency of heat waves, which have claimed human and animal lives (Azhar et al. 2014; Witze 2022; Zachariah et al. 2022). Extremely low and high rainfall events have been recorded (Fowler and Ali 2022), leading to decreased or total failure in crop production (Li et al. 2019; Brás et al. 2021; Schmitt et al. 2022). Similarly, over the last 35 years, severities of typhoons and tornadoes (Fig. 1) and associated risks to human livelihoods and environment, have progressively increased in many areas (Kossin et al. 2020; Bhatia et al. 2022) all attributable to climate change with a positive contribution from anthropogenic forcing (Table 1).

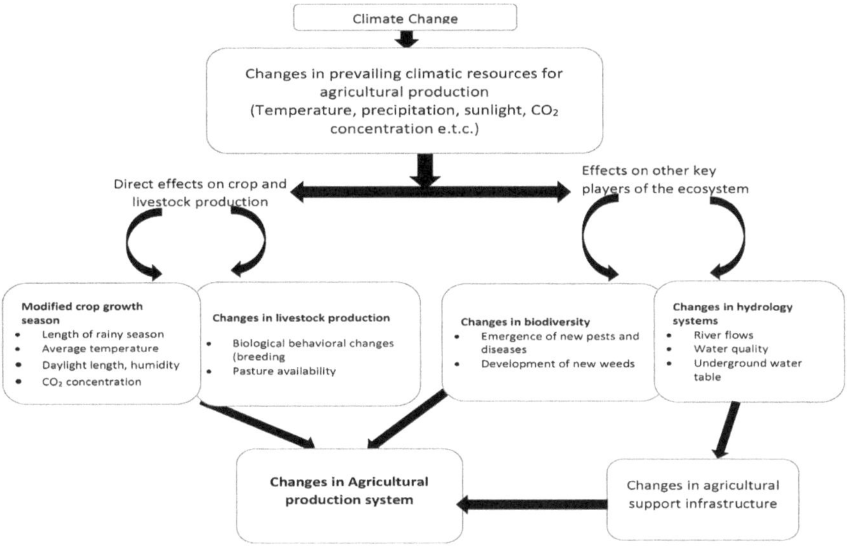

Fig. 1 Flow of climate change effects on agricultural production

2 Effects of Climate Change on the Environment and Agriculture

Climate change has traditionally been tracked using a couple of parameters including temperature, precipitation and wind patterns. This is because the magnitudes and trends of these components of the climate do change over time.

Variation in temperature due to climate change: While the Earth has warmed by approximately 0.8 °C since the late nineteenth century, the Arctic and Antarctic poles have experienced differential patterns of warming in the same period. In the last two decades, the Arctic has warmed by about 3° while the Antarctic has shown no obvious upward trend in temperature (Post et al. 2019). Continued polar warming has resulted in melting land ice and a rise in sea levels (Church et al. 2013). The melting of polar ice can result in the submergence of some islands and coastal strips in some countries. Records show for example, that five tiny Pacific islands have completely disappeared due to rising seas and erosion (Albert et al. 2016), and coasts in the Maghreb have disappeared due to coastal erosion (World Bank 2021). Rises in temperature are also associated with increased frequency of droughts. More frequent and longer-lasting droughts are a direct result of rising global temperatures. Research shows that even with a minimum of 1.5 °C warming scenario, the drought probability is projected to triple in some countries such as Brazil and China, nearly double in Ethiopia and Ghana, increase slightly in India and substantially increase in Egypt (Price et al. 2022).

Variation in global rainfall events due to climate change: Extreme precipitation events have become more frequent now than back in time. While some places

Table 1 Selected events of intense typhoons and tornadoes in the world

	Typhoon or tornado event	Place and time of occurrence	Detail of destruction of life and property	Reference(s)
1	Daulatpur–Saturia tornado	Manikganj District, Bangladesh, April 26, 1989	Holds the record of deadliest tornado in history, killed around 1300 people and over 12,000 injuries	Johnson (2021)
2	2011 super outbreak tornadoes	April 25–28, 2011 spread in over 20 states including Alabama	Most devastating tornado outbreaks in U.S. history. It produced a total of 362 tornadoes that killed at least 324 people with thousands of injuries and over $12 billion of property damage	Matthews and Sanchez (2022)
3	Typhoon Mangkhut	6 September 2018, Hong Kong, China	It is the fiercest typhoon on record to hit Hong Kong. It injured around 500 people, caused blackout and widespread damage to property	Choy et al. (2020)
4	Typhoon Morakot	August 2009, Taiwan	Caused record breaking rainfall in Taiwan leading to 677 deaths mainly from landslides and agricultural losses amounting to almost US$300 million	Wu (2013), Lo and Fan (2022), Tu and Chen (2017)
5	Typhoon Maemi	6 September 2003, South Korea	With sustained winds of up to of 195 km/h, Maemi killed at least 117 people, destroyed 5000 homes and caused economic losses of up to US$4.8 billion	Kim et al. (2019)
6	Typhoon Hagibis	October, 2019	Japan's most devastating typhoon on record claiming 98 lives and costing the country at least US$15 billion. Violence from it caused cancellation of several Rugby work cup tournaments	The Japan Times (2019)

in the world experience below-average rainfall, others receive excessive rainfall events (Mirhosseini et al. 2012). Rainfall events have changed in terms of intensity, frequency and seasonality (Mirhosseini et al. 2012; Dourte et al. 2015). High-intensity rain events due to changing climate are becoming a common reality in many parts of the world (Higgins and Kousky 2013; Dourte et al. 2015; Blenkinsop et al. 2021).

Global warming is associated with an increase in precipitation in northern Europe, Canada, and northern Russia, but less in swaths of sub-Saharan Africa, southern India, and southeast Asia (WMO 2020; Mamalakis et al. 2021). In some parts of the world such as the boreal high latitudes, precipitation seasons have become longer in the last 50 years due to an earlier onset of spring (Trenberth et al. 2007; Trenberth 2011).

Variation in wind and wind patterns due to climate change: Winds and wind patterns, which have a bearing on rains, have also changed in strength and speed to the point of resulting in strong tornadoes, hurricanes and typhoons that have ravaged many countries. Overall, studies suggest that with the changing climate, there is a parallel change in persistent wind patterns, which in turn can lead to large changes in net precipitation (Wills et al. 2016). Westerlies or westerly winds, which typically blow from west to east across the planet's middle latitudes, are changing, migrating poleward mainly due to climate change (Abell et al. 2021). The global average wind speed has been increasing since about 2010 at a faster rate from about 7 to 7.4 mph, especially in North America, Europe and Asia (Zeng et al. 2019). This reversal in trend happened after a period of steady decrease in global wind speed that spanned from 1970 to around 2010 (Zeng et al. 2019). Over that period, surface wind speeds over all the continental areas of the northern mid-latitudes had declined by 5–15% and strong winds had slowed faster than weak winds (Xu et al. 2006; Vautard et al. 2010). Slowing of wind speeds or atmospheric stilling- as it is known, is partly due to increased surface roughness which in turn, is a result of increases in biomass and land-use change across the globe.

3 Measures to Ensure Adaptability and Resilience to Climate Change

Climate change is now an unpleasant fact. Therefore, the world needs to operate in ways that ensure an acceptable level of biological productivity to sustain livelihoods even in the presence of this current and continuing change of climate. Before climate changed (Abram et al. 2016), temperatures were relatively lower and optimal for plant growth; rainfall levels and distribution were likewise favourable. As climate change progressively intensified, temperatures and rainfall patterns changed for the worse (Abram et al. 2016; Price et al. 2022), and agricultural productivity correspondingly decreased (Nelson et al. 2009). Scientific projections show that, compared to a no-climate-change future, an additional 78 million people, half of them in Africa south of the Sahara, will face chronic hunger by 2050 (Sulser et al. 2021). Overall, even in this era of climate change, livelihoods have to be sustained. To that end, there are a variety of actions that can be adopted to lessen the effects or impacts of climate change on agricultural productivity. Some of the most noteworthy of these are explained here:

3.1 Improve Weather Forecasting to Know the Length of Rains for Crop Choice

In many developing countries, weather forecasting is still at rudimentary levels in comparison to those in developed countries. In many of these countries, including those in Sub-Saharan Africa (SSA), farmers' crop yields have been drastically reduced, or sometimes complete crop failures have happened, due to absence of accurate forecasts to guide agricultural operations (Caparas et al. 2021). Crop failures have been associated with extreme weather events (Li et al. 2019; Beillouin et al.

2020; Caparas et al. 2021; Schmitt et al. 2022), many of which were not forecasted prior to their occurrence, and thus deterrence and avoidance measures could not be taken beforehand. Climate forecast data can help farmers make climate-sensitive decisions which may include time for planting, the selection of late- and/or early-maturing seed varieties, and crop mixtures that suit the forecast scenarios (Guido et al. 2020). However, most of the developing world has a shortage of forecast capability to inform crop-related decisions (Guido et al. 2020). Therefore, there is an urgent need to improve weather and climate forecasting to aid farming decisions, which can be achieved through localized seasonal and sub-seasonal weather forecasts.

3.2 Adopting Climate-Resilient Crops/Varieties

In the era of climate change that has resulted in a shortening of the crop growth season, farmers should embrace early-maturing crop varieties that will fit into the shortened season. Some levels of yields may be compromised, but this is not the same as losing an entire crop. In these scenarios, there is a need for climate adaptation strategies, which would help farmers adjust to using the right crop varieties, and ultimately minimize the climate risk (van Ettena et al. 2019). Sometimes, this is coupled with the concept of Crop adaptation—a practice of switching to the use of more climate-resilient crop species or varieties. Such varieties may include those with traits for early maturing, heat tolerance, drought tolerance or enhanced salinity tolerance (Westengen and Brysting 2014). Where indigenous varieties exist, they should be encouraged, as they will have developed resilience to climate change within those locations (FAO 2007). Otherwise, crop-breeding efforts should be mounted or expanded to acquire early-maturing ones suited to a particular area. If successfully adopted, all the practices mentioned above can boost harvests and, therefore, help farmers to cope with the negative effects of climate change on agricultural production (Acevedo et al. 2020). Where drought is one of serious problems brought about by climate change, the proposed strategy may be to broaden the range of crops cultivated to include drought-tolerant varieties. This is mainly because, in some scenarios, especially where there is widespread use of highly bred varieties, the latter may be more vulnerable to climate change, and yields may decline substantially (Kusmec et al. 2023), or a complete crop loss experienced. Highly bred varieties may have been bred to accommodate a narrow band of suitable weather conditions outside of which their performance is poor.

3.3 Water Harvesting and Soil Moisture Conservation

Climate change is associated with increased frequency of extreme rainy events in the world. Water may be too much in some areas during the rainy seasons, and the opposite has been experienced in other regions resulting in failures of crops to mature. In such circumstances, supplemental irrigation could carry the crops through. However, in most developing countries, such as those of the SSA region, water harvesting is generally not practiced, and thus there is no water available for irrigation once the rainy season has ended. There are no dams/reservoirs to impound water, and rivers dry up within a short period of time after the rains. In most of these countries, water can be harvested on a small scale using ponds or using other in-field techniques of rainwater harvesting (Ndlovu et al. 2020). Appropriate technologies may include dead level contours, ephemeral stream diversion, ridges/furrows and hillside sheet runoff (Ndlovu et al. 2020) or *fanya juu* and *fanya chini* contour systems of farming (Mwanyoka and Lopa 2016). Additionally, suitable agronomic techniques need to be brought to bear on the process, for example, mulching, and turning under crop residues after harvest. It is known that crop residues are oftentimes subjected to other uses, e.g. grazing of livestock. These practices should be discouraged for the sake of improving crop production soils in terms of soil moisture conservation.

At a higher level is the construction of dams—both small and large dams. These are costly, but necessary undertakings, especially now in the era of climate change. With big dams, large-scale irrigation could be undertaken, although this comes with additional costs. The possibility of multiple damming of a given river should be explored. Such dams would then impound large quantities of water to cater for the different needs.

3.4 *Afforestation/Reforestation*

Well conserved forested lands conserve moisture in their soils and act as sources of the water that flows in rivers and streams. They enable slow release of rainwater, making the rivers flow throughout the year without drying. This is the opposite of bare lands whereby rainwater runs on the surface and is eventually lost.

Therefore, afforestation of bare lands and reforestation of cleared forest lands need to be steadily undertaken so that over time the new forests can fulfil this role of water conservation. In Tanzania, for examples, in the 1970s the government undertook the afforestation of 135,903 ha of bare land to result in the Sao Hill Forest Plantation in the Iringa region (Kangalawe 2019). The microclimate in this forest has improved compared to the surrounding non-forested lands. It is a perennial water source for surrounding areas. Similarly, in Egypt, the government via the Egyptian Environmental Affairs Agency is overseeing the execution of afforestation projects using sewage water for irrigation of the trees. One such project is the Serapium forest project in which land as big as 200 ha of desert has been turned into a forest using sewage water for irrigation (FAO 2012). Much more can be done in non-desert countries.

3.5 *Controlling Industrial Gaseous Emissions*

Climate change has been linked to the presence in the atmosphere of high levels of greenhouse gases (GHGs), mainly carbon dioxide (CO_2) and methane (CH_4) among others, as a result of emissions from industrial activities and products thereof. The main emitters of these gases are the industrialized countries of the world. Once emitted, the GHGs are not confined to the atmospheres of emitting countries, they spread globally due to atmospheric transport of such gases over long distances. GHGs emitted at one location can, therefore, cause climate change in receiving locations far away. It is for this reason for example, that glaciers in the north and south poles are melting as a result of climate change, despite the fact that the poles are not the source of GHGs. Technologies for reducing gaseous emissions are available but expensive. This may probably account for the reluctance encountered in industrial countries to substantially reduce emissions. However, this is a necessary and indispensable option for environmental protection and reversing climate change in the long term.

3.6 *Strengthening Environmental/Weather Institutions, Legislation and Supervision*

Most countries of the world have in place comprehensive laws and policies that are meant to put legal limits on gaseous emissions from industrial operations. An example is laws that are in existence in the United States and in Europe (UK Climate Change Act 2008). The problem lies in non-compliance with those laws. This is because compliance to reduce gaseous emissions requires enforcement of legal penalties and monitoring through the installation of pollution control equipment—which is expensive and eats into profits. The industry has for long been unwilling to comprehensively engage in compliance because of the profit factor. In these cases, governments should continue to develop ways, including legislation, that are informed by sound research, education, and public opinion and advocacy, to improve compliance. Where legislation on a particular aspect is weak or lacking, this should be addressed by enacting new legislation. After enacting the requisite laws/by-laws, compliance should be ensured through effective supervision.

4 Climate Change and Effect on Global Food Security

4.1 *Current and Future Climate Change Scenarios*

Climate change is a threat to both current and future global security and livelihood. Recent scientific evidence suggests that climate change invariably leads to consequences that trigger three major issues—alternate drought and flooding, diseases and famine. Flooding episodes result into unprecedented scales of mass migration into areas of already high tension as well as significant crop-failure and/or losses, leading to intense competition for food, water and energy. On the other hand, changing climate can lead to increased global warming or increased rainfall episodes, both of which have the tendency of enabling certain disease vectors like mosquitoes, ticks and birds to expand their geographical range, putting human populations previously outside the range into risk of contracting the diseases they spread. Furthermore, a recent study has reported that up to 58% of infectious diseases that affect humanity are aggravated by climate change through increased greenhouse house gas emission-related global warming (Mora et al. 2022).

Climate science shows that there is a steady increase in extreme weather events attributable to changing climate. Examples include record-breaking episodes of Heatwaves in Europe, colossal floods in Pakistan and Prolonged and severe droughts in parts of China, the Horn of Africa and the United States (WMO 2022). Data further shows that currently, emissions of GHGs including carbon dioxide (CO_2), methane (CH_4) and nitrous oxide (N_2O) are on the rise. Similarly, the WMO has reported that the most recent seven years of 2015–2021 were the warmest on record with a global mean temperature average (based on data of up to June 2022) for the period 2018–2021 standing at a record of 1.17 ± 0.13 °C above that of the period from 1850 to 1900. Since around 90% of the accumulated heat in the earth's systems is stored in the oceans, the Ocean Heat Content for the 2018–2022 period was higher than in any other 5-year period in history (UNFCCC 2022).

4.2 *Effect of Climate Change on Agricultural Production and Food Security*

Climate change increases the frequency and intensity of extreme events like droughts, flooding and storms. This makes it more difficult to grow crops or raise animals in ways and places that communities used to do before changes in climatic conditions. Conceptually, the effects flow through a number of contributing sectors as described in Fig. 1.

Increases in the frequency and severity of droughts and floods, for example, can harm crops and reduce yields due to direct interference in the crops' biochemical processes. Reduction in water supply, and hence the inability to suffice crop requirements affecting yield, is a direct consequence of climate change-induced droughts. Reports show that the increased frequency of drought episodes does significantly reduce renewable surface—and groundwater—in most subtropical regions (Jiménez Cisneros et al. 2014), leading to crop failures. Increased precipitation frequency and intensity because of climate change may lead to flooding, which in turn may create waterlogging conditions, and also lead to catastrophic soil erosion and associated loss of property.

Similarly, changes in average temperature can result in negative or positive consequences on crop and livestock production, depending on the direction of change. Rises in global temperature due to global warming will likely increase growth rates and productivity of particular crops; at the same time, it will reduce the productivity of other crops that prefer low temperature for their growth and reproduction. Such changes in temperature can also lead to the emergence of new pest and weed species due to shifts in their respective temperature optima for growth and reproduction. On the other extreme, rising global temperatures may cause the melting of polar ice reserves and lead to rising sea levels and floods. This may in turn cause the displacement of brackish and freshwaters in river deltas, wiping out some coastline agriculture and aquaculture practices and destruction of wetlands (FAO 2015).

Carbon dioxide additions into the atmosphere can occur due to natural factors, such as the respiration by organisms or decomposition (decay) of organic matter, weathering of carbonate rocks, occurrence of forest fires, and eruption of volcanoes. These factors can result in increased annual CO_2 flux to the atmosphere. Such increases in atmospheric carbon dioxide gas can have both negative and positive consequences on agricultural production and food security. Research has shown that many food crops, when grown under elevated CO_2 concentrations (546–586 ppm) tend to accumulate lesser nutrients, such as iron, zinc, and protein, which in turn may be directly linked to the risk of hidden hunger to consumers (Myers et al. 2014). Furthermore, C3 grains and legumes are sensitive to CO_2 concentrations, and thus have lower concentrations of zinc and iron when grown under field conditions at elevated atmospheric CO_2 concentrations, whereas C4 crops seem to be less affected.

5 Soil Carbon Sequestration and Adaptability to Climate Change

Carbon sequestration is a process of capturing and storing atmospheric carbon dioxide. It is the uptake and conversion of atmospheric CO_2 to organic matter through photosynthesis by plants, and the subsequent transfer of the organic matter into the soil reservoir for storage in a way that prolongs its mean residence time (Lal et al. 2015; Were et al. 2020). It is one way of reducing atmospheric concentrations of carbon dioxide and hence reducing global warming. Soil carbon sequestration happens when the captured carbon is stored in the soil, usually as organic soil carbon. The soil carbon pool, also referred to in some literature as the soil pedological pool, is divided into two major forms—the soil organic carbon (SOC) and soil inorganic carbon (SIC). The soil holds over 2500 Gigatons of carbon as SOC and SIC combined, which is equal to over 80% of the total carbon held in the world's terrestrial ecosystem (Lal 2008; Ontl and Schulte 2012; Were et al. 2020). SOC, which is approximately 1500–1550 GT, and is derived from organic materials such as leaf litter, branches, plant roots and manure, is mainly held within the upper one meter of the soil profile. Additionally, about 650 GT of carbon forms the biotic pool-carbon held in living plants and animals. On the other hand, about 950–1000 GT of carbon are in the form of SIC and consists mainly of elemental carbon and carbonate materials such as calcite, dolomite, and gypsum (Ontl and Schulte 2012; Were et al. 2020). Exchanges of carbon between the atmosphere and the soil happen either as the microbial community (plants and animals) respire, releasing C in the form of CO_2 to the atmosphere or as the carbon locked up in the dead plants and animal tissues is released into the soil by the action of decomposer microorganisms. Climate change has swayed out of balance the carbon concentrations between these pools. Atmospheric CO_2 concentrations have risen from approximately 280 parts per million (ppm) prior to 1850, to 414.72 ppm in 2021 (Lindsey 2023). A significant proportion of this increase is due to burning of fossil fuels, with the remainder coming from SOC loss due to land use change such as the clearing of forests and the cultivation of land for food production (Ontl and Schulte 2012; Were et al. 2020).

The conversion of natural ecosystems to agricultural use—a key attribute of industrial revolution—has resulted in depletion of SOC levels worldwide. This conversion is also associated with reductions in amounts of plant residues returned to soil, increased decomposition from soil tillage, and increased soil erosion, which together have created a soil carbon deficit (Lemus and Lal 2005; Lal 2009; Ontl and Schulte 2012). This carbon deficit in the soil creates an opportunity to store more carbon captured back from the atmosphere through a variety of land management approaches amounting to soil carbon sequestration. Soil carbon sequestration is realized when a positive carbon balance in soil is attained (Were et al. 2020). Ideally, there are two main groups of techniques through which the positive C balance can be attained as summarized in Fig. 2.

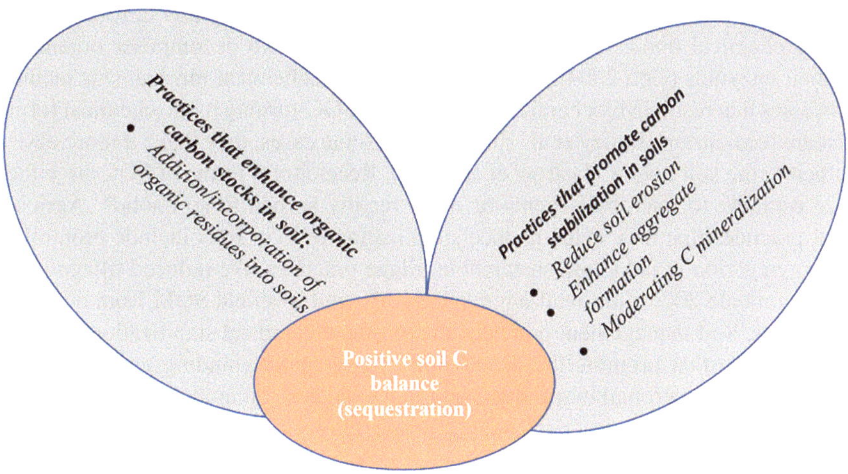

Fig. 2 Main techniques leading to positive soil carbon balance

5.1 Enhancing Soil Carbon Sequestration in Agricultural Soils

(A) **Capitalizing on agricultural practices that increase organic carbon stocks in soils**

The increase in soil organic matter and consequently soil organic carbon (SOC) is directly influenced by the amounts of carbon input into the soil. In such inputs, which mainly come from plant residues, the role of plants in increasing organic carbon stocks in soils is two-fold: (a) plants are the primary and main source of soil organic C through production of litter (shoots and roots), and (b) plants contribute to OM stabilization by producing poorly-degradable compounds and promoting stable aggregate formation (Dignac et al. 2017). Plant species differ in their ability for above-ground biomass production, which in turn may impact how much organic carbon is added into the soil. Generally, crop varieties selected for their abilities to allocate more resources to the harvested parts and increase yield (such as grain) tend to decrease biomass allocation to vegetative parts and roots, hence reducing prospects for more biomass additions to soils (Dignac et al. 2017). This means that when the intention is to promote biomass additions to the soil, then crop varieties that produce more biomass for eventual incorporation into the soil are encouraged. Other practices that help increase OM additions in the soil are summarized in Table 2.

(B) **Capitalizing on processes that promote carbon stabilization in soils**

Carbon stabilization in soils is a consequence of biochemical alteration and physicochemical protection mechanisms. Physicochemical protection mechanisms are those that physically shield the soil carbon from decomposer communities and/or their enzymes. The interactions of SOC with soil mineral matrix results in formation of strong chemical bonds and physical shielding of SOC to decomposer organisms or their enzymes (Goh 2004). On the other hand, biochemical mechanisms include processes that result in biochemical alterations of SOC-turning it into chemical forms that are recalcitrant (Bailey et al. 2019); and in some cases, they make it more easily sorbed to the soil matrix (Jastrow et al. 2007). Recalcitrant forms of SOC are either less available to microbial agents or more readily sorbed to soil solids. Agricultural practices that may help enhance stabilization of C in soils include promoting soil aggregation by adopting sustainable tillage practices like reduced tillage (Raji and Ogunwole 2006); for, well-aggregated soils tend to shield SOM from degradative forces. Soil management practices that promote chemical stabilization of SOM include those that promote the production of aggregating/binding agents such as fungal hyphae, microbial by-products and root exudates (Zotarelli et al. 2005; Singh and Benbi 2021). Such practices include proper choices on prevalent crop/cropping system and cropping intensity.

Table 2 Agricultural practices that enhance organic carbon stocks in soils

Practice	Influence by increased gains in SOM	Influence by decrease in losses of SOM	Reference(s)
Reduced or no tillage	Encourages increased C inputs from weed biomass resulting from a higher weed incidence in reduced tillage	Reduces surface runoff and erosion and encourage more surface residue accumulation	Krauss et al. (2022), Whitehead et al. (2018)
Crop rotations	More litter produced when rotations involve species with high biomass production index which upon decomposition adds SOM to soil	Reduced removal of top soil and thus loss of SOM due to efficient cover by produced biomass More cover of surface soil reducing loss by erosion such as that caused by raindrop impacts, when rotations involve cover cropping	Rigon and Calonego (2020), Land et al. (2017)
Cover cropping	Results into increased production of biomass	–	Moukanni et al. (2022)
Use of mulch and other organic amendments	Increased SOC accumulation due to frequent addition of organic mulch	–	Sun et al. (2021)
Irrigation, liming and fertilization	Increase primary production, input to soil and modify plant rooting	–	Chenu et al. (2019)

5.2 *Major Challenges to Carbon Sequestration in Agricultural Soils*

Estimates show that the nearly 10,000 years of cultivated agriculture is responsible for around 116 gigatons of soil carbon reductions globally (Sanderman et al. 2017). This creates a carbon deficit in the soil, and is thus a convenient overall carbon sink. This deficit, however, can be corrected by carbon sequestration techniques that capture carbon dioxide from the atmosphere, and store it as organic carbon stocks in the soil. Soil organic carbon (SOC) sequestration mainly includes processes of soil organic matter (SOM) addition, stabilization of SOM into recalcitrant SOC pools and conserving SOC pools by minimizing C-loss pathway (Sakar et al. 2020).

Factors affecting these processes in agricultural soils are divided into several categories; and the major ones include social, economic, political, organizational, cultural and biophysical challenges. Studies show, for example, that if the world increases the quantity of carbon contained in the soil by 4% annually, then the annual increase in atmospheric carbon dioxide can be halted, hence reducing the risks of climate change. For this to be practically possible, it must be implemented immediately, and sustained without change for decades, all over the world. This means, political authorities all over the world must show and oversee its implementation; this is a political challenge that is hard to overcome. Conservation agriculture as a tool for soil carbon sequestration may not be sound in all areas, especially those with barriers such as the inherently low capacity of soil carbon storage in sandy soils, unavailability of organic matter, monsoon variability, fire and erosion—prone areas. These are collectively referred to as biophysical challenges (Demenois et al. 2020).

Similarly, most techniques that prove very effective at sequestering organic carbon in agricultural soils do not present any additional income for soil carbon sequestration and improvement of the quality of the products (e.g., nutritional quality). This disincentivizes farmers from practicing and investments in such techniques. While creating financial policy mechanisms that encourage farmers to sequester carbon in agricultural soils is a useful approach, the revenues that may be received by farmers who may deploy such techniques in their farms may be small relative to the costs and revenues of not practicing, which directly impacts on their decision making (Pannell and Crawford (2022). As a result, in most developing countries soil carbon sequestration as a means to improve soil fertility, and indeed manage climate change, is usually not a priority because getting immediate benefits from farming to ensure income or food security comes first (Luedeling and Neufeldt 2012; Jerneck and Olsson 2014).

5.3 Techniques for Quantifying Soil C Sequestration

Estimating soil carbon sequestration can be accomplished through two major approaches—(i) the conventional approach in which quantification of soil carbon stocks is done using specified soil sampling methods or (ii) using a hybrid approach of measurements from soil sampling and soil carbon model estimates. The most common methods (conventional or hybrid) used to estimate soil carbon stocks and their respective changes over time are summarized in Table 3. They range from traditional invasive and destructive sampling techniques to modern, non-destructive methods of carbon estimation techniques. One common association is that the traditional core sampling and lab-based analysis method has always formed the basis and calibration reference method for modern techniques.

6 Concluding Remarks

Climate change has a significant influence on agricultural production. It affects the ability of agricultural systems to produce, through changes in ambient temperature and water supply, increased incidences of pests and diseases, as well as increases in the frequencies of catastrophic events such as typhoons and tornados that destroy agricultural produce, infrastructure and lives. There is no single approach to curtailing the effects of climate change in Agriculture, but healthy soils have the ability to store

Table 3 Available techniques for measuring soil carbon

Category	Technique	Detail and strength of approach	Shortfalls	References
Conventional techniques	(a) Soil coring and furnace technology for analysis	Uses core samples individually collected over an area/site with spatial and temporal variability consideration Data from this method is usually used as basis for soil carbon inventory of any geographical area Precise analytical protocols are well developed and readily available	• Sampling is destructive, • Highly costly • Large number of samples are required to account for spatial variability	Whitehead et al. (2010), Smith et al. (2019)
	(b) Estimation of ecosystem carbon exchange rates	It is based on the use of biophysical drivers to estimate carbon exchange of the ecosystem • Can facilitate diurnal and seasonal exchange dynamics • Enables continuous measurements of net input and loss of carbon at high temporal resolution	Difficult to execute over a large study area	Smith et al. 2019

(continued)

Table 3 (continued)

Category	Technique	Detail and strength of approach	Shortfalls	References
Spectroscopy techniques	(a) Near infrared reflectance spectroscopy (NIRS)	It is based on the vibration of covalent bonds of small atoms such as C–H, O–H, N–H etc. • It provides a non-disruptive sensor aided technology for mapping spatial variability in soil carbon stock • It requires less amount of samples compared to standard core sampling technique • Low cost high resolution measurements	-Requires a large amount of samples in the laboratory (destructive sampling) to account for spatial variability	Workman and Shenk (2004); Kusumo and Bustan (2018)
	(b) Geo-statistics coupled near infrared reflectance spectroscopy	The NIRS technique is coupled with GIS and geo-statistical interpretation technology to allow rapid mapping of soil carbon spatial distribution in real time	Need for calibration for every type of soil Requires high software expertise capability	Odlare et al. (2005), Whitehead et al. (2010)

(continued)

Table 3 (continued)

Category	Technique	Detail and strength of approach	Shortfalls	References
	(c) Laser-induced breakdown spectroscopy (LIBS)	It is an atomic emission spectroscopy technique, which uses a high-energy laser pulse focused on a sample surface to cause ablation, evaporation, atomization and ionization It is less-destructive and can be applied in situ	Common overlap of the emission lines that cannot be spectrally resolved It also suffers interference from a wide range of objects (roots, stones, variability in soil texture or water content)	Yu et al. (2020), Ma et al. (2021)
Non-invasive soil carbon measurement techniques	(a) Inelastic Neutron Scattering (INS) technique	It is based on the registration and analysis of gamma rays created when neutrons interact with soil elements It is non-destructive in situ method of carbon measurement It can allow simultaneous analysis of carbon and multiple other elements	It is relatively insensitive and restricted to the solid state It has much lower flux compared to X-rays that requires longer exposure for comparable sensitivity	Whitehead et al. (2010), Yakubova et al. (2016)

terrestrial carbon. Through increases in carbon sequestration, the capacity of soils to store large amounts of carbon will be boosted, and this will decrease greenhouse gas emissions to the atmosphere. On the other hand, enhancing carbon sequestration increases the ability of soils to hold water, supply more nutrients and increase agricultural produce.

Abbreviations and Acronyms

GHGs Green House Gases

CO_2	Carbon dioxide
IPCC	Intergovernmental Panel for Climate Change
WMO	World Meteorological Organization
SSA	Sub-Saharan Africa
FAO	Food and Agriculture Organization
CH_4	Methane
SO_2	Sulfur dioxide
UK	United Kingdom
UNFCCC	United Nations Framework Convention on Climate Change
N_2O	Dinitrogen Oxide
GT	Gigaton
SOC	Soil Organic Carbon
OM	Organic Matter
ppm	Parts per Million

References

Abell JT, Winckler G, Anderson RF, Herbert TD (2021) Poleward and weakened westerlies during Pliocene warmth. Nature 589:70–75. https://doi.org/10.1038/s41586-020-03062-1

Abram N, McGregor H, Tierney J, Evans MN, McKay NP, Kaufman DS (2016) Early onset of industrial-era warming across the oceans and continents. Nature 536:411–418. The Pages 2k Consortium. https://doi.org/10.1038/nature19082

Acevedo S, Mrkaic M, Novta N, Pugacheva E, Topalova P (2020) The effects of weather shocks on economic activity: what are the channels of impact? J Macroecon 65:103207. https://doi.org/10.1016/j.jmacro.2020.103207

Albert S, Leon JX, Grinham AR, Church JA, Gibbes BR, Woodroffe CD (2016) Interactions between sea-level rise and wave exposure on reef Island dynamics in the Solomon Islands. Environ Res Lett 11:054011. https://doi.org/10.1088/1748-9326/11/5/054011

Anonymous. (1912) Coal consumption affecting climate. Rodney and Otamatea Times, Waitemata and Kaipara Gazette, Warkworth, New Zealand, p 7

AO (2012) Serapium forest plantation, Ismailia, Egypt: Spatial analysis report of the plantation area for the forest management plan 2013–2022. Available at: https://www.fao.org/forestry/35800-094f4eedb0c86a4a33be1a3d906d80d3f.pdf. Accessed on 9th Feb 2023

Azhar GS, Mavalanka D, Nori-Sarma A, Rajiva A, Dutta P, Jaiswal A, Sheffield P, Knowlton K, Hess JJ (2014) Heat-related mortality in India: excess all-cause mortality associated with the 2010 Ahmedabad heat wave. PLoS One 9(3):e91831. https://doi.org/10.1371/journal.pone.0091831

Bailey VL, Pries CH, Lajtha K (2019) What do we know about soil carbon destabilization? Environ Res Lett 14:083004. https://doi.org/10.1088/1748-9326/ab2c11

Bamber JL, Oppenheimer M, Kopp RE, Aspinall WP, Cooke RM (2019) Ice sheet contributions to future sea-level rise from structured expert judgment. Proc Natl Acad Sci USA 116(23):1195–1120. https://doi.org/10.1073/pnas.1817205116

Beillouin D, Schauberger B, Bastos A, Ciais P, Makowski D (2020) Impact of extreme weather conditions on European crop production in 2018. Phil Trans Royal Soc B: Biol Sci 375(1810):20190510. https://doi.org/10.1098/rstb.2019.0510

Bhatia K, Baker A, Yang W, Vecchi G, Knutson T, Murakami H, Kossin J, Hodges K, Dixon K, Benjamin BB, Whitlock C (2022) A potential explanation for the global increase in tropical

cyclone rapid intensification. Nat Commun 13:6626. https://doi.org/10.1038/s41467-022-34321-6
Blenkinsop, S, Muniz Alves, L, Smith, AJP. (2021) Science brief review: climate change increases extreme rainfall and the chance of floods. In: LeQuéré C, Liss P, Forster P (eds) Critical issues in climate change science. Science Brief News. https://doi.org/10.5281/zenodo.4779119
Braddock S, Hall BL, Johnson JS, Balco G, Spoth M, Whitehouse PL, Campbell S, Goehring BM, Rood DH, Woodward J (2022) Relative sea-level data preclude major late Holocene ice-mass change in Pine Island Bay. Nat Geosci 15:568–572. https://doi.org/10.1038/s41561-022-00961-y
Brás TA, Seixas J, Carvalhais N, Jägermey J (2021) Severity of drought and heatwave crop losses tripled over the last five decades in Europe. Environ Res Lett 16(6):065012. https://doi.org/10.1088/1748-9326/abf004
Caparas M, Zobel Z, Castanho ADA, Schwalm CR (2021) Increasing risks of crop failure and water scarcity in global breadbaskets by 2030. Environ Res Lett 16:104013. https://doi.org/10.1088/1748-9326/ac22c1
Chenu C, Angers DA, Pierre Barré P, Derrien D, Arrouays D, Balesdent J (2019) Increasing organic stocks in agricultural soils: knowledge gaps and potential innovations. Soil Tillage Res 188:41–52. https://doi.org/10.1016/j.still.2018.04.011
Choy CW, Lau DS, He Y (2020) Super typhoons Hato and Mangkhut, part I: analysis of maximum intensity and wind structure. Weather 77(9):314–320. https://doi.org/10.1002/wea.3797
Church JA, Clark PU, Cazenave A, Gregory AJ, Jevrejeva S, Levermann A, Merrifield MA, Milne GA, Nerem RS, Nunn PD, Payne AJ, Pfeffer WT, Stammer D, Unnikrishnan AS (2013) Sea level change. In: Stocker TF, Qin D, Plattner G-K, Tignor M, Allen SK, Boschung J, Nauels A, Xia Y, Bex V, Midgley PM (eds) Climate change 2013: the physical science basis. Contribution of working group I to the fifth assessment report of the intergovernmental panel on climate change. Cambridge University Press, Cambridge, United Kingdom and New York, NY, USA
Climate Change Act (2008) (legislation.gov.uk). Available at https://www.legislation.gov.uk/ukpga/2008/27. Accessed on 10 Feb 2023
Demenois J, Torquebiau E, Arnoult MH, Eglin T, Masse D, Assouma MH, Blanfort V, Chenu C, Chapuis-Lardy L, Medoc JM, Sall SN (2020) Barriers and strategies to boost soil carbon sequestration in agriculture. Front Sustain Food Syst 4:37. https://doi.org/10.3389/fsufs.2020.00037
Dignac MF, Derrien D, Barré P, Barot S, Cécillon L, Chenu C, Chevallier T, Freschet GT, Garnier P, Guenet B, Hedde M, Klumpp K, Lashermes G, Maron PA, Nunan N, Roumet C, Basile-Doelsch I (2017) Increasing soil carbon storage: mechanisms, effects of agricultural practices and proxies: a review. Agron Sustain Dev 37:14. https://doi.org/10.1007/s13593-017-0421-2
Dourte DR, Fraisse CW, Bartels WL (2015) Exploring changes in rainfall intensity and seasonal variability in the Southeastern U.S.: stakeholder engagement, observations, and adaptation. Clim Risk Manage 7:11–19. https://doi.org/10.1016/j.crm.2015.02.001
FAO (2007) Adaptation to climate change in agriculture, forestry and fisheries: Perspective, framework and priorities. Available at https://www.fao.org/3/au030e/au030e.pdf. Accessed on 7 Jan 2023
FAO (2015) Climate change and food security: risks and responses. Available at https://www.fao.org/3/i5188e/I5188E.pdf. Accessed on 19 Nov 2022
Fowler H, Ali H (2022) Analysis of extreme rainfall events under the climatic change. In: Morbidelli R (ed). Elsevier, Amsterdam, The Netherlands, pp 307–326
Goh KM (2004) Carbon sequestration and stabilization in soils: Implications for soil productivity and climate change. Soil Sci Plant Nutr 50(4):467–476. https://doi.org/10.1080/00380768.2004.10408502
Guido Z, Zimmer A, Lopus S, Hannah C, Gower D, Waldman K, Evans T (2020) Farmer forecasts: impacts of seasonal rainfall expectations on agricultural decision-making in Sub-Saharan Africa. Clim Risk Manage 30:100247. https://doi.org/10.1016/j.crm.2020.100247

Higgins RW, Kousky VE (2013) Changes in observed daily precipitation over the United States between 1950–1979 and 1980–2009. J Hydrometeorol 14:105–121. https://doi.org/10.1175/JHM-D-12-062.1

Holland D, Nicholls K, Basinski A (2020) The southern ocean and its interaction with the Antarctic ice sheet. Science 367:1326–1330

Jastrow JD, Amonette JE, Bailey VL (2007) Mechanisms controlling soil carbon turnover and their potential application for enhancing carbon sequestration. Clim Chang 80:5–23. https://doi.org/10.1007/s10584-006-9178-3

Jerneck A, Olsson L (2014) Food first! Theorising assets and actors in agroforestry: risk evaders, opportunity seekers and 'the food imperative' in sub-Saharan Africa. Int J Agric Sustain 12(1):1–2. https://doi.org/10.1080/14735903.2012.751714

Jiménez Cisneros BE, Oki T, Arnell NW, Benito G, Cogley JG, Döll P, Jiang T, Mwakalila SS (2014) Freshwater resources. In: Field CB, Barros VR, Dokken DJ, Mach, KJ, Mastrandrea MD, Bilir TE, Chatterjee M, Ebi KL, Estrada YO, Genova RC, Girma B, Kissel ES, Levy AN, MacCracken S, Mastrandrea PR, White LL (eds) Climate change 2014: impacts, adaptation, and vulnerability. Part a: global and sectoral aspects. Contribution of working Group II to the fifth assessment report of the intergovernmental panel on climate change. Cambridge University Press, Cambridge, United Kingdom and New York, NY, USA, pp 229–269

Johnson B (2021) The world's 5 deadliest tornadoes. Available at: https://www.thoughtco.com/worlds-worst-tornadoes-3555048. Accessed on 20 Feb 2023

Kangalawe HR (2019) Who took our land—the world bank or the state?" the land question in the state plantation forests in Tanzania: the case of the Sao Hill forests, 1962–2012. Tanzania J Dev Stud 17(2):77–96

Kim JM, Kim T, Son K, Yum SG, Ahn S (2019) Measuring vulnerability of typhoon in residential facilities: focusing on typhoon Maemi in South Korea. Sustainabil. 11:2768. https://doi.org/10.3390/su11102768

Kniivilä M (2007) Industrial development and economic growth: Implications for poverty reduction and income inequality. In: Industrial development for the 21st century: sustainable development perspectives. United Nations, Department of Social and Economic Affairs. New York, USA, pp 295–332

Kossin JP, Knapp KR, Olander TL, Velden CS (2020) Global increase in major tropical cyclone exceedance probability over the past four decades. PLoS 117(22):11975–11980. https://doi.org/10.1073/pnas.1920849117

Krauss M, Wiesmeier M, Don A, Cuperus F, Gattinger A, Gruber S, Haagsma WK, Peigné J, Chiodelli Palazzoli M, Schulz F, van der Heijden MGA, Vincent-Caboud L, Wittwer RA, Zikeli S, Steffens M (2022). Reduced tillage in organic farming affects soil organic carbon stocks in temperate Europe. Soil Tillage Res 216: 105262

Kusmec A, Attigala L, Dai X, Srinivasan S, Yeh CT, Schnable PS (2023) A genetic trade-off for tolerance to moderate and severe heat stress in US hybrid maize. PLoS Genet 19(7):e1010799. https://doi.org/10.1371/journal.pgen.1010799

Kusumo BH, Bustan S (2018) The rapid measurement of soil carbon stock using near-infrared technology. IOP Conf Ser: Earth Environ Sci 129:012023. https://doi.org/10.1088/1755-1315/129/1/012023

Lal R (2008) Sequestration of atmospheric CO_2 in global carbon pools. Energy Environ Sci 1:86–100. https://doi.org/10.1039/b809492f

Lal R (2009) Challenges and opportunities in soil organic matter research. Eur J Soil Sci 60:158–169. https://doi.org/10.1111/j.1365-2389.2008.01114.x

Lal R, Negassa W, Klaus Lorenz K (2015) Carbon sequestration in soil. Curr Opin Environ Sustain 15:79–86. https://doi.org/10.1016/j.cosust.2015.09.002

Land M, Haddaway NR, Hedlund K, Jørgensen HB, Bracht H, Kätterer T, Isberg PE (2017) How do selected crop rotations affect soil organic carbon in boreo-temperate systems? A systematic review protocol. Environ Evid 6:9. https://doi.org/10.1186/s13750-017-0086-y

Lemus R, Lal R (2005) Bioenergy crops and carbon sequestration. Crit Rev Plant Sci 24:1–21. https://doi.org/10.1080/07352680590910393
Li Y, Guan K, Schnitkey GD, DeLucia E, Peng B (2019) Excessive rainfall leads to maize yield loss of a comparable magnitude to extreme drought in the United States. Glob Chang Biol 25:2325–2337. https://doi.org/10.1111/gcb.14628
Lindsey R (2023) Climate change: atmospheric carbon dioxide. Available at https://www.climate.gov/news-features/understanding-climate/climate-change-atmospheric-carbon-dioxide. Accessed on 12 Feb 2023
Lo MCM, Fan Y (2022) How narratives of disaster impact survivors' emotionality: THE case of Typhoon Morakot. Poetics 93(Part A). https://doi.org/10.1016/j.poetic.2021.101579
Luedeling E, Neufeldt H (2012) Carbon sequestration potential of parkland agroforestry in the Sahel. Climatic Chang 115:443–461. https://doi.org/10.1007/s10584-012-0438-0
Ma F, Mao X, Zorba V, Russo RE, González J, Du C (2021) Combination of high-resolution laser-induced breakdown spectroscopy and least square method for reducing soil carbon overestimation due to iron interference. Geoderma 385:114881. https://doi.org/10.1016/j.geoderma.2020.114881
Mamalakis A, Randerson JT, Yu JY, Pritchard MS, Magnusdottir G, Smyth P, Levine PA, Yu S, Foufoula-Georgiou E (2021) Zonally contrasting shifts of the tropical rain belt in response to climate change. Nat Clim Chang 11:143–151. https://doi.org/10.1038/s41558-020-00963-x
Matthews T, Sanchez R (2022) The deadliest tornadoes, tornado outbreaks of the 21st century. Available at: https://www.foxweather.com/learn/7-deadliest-tornadoes-outbreaks-of-21st-century. Accessed on 20 Feb 2023
Mauritsen T, Pincus R (2017) Committed warming inferred from observations. Nat Clim Chang 7:652–655. https://doi.org/10.1038/nclimate3357
Mirhosseini G, Srivastava P, Stefanova L (2012) The impact of climate change on rainfall Intensity–duration–frequency (IDF) curves in Alabama. Reg Environ Chang. https://doi.org/10.1007/s10113-012-0375-5
Mora C, McKenzie T, Gaw IM, Dean JM, Hammerstein HV, Knudson TA, Setter RO, Smith CZ, Webster KM, Patz JA, Franklin EC (2022) Over half of known human pathogenic diseases can be aggravated by climate change. Nat Clim Chang 12:869–875. https://doi.org/10.1038/s41558-022-01426-1
Moukanni N, Brewer KM, Gaudin ACM, O'Geen AT (2022) Optimizing carbon sequestration through cover cropping in mediterranean agroecosystems: synthesis of mechanisms and implications for management. Front Agron 4:844166. https://doi.org/10.3389/fagro.2022.844166
Mwanyoka I, Lopa D (2016) Communities' perception on the contribution of soil and water conservation measures in improving land productivity in the dry-land areas of tanzania: the case of terrace, "Fanya Juu" and double digging in same district. Int J Agric Environ Res 2(6):1679–1695
Myers S, Zanobetti A, Kloog I, Huybers P, Leakey A, Bloom A, Carlisle E, Dietterich LH, Fitzgerald G, Hasegawa T, Holbrook NM, Nelson RL, Ottman MJ, Raboy V, Sakai H, Sartor KA, Schwartz J, Seneweera S, Tausz M, Usui Y (2014) Increasing CO_2 threatens human nutrition. Nature 510:139–142. https://doi.org/10.1038/nature13179
Ndlovu S, Mathe B, Phiri K, Nyathi D (2020) Factoring water harvesting into climate change adaptation: endogenous responses by smallholder farmers in Gwanda district. Zimbabwe Cog Soc Sci 6(1):1784652. https://doi.org/10.1080/23311886.2020.1784652
Nelson GC, Rosegrant MW, Koo J, Robertson RD, Sulser T, Zhu T, Ringler C, Msangi S, Palazzo A, Batka M, Magalhaes M, Valmonte-Santos R, Ewing M, Lee DR (2009) Climate change impacts on agriculture and costs on production. Food Policy Rep 30. https://doi.org/10.2499/0896295354
Odlare M, Svensson K, Pell M (2005) Near-infrared reflectance spectroscopy for assessment of spatial soil variation in an agricultural field. Geoderma 126(3–4):193–202. https://doi.org/10.1016/j.geoderma.2004.09.013
Ontl TA, Schulte LA (2012) Soil carbon storage. Nat Educ Knowl 3(10):35

Pannell D, Crawford M (2022) Challenges in making soil-carbon sequestration a worthwhile policy. Farm Pol J FPJ1901

Post E, Alley RB, Christensen TR, Macias-Fauria M, Forbes BC, Gooseff MN, Iler A, Kerby JT, Laidre KL, Mann ME, Olofsson J, Stroeve JC, Ulmer F, Virginia RA, Wang M (2019) The polar regions in a 2°C warmer world. Sci Adv 5(12):eaaw9883. https://doi.org/10.1126/sciadv.aaw9883

Price J, Warren R, Forstenhäusler N et al (2022) Quantification of meteorological drought risks between 1.5 °C and 4 °C of global warming in six countries. Clim Chang 174:12 (2022). https://doi.org/10.1007/s10584-022-03359-2

Raftery A, Zimmer A, Frierson DMW, Startz R, Liu P (2017) Less than 2°C warming by 2100 unlikely. Nat Clim Chang 7:637–641 (2017). https://doi.org/10.1038/nclimate3352

Raji BA, Ogunwole JO (2006) Potential of soil carbon sequestration under various land use in the sub-humid and semi-arid savannah of Nigeria: lessons from long-term experiments. Int J Soil Sci 1(1):33–43

Rigon JPG, Calonego JC (2020) Soil carbon fluxes and balances of crop rotations under long-term no-till. Carbon Balance Manage 15:19. https://doi.org/10.1186/s13021-020-00154-3

Sanderman J, Hengl T, Fiske GJ (2017) Soil carbon debt of 12,000 years of human land use. Proc Natl Acad Sci USA 114:9575–9580

Sarkar R, Corriher-Olson V, Long C, Somenahally A (2020) Challenges and potentials for soil organic carbon sequestration in forage and grazing systems. Rangel Ecol Manage 73(6):786–795

Schmitt J, Offermann F, Söder M, Frühauf C, Finger R (2022) Extreme weather events cause significant crop yield losses at the farm level in German agriculture. Food Policy 112:102359

Singh P, Benbi DK (2021) Physical and chemical stabilization of soil organic matter in cropland ecosystems under rice–wheat, maize–wheat and cotton–wheat cropping systems in northwestern India. Carbon Manage 12(6):603–612. https://doi.org/10.1080/17583004.2021.1992505

Smith P, Soussana J, Angers D, Schipper L, Chenu C, Rasse DP, Klumpp K (2019) How to measure, report and verify soil carbon change to realise the potential of soil carbon sequestration for atmospheric greenhouse gas removal. Glob Chang Biol. https://doi.org/10.1111/gcb.14815

Sulser, T, Wiebe, KD, Duston, S, Cenacchi, N, Nin-Pratt, A, Mason-D'Croz, Robertson, RD, Willenbockel, D, Rosegrant, MW. (2021). Climate Change and hunger: Estimating costs of adaptation in the agrifood system. Food Pol. Rep. 62: https://doi.org/10.2499/9780896294165

Sun X, Wang G, Ma Q, Liao J, Wang D, Guan Q, Jones DL (2021) Organic mulching promotes soil organic carbon accumulation to deep soil layer in an urban plantation forest. For Ecosyst 8:2. https://doi.org/10.1186/s40663-020-00278-5

The Japan Times (2019) Extremely large typhoon draws bead on Japan, could hit main islands this weekend. Available at Extremely large typhoon draws bead on Japan, could hit main islands this weekend | The Japan Times. Accessed 31 March 2023

Trenberth KE, Jones PD, Ambenje P, Bojariu R et al (2007) Observations: surface and atmospheric climate change. In: Solomon S, Qin D, Manning M, Chen Z et al (eds) Climate change 2007. The physical science basis. Intergovernmental Panel on Climate Change 4th Assessment Report.

Trenberth KE (2011) Changes in precipitation with climate change. Clim Res 47:123–138. https://doi.org/10.3354/cr00953

Tu HM, Chen HM (2017) Slopeland hazard and respiratory health: the example of Typhoon Morakot in Taiwan. Landsc Urban Plann 157:375–382. https://doi.org/10.1016/j.landurbplan.2016.08.008

UNFCCC (2022) United in science: we are heading in the wrong direction. External Press Release. Available at https://unfccc.int/news/We%20are%20Heading%20in%20the%20Wrong%20Direction%20&C%20UNFCCC. Accessed on 3 Oct 2022

Van Ettena J, De Sousa K, Aguilarc A, Barriosc M, Cotoa A, Dell'Acquad M, Faddae C, Gebrehawaryate Y, Van de Gevelf J, Guptag A, Kirosh AY, Madriza B, Mathurg P, Mengistueh DK, Mercadoi L, Mohammedh JN, Paliwalg A, Ped ME, Quirosa CF, Rosasj JC, Sharmag N, Singh SS, Solankil IS, Steinkea J (2019) Crop variety management for climate adaptation supported by citizen science. PNAS 116(10):4194–4199

Vautard R, Cattiaux J, Yiou P, Thepaut JN, Ciais P (2010) Northern hemisphere atmospheric stilling partly attributed to an increase in surface roughness. Nat Geosci 3:756–761. https://doi.org/10.1038/ngeo979

Were K, Singh BR, Ayaga G (2020) Land use changes and sustainable land management practices for soil carbon sequestration in sub-saharan african agro-ecosystems. In: Singh B, Safalaoh A, Amuri N, Eik L, Sitaula B, Lal R (eds) Climate impacts on agricultural and natural resource sustainability in Africa. Springer, Cham. https://doi.org/10.1007/978-3-030-37537-9_3

Westengen OT, Brysting AK (2014) Crop adaptation to climate change in the semi-arid zone in Tanzania: the role of genetic resources and seed systems. Agric Food Secur 3:3. https://doi.org/10.1186/2048-7010-3-3

Whitehead D, Baisden T, Beare M, Campbell D et al (2010) Review of soil carbon measurement methodologies and technologies, including nature and intensity of sampling, their uncertainities and costs. Landcare Research Report. Available at Review of soil carbon measurement methodologies and technologies, including nature and intensity of sampling, their uncertainties and costs (mpi.govt.nz). Accessed on 20 June 2022

Whitehead D, Schipper LA, Pronger J, Moinet GYK, Mudge PL, Calvelo Pereira R, Camps-Arbestain M (2018) Management practices to reduce losses or increase soil carbon stocks in temperate grazed grasslands: New Zealand as a case study. Agric Ecosyst Environ 265:432–443. https://doi.org/10.1016/j.agee.2018.06.022

Wills RC, Byrne MP, Schneider T (2016) Thermodynamic and dynamic controls on changes in the zonally anomalous hydrological cycle. Geophys Res Lett 43(9):4640–4649. https://doi.org/10.1002/2016gl068418

Witze A (2022) Extreme heatwaves: surprising lessons from the record warmth: unprecedented temperatures are coming faster and more furiously than researchers expected, raising questions about what to anticipate in the future. Nature 608:464–465. https://doi.org/10.1038/d41586-022-02114-y

Workman JJ, Shenk J (2004) Understanding and using the near-infrared spectrum as an analytical method. In: Al-Amoodi L, Roberts CA, Workman JJ et al. (eds.) Near-infrared spectroscopy in agriculture. Agronomy Monograph no. 44. Madion, Wisconsin, USA, American Society of Agronomy, Inc., Crop Science Society of America, Inc., and Soil Science Society of America, Inc. 44 Agronomy series: 3–10, Madion, Wisconsin, USA

World bank (2021) Disappearing coasts in the Maghreb: coastal erosion and its costs. Available at https://www.worldbank.org/en/country/morocco/publication/disappearing-coasts-in-the-maghreb-coastal-erosion-and-its-costs. Accessed on 3rd Nov 2025

World Meteorological Organization (WMO) (2020) State of the climate in Africa 2019 (WMO-No. 1253). https://library.wmo.int/index.php?lvl=notice_display&id=21778#.Y6viBhRBzIW

WMO (2022) State of global climate 2022. Available at https://wmo.int/publication-series/state-of-global-climate-2022. Accessed on 3rd Nov 2025

Wu CC (2013) Typhoon Morakot: key findings from the journal TAO for improving prediction of extreme rains at landfall. Bull Am Meteorol Soc 94(2). https://doi.org/10.1175/BAMS-D-11-00155.1

Xu M, Chang CP, Fu C, Qi Y, Robock A, Robinson D, Zhang H (2006) Steady decline of east Asian monsoon winds, 1969–2000: evidence from direct ground measurements of wind speed. J Geophys Res 111(D24). https://doi.org/10.1029/2006jd007337

Yakubova G, Kavetskiy A, Prior SA, Torbert HA (2016) Benchmarking the inelastic neutron scattering soil carbon method. Vad Zon J. https://doi.org/10.2136/vzj2015.04.0056.

Yu K, Ren J, Zhao Y (2020) Principles, developments and applications of laser-induced breakdown spectroscopy in agriculture: a review. Artif Intel Agric 4:127–139. https://doi.org/10.1016/j.aiia.2020.07.001

Zachariah M, Arulalan T, Rao KA, Saeed F, Jha R, Dhasmana MK, Mondal A, Bonnet R, Heinrich D, Thalheimer L, Marghidan CP, Kapoor A, Aalst M, Raju E, Li S, Sun J, Vecchi G, Yang W, Hauser M, Schumacher DL, Seneviratne SI, Harrington LJ, Otto FEL (2022) Climate change made devastating early heat in India and Pakistan 30 times more likely. World Weather Attribution

(2022). Available at https://www.worldweatherattribution.org/wp-content/uploads/India_Pak-Heatwave-scientific-report.pdf. Accessed on 5 Dec 2022
Zeng Z, Ziegler AD, Searchinger T, Yang L, Chen A, Ju K, Piao S, Li LZX, Ciais P, Chen D, Liu J, Azorin-Molina C, Chappell A, Medvigy D, Wood EF (2019) A reversal in global terrestrial stilling and its implications for wind energy production. Nat Clim Chang 9:979–985. https://doi.org/10.1038/s41558-019-0622-6
Zotarelli L, Alves BJR, Urquiaga S et al (2005) Impact of tillage ad crop rotation on aggregate-associated carbon on two oxisols. Soil Sci Soc Am J 69(2):482–491. https://doi.org/10.2136/sssaj2005.0482

Open Access This chapter is licensed under the terms of the Creative Commons Attribution 4.0 International License (http://creativecommons.org/licenses/by/4.0/), which permits use, sharing, adaptation, distribution and reproduction in any medium or format, as long as you give appropriate credit to the original author(s) and the source, provide a link to the Creative Commons license and indicate if changes were made.

The images or other third party material in this chapter are included in the chapter's Creative Commons license, unless indicated otherwise in a credit line to the material. If material is not included in the chapter's Creative Commons license and your intended use is not permitted by statutory regulation or exceeds the permitted use, you will need to obtain permission directly from the copyright holder.

Agroforestry Innovations for Improved Soil Fertility and Resilience to Climate Change

Joyce P. B. Njoloma, Robertson R. B. Khataza, Isaac B. Nyoka, Anthony Kimaro, Tadala Makuluni, Chisomo Haswell, Batson Mazungwi, Franklin Chilomba, and Gift Kamanga Thole

Abstract Agroforestry practices involve a wide range of trees and woody shrubs that are protected or planted and managed on agricultural landscapes. Integrating trees in cropping and livestock systems has led to the development of various agroforestry innovations. These innovations are considered to be low-cost, agro-ecologically sustainable, and can be relied upon to build smallholder farmer's resilience to climate change. These agroforestry innovations have a great potential of contributing to both climate change adaptation and mitigation, thereby making considerable positive impact to sustainable agriculture and natural resources productivity leading to increased crop yields, better livelihoods, and a more resilient environment.

The soil fertility enhancing agroforestry technologies include intercropping with *Gliricidia sepium* coppicing trees, short rotation and relay fallowing with non-coppicing trees (*Tephrosia vogelii*), high canopy tree species of *Faidherbia albida*-based agroforestry system. It must be noted that while these agroforestry systems have the potential to enhance soil fertility for agricultural productivity, they also enhance biodiversity and resilience to climate change associated livelihood shocks. Proper management of the trees in agroforestry systems have resulted into provisioning of more benefits such as increased soil organic matter, soil moisture retention as well as enhanced soil water infiltration, and nutrient cycling with high levels of nitrogen that supports crop's vegetative growth. The associated agroforestry trees also do provide shade, reduce wind speed, and control soil erosion. Other potential benefits include

J. P. B. Njoloma (✉) · R. R. B. Khataza · T. Makuluni · C. Haswell · B. Mazungwi · F. Chilomba · G. Kamanga Thole
Lilongwe University of Agriculture and Natural Resources (LUANAR), Bunda College, Lilongwe PO Box 219, Malawi
e-mail: jnjoloma@luanar.ac.mw; jnjoloma@gmail.com

I. B. Nyoka
CIFOR-ICRAF, Lilongwe, Malawi

A. Kimaro
CIFOR-ICRAF, Dar es Salaam, Tanzania

© The Author(s) 2026

B. R. Singh et al. (eds.), *Sustainable Food Systems in Sub-Saharan Africa*,
https://doi.org/10.1007/978-3-032-11288-0_13

removal of significant amounts of carbon as a greenhouse gas through increased carbon storage in above-ground and below-ground biomass.

However, despite the many acknowledged benefits, adoption of agroforestry innovations remains low and slow. Considering the complexities associated with combination of the agroforestry components and longer gestation periods for the expected benefits requires tailor made policies to support the adoption at both household and landscape levels. Additionally, it is known that adoption rates are influenced by a variety of factors, which include agroforestry related extension services and access to markets for the agroforestry product, availability of land and labour, and farmers' perceptions of the benefits and risks of managing the agroforestry innovations.

1 Introduction

In the recent decade, climate change-induced weather variability has become apparent, thereby raising global concern largely because of its negative impacts, and this has led to severe loss and damage (Garrity et al. 2010; Mechler et al. 2019; Shimada 2022). Generally, developing countries are more prone to adverse impacts of global warming and climate change; thus necessitating the need for promoting viable adaptation and mitigation strategies (Seghieri et al. 2021). The agricultural impacts of climate change are of greatest concern to developing countries because of their increased vulnerability, resulting from lack of adaptive capacity. This is largely due to overreliance on rain-fed agriculture, subsistence level of operations, and limited institutional support (Sorgho et al. 2020). Thus, agroforestry remains one of the low-cost, sustainable agro-ecological innovations extensively relied upon to build resilience to climate change (Sheppard et al. 2020; Seghieri et al. 2021). As an agro-ecological innovation, agroforestry is increasingly recognized as a land management system that can contribute to both climate change adaptation and mitigation, while enhancing livelihood resilience for smallholder farmers (Mbow et al. 2019). Agroforestry can generate multiple livelihood benefits, including reducing poverty, improving food and income security as well as sustaining environmental quality (Pantera et al. 2021; Lelamo 2021).

In sub Saharan Africa, ambitious targets have been set for member countries to increase their fertilizer usage, but there is still low mineral fertilizer use and use efficiency. Currently an average farmer cannot afford to timely buy and apply chemical fertilizers at the recommended rates, resulting in poor fertilizer use efficiency. Poor soil health is another factor that has led to poor fertilizer use efficiency of the very low amounts of fertilizers that a farmer can afford (Vanlauwe et al. 2023). With proper implementation and support, agroforestry practices can improve soil health, increase soil fertility and reduce dependence on mineral fertilizers, thereby improving food security and livelihoods for smallholder farmers. Agroforestry innovations provide a suite of climate-smart practices that can contribute to sustainable agriculture by improving agricultural and natural resources productivity, leading to increased crop yields, better livelihoods, and a more resilient environment (Garrity et al. 2010).

Promoting the adoption of these climate-smart agroforestry innovations will ensure increased agricultural production and adaptive capacity to climate change and its associated livelihood shocks.

2 Development of Soil Fertility Enhancing Agroforestry Technologies

The need for application of inorganic fertilizers for soil reconditioning and to raise crop productivity in sub-Saharan Africa (SSA) is well established. However, the high cost of chemical fertilizers and low incomes among smallholder farmers pose significant challenges to increasing fertilizer use in the region (Sheahan and Barrett 2017). The high fertilizer costs and distribution inefficiencies has led to low fertilizer usage of less than 10 kg ha^{-1} of fertilizer compared to 90 kg ha^{-1} in Latin America and over 170 kg ha^{-1} in Asia (Tittonell et al. 2016; Mafongoya et al. 2007). On the other hand, organic soil amendments including agroforestry do increase soil organic matter (SOM) content, which in turn increases fertilizer-use efficiency. Thus, the best strategy to improve productivity and maintain soil fertility in SSA is a combination of both inorganic and organic fertilizers (Mafongoya et al. 2007).

The need to complement inorganic and organic fertilizers has compelled researchers to develop adaptable and viable agroforestry technologies as part of organic soil amendments. Typically, farmers can integrate trees in agricultural landscapes through three processes; namely, retention of trees present on the landscape being converted to agriculture, natural tree regeneration after farm establishment, and active tree planting in preferred locations and patterns within the agricultural landscapes (Garrity et al. 2010). Studies have shown that growing of leguminous trees that are associated with nitrogen-fixing bacteria in agroforestry systems and retention of naturally regenerating trees on crop fields do enhance soil fertility (Fahad et al. 2022). For example, one of the most successful agroforestry practices in Africa is farmer-managed natural regeneration (FMNR), which involves selective pruning and managing trees that naturally regenerate on farms. FMNR has been shown to increase soil fertility, water retention, and crop yields (Opoku Mensah et al. 2024). Another effective agroforestry practice is the use of agroforestry-based fallows, which involve planting trees and shrubs on short and relay fallows to improve soil fertility and reduce soil erosion. Studies on alley cropping systems of *Gliricidia sepium* and the systematic planting of *Faidherbia albida* have shown improvements in soil fertility and increases in maize yields by 0.7–2.5 Mg ha^{-1} over the no-input control, which translates to an 89–318% increase yields across Africa (Sileshi et al. 2014). Studies have further shown that Faidherbia trees in parklands can provide significant amounts of agriculturally important nutrients to the soil, including nitrogen, potassium, and organic carbon. Fully grown *Faidherbia* trees, at a density of 100 trees per hectare, can supply the equivalent of 300 kg per year of complete fertilizer, resulting in an estimated 250–400% increase in maize yields. This makes *Faidherbia* trees a valuable resource for resource-constrained households struggling to replenish soil nutrients through mineral fertilizer amendments (Sileshi 2016). In Kenya, it has been reported that planting *Faidherbia albida* trees at a density of 50 trees per hectare was sufficient for significant crop yield gains compared to sole maize cropping. In smallholder farming systems, promising soil fertility replenishment technologies also include intercropping with coppicing tree species, short rotation and relay fallowing with non-coppicing trees (*Cajanus cajan, Sesbania sesban* and *Tephrosia vogelii*), and high canopy tree species of *Faidherbia albida*-based agroforestry system. These agroforestry systems are widely used and revered by farmers in various regions and have the potential to not only enhance soil fertility for agricultural productivity but also increase biodiversity and resilience to climate change (Kuyah et al. 2020).

3 Agroforestry Technologies and Soil Fertility Management

Agroforestry service functions are diverse, but the notable ones include soil fertility management, usually achieved through Biological Nitrogen Fixation (BNF) process, microclimate regulation and erosion control. Among the various organic soil amendment technologies, agroforestry fertilizer trees (when incorporated in the soils) have the potential to address the low SOM levels. As a result, they benefit smallholder farmers through enhanced soil fertility and improved crop productivity (Giller et al. 2019; Tittonell et al. 2022). The depletion of SOM due to poor agricultural practices, such as traditional fallows and mono-cropping, causes nutrient deficiency because soils become inept to hold nutrients and water, resulting in poor crop growth during dry spells. SOM is an important factor in retaining nutrients and moisture in highly weathered soils, which are common in SSA. A study by Akinnifesi et al. (2010) in Malawi found that intercropping maize with pigeon peas resulted in a 25% increase in SOM content and a 50% increase in nitrogen content compared to monoculture maize. Similarly, Beedy et al. (2010) reported that intercropping maize with trees, such as *Sesbania sesban, Calliandra calothyrsus*, and *Gliricidia sepium* resulted in a 30–50% increase in SOM content and a 20–40% increase in nitrogen content compared to monoculture maize. In addition, a study by Njoloma et al. (2016) in Malawi showed that agroforestry-based soil fertility management practices, particularly using *Faidherbia albida* and other legume tree species such as Gliricidia and Tephrosia, can improve soil organic carbon (SOC) content. Gliricidia tends to be superior to other legume tree species in subhumid regions due to its drought tolerance (Kerr 2012) and particularly high biomass production with high foliage N content (Kuyah et al. 2020). Legume tree intercropping studies in Malawi and Kenya have revealed a moderate build-up of organic carbon after 10–12 years (Eze et al. 2020; Makumba et al. 2007; Martinsen et al. 2017). However, the leguminous tree species biomass enhance soil fertility through decomposition and subsequent nutrient release.

Over the years, several studies have provided evidence on the extent to which agroforestry systems have contributed towards improving soil health and crop productivity (Misra 2011). A meta-analysis review of 87 studies by Nair et al. (2019) revealed that agroforestry systems supply higher SOM, nitrogen, and phosphorus levels than monoculture systems. For instance, use of green biomass manure can enhance soil health by boosting SOM, soil nutrients, soil microbial activities and enhancing soil structure and hence increased soil fertility. Further, research evidence shows that incorporating leguminous trees, which produce large quantities of leaf biomass, can significantly increase soil nitrogen and organic matter content (Kuyah et al. 2020). Ikerra et al. (1999) showed that Maize/Gliricidia intercrop supplied nitrogen in the ranges from 60 to 120 kg ha^{-1} yr^{-1} in Malawi and western Kenya where sizes of land holdings preclude fallows. Trees were managed through repeated cutting back so that they do not interfere with the crop. In western Kenya, maize yield following improved fallows averaged 4.1 t ha^{-1}, which was much higher than maize yield from continuously planted non-fertilized plots (1.7 t ha^{-1}). In Zambia, maize yield averaged to over 3.5 t ha^{-1} under *Gliricidia* alley system where Gliricidia green biomass manure were incorporated under sandy soil conditions (Mehreteab et al. 2022). In Malawi, Makumba et al. (2006) evaluated the impacts of an agroforestry system incorporating *Gliricidia sepium* trees on soil fertility and maize yields. The study found that the agroforestry system significantly improves soil fertility and yields compared to conventional maize cropping. The contribution of agroforestry has also been reported in many other studies, including that of (Rusinamhodzi et al. 2012). According to Rusinamhodzi et al. (2012), for a farmer to attain tangible benefits in maize-based agroforestry, the intercropping of maize with legumes must match with both the biophysical conditions and socio-economic status of the smallholder farmers.

4 Agroforestry Technologies and Resilience to Climate Change

Climate change remains one of the most critical threats to the livelihoods of agriculture-dependent households in low income countries. Consequently, agroforestry technologies are considered as an important component for sustainable agriculture in the context of climate change adaptation and mitigation (Kalele et al. 2021). Climate change risks, including severe droughts, flooding, and disease outbreaks can have extensive impacts on agricultural systems, triggering soil erosion, crop failure, loss of biodiversity, pest damages and economic losses. Frequent extreme events and greater occurrence of drier and wetter conditions are already making it difficult for farmers to plan planting and harvesting (SIWI 2018; Kalele et al. 2021). On the other hand, about 23% of the world's total emission of greenhouse gases (GHGs), such as carbon dioxide (CO_2), methane (CH_4) and nitrous oxide (N_2O) come from agriculture, forestry and other land uses; for example, deforestation, livestock production and soil and biodiversity degradation (IPCC 2019). Farming activities may generate significant amounts of GHGs, for instance from inorganic fertilizers, pesticides, fossil fuels, manure, and crop residues, as well as increased numbers of livestock. At the same time, impacts of climate change make farming more difficult and unpredictable for smallholder farmers, while the demand for food increases as populations grow (Agroforestry Network 2019). Degraded landscapes mean lost opportunities and negative impacts for people, society, economies, as well as for ecosystem services and environmental flows. Also, allowing degradation to continue increases the difficulty and cost of land restoration (Molin et al. 2018); and greenhouse gases continue to increase, causing adverse climate change impacts on livelihoods (Ripple et al. 2020).

Agroforestry for climate adaptation at the farm level, and enhanced resilience at the landscape level, can take many forms. The simultaneous cultivation of trees and crops has shown that trees on agricultural landscapes play a vital role in water conservation, soil erosion prevention, temperature moderation, and the improvement of SOC and fertility, leading to better land use and higher crop productivity (Padmavathy and Poyyamoli 2013). Agroforestry practices can also reduce or remove significant amounts of GHGs through increased carbon storage in above-ground and below-ground biomass and in SOC (IPCC 2019). Agroforestry technologies can reduce air pollution, moderating both warming and cooling of the atmosphere, thereby creating a resilient microclimate for crops and livestock (Ellison et al. 2017; Ripple et al. 2020). Integrating agroforestry in cropping and livestock keeping systems can enhance carbon sequestration by significant amounts, thus enabling farming systems to be more resilient to the effects of climate change (Verchot et al. 2007). Agroforestry stores more carbon than pastures and fields with annual crops, but less than forested areas (Agroforestry Network 2019).

Recent studies have shown that agroforestry systems provide multiple benefits, including reducing GHG emissions. In their analysis, Nair et al. (2019) indicated, for instance, that agroforestry can store considerable amounts of carbon in soil and biomass. Thus, agroforestry remains a better option for lowering levels of atmospheric carbon dioxide and reducing GHG emissions. Using trees in agricultural landscapes can significantly reduce GHG emissions and enhance carbon sequestration by accumulating biomass in trees and soils than in traditional agriculture systems (Nair et al. 2019).

Agroforestry techniques offer a variety of ecosystem services that are helpful in the adaptation and mitigation of climate change by improving soil moisture retention and reduce soil erosion; thereby, increasing the resilience of farming systems to climate change (Mngoma et al. 2021). Thus, the integration of trees in agricultural systems ensure enhanced sustainable land management practices, enhanced resilience to climate change, and improved livelihoods for smallholder farmers. A study by Makumba et al. (2006) in Malawi found that using fertilizer trees, such as *Tephrosia vogelii* and *Cajanus cajan*, improved maize yields by up to 61% compared to fields without fertilizer trees. These fertilizer trees increased SOC and nitrogen content and soil moisture retention. Maize yields are also more stable in agroforestry fields, especially maize crop intercropped with *Leucaena* and *Gliricidia*, than in a fully fertilized sole maize field. In Cameroon, agroforestry-based soil fertility management practices enhanced crop productivity, reduced soil erosion, and promoted carbon sequestration (Awazi and Tchamba 2019). Basically, fields under agroforestry systems have higher SOM, soil moisture, and nitrogen levels, and hence increased soil fertility and moisture retention, which is attributable to the presence of trees responsible for providing shade, reduced wind speed, and enhanced soil water infiltration and nutrient cycling (Awazi and Tchamba 2019).

5 Other Benefits of Agroforestry Systems

Besides low crop productivity challenges, low-income countries experience the problem of natural resource degradation. Scarcity of environmental resources has serious implications for gender inequality, whereby women and girls spend long time fetching firewood and water for the entire family. Emerging evidence suggests that poverty levels are high in the periphery of forests or biodiversity conservation zones (Verde Selva et al. 2019). As a result, poverty fuels excessive and illegal resource extraction from proximate nature conservation areas (Verde Selva et al. 2019; MacKenzie and Hartter 2013). Thus, it is important to protect the soil nutrient-cycling benefits of trees from the short-term pressure of household energy and income needs. Existing literature highlights multiple benefits of agroforestry systems in favour of forests and biodiversity conservation. A study by Sibelet et al. (2019) shows that the agroforestry system contributes 65% of wood energy compared to 7% for the conventional system. Families using the conventional agricultural system extracted 81% of their wood energy from the forest, while families using the enriched agroforestry system only extracted 32%. The wood energy produced by the enriched agroforestry system offers an additional benefit to women, children and young people, who are responsible for transporting the bundles, because they do not have to travel long distances to supply the household (Sibelet et al. 2019). Collecting firewood from agroforestry systems reduces pressure put on protected areas which eventually save their biodiversity and mitigate the climate change associated impacts.

Tchoundjeu et al. (2016) showed that when *Faidherbia albida* is integrated into agroforestry systems for soil fertility improvement, the fields involved are less labor-intensive and more beneficial to women and children than other tree-based soil fertility-enhancing technologies. Pilote et al. (2017), in a study in southern province of Rwanda, showed that agroforestry practices are contributing to an increase in the income of agroforestry adopters compared to non-agroforestry adopters, and are improving soil fertility, reducing deforestation, and conserving soil and water in the district. Agroforestry systems incorporating and utilizing some of the legume trees (eg. Neem) or shrubs (Tephrosia/Tithonia) have potential to reduce pest infestations and weeds, leading to increased crop yields and reduced use of pesticides

(Kerebba et al. 2019). This alone suggest that agroforestry systems provide small-scale farmers with a more sustainable and resilient approach to pest management than monocropping.

6 Scaling and Incentivizing Adoption of Agroforestry Technologies

The adoption of agricultural technologies differs from one place to another (Tafere and Nigussie 2018). In the case of agroforestry innovation, its adoption is a necessary condition for the realization of benefits at both household and landscape levels. However, the uptake of agroforestry technologies is more complicated than those of annual crops because of the associated multiple components and longer gestation periods for the expected benefits (Pritha et al. 2024). As a result, adoption rates in SSA range from less than 5% in some areas to over 50% in others, with a median adoption rate of 20%. These adoption rates are influenced by a variety of factors, which include access to markets and extension services, availability of land and labour, the political and socio-economic environment, and farmers' perceptions of the benefits and risks of agroforestry (Mbow et al. 2013; Ullah et al. 2023). Among the maize-coppicing tree intercrops and improved fallows, adoption of *Tephrosia* and *Sesbania* has been limited due to poor supply of germplasm and perceived high costs. Some constraints are related to the interaction between technology and farm characteristics. For example, practices which require significant labour for both seasonal and off season tree management and protection, land or cash are deemed less attractive on farms with relatively low endowments of those factors (Ullah et al. 2023).

In order to increase adoption levels, the importance of policy support for developing and adopting agroforestry practices that enhance soil fertility cannot be overemphasized. Policies that promote adoption of agroforestry practices for soil fertility improvement can significantly impact on food security and livelihoods as the policies play crucial roles in providing incentives and disincentives for farmers to invest in agroforestry systems (Place and Dewees 1999). Thus, such policies must be tailored to local contexts, and must consider the diverse social, economic, and environmental factors that influence adoption of the respective agroforestry practices. Notable associated policies include the development of community bylaws related to tree planting. These bylaws affect the farmers' decision-making in adopting agroforestry farming systems. Furthermore, the bylaws also affect tenure-related security for timely adoption of agroforestry innovation (Tafere and Nigussie 2018; Ullah et al. 2023). This means that there is need to formulate necessary policies or bylaws that will encourage participation of farmers in various agricultural activities, which include the adoption of various agroforestry innovations.

7 Conclusion

Agroforestry innovations provide a variety of cost-effective means to enhance food security and livelihoods among smallholder farmers; at the same time, they contribute to climate change adaptation and mitigation. Studies have shown that soil fertility enhancing agroforestry technologies such intercropping with *Gliricidia sepium* a coppicing tree species, short rotation and relay fallowing with non-coppicing trees (*Cajanus cajan*, *Sesbania sesban* and *Tephrosia vogelii*), as well as high canopy tree species such as the *Faidherbia albida*-based agroforestry system with good tree management, can contribute to soil fertility and improved crop productivity. As a result, they enhance resilience of smallholder farmers against the negative impact of climate change. However, it has also been noted that despite the reported benefits, several factors, including lack of by-laws enforcement, have contributed to low adoption of these technologies. Thus, promoting localized policy support for developing and adopting agroforestry practices should be encouraged. Such policies should be tailored to local contexts, considering the diverse social, economic and environmental factors that influence adoption of the respective agroforestry innovations.

References

Agroforestry Network (2019) Agroforestry for adaptation and mitigation to climate change—2019. Agroforestry-for-adapation-and-mitigation-to-climate-change_web.pdf. https://siwi.org/wp-content/uploads/2020/02/

Akinnifesi FK, Ajayi O, Sileshi G, Chirwa PW, Chianu J (2010) Fertiliser trees for sustainable food security in the maize-based production systems of East and Southern Africa. A review. Agron Sustain Dev 30:615–629

Awazi NP, Tchamba MN (2019) Enhancing agricultural sustainability and productivity under changing climate conditions through improved agroforestry practices in smallholder farming systems in Sub-Saharan Africa. Afr J Agric Res 14(7):379–388. https://doi.org/10.5897/AJAR2018.12972

Beedy TL, Snapp SS, Akinnifesi FK, Sileshi GW (2010) Impact of Gliricidia sepium intercropping on soil organic matter fractions in a maize-based cropping system. Agric Ecosyst Environ 138:139–146

Ellison D, Morris CE, Locatelli B (2017) Trees, forests and water: cool insights for a hot world. Glob Environ Chang 43(51):51–61. https://doi.org/10.1016/j.gloenvcha.2017.01.002

Eze S, Dougill AJ, Banwart SA, Hermans TDG, Ligowe IS, Thierfelder C (2020) Impacts of conservation agriculture on soil structure and hydraulic properties of Malawian agricultural systems. Soil Tillage Res 201:104639. https://doi.org/10.1016/j.still.2020.104639

Fahad S, Chavan S, Chichahgare A, Uthapa A, Kumar M, Kakade V, Poczai P (2022) Agroforestry systems for soil health improvement and maintenance. Sustainability 14:14877. https://doi.org/10.3390/su142214877

Garrity DP, Akinnifesi FK, Ajayi OC, Weldesemayat SG, Mowo JG, Kalinganire A, Larwanou M, Bayala J (2010) Evergreen agriculture: a robust approach to sustainable food security in Africa. Food Secur 2:197–214

Giller KE, Tittonell P, Rufino MC, Wijk MTV, Zingore S, Mapfumo P, Adjei-Nsiah S (2019) Communicating complexity: integrated assessment of trade-offs concerning soil fertility

management within African farming systems to support innovation and development. Agric Syst 171:65–75
Ikerra ST, Maghembe JA, Smithson PC, Buresh RJ (1999) Soil nitrogen dynamics and relationships with maize yields in a gliricidia–maize intercrop in Malawi. Plant Soil 211:155–164
IPCC (2019) IPCC special report on climate change, desertification, land degradation, sustainable land management, food security, and greenhouse gas fluxes in terrestrial ecosystems. Intergovernmental panel on climate change. IPCC. (2019). Refinement to the 2006 IPCC Guidelines for National Greenhouse Gas Inventories. IPCC's 49th Session (Kyoto, Japan) and adopted/accepted on 12 May 2019
Kalele DN, Ogara WO, Oludhe C, Onono JO (2021) Climate change impacts and relevance of smallholder farmers' response in arid and semi-arid lands in Kenya. Sci Afr. https://doi.org/10.1016/j.sciaf.2021.e00814
Kerebba N, Oyedeji AO, Byamukama R, Kuria SK, Oyedeji OO (2019) Pesticidal activity of Tithonia diversifolia (Hemsl.) A. Gray and Tephrosia vogelii (Hook f.); phytochemical isolation and characterization: a review. S Afr J Bot 121:366–376
Kerr AC (2012) Drought resilience of maize-legume agroforestry systems in Malawi. UC Berkeley Electron. Theses Diss. UC Berkeley
Kuyah S, Sileshi G, Luedeling E, Akinnifesi F, Whitney C, Bayala J, Kuntashula E, Dimobe K, Mafongoya P (2020) Potential of agroforestry to enhance livelihood security in Africa. In: Agroforestry for degraded landscapes. Springer, Singapore, pp135–167. https://doi.org/10.1007/978-981-15-4136-0_4
Lelamo L (2021) A review on the indigenous multipurpose agroforestry tree species in Ethiopia: management, their productive and service roles and constraints. Heliyon 7(9). https://doi.org/10.1016/j.heliyon.2021.e07874
MacKenzie C, Hartter J (2013) Demand and proximity: drivers of illegal forest resource extraction. Oryx 47(2):288–297. https://doi.org/10.1017/S0030605312000026
Mafongoya PL, Bationo A, Kihara J, Waswa BS (2007) Appropriate technologies to replenish soil fertility in southern Africa. In: Bationo A, Waswa B, Kihara J, Kimetu J (eds) Advances in integrated soil fertility management in sub-saharan africa: challenges and opportunities. Dordrecht Springer, Netherlands, pp 29–43
Makumba W, Janssen B, Oenema O, Akinnifesi FK, Mweta D, Kwesinga F (2006) The long-term effects of a gliricidia–maize intercropping system in Southern Malawi, on gliricidia and maize yields, and soil properties. Agr Ecosyst Environ 116(1–2):85–92. https://doi.org/10.1016/j.agee.2006.03.012
Makumba W, Akinnifesi FK, Janssen B, Oenema O (2007) Long-term impact of a gliricidia-maize intercropping system on carbon sequestration in southern Malawi. Agric Ecosyst Environ 118:237–243. https://doi.org/10.1016/J.AGEE.2006.05.011
Martinsen V, Shitumbanuma V, Mulder J, Ritz C, Cornelissen G (2017) Effects of hand-hoe tilled conservation farming on soil quality and carbon stocks under onfarm conditions in Zambia. Agric Ecosyst Environ 241:168–178. https://doi.org/10.1016/J.AGEE.2017.03.010
Mbow C, Van Noordwijk M, Luedeling E, Neufeldt H, Minang PA, Kowero G (2013) Agroforestry solutions to address food security and climate change challenges in Africa. Elsevier
Mbow C, Rosenstock TS, Barioni LG, Benton TG, Herrero M, Krishnapillai M, Liwenga E, Pradhan P, Rivera-Ferre MG, Sapkota T, Tubiello FN, Xu Y, Yue SC (2019) Chapter 5: food security. In: IPCC special report on climate change, desertification, land degradation, sustainable land management, food security, and greenhouse gas fluxes in terrestrial ecosystems. Retrieved from https://www.ipcc.ch/srccl/chapter/chapter-5/
Mechler R, Bouwer LM, Schinko T, Surminski, S, Linnerooth-Bayer, J (2019) Loss and damage from climate change: concepts, methods and policy options. Springer Nature, p 557
Mehreteab T, Alamu EO, Njoloma, JB, Udaya SN, Ngumayo J (2022) Agroecological farming approaches that enhance resilience and mitigation to climate change in vulnerable farming systems. In: Nagothu US (ed) Climate neutral and resilient farming systems: practical solutions

for climate mitigation and adaptation, 1st edn. Routledge. https://doi.org/10.4324/9781003273172
Misra P (2011) Soil fertility management in agroforestry system. Int J Biotechnol Biochem 7(5):637–644. Retrieved from https://www.ripublication.com/Openaccess/ijbbv7n5_11.pdf
Mngoma P, Shackleton CM, Naidoo R (2021) Agroforestry and ecosystem services: a review of the science and practice in South Africa. Forest Policy Econ 128:102468
Molin PG, Chazdon R, de Barros F, Ferraz S, Brancalion PH (2018) A landscape approach for cost effective large-scale forest restoration. J Appl Ecol 55(6):2767–2778
Nair PKR, Kumar R, Nair EM (2019) Agroforestry as a strategy for carbon sequestration. J Soil Water Conserv 74(1):11–16
Njoloma JP, Sileshi GW, Sosola BG, Nalivata PC, Nyoka BI (2016) Soil fertility status in smallholder farmers' fields in Malawi. Afr J Agric Res (AJAR) 11(19):1679–1687. https://doi.org/10.5897/AJAR2015.10018
Opoku Mensah S, Ibrahim SK, Jacobs B, Cunningham R, Owusu-Ansah D, Adjei E (2024) Benefits of farmer managed natural regeneration to food security in semi-arid Ghana. Agric Hum Values. https://doi.org/10.1007/s10460-024-10546-7
Padmavathy A, Poyyamoli G (2013) Role of Agro-forestry on organic and conventional farmers' livelihood in Bahour, Puducherry-India. Int J Agric Sci 2(12):400–409
Pantera A, Mosquera-Losada MR, Herzog F, den Herder M (2021) Agroforestry and the environment. Agrofor Syst 95:767–774
Pilote K, Jewel A, Yohan L, Don Koo L (2017) Benefits and challenges of agroforestry adoption: a case of Musebeya sector, Nyamagabe District in southern province of Rwanda. For Sci Technol 13(4):174–180. https://doi.org/10.1080/21580103.2017.1392367
Place F, Dewees P (1999) Policies and incentives for the adoption of improved fallows. Agrofor Syst 47:323–343. https://doi.org/10.1023/A:1006248304024Pritha
Pritha D, Bhagirath B, Dil Bahadur R (2024) Assessing the role of agriculture-forestry–livestock nexus in improving farmers' food security in South Asia: a systematic literature review. Agric Syst 213:103807. ISSN 0308-521X (2024). https://doi.org/10.1016/j.agsy.2023.103807
Ripple WJ, Wolf C, Newsome TM, Barnard P, Moomaw WR (2020) World scientists' warning of a climate emergency. Bioscience 70(1):8–12. https://doi.org/10.1093/biosci/biz088
Rusinamhodzi L, Corbeels M, Nyamangara J, Giller KE (2012) Maize–grain legume intercropping is an attractive option for ecological intensification that reduces climatic risk for smallholder farmers in central Mozambique. Field Crop Res 136:12–22
Seghieri J, Droy I, Hadgu K (2021) Introduction to the special issue "scaling up of agroforestry innovations: enhancing food, nutrition and income security." Agroforest Syst 95:1245–1249. https://doi.org/10.1007/s10457-021-00689-5
Sheahan M, Barrett CB (2017) Ten striking facts about agricultural input use in Sub-Saharan Africa. Food Policy 67:12–25
Sheppard JP, Bohn Reckziegel R, Borrass L, Chirwa PW, Cuaranhua CJ, Hassler SK, Hoffmeister S, Kestel F, Maier R, Mälicke M (2020) Agroforestry: an appropriate and sustainable response to a changing climate in southern Africa? Sustainability 12:6796. https://doi.org/10.3390/su12176796
Shimada G (2022) The impact of climate-change-related disasters on Africa's economic growth, agriculture, and conflicts: can humantarian aid and food assistance offset the damage? Int J Environ Res Public Health 19(1):467. https://doi.org/10.3390/ijerph19010467
Sibelet N, Posada KE, Gutiérrez-Montes IA (2019) Agroforestry systems provide wood energy that improves livelihoods in Guatemala. Bois et Forêts des Tropiques 340:91–102. https://doi.org/10.19182/bft2019.340.a31692
Sileshi GW (2016) The magnitude and spatial extent of influence of Faidherbia albida trees on soil properties and primary productivity in drylands. J Arid Environ 132:1–14. https://doi.org/10.1016/j.jaridenv.2016.03.002
Sileshi G, Mafongoya P, Akinnifesi F, Phiri E, Chirwa P, Beedy T, Makumba W, Nyamadzawo G, Njoloma J, Wuta M, Nyamugafata P, Jiri O (2014) Agroforestry: fertilizer trees. In: Alfen NV

(ed) Encyclopedia of agriculture and food systems, vol 1, pp 222–234. https://doi.org/10.1016/B978-0-444-52512-3.00022-X
SIWI (2018) Water for productive and multifunctional landscapes. Stockholm International Water Institute, Report no. 38
Sorgho R, Quinonez C, Louis V, Winkler V, Dambach P, Sauerborn R, Horstick O (2020) Climate change policies in 16 west African countries: a systematic review of adaptation with a focus on agriculture, food security, and nutrition. Int J Environ Res Public Health 17(23):8897. https://doi.org/10.3390/ijerph17238897
Tafere SM, Nigussie ZA (2018) The adoption of introduced agroforestry innovations: determinants of a high adoption rate—a case study from Ethiopia. For Trees Livelihoods. https://doi.org/10.1080/14728028.2018.1493954
Tchoundjeu Z, Degrande A, Leakey RRB, Ngo-Mpeck ML, Asaah E, Tsobeng A (2016) An overview of the contribution of trees and agroforestry to sustainable development goals in Africa. For Trees Livelihoods 25(1):1–13
Tittonell P, Muriuki A, Klapwijk CJ, Shepherd KD, Coe R, Vanlauwe B, Giller KE (2016) The diversity of rural livelihoods and their influence on soil fertility in agricultural systems of East Africa—a typology of smallholder farms. Agric Syst 143:106–118
Tittonell P, El Mujtar V, Felix G, Kebede Y, Laborda L, Luján Soto R, de Vente J (2022) Regenerative agriculture—agroecology without politics? Front Sustain Food Syst 6:844261. https://doi.org/10.3389/fsufs.2022.844261
Ullah A, Mishra A, Bavorova M (2023) Agroforestry adoption decision in green growth initiative programs: key lessons from the billion trees afforestation project (BTAP). Environ Manage 71(5):950–964. https://doi.org/10.1007/s00267-023-01797-x
Vanlauwe B, Amede T, Bationo A, Bindraban P, Breman H, Cardinael R, Couedel A, Chivenge P, Corbeels M, Dobermann A, Falconnier G, Fatunbi W, Giller K, Harawa R, Kamau M, Merckx R, Palm C, Powlson D, Rusinamhodzi L, Johan S, Singh U, Zachary S, van Ittersum M, Witt C, Zingore S, Groot R (2023) Fertilizer and soil health in Africa: the role of fertilizer in building soil health to sustain farming and address climate change. International Fertilizer Development Center, Muscle Shoals, AL 35662 | US
Verchot LV, van Noordwijk M, Kandji S, Tomich T, Ong C, Albrecht A, Mackensen J, Bantilan C, Anupama KV, Palm C (2007) Climate change: linking adaptation and mitigation through agroforestry. Mitig Adapt Strat Glob Chang 12(5):901–918
Verde Selva G, Pauli N, Kiatkoski Kim M, Clifton J (2019) Can environmental compensation contribute to socially equitable conservation? The case of an ecological fiscal transfer in the Brazilian Atlantic forest. Local Environ 24(10):931–948

Open Access This chapter is licensed under the terms of the Creative Commons Attribution 4.0 International License (http://creativecommons.org/licenses/by/4.0/), which permits use, sharing, adaptation, distribution and reproduction in any medium or format, as long as you give appropriate credit to the original author(s) and the source, provide a link to the Creative Commons license and indicate if changes were made.

The images or other third party material in this chapter are included in the chapter's Creative Commons license, unless indicated otherwise in a credit line to the material. If material is not included in the chapter's Creative Commons license and your intended use is not permitted by statutory regulation or exceeds the permitted use, you will need to obtain permission directly from the copyright holder.

Importance, Impacts and Farmer Perception of In-Situ Rainwater Harvesting for Climate Adaptation in Malawi

Vernon H. Kabambe, Amos R. Ngwira, and Bishal K. Sitaula

Abstract Sustainable food production in the face of climate change and variability is highly dependent on the ability of the soil to hold water, sustain nutrient retention and microbial life for improved crop growth and development. In Malawi, crop yields are low, mainly due to poor soil fertility caused by many factors, including soil erosion, and crop nutrient mining. In-situ rain water harvesting technologies (IRWHT) provide practical approaches that are within reach of the resource-constrained farmers, to notably reduce soil erosion, improve water and nutrient availability, and serve as a platform to improve soil microbial diversity. This chapter provides a review of the soil fertility status, the extent of climate variability, soil erosion and the impacts of IRWHTs in Malawi and southern Africa. The chapter finally presents results of farmer attitudes and perceptions to IRWHTs and yield results on maize (Zea mays {L.}) in a case study conducted in 2017/18 season in Balaka district, southern Malawi. The study evaluated simple IRWHT's (mulching, box ridges, animal manure and manure and box ridges) in comparison to no intervention, all under a recommended fertilizer package and good agricultural practices (GAP). The review shows that a wide range of IRWHTs are available and suitable for smallholder farmers. The evidence in the literature shows that these are effective in improving water use efficiency, rainfall use efficiency and crop yields. For the best results, IRWHTs must be accompanied by GAP, including proper use of fertilizers and improved varieties. In the perceptions study, farmers were asked to rate the field performance of the IWRHTs (using a scale of 1 = most liked and 4 = least liked).

V. H. Kabambe (✉)
Lilongwe University of Agriculture and Natural Resources, Lilongwe, Malawi
e-mail: kabambev@mau.adventist.org

Malawi Adventist University, Ntcheu, Malawi

A. R. Ngwira
Intrenational Crops Research Institute for the Semi-Arid Tropics, Lilongwe, Malawi
e-mail: ANgwira@agra.org

B. K. Sitaula
Department of International Environment and Development Studies, Norwegian University of Life Sciences, Ås, Norway
e-mail: bishal.sitaula@nmbu.no

© The Author(s) 2026

B. R. Singh et al. (eds.), *Sustainable Food Systems in Sub-Saharan Africa*,
https://doi.org/10.1007/978-3-032-11288-0_14

The results showed that farmers had good comprehension of the IRWHTs, were aware of advantages and disadvantages, and were able to make informed choices based on their circumstances. The manure and box ridge treatment was very well rated for soil conservation (1.15) and drought resistance (1.30), while the mulching treatment was rated highly based on access (1.24) and labour requirement (2.0). The control plot was consistently rated the least. No significant differences in yield were observed amongst the treatments, but yield significantly varied between sites (range 3.9–3.8 t ha^{-1}). It is concluded that stakeholders should aim at working with farmers to facilitate access and adoption of these technologies, in combination with good agricultural practices, to prevent further land degradation (through erosion) and increase crop productivity.

Keywords Climate resilience · Conservation agriculture · Rain use efficiency · Water use efficiency

1 Introduction

1.1 Importance of State of Crop Production in Malawi

In Malawi, maize (Zea mays {L.}) is the staple food crop, accounting for more than 65% of land under cultivation, predominantly under rain-fed smallholder based cropping systems (MoA, 2020). In the 2019/20 season, maize was grown on 1.5 M ha of land. Table 1 shows the average smallholder farmers' yields of maize, and other commonly grown crops between 2015/16 and 2019/20. These crops included rice (*Oryza sativa*) groundnuts (*Arachis hypogea*), soybean (*Glycine max*), pigeon peas (*Cajanus cajan*), common beans (*Phaseolus vulgaris*), cowpeas (*Vigna unguiculata*), sorghum (*Sorghum bicolor* {L.} Moench), sunflower (*Helianthus annuus*), cassava (*Manihot esculanta*), potatoes (*Solanum tuberosum*) and sweet potatoes (*Ipomoea batatas*). Other crops not included in the table include cotton (*Gossypium hirsutum*), dolichos beans (*Dolichos lablab*), chickpeas (*Cicer arientinum*), ground beans (*Vigna subterranea*), green grams (*Vigna aureus*), finger millet (*Eleusine coracana*), sesame (*Sesamum indicum L.*), chillies, fruits and vegetables. These yields are low compared to their potential yields. For example, the yields of maize are 20–40% of potential yields, the trend being similar with other crops. The wide range of crops grown by farmers gives opportunity for crop rotation, for food and dietary diversification, income diversification, and for pest management and soil fertility improvement purposes. Based on Malawi Vision 2063 national strategic plan, agriculture commercialization is pillar number one of the government (NPC, 2020); thus, it is imperative to find strategic production approaches that can sustainably increase yields under the prevailing climate variability events.

Table 1 National average crop yields and potential yields (kg ha^{-1}) of commonly grown crops in Malawi between 2015/16 and 2019/20 seasons

Crop	Season and yields, kg ha^{-1}					Yield potential
	2015/16	2016/17	2017/18	2018/19	2019/20	–
Maize	1415	1937	1601	1468	1800	5000–10,000
Groundnuts	743	742	987	848	1011	2000–2500
Soybean	890	885	1099	977	1093	3500
Pigeon peas	1506	1507	1731	1743	1624	2500
Rice	1560	1414	1744	1,624	1759	4000
Common beans	481	390	509	431	598	2–2500
Cowpeas	341	335	473	434	503	2000
Sunflower	974	975	1108	1058	1158	3000
Cassava	21,889	21,413	22,694	23,490	23,543	45,000
Potatoes	15,321	18,416	15,216	12,602	15,831	15,000–20,000
Sweet potatoes	17,401	20,159	19,721	21,258	21,950	40,000
Sorghum	589	589	870	780	11,677	2500

Source MoA 2020

1.2 *Constraints to Food Production in Malawi and Southern Africa*

Poor and declining soil fertility as well as rainfall variability are some of the major reasons contributing to low crop yields in southern Africa (Kumwenda et al. 1997; Blackie and Mann 2005; MoA 2020). Other constraints include insects, pests, parasitic weeds and diseases (MoA 2020). Poor soil fertility prompted the Government of Malawi to initiate small fertilizer support programs (Blackie and Mann 2005; MoAIWD 2012). Poor management practices, such as late planting, fertilizer application and weeding, and lack of or poor pest control practices result in crops not using fertilizer and water efficiently, hence large yield gaps. World-wide, 80% of cropland is under rain-fed production (Chen et al. 2018), such that irregular rainfall patterns are responsible for water shortage for crop production in arid regions (Qin et al. 2015; Li et al 2017). As prices of inputs increase sharply, and with rapid population growth of 2.59% (WorldData.infor 2022), it is important to focus on crop management practices that improve efficient use of resources such as water and nutrients. In-situ rain water harvesting practices help to retain water, thereby reducing erosion, improving nutrient and water use efficiencies, thereby providing adaptation and mitigation to inter-annual rainfall variations.

Malawi is one of the countries that has been experiencing adverse effects of climate change. In a review, Bei et al. (2008) reported that Intergovernmental Panel on Climate Change (IPCC) prediction models show increased uncertainty on regional and national scales. Environmental Affairs Department (EAD 2002) noted that the vulnerability of Malawi to climate change also mainly arises from socio-economic, demographic and climatic factors, which include a slim economic base, limited agro-processing facilities, over-dependence on rain-fed agriculture and fuel wood for energy. Fuji et al. (2020) described described both the impacts and extend of climate change in Malawi.

CIMMYT (2013) reported that about 40% of area grown under maize in sub-Saharan Africa experiences drought stress, causing 10–25% yield loss. CIMMYT (2013) also noted that around 25% of the maize crop suffers from frequent drought leading to losses of about half of the harvest. In addition, Edmeandes et al. (2000) noted that 20–25% of the yield gap between well-watered and water-limited crops can be closed by water-conserving agronomic practices. In groundnuts, drought stress reduces nodulation, pod numbers, size and yield (Dakora 1984). It also enhances Aspergillus invasion and aflatoxin production by promoting favourable carbohydrate compounds.

While some rainfall reports tend to show non-significant rainfall variations in annual seasonal trends in Malawi (FAO 2017), irregular rainfall is common (Kabambe et al. 2018; FAO 2017). In temporal trends study between 1981 and 2021, Demissie and Gebrechorkos (2024) reported non-significant increase in precipitation off 3 mm/year in Zambia and Malawi Malawi, whereas they reported significant increase trend in temperature of 0.06 °C/decade. In Zimbabwe, Mupangwa et al. (2006) reported that annual rainfall for a single site can vary by up to 1000 mm from year to year.

One main purpose of IRWHT is to reduce soil erosion, which is a major form of environmental degradation in Malawi. Soil erosion takes with it soil nutrients, thus further incapacitating the crops from growing well enough to provide soil cover. In Malawi, a country-wide study by Vargas and Omuto (2016) classified soil erosion as moderate in the northern region, and light elsewhere. They reported soil loss (t ha^{-1} $year^{-1}$) of 0 to < 5 in 13 districts, 5 to < 10 in 7 districts and > 10 in 5 districts.

2 Opportunities of Restoring and Maintaining Soil Productivity by Means of IRWHT

2.1 Definition and Range of IRWH Technologies

In order to increase adoption of some of the practices, it is important that development and extension actors must be well abreast with the range of options available and their impacts on productivity. FAO (2017) noted that targeting extension messages to specific populations and regions and household needs is important as people have different challenges and due to their location and socioeconomic conditions.

In rain-fed agricultural production, Rockstrom et al. (2010) indicated that the two main strategies in drought prone agriculture should be to increase plant water availability and increase plant water uptake. The approaches that increase plant water availability are sometimes referred to as *in-situ* water harvesting technologies. These are field practices that help to store water in areas where water it will be utilized by the crops (Rockstrom et al. 2010). A non-exhaustive list of IRWHT is shown in Box 1.

Box 1. Non-exhaustive list of in-situ rain water harvesting technologies

Agroforestry practices	Manure application—all types
Grass strips including vetiver grass	Box and tied ridges
Mulching—all types	Contour ridges and bunds
Pit planting	Swales and pot holes
Cover crops	Terracing
Intercropping	Zero and reduced tillage practices
Green manures all types	Fertilizer application
Deep tillage practices—all types	Good Agricultural Practices
Compost and animal manures	Micro-basins

Most of the IRWHT techniques fall within the broader range of practices under sustainable agriculture, climate smart agriculture and agroecology. In Malawi these practices are well described in 'Guide to Agriculture Production and Natural Resource Management' (MOA 2020), and 'National Guidelines for Implementing Conservation Agriculture in Malawi' (NCATF 2016). Most recently Southern Africa Development Community (SADC) has released the Climate Smart Agriculture Handbook (CCARDESA 2023). There is also the Malawi Climate Smart Agriculture Handbook for Frontline Extension Staff.

2.2 *Impacts of IRWHT as Means of Adaptation Under Inter-Annual Dry Spells Through Water Retention, on Soil Productivity, Crop and Growth and Yield*

In this chapter, the focus is on the impacts of those practices within affordable reach of farmers, including those involving physical structure such as box or tied ridging, mulching, use of animal or green manure, indirect practices that improve rainfall use efficiency (RUE), water use efficiency (WUE) & fertilizer use efficiency (FUE), such as time of planting, type and timing of fertilizer application. The chapter then presents the results of a case study which evaluated simple IRWHT with farmers in a semi-arid area of Malawi. After a review, Turner (2004) concluded that the major impact of agronomic management on RUE has not arisen from increasing water use by the crop in evapotranspiration, but from increasing water use by the crop in itself at the expense of water loss by weeds or from the soil by evaporation.

2.2.1 Mulching

In the context of smallholder farmers, mulching involves placing usually dead plant litter on the soil surface to achieve soil cover, and it is a key component of conservation agriculture packages (NCATF 2016; MoA 2020). A review by Fairhurst (2012) showed that straw mulch of 2.0 t ha^{-1} resulted in 60% reduction in runoff and 90% reduction in erosion. With 6 t/ha of straw mulch there was 90% reduction in runoff and erosion reduced to zero. Mulching has also been reported to conserve soil moisture (Thierfelder and Wall, 2009; Lal, 1978). Mulch amounts of 2–5 t ha^{-1} improved water infiltration even after the first season, particularly under drier conditions (Thierfelder and Wall 2009; Roth et al. 1988). Lal (1978) reported increased soil moisture conservation, reduced bulk density and enhanced maize leaf area in mulched soil compared to unmulched soil. However, intensive rainfall in unmulched fields could also result in waterlogging. In Zambia, Gill et al. (1992) demonstrated a significant decline in weed density as a function of mulching, while Ngwira et al. (2014) reported exponential decline in weed biomass as a function of mulch density in Malawi. Giller et al. (2009) noted that residues with a wide C/N ratio can reduce N supply in soil to lock up the N by microbial action. Such effects can be of minimum concern when nominal rates of 30 kg N ha^{-1} are applied or in N rich soils. Coppens et al. (2007) reported a lower decomposition rate of 35% for surface placed residues compared with incorporated residues. In Malawi, Sanga and Kabambe (2014) reported higher maize grain yields with pigeon residues incorporation than surface placement. In a review, Zhang et al. (2017) reported that overall, mulch increased WUE by 61% because of the change in water balance and the productivity of the maize crop. Ali et al. (2018) reported that the presence of wheat residue increased RUE by 50%.

2.2.2 Box or Tied Ridges, Basins and Contour Bands

Use of box ridges, tied ridges, contour bands or contour hedgerows are all recommended and well described in Malawi (MoA 2020). A study by Mudatengubha et al. (2005) in a semi-arid area in Rwanda compared use of pot holes, tied ridging, and mulching to flat planting (control). The authors reported significant increase of soil moisture, maize plant dry weight by all the treatments compared to the control (flat planting) at 8 weeks after planting, as well as in yield and yield components. Wiyo et al. (1999) conducted a study in which mean yields due to tie-ridging in drought, dry, normal and wet years at different N levels were derived from observed smallholder maize yield data using a linear nitrogen response model and field-observed retained rainwater amounts in tie-ridged fields. Their simulation results indicated that tie-ridging will result in hybrid maize yield gain of 1050 kg ha^{-1} in a drought year, and of 560 kg ha^{-1} in a dry year. There will be a hybrid maize yield loss of 350 kg ha^{-1} in a normal year, and of 700 kg ha^{-1} in a wet year. The results suggest suppressive effects of excess moisture in wet periods. The designs of physical water conservation structures must allow excess water to be drained away slowly (MOA 2020). In Tanzania, Kizito et al. (2022) reported maize yield increases of 58 and

33% with tied ridging and ripping compared to conventional tillage. The authors also reported 86 and 87% improvement in moisture retention, 88 and 31% reduction in erosion for the same treatment comparisons.

2.2.3 Organic Soil Inputs

Organic soil inputs encompass all types of manure, such as compost, animal manure, crops residues, and more recently arbuscular mychorrizal fungi, AMF, (Njira 2014). The benefits of organic inputs, apart from supply of mineral nutrients to crops, are well documented. Organic inputs directly supply available forms of carbon that stimulate soil biological activity, which contributes to soil organic matter formation, SOM (Fairhurst 2012). The SOM also improves soil productivity through regulation of the rates and amount of nutrients released, improvement of soil water infiltration rate and soil water holding capacity, increasing cation exchange capacity, enhancing soil aggregation of organic matter, improving soil structure, reducing bulk density and promoting good aeration; buffering the soil against changes in acidity, alkalinity and salinity, and improving availability of other nutrients such as P in acidic soils (Fairhust 2012; Tisdale et al. 1985). The best benefits of manure are often obtained when combined with inorganic fertilizers (Mloza-Banda and Kabambe 2014), usually because manure contains lower amounts of nutrients. Kabambe et al. (2012) reported that application of compost manure, together with the recommended inorganic fertilizer, increased yield from 4.25 to 5.84 t ha^{-1} (37%) in acidic soils. The best options for furnishing organic input is through growing crops and plants with high forage nitrogen and returning the forage to the soil. Grain legume crops such as groundnuts, cowpeas and pigeon peas and soybeans are widely used for this purpose. In addition, further techniques such as doubling up of legumes as intercrops, have been shown to further enhance legume biomass and biological nitrogen fixation (57–92 kg ha^{-1} fixed) and improve yields in subsequent seasons (Njira et al. 2017, 2020; Mhango et al. 2017).

2.2.4 Fertilizers and Soil Fertility Management

Soil fertility can be defined as the capacity of the soil to support the production of crops and livestock (Fairhust 2012). The support must necessarily include adequate and timely supply of water, air, nutrients and anchorage. Thus practices to improve soil fertility can also be regarded as IRWHTs (Rockstrom et al. 2010; Tittonel et al. 2007). The emphasis in this section is on those practices that are not already mentioned in other sections. In connection with fertilizer management, there are 'Five Rights' (5R's) of fertilizer management recommended. These are the right amount, right placement, right type, right time and the final R is targeting of the most remunerative options (Fairhust, 2012). This principle serves to encourage farmers to have continuous and long term plans to improve the soil's physical and chemical properties. Application of lime, manure or any other soil reaction amendment all constitute measures that improve crop growth and WUE. From a study under dryland conditions in Kenya, Oduor et al (2023) reported that in comparison to 30 kg ha^{-1}, 120 and 60 kg ha^{-1} N significantly enhanced WUE by 66 and 25%. Kunrath et al. (2018) reported that limiting nitrogen to tall fescue had a negative effect on WUE because the transpiration decreased relative soil water evaporation. Gaiser et al. (2004) reported that WUE of a maize/cowpea intercrop on a highly acid soil was positively affected by liming and fertilizer application.

2.2.5 Tillage Practices and Crop Rotations

One of approaches that helps to improve crop water use in dry environments is to increase the depth of rooting, since crops may fail to use available water due to restricted root growth. Deep ripping to about 30 cm was shown to increase yields and RUE on deep sandy soils (Delroy and Bowden 1986). In southern Africa, many crops such as pigeon peas, cassava, cotton and agroforestry tree species provide 'biological plows" or biopores which achieve the same purpose (Cresswell and Kirkegaard 1995), thus making purposeful rotations important. Sometimes soils are compacted due to acid reaction. Such constraint can be alleviated through deep placement of lime (gypsum) application to flocculate the soil particles and increase water penetration and root growth (Hamza and Anderson 2003). In Zimbabwe, Mupangwa et al. (2006) reported that mulch ripping and other minimum tillage practices consistently increased soil water content and crop yields compared to traditional spring ploughing. They further reported higher soil loss from conventional tillage compared to plots under different tillage practices. Rotations are sometimes useful in the management of diseases and pests which attack roots, stems and leaves. Healthier plants have much better WUE due to better water, nutrients, sunlight and water capture.

3 Case Study: Evaluation of Selected IRWHT in Balaka District, Malawi—Farmers' Assessment Rating and Yield

3.1 Study Background and Site Description

This case study was conducted in 2017/18 season in Balaka district, Ulongwe Extension Planning Area (EPA) area in Malawi. In the Malawi extension system, districts are divided into Extension Planning Areas (EPAs) and sections. The study was prompted by rainfall records and observations in the previous years, and was a follow up to other activities in the area. In 2015/16 season, the total rainfall in Ulongwe EPA ranged from 326 mm at Chombe village in Chibwana Msaamala section to 527 mm at Kalembo 1 village in Chitseko section. The long term annual rainfall for the EPA is 840 to 1000 mm. The rainfall was erratic, with up to 3 dry spells (periods with at least 10 days of no rain or trace rainfall) recorded in two sections, and 2 dry spells in another. Kalembo 1 village in Chitseko section had no dry spells as well as the highest rainfall (Kabambe et al. 2018). Based on the above, it is clear that farmers in Balaka are exposed to high risk of dry spells; thus researchers were prompted to initiate this study. The objective was to provide a platform for co-learning between farmers and stakeholders on roles, farmers' assessment and impacts of IRWHT with maize as a test crop.

3.2 *Study Description and Designs*

Five treatments were evaluated on farmers' fields as detailed in Table 2. Each set of the five treatments was allocated to one farmer, in a randomized complete block design with the farmer as replication. Due to the learning objective, the set of plots was termed learning centers. The spread of treatments within each section is shown in Table 3.

Table 2 Treatments for 2017/18 maize IRWHT learning centres. Maize variety = DK8033

No	Treatment description	
	Standard good agricultural practices	In-situ water harvesting technology
1	Ridges at 0.75 m × 0.5 m × 2 plants. Fertilizer:- 69:21:04S as a basal and top dressing from the compound 23:21:0 + 4S and urea. Kept weed free by weeding 3 times. Plant with first rains	Control plot. No IRWH applied
2	Same as in 1	Mulching with maize stover or alternative at 5 t/ha or 9 kg per plot
3	Same as in 1	Make box ridges
4	Same as in 1	5.0 t ha^{-1} animal manure (2.25 kg/ridge) in a 15 cm groove made of the ridge, then bury
5	Same as in 1	A combination of manure and box ridges

Table 3 Distribution five- plot host farmers for in-rain water harvesting trials, 2917/18 season

Section	Number of villages	Number of farmers (replicates)
Chibwana	3	7
Hindahinda	3	7
Chitseko	3	9
Mulambe	2	6

Plot sizes were 4 ridges × 0.75 m × 6 m. Basal dressing fertilizer was applied at a rate of 100 kg of 23:21:0 + 4S for all maize plots at planting or soon after crop emergence. For top-dressing, 100 kg/ha urea (46 kg N) was applied to all maize plots at 3 weeks after planting. Participating farmers were encouraged to weed the plots 2–3 times, following the recommendations. The recommended agronomic practices were supervised by lead farmers, district field officers and project staff. Yield was recorded from the middle two ridges, and adjusted to a storage moisture of 12.5%.

Farmer evaluation on the learning centre (LC) treatments

Field days were held in some of the learning centres (LCs). During some field days and visits, discussions on advantages and disadvantages of the technologies were held with LC host farmers and non-host farmers as well, and were recorded. In some cases one farmer, one vote system was used to identify most and least preferred treatments. In addition, a rating system was used with LC host farmers only. On a rating of 1–4, farmers were asked to rate each treatment in terms of labour requirements, access to required resources, cost, drought resistance of the crop, success with water conservation.

3.3 *Results and Discussion*

3.3.1 Early Season Farmer Assessment of IRWH Learning Centres

Farmers' listing of advantages and disadvantages of the in-situ water harvesting technologies in the 2017/18 LCs, compared to control, is shown in Table 4. The discussions were conducted in January 2018 when the crop was in advanced vegetative stage. The discussions are a summary from 5 villages, involving a total of 76 farmers (52 women, 24 men). All farmers were host farmers to an LC. The results of the discussion are interesting in that the farmers were able to mention advantages relating to mechanisms or technical aspects of the technology, such as mulch and manure being source of nutrients, and the fact that preventing run-off also preserves nutrients. The farmers were also able to commend that the design or choice of treatments, noting that it allows fair comparison and appreciation of some factors; for example, having a control allows direct comparison with mulching, box ridges or manure. The farmers appreciated that the combination of box ridges and manures will help them determine if using both is necessary. Farmers were also able to foretell disadvantages, such as potential for waterlogging with box ridges.

3.3.2 Preference Ranking by Combination of Host and Non-Host Farmers During Field days

Field days were held in mid-February 2018, when the maize crop was nearing maturity. It was a good time to assess the performance of the treatments. Farmers were

Table 4 Summary of advantages and disadvantages mentioned by five groups of LC host farmers in Chitseko and Mulambe section, January 2018 at the vegetative stage. The + shows number of times mentioned out of 5 group discussions

Technology type: control plot	
Advantages/ & + ve experience gained	Disadvantages/problems & -ve experiences
No time wasted in field preparation + Less work involved +	Encourages erosion + + . No moisture conserved leading to wilting + + +
Technology type: mulching approx 5t/ha	
Conserves moisture + + + + + Mulch becomes manure after decomposition + + + + Less work during weeding + + + Less weeds, just hand weeding +	Availability of mulch problem for large fields + + + Mulch brings fall army worms + Mulch attract termites + Mulch damaged by goats +
Technology type: box ridges	
Holds water + + + + Prevents run-off + Prevents erosion +	Possible leaching under excess rain + + +
Technology type: animal manure, 5 t/ha	
Provides nutrients + + + + + Improves soil texture + + + + Improves soil water holding capacity + + + + +	Difficult to access + + Soil prone to run-off + Brings white grubs +
Technology type: manure and box ridges	
Both box ridges and manure hold water + + + + Manure provides nutrients + + + +	Manure not available to some farmers + Possible leaching with box ridges under excess rain +

asked to identify their best plot in terms of how it performed on the site. They were also asked to identify a treatment which they would actually adopt based on their own situation. This was a *one person, one vote* arrangement. The results are shown in Table 5. The results show variability of first choice by section, crop and village. Overall mulching was most popular. The reasons as to why farmers made the choice as to which performance was most suitable were apparently the same. While manure and box ridges were not popular on their own, they were popular when combined.

3.3.3 Individual Rating of IWRHTs Treatments by LC Host Farmers

At the end of March 2018, when maize was at maturity and heading towards drying off, an individual host famer rating exercise was undertaken. The rating asked each famer to weigh each technology with respect to labour, cost, drought resilience, effectiveness on water conservation, and expected yield. The weighing was done on a scale of 1–4, where 1 was most acceptable and 4 was least acceptable or liked. The purpose of these ratings was to further understand the reasons for the choices made earlier. Only farmers hosting the five plot learning centres were asked to rank, as they

Table 5 First choice ranking of IRWH technologies on basis of observed performance and suitability for adoption in sections (sect) and villages (vge) x basis of ranking, February 2018. Assessment by a mixture LC host and non-host farmers

Treatment	Chitseko sect, Chakwita vge		Mulambe sect, Namunde vge		Chibwana sect Chibwana vge		Hinda hinda sect Chipyali vge
	Performance	Most suited	Performance	Most suited	Performance	Most suited	Performance
Control	0	0	0	0	0	0	0
Mulching	25	27	10	10	12	14	23
Manure	2	5	5	6	12	15	21
Box ridges	3	4	7	7	19	19	0
Manure & Box ridges	18	19	8	7	23	16	1
Total	86	86	30	30	66	66	44

had experience with all treatments. The results are shown in Table 6. The control was least rated throughout. It is most interesting that the best rated treatments varied according factor of consideration. For example manure and box ridge treatment was very well rated for soil conservation and drought resistance; the mulching treatment best rated or liked for cost, access and labour.

Table 6 Individual rating of treatments by maize host farmers conducting at crop maturity period based on specified criteria

Treatment	Aspect rated, 1 = most liked, acceptable; 4 = least acceptable or liked (n = 100)					
	Labour	Access	Cost	Drought resistance	Water conservation	Expected yield
Control	3.16	2.08	3.0	3.72	3.88	3.48
Mulching	2.00	1.24	2.12	1.56	1.40	1.40
Box ridges	2.08	1.32	2.36	1.88	1.52	1.72
Manure	2.04	1.72	2.32	1.64	2.00	1.76
Manure + box	2.23	1.48	2.08	1.30	1.15	1.59
Mean	2.29	1.57	2.40	2.05	2.00	2.01
P	< 0.001	0.002	0.014	< 0.001	< 0.001	0.148
LSD	0.55	0.43	0.57	0.40	0.32	0.38
P Sect x Treat	0.81	0.64	0.89	0.8	0.35	0.036
CV	42	50	42	34	28	36

Table 7 Maize grain yield (kg ha^{-1}) results from IRWH learning centres

Treatment	Section				Mean
	Chibwana Nsamala	Hindahinda	Chitseko	Mulambe	
Control	3104	3003	2805	3129	2993
Mulching	3779	4070	3532	2672	3557
Box ridges	3234	4059	2899	2703	3221
Manure	2956	4258	3015	3679	3302
Manure + box	3385	3824	3371	2631	3337
Mean	3292	3843	3122	2963	
F Prob section	0.017				
F Prob treat	0.54				
P sect x Treat	0.92				
CV %	83				

3.3.4 Maize Yield Results

Maize grain yield results (Table 7) were not significantly affected by treatment effects. However, there were significant differences for yield from the sections. The highest yield was obtained from Hindahinda.

4 Lessons and General Conclusions and Recommendations

The review shows that there is a wide variety of options available to farmers to conserve moisture in their fields and get better yields. Rainfall use efficiency is achieved not from increasing water use by the crop in evapotranspiration, but through increasing water use by the crop in itself at the expense of water loss by weeds or from the soil by evaporation (Turner 2004). Water use efficiency, defined as crop biomass gain per unit of water used, is achieved when water in the field is used by the crop to gain growth. Other factors such as nutrients, efficient cultivar, freedom from weeds and pests must be available and in good balance. The review has shown that farmers can moderate or avoid harm from dry spells, and they can exploit rain water by making it more available for crop use. By improving rainfall use, IRWHTs helps to prevent erosion (Fairhurst 2012); and when integrated with organic residue retention, soil fertility can be restored or increased. The case study has shown that farmers are aware of the benefits IRWHTs, and that their impacts may vary based on other circumstances such rainfall amount and intensities. However, no significant effects on yield were detected, and this is in agreement with Wiyo et al. (1999) who reported similar results. Use of biosystems such as intercropping with legumes such as pigeon peas, and agroforestry, has strong advantages for smallholder farmers in that other tangible benefits such as legume yield and fuel can be realized (Njira et al

2017, 2020; Mhango et al 2017; Ngwira et al 2012), in addition to biological plowing benefits (Cresswell and Kirkegaard 1995). To avoid the problem of water logging, designs of physical structures are made to allow for gentle drainage when water is in excess (MoA 2020). We recommend that stakeholders should adopt an integrated approach that embraces IRWHTs and soil fertility enhancement, to prevent erosion and enhance rain, water and nutrient use efficiency.

Acknowledgements The authors wish to thank the Norwegian government for supporting this work through grant support provided to Lilongwe University of Agriculture and Natural Resources (LUANAR) under the project titled "Capacity Building for Managing Climate Change in Malawi." We also thank management and staff of LUANAR for their support in the smooth running of this work. Likewise, we thank all our collaborating staff and farmers in Balaka district without whom this work would not have been a success.

References

Ali S, Manzoor AJ, Sohail A, Khan A, Khan MI et al (2018) Soil amendment strategies to improve water use efficiency and productivity of maize under different irrigation conditions. Agric Water Manage 210:88–95

Bei SW, Mkwambisi D, and Gomani M (2008) Climate and rural Livelihoods in Malawi. Review study report of Norwegian support to FAO and SCC in Malawi, with a note on some regional implications. Noragic Report No. 41, Noragic University of Life Sciences (UMB), Norway

Blackie MJ, Mann CK (2005) The origin and concept of the starter pack. In: Levy S (ed) Starter packs: strategy to fight hunger in developing countries? CABI Publishing, Reading, UK, Lessons from the Malawi Experience, pp 15–27

CCARDESA Centre for Coordination of Agriculture Research and Development for Southern Africa (2023) Climate smart agriculture handbook. SADC Secretariat, Gaberone, Botswana

Chen B, Liu E, Mei X, Yan C, Garre S (2018) Modelling soil water dynamics in rain-fed spring maize with plastic mulching. Agric Water Manage [Internet] 198:19–27. Available from:http://linkinghub.elsevier.com/retrieve/pii/SO378377417303980. Retrieval date?

CIMMYT (2013) Drought tolerant maize for Africa project. DTMA brief, Sept 2013. http://dtma.cimmyt.org/index.php/about/background

Coppens F, Garnier P, Findeling A, Merckx R, Recous S (2007) Decomposition of mulched versus incorporated crop residues: modelling with PASTIS clarifies interactions between residue quality and location. Soil Biol Biochem 39(9):2339–2350

Cresswell HP, Kirkegaard JA (1995) Subsoil amelioration by plant roots: the process and evidence. Aust J Soil Res 33:221–239

Dakora FD (1984) Nodulation and nitrogen fixation by groundnut in amended and un-amended field soils in Ghana. In: Ssali H, Keya SP (eds) Biological nitrogen fixation in Africa, pp 324–339

Delroy ND, Bowden JW (1986) Effect of deep ripping, the previous crop, and applied nitrogen on growth and yield of a wheat crop. Aust J Exp Agric 26:469–479

Demissie T, Gebrechorkos AH (2024) Spatio-temporal trends in precipitation, temperature and extremes: study of Malawi and Zambia (1981–2021). Sustainability 16 (10):3885. https://hdl.handle.net/10565/148735

EAD (Environmental Affairs Department) (2002) Initial communication under the united nations framework on climate Change. Ministry of Natural Resources and Environment, Lilongwe, Malawi, p 35

Edmeades GO, Bolaños J, Elings A, Ribaut JM, Bänziger M, Westgate ME (2000) The role and regulation of the anthesis-silking interval in maize, p 43–73. In: Westgate ME, Boote KJ (eds) Physiology and modelling kernel set in maize. CSSA Special Publication No. 29. CSSA, Madison, WI, pp 43–73
Fairhurst T (ed) (2012) Handbook for integrated soil fertility management. Africa Soil Health Consortium, Nairobi
FAO, Food and Agriculture Organisation of the United Nations (2017) Tackling climate change in Zambia and Malawi. Bringing together evidence and policy insights. www.fao.org/climatechange/epic
Fujisawa, M Gordes A, Heureux A (2020) Assessing the impacts of climate change on the agriculture sectors in Malawi. The MOSSAIC methodology for national adaptation planning. Rome, FAO. https://doi.org/10.4060/ca8624en
Gaiser T, de Barros L, Lange FM, Williams (2004) Water use efficiency of a maize/cowpea intercrop on highly acidic soil as affected by liming and fertilizaer application. Plant Soil 263:165–171
Gill KS, Arshd MA, Chivunda BK, Phiri B, Gumbo M (1992) Influence of residue mulch, tillage and cultural practices on weed mass and corn yield from three experiments. Soil Tillage Res 24(30):211–223
Giller KE, Witter E, Crobeels M, Tittonell P (2009) Conservation agriculture and smallholder farming in Africa: the heretics' view. Field Crop Res 114(1):23–34
Hamza MA, Anderson WK (2003) Responses of soil properties and grain yields to deep ripping and gypsum application in compacted loamy sand soil contrasted with a sandy clay loam soil in Western Australia. Aust J Agric Res 54:273–282
Kabambe VH, Ngwira AR, Aune JB, Sitaula BK, Chilongo T (2018) Productivity and profitability of pigeon pea and maize rotation system in Balaka district, southern Malawi. Agrican J Agric Res 13:2392–2398
Kabambe VH, Chilimba ADC, Ngwira A, Mbawe M, Kambauwa G, Mapfumo P (2012) Using innovation platforms to scale out soil acidity ameliorating technologies in Dedza district, central Malawi. Afr J Biotech 11(3):561–569. https://doi.org/10.5897/AJB10.2227
Kizito F, Chikowo R, Kimaro A, Swai E (2022) Soil and water conservation for climate resilient agriculture. In: Bekunda M, Hoeschie-Zeledon, Odhong J (eds) Sustainable agriculture intensification. CAB International
Kumwenda JDT, Waddington SR, Snapp SS, Jones RB, Blackie MJ (1997) Soil fertility management in southern Africa. In: Byerlee D, Eicher CK (eds) Africa's emerging maize revolution. Lynne Reiner Publishers, Colorado
Kunrath TR, Lemaire G, Sadras VO, Gastal F (2018) Water use efficiency in perennial species: interactions between nitrogen nutrition and water deficit. Field Crops Res 222:1–11
Lal R (1978) Influence of within and between row mulching on soil temperature, soil moisture, root development and yield of maize (Zea mays L) in a tropical soil. Field Crops Res 1:127–139
Li C, Wang C, Wen X, Qin X, Liu Y, Han J, Li Y, Llao Y, Wu W (2017). Ridge-furrow with plastic film mulching practice improves maize productivity and resource use efficiency under the wheat—maize double—cropping system in dry semi-humid areas. Field Crop Res 203201–211. https://doi.org/10.1016/jfcr2016.12029
McSweeney C, New M, Lizcano G (2025) UNDP Climate Change Profiles, Malawi. https:// Un-gsp.org/sites/ default/ files/ documents/ malawi.oxford. reportpdf
McSweeney M, New M, Lizcano G, Lu X (2010) Improving accessibility of observed and projected climate change in developing countries. Article Bull Am Meteorol Soc. https://doi.org/10.1175/2009BAMS2826.1
MoA—Ministry of Agriculture (2020) A guide to agriculture and natural resources management. Government of Malawi. Lilongwe
MoAIWD (2012) 2011/12 Annual agricultural statistical bulletin. Ministry of Agriculture, Irrigation and Water Development. Lilongwe, Malawi

Mhango WG, Snapp S, Kanyama Phiri G (2017) Biological nitrogen fixation and yield of pigeon peas and groundnut: quantifying response on smallholder farms in northern Malawi. Afr J Agric Res 12(16):1385–1394

Mloza Banda C, Kabambe VH (2014) Agronomic evaluation of cattle manure and inorganic fertilizer application on maize yield at Bunda College, Lilongwe, Malawi. In: Kabambe v, Tembo Y, Ngwira AR (eds) Proceedings of the first integrated soil fertility management symposium held at sunbird Lilongwe Hotel. Lilongwe University of Agriculture and Natural Resources, Bunda College Campus and Health Consortium of Malawi, pp 20–26

Mudatenguha F, Anena F, Kiptum CK, Mashingaidze AB (2005) In Situ rain water harvesting techniques increase maize growth and grain yield in a semi-arid agro-ecology of Nyagatare, Rwanda. Int J Agric Biol 16:996–1000

Mupangwa W, Love D Tromlow S (2006) Soil-water conservation and rainwater harvesting strategies in the semi-arid Mzingwane Catchment, Lilmpopo basin, Zimbabwe. Phys Chem Earth 31:893–900

NPC National Planning Commission (2020) Malawi 2063. Lilongwe, Malawi

NCATF National Conservation Agriculture Task Force (2016) Guidelines for implementing conservation agriculture in Malawi. Ministry of Agriculture, Irrigation and Water Development

Ngwira A, Aune J, Thiefelder C (2014) On-farm evaluation of the effects of principles and components of conservation agriculture on maize yield and weed biomass in Malawi. Exp Agric 50(4):591–690

Ngwira AR, Kabambe VH, Kambauwa G, Mhango WG, Mwale CD, Chimphero L, Chimbizi A, Mapfumo P (2012) Scaling out best fit legume technologies for soil fertility enhancement among smallholder farmers in Malawi. Afr J Agri Res 7(6):918–928

Njira KOW, Semu E, Mrema JP, Nalivata PC (2020) Integrated soil fertility management based on pigeon pea and cowpea cropping systems influences nitrogen use efficiency, yields and quality of subsequent maize on Alfisols in central Malawi. In: Singh B, Safalaoh A, Amuri N, Eik L, Sitaula B, Lal R (eds) Climate impacts on agricultural and natural resources sustainability in Africa. Springer https://doi.org/10.1007/978-3-030-37537-9.

Njira KOW, Semu E, Mrema JP, Nalivata PC (2017) Biological nitrogen fixation by pigeon pea and cowpea in doubled-up and other cropping systems on the Luvisols of Central Malawi. Afr J Agric Res 12(15):1341–1352. https://doi.org/10.5897/AJAR2017.12167

Njira KOW (2014) Whither biofertilizer technologies in Malawi?. In: Kabambe V, Tembo Y, Ngwira AR (eds) Proceedings of the first integrated soil fertility management symposium held at sunbird Lilongwe hotel. Lilongwe University of Agriculture and Natural Resources, Bunda College Campus and Health Consortium of Malawi, pp 27–40

Oduor O, Muchena-Muna M, Mugwe NJ Sijali I, Nyabuga G, Mugendi ND (2023) Soil water use efficiency under integrated soil fertility management practices in the drylands of Kenya. Heliyon 9(5):e16145

Qin W, Hu C, Oenema O (2015) Soil mulching significantly enhances yields and water and use efficiencies of maize and wheat; a meta analysis. Sci Rep (Internet) 5:16210. http//www.nature.com/articles/srep16210.

Rockstrom J, Kariberg L, Wan SP, Barron J (2010) Managing water in rainfed agriculture. The need for paradigm shift. Agric Water Manage 97:543–550

Roth CH, Meyer B, Frede HG, Derpsch R (1988) Effect of mulch rates and tillage systems on infiltration and other soil physical properties of an Oxisol in Parana, Brazil. Soil Tillage Res 11:81–91

Sanga F, Kabambe VH (2014) Evaluation of agronomic management of leaf litter and inorganic fertilizer on maize *(Zea mays* L.) grown after two-year pigeon (*Cajanus cajan*) fallow in Malawi. Int J Agric Science 46:307–312

Tittonell P, Zingore S, van Wijk MT, Corbeels M, Giller KE (2007) Nutrient use efficiencies and crop responses to N, P and manure applications in Zimbabwean soils: exploring management strategies across soil fertility gradients. Field Crop Res 100:348–368

Tisdale SL, Nelson WL, Beaton JD (1985) Soil fertility and fertilizers. McMillan Publishing Co., New York

Thierfelder C, Wall P (2009) Effects of conservation agriculture techniques on infiltration and soil water content in Zambia and Zimbabwe. Soil Tillage Res 105:217–227

Turner NC (2004) Agronomic options for improving rainfall-use efficiency of crops in dryland farming systems. J Exp Bot 55:2413–2425

Vargas R, Omuto C (2016) Soil loss assessment in Malawi. FAO, UNDP-UNEP, Malawi Ministry of Agriculture, Irrigation and Water Development

Wiyo KA, Kasomekera ZM, Feyen J (1999) Effect of tied-ridging on soil water status of a maize crop under Malawi conditions. Agric Water Manag 45:101–125

WorldData.infor (2022) WorldData.infor/Africa/Malawi/populationgrowth.php

Zhang Q, Wang Z, Miso F, Wang G (2017) Dryland maize yield and water use efficiency responses to mulching and tillage practices. Where published?

Open Access This chapter is licensed under the terms of the Creative Commons Attribution 4.0 International License (http://creativecommons.org/licenses/by/4.0/), which permits use, sharing, adaptation, distribution and reproduction in any medium or format, as long as you give appropriate credit to the original author(s) and the source, provide a link to the Creative Commons license and indicate if changes were made.

The images or other third party material in this chapter are included in the chapter's Creative Commons license, unless indicated otherwise in a credit line to the material. If material is not included in the chapter's Creative Commons license and your intended use is not permitted by statutory regulation or exceeds the permitted use, you will need to obtain permission directly from the copyright holder.

Integrated Landscape Management Approach for Scaling Sustainable Land Management Practices and Improving Food Production in a Changing Climate: Experiences from Western Kenya

K. O. Were, G. O. Ayaga, R. Orlale, H. Agevi, F. N. Muyekho, J. K. Mwenesi, S. O. Ojung'a, J. M. Otuoma, I. S. Adolwa, and A. M. Kiwia

Abstract Landscape degradation is widespread with significant ramifications on ecosystem services. Currently, more than 50% of all fertile soils globally are severely degraded, with 25–40 billion megagrams (Mg) being lost annually. Similarly, about 20%, 14% and 12% of the total land area in Sub-Saharan Africa, East Africa and Kenya, respectively, are already suffering from severe to very severe degradation. In Kenya, productive landscapes in the western region have been the worst hit, losing an average of 0.5 Mg of soil per hectare per year (Mg ha^{-1} yr^{-1}) in 2017. The annual cost of land degradation in the region was estimated at USD 31 million between 2001 and 2007 using land use change as an indicator. Cognizant of the associated ecological and socio-economic impacts, several Sustainable Land Management (SLM) initiatives targeting landscape restoration have been implemented in the area. Most of the SLM initiatives adopted sectoral and isolated approaches, focusing on specific farm fields, pilot sites and production systems to demonstrate best-fit SLM practices for adoption. The impact of such narrow, localized and piecemeal approaches on the uptake of SLM practices, food security and livelihoods has, however, been low. This is mainly due to the complexity of land degradation processes, which transcend the

K. O. Were (✉) · G. O. Ayaga · R. Orlale · I. S. Adolwa
Kenya Agricultural and Livestock Research Organization, Nairobi, Kenya
e-mail: kennedy.were@kalro.org

H. Agevi · F. N. Muyekho
Department of Biological Sciences, Masinde Muliro University of Science and Technology, Kakamega, Kenya

J. K. Mwenesi
Anglican Development Services—Western, Kakamega, Kenya

S. O. Ojung'a · J. M. Otuoma
Kenya Forestry Research Institute, Nairobi, Kenya

A. M. Kiwia
Alliance for a Green Revolution in Africa, Nairobi, Kenya

© The Author(s) 2026

B. R. Singh et al. (eds.), *Sustainable Food Systems in Sub-Saharan Africa*,
https://doi.org/10.1007/978-3-032-11288-0_15

traditional management boundaries and sectors. Effective management and unravelling of this complexity to enhance SLM adoption, agricultural productivity and livelihoods requires a new thinking that looks at ecosystem services beyond individual farms and plots, and integrates perspectives from multiple disciplines, stakeholders and institutions. In this chapter, we have presented an innovative and replicable Integrated Landscape Management approach, which was developed and applied by a model SLM Project that aimed at reducing land degradation and improving productivity and livelihoods through rapid adoption of proven SLM practices across the productive landscape of the Kakamega-Nandi ecosystem.

Keywords Landscape approach · Land degradation · Land restoration · Land productivity · Ecosystem services · Sustainable land management · Kakamega forest · Nandi forest

1 Introduction

Landscapes are the basis of many life support systems through the performance of multiple ecological and socio-economic functions, including the supply of food, fodder, fibre, fuel, timber and other ecosystems goods and services for human use. In spite of their multi-functionality, many landscapes are being degraded and their integrity disrupted owing to overexploitation of their natural resources. In many places, soils are deteriorating, biological diversity is declining, and ecosystem services and productivity are diminishing. Recent studies have shown that more than 50% of all fertile soils globally are now classified as severely degraded, with 25–40 billion megagrams (Mg) being lost annually (UNCCD 2015; FAO and ITPS 2015). Sub-Saharan Africa (SSA) is the worst hit region, with up to 65% of productive land having been degraded and desertification affecting 45% of the total land area (Mansourian and Berrahmouni 2021). This is largely because more than 70% of the people in SSA directly depend on land resources for their livelihoods, which exerts undue pressure on them and increases land consumption (Mbow, 2020). The situation is expected to worsen in the future with the projected demographic changes. The official United Nations (UN) population projections show that Africa's population will reach 1.7 billion in 2030, 2.5 billion in 2050, and 4.5 billion in 2100 (UN, 2017).

Kenya has not been spared either: the challenge looms large. Recent studies revealed that about 12% of the total land area in Kenya, supporting about 27% of the population, is degraded, with the hotspots distributed across the western and southern parts (Kizito et al. 2018). In western Kenya, degradation of the productive landscapes has been occurring at alarming rates. Kogo et al. (2020) reported that the average soil loss was 0.5 Mg per hectare per year (Mg ha^{-1} yr^{-1}) in 2017 compared to 0.3 Mg ha^{-1} yr^{-1} in 1995. Using land use change as an indicator, Mulinge et al. (2016) estimated the annual cost of land degradation in the region at USD 31 million between 2001 and 2007. The associated impacts on the ecological and socio-economic landscape functions can be devastating, and this has provided a compelling reason for the implementation of several Sustainable Land Management (SLM) initiatives targeting landscape restoration in the region over the years (Dallimer et al. 2016). Most of these initiatives adopted approaches that focused on specific farm fields, pilot sites or production systems to demonstrate and scale out best-fit SLM practices within the landscape and, mostly, neglecting the concerns of stakeholders. However, the impact of such narrow, localized and piecemeal approaches on the uptake of proven SLM practices, food security and livelihoods has been quite low. This is due to the complexity and multi-faceted nature of land degradation processes and problems, which transcend the conventional management boundaries, scale and sectors (Musafiri et al. 2022). Effective management and unravelling of this complexity with a view to enhancing SLM adoption, agricultural productivity and livelihoods calls for a more holistic, integrative thinking that looks at ecosystem services and livelihoods on a wider scale beyond individual farms, pilot sites and demonstration plots, and assimilates perspectives from multiple disciplines, stakeholders and institutions.

This paper presents an innovative and replicable Integrated Landscape Management (ILM) approach, which was developed and applied by a model SLM Project that aimed at reducing land degradation and improving productivity and livelihoods through rapid adoption of proven SLM practices across the productive landscape of the Kakamega-Nandi ecosystem. It systematically describes the strategies, processes and tools that were used to reconcile, synergize and integrate the diverse elements of this landscape, including the land resources, production systems, and stakeholders to achieve a shared vision. The paper also captures and discusses the impact, challenges and lessons learnt from application of the new ILM approach.

2 The Approach

2.1 Description of the Project Implementation Area

The SLM Project was implemented in Kakamega, Nandi and Vihiga counties, which lie between the Latitudes 0° 5′ and 0° 30′ N, and the Longitudes 34° 45′ and 35° 0′ E (Fig. 1). These three counties cover approximately 6402 km^2, with a population of about 3.3 million as at 2019 (KNBS 2019). The huge population, coupled with notable trends of encroachment into protected resources, such as Kakamega and South Nandi forests made western region of Kenya an ideal implementation area for the Project.

Agriculture is the main economic activity and land use in the area, with maize, beans, sweet potatoes, cassava, sorghum, vegetables and finger millet being cultivated as the major food crops, and tea, sugarcane and coffee being grown as the major cash crops. Farmers in the counties also keep livestock, such as Zebu cattle, dairy cattle, goats, sheep, pig, and poultry. These agricultural activities take place in the Upper Highland (UH1 and 2), Lower Highland (LH1, 2 and 3), Upper Midland (UM0, 1, 2, 3 and 4) and Lower Midland (LM1 and 2) (Jaetzold et al. 2005, 2010). There is

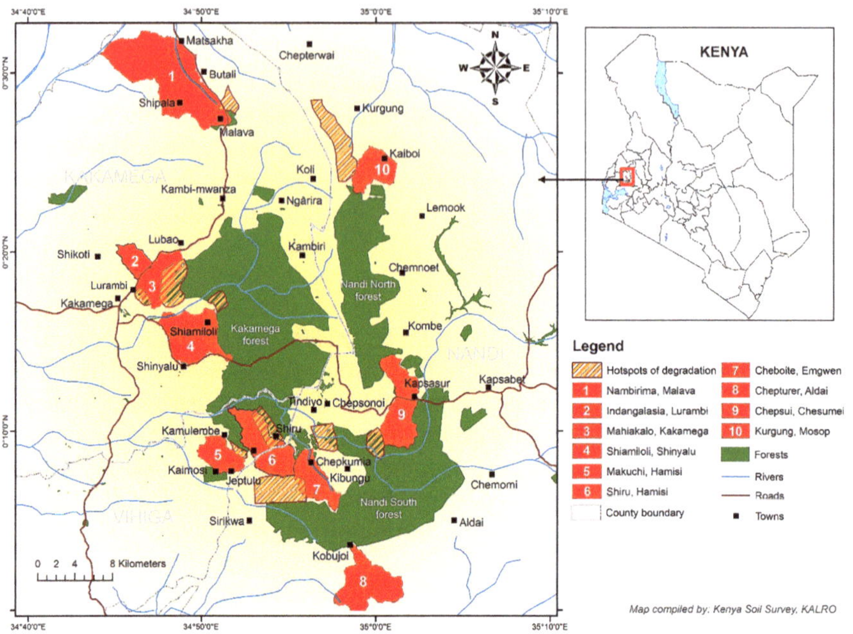

Fig. 1 The project implementation area in western Kenya

also a significant tourism industry centred on Kakamega forest, which is Kenya's last remnant of the ancient Guineo-Congolian rainforest (Waswa 2012).

There is substantial variation in topography, climate, soils, and geomorphology. The altitudes range from 1200 to 2600 m above sea level, while the annual mean rainfall average is between 1200 and 2200 mm. In addition, potential evaporation varies from 1200 to 2100 mm per year, whereas annual mean temperature ranges from 15 to 22.5 °C. With regard to physiography, the area is characterized by mountain slopes, hills, uplands, bottomland, valleys, foot slopes, plateaus and alluvial floodplains (Jaetzold et al. 2005, 2010). Finally, several streams, such as the Isiukhu, Nambirima, Makuchi and Cheboite drain the intensively cultivated landscape, joining the Yala and Nzoia Rivers, which discharge into Lake Victoria.

2.2 *Conceptual Framework of the ILM Approach*

As conceptualized in Fig. 2, the ILM approach was about wide-scale adoption of SLM technologies through involvement of local communities and a consortium of supportive institutions and partners in the agricultural value chain, including local institutions or innovation platforms (IPs). It was a new way of thinking on how the uptake of SLM technologies by farmers could be accelerated to improve agricultural productivity, food security, livelihoods, and environmental sustainability.

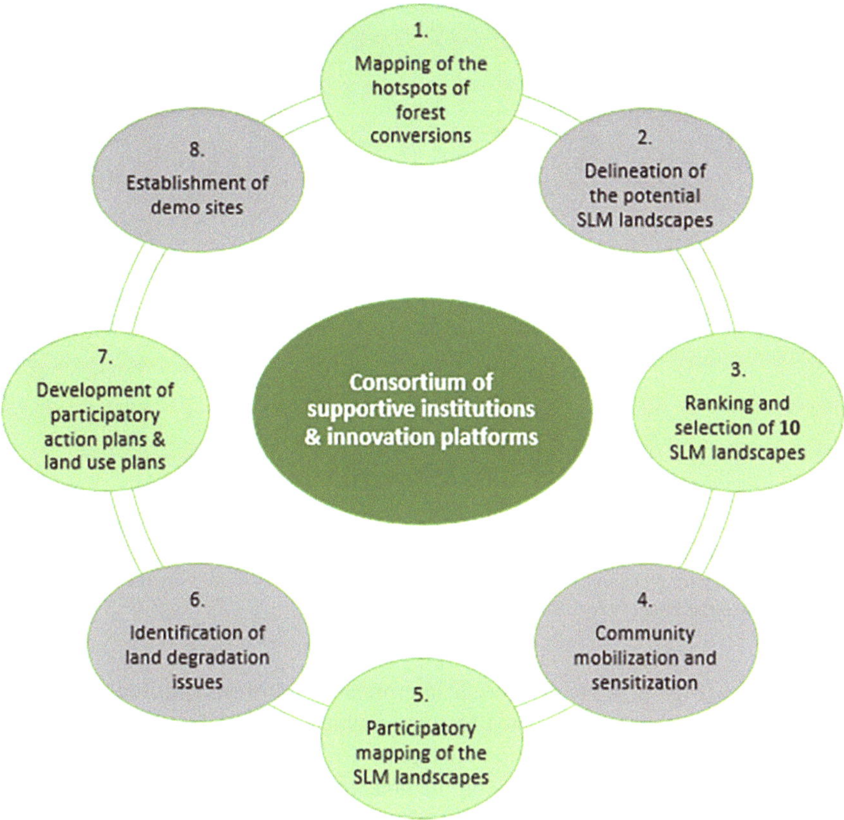

Fig. 2 Conceptual framework of the ILM approach

2.3 Delineation, Mapping, and Prioritization of SLM Landscapes

At the outset, a Shuttle Radar Topographic Mission Digital Elevation Model (SRTM DEM, https://earthexplorer.usgs.gov/) was analyzed using geographical information systems (GIS) techniques to delineate the potential micro-catchments (i.e., SLM landscapes) for project implementation. The output was a thematic map, depicting the boundaries of all the SLM landscapes with a threshold area of 25 km^2 in the Kakamega-Vihiga-Nandi landscape. After delineating the SLM landscapes, time-series satellite images were analyzed to extract the hotspots of forest conversions. Specifically, Landsat Multi-spectral Scanner and Thematic Mapper images acquired in 1973, 1980, 1983, 1986, 1995, 2003 and 2017 were obtained from the United States Geological Survey geoportal (https://earthexplorer.usgs.gov/), pre-processed and subsets of the project area made. The image subsets were then juxtaposed and visually interpreted to identify the hotspots of conversion from forest to crop lands

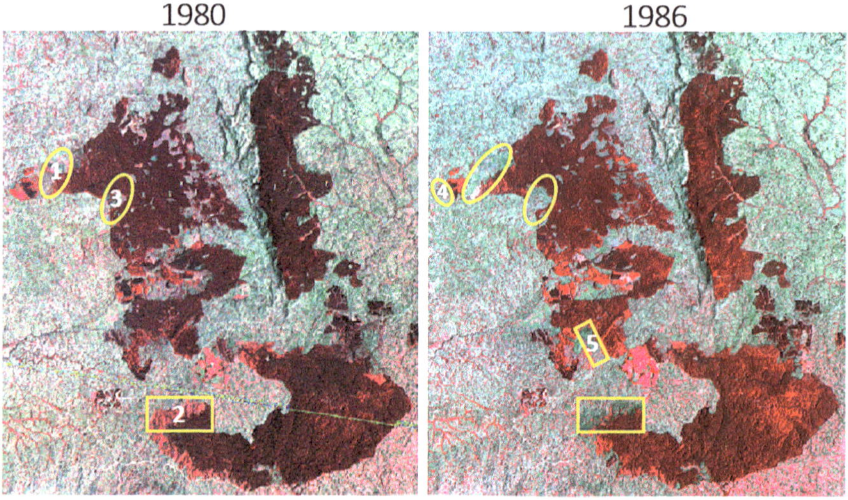

Fig. 3 The hotspots of forest degradation identified through comparison of Landsat imagery acquired in 1980 and 1986. Numbers 1, 2 and 3 are the hotspots of degradation that were identified when the imagery for 1973 and 1980 were compared, while Numbers 4 and 5 are the ones that were identified when the imagery for 1980 and 1986 were compared

over the years (Figs. 3, 4, 5 and 6). Visual interpretation of forests and croplands on the satellite imagery was supplemented by topographical sheets, Open Street Map and Google Earth imagery. The identified hotspots of forest conversions were manually digitized and a thematic map showing the hotspots of forest conversion from 1973 to 2017 produced. The overall thematic map was a combination of the delineated SLM landscapes and hotspots of deforestation (Fig. 7), which guided the scoring and prioritization of SLM landscapes by the Sub-county Technical Committees (STCs) in each County.

The STCs identified the SLM landscapes that were within five kilometres radius from the edge of the hotspots of forest land conversions and scored them based on a nine-point selection criteria: (a) Within 5 km radius from the edge of the hotspots of forest conversions; (b) A micro-catchment; (c) Area of the site (Ha); (d) Extent of degradation; (e) Vulnerability of the community (i.e., poverty index); (f) Extent of forest encroachment; (g) Willingness of the community to accept the project, or availability of registered and active groups; (h) Evidence of other stakeholders' activities with similar objectives; and (i) Existence of other economic activities apart from farming. The SLM landscapes that ranked highly in each County were selected as the project implementation sites (See Table 1). Of the 10 SLM landscapes that were selected (Fig. 1), 4 were in Kakamega County (i.e., Shiamiloli, Indangalasia, Nambirima and Mahiakalo), 4 in Nandi (i.e., Cheboite, Kurgung, Chepsui and Chepturer) and 2 in Vihiga (i.e., Makuchi and Shiru).

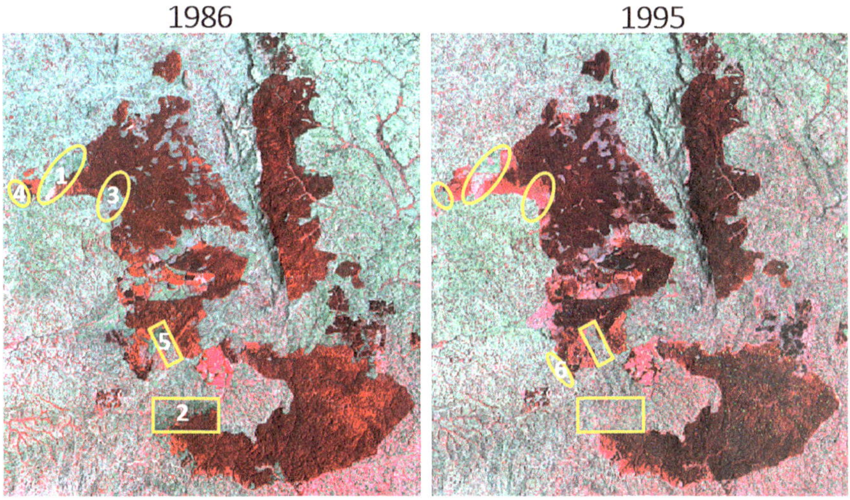

Fig. 4 The hotspots of forest degradation identified through comparison of Landsat imagery acquired in 1986 and 1995. Numbers 1–5 are the hotspots of degradation that had been identified in the previous images, while Number 6 is the one that was identified when the imagery for 1986 and 1995 were compared

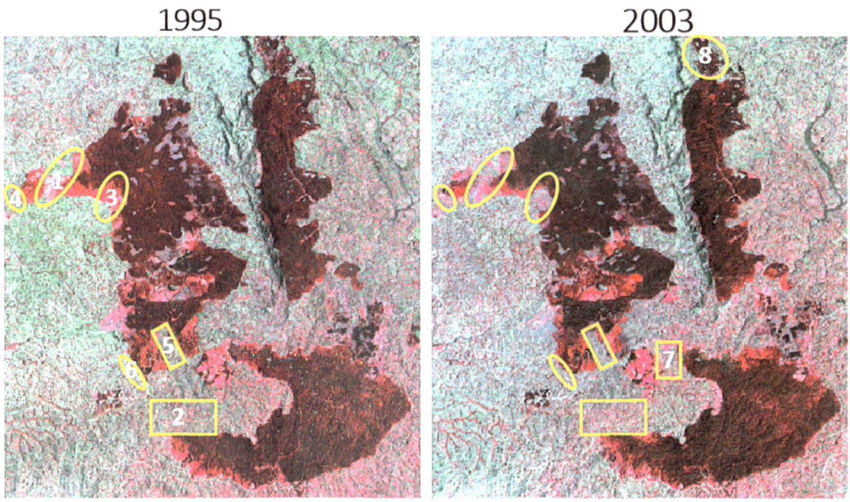

Fig. 5 The hotspots of forest degradation identified through comparison of Landsat imagery acquired in 1995 and 2003. Numbers 1–6 are the hotspots of degradation that had been identified in the previous images, while Number 7 and 8 are the ones that were identified when the imagery for 1995 and 2003 were compared

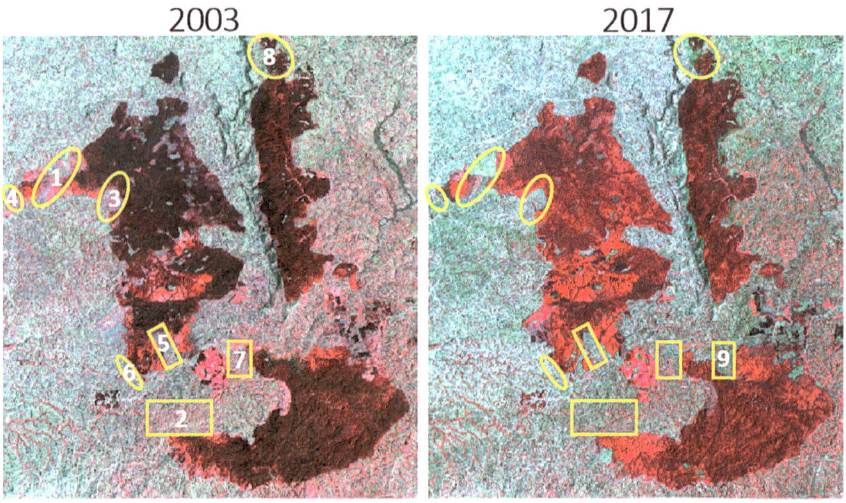

Fig. 6 The hotspots of forest degradation identified through comparison of Landsat imagery acquired in 2003 and 2017. Numbers 1–8 are the hotspots of degradation that had been identified in the previous images, while Number 9 is the one that was identified when the imagery for 2003 and 2017 were compared

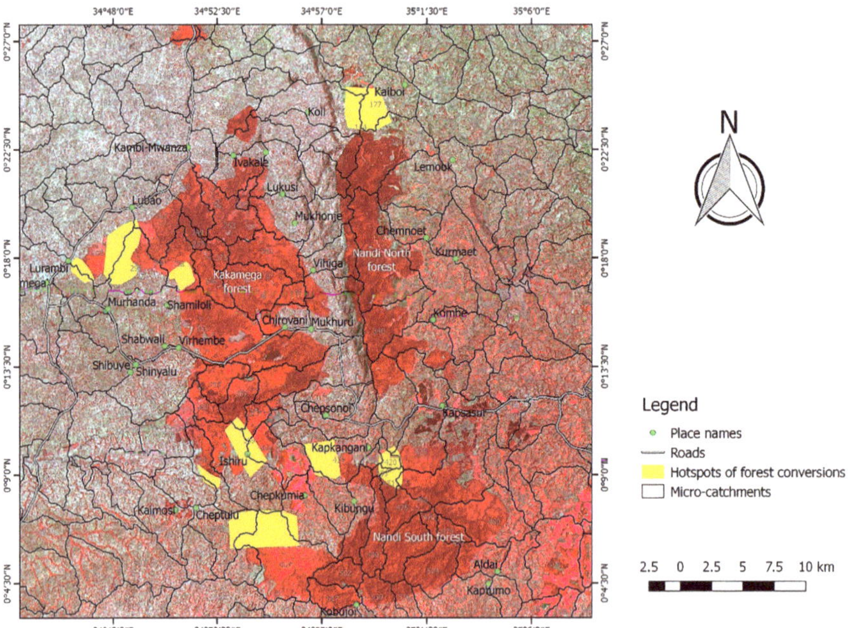

Fig. 7 DEM-extracted SLM landscapes (with black boundaries) overlaid on the hotspots of forest degradation and Landsat imagery

Table 1 Site selection score sheet for SLM landscapes in Vihiga county

Selection criterion	Candidate SLM landscape					
	Makuchi	Shiru	Muhudu	Musunju	Jeptulu	Musasa
A	3	3	3	3	1	3
B	3	3	3	3	2	3
C	3	2	3	3	3	1
D	3	3	2	2	3	1
E	3	2	2	2	2	1
F	1	2	1	1	2	1
G	2	2	2	3	2	1
H	2	2	2	1	2	1
I	2	2	2	2	2	1
J	2	2	2	1	2	1
Total score	**24**	**23**	22	21	21	14
Rank	**1**	**2**	3	4	5	6
Selected sites	Makuchi & Shiru					

Notes Scores: 3 = high; 2 = moderate; 1 = low
Selection criterion: A = 5 km radius from forest edge; B = Micro-catchment area; C = Area of the site in Ha; D = Extent of degradation; E = Vulnerability (poverty index); F = Extent of forest encroachment; G = Willingness of the community, or availability of registered and active groups; H = Existence of other stakeholders' activities with similar objectives; I = Existence of other economic activities apart from farming; J = Judging team score after assessment.

2.4 Community Participation

At the core of the ILM approach was the involvement of local communities at all stages for maximum sustainability. This was important because the local communities are the ones who suffered most from the negative effects of environmental degradation. At the same time, they are the ones who influenced the nature and severity of the degradation processes through exploitation of the land resources.

2.4.1 Mobilization and Sensitization

Over 5000 smallholder farmers living within the 10 selected SLM landscapes were mobilized by extension service providers and sub-county technical leaders, who invited them to meetings in central venues through announcements in farmer groups, churches, and schools. In the meetings, the communities were sensitized to fully understand the extent and impact of land degradation, including the available natural resources, problems that affected their livelihoods, and implications of land resource degradation. In addition, functional farmer groups for project implementation were identified, community and project expectations were outlined, the prioritized SLM

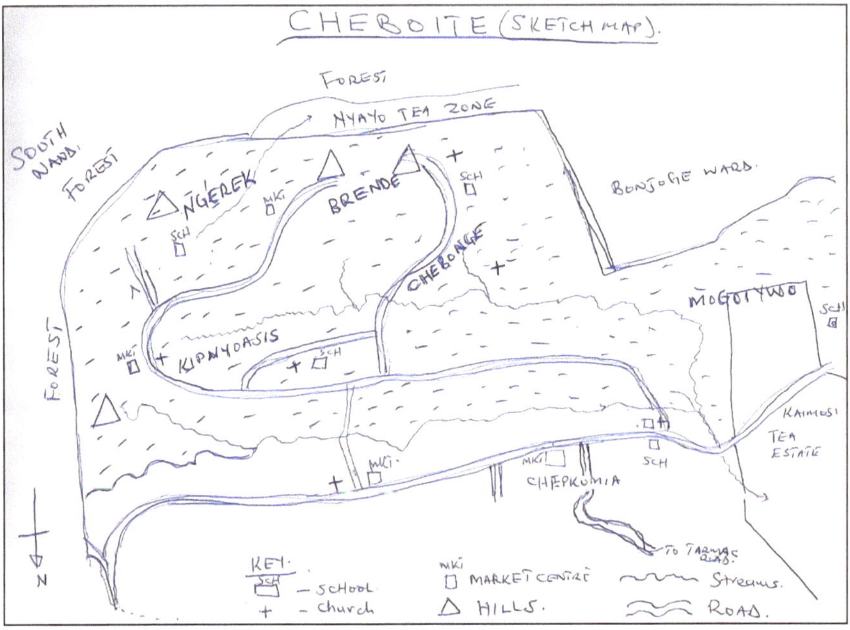

Fig. 8 A sketch map of Cheboite SLM landscape in Emgwen sub-county

landscapes were mapped and named, land resource problems were identified, listed and prioritized, and participatory action plans (PAPs) were developed.

Participatory Mapping of the Prioritized SLM Landscapes

Local communities were actively involved in the mapping of the SLM landscapes. The SLM landscapes were visited by the Project coordination team and partners who, together with the local communities, conducted transect walks to gather information and isolate unique characteristics, such as location and distribution of natural resources, land use, vegetation, physical features, topography, production enterprises, land management technologies and practices, and major land degradation problems. After the transect walks, transect diagrams, and sketch maps of the resources and landscape boundaries were drawn in a participatory fashion (Fig. 8). The drawings were made on flip charts and, in some instances, on Google Earth platform.

The final step involved matching the GIS-generated SLM landscape maps with the community-generated sketch maps. That is, the sketch maps of the landscapes were compared with those extracted from the SRTM DEM both visually and through GIS overlay. The local communities validated the boundaries using their local experiences, perceptions, and observations. This ensured that the defined boundaries of the selected SLM landscapes were not only accurate and founded on the hydrological tenets of watershed delineation, but also integrated the local communities'

Table 2 Pairwise ranking of the land degradation issues in Shiru SLM landscape

No	Problems	1	2	3	4	5	6	7	8	Total	Rank
1	Small land size	x	2	3	4	5	6	7	8	0	8
2	Market access		x	3	4	5	6	7	8	1	7
3	Lack of knowledge			x	3	3	3	3	3	7	1
4	Low crop yields				x	5	6	7	8	2	6
5	Soil infertility					x	6	7	8	3	5
6	Soil erosion						x	7	6	5	3
7	Deforestation							x	7	6	2
8	Pest and diseases								x	4	4

perceptions of their spatial extents and the associated resources. The whole process contributed to empowerment and ownership of the Project by the local communities.

Identification and Prioritization of Land Degradation Issues

The specific land degradation problems in the 10 SLM landscapes were identified and listed through various participatory strategies, including reconnaissance surveys, transect walks and stakeholder engagement meetings. Biophysical information (e.g., land use, climate, soils, vegetation, and landforms) was compiled and analyzed to determine the baseline land conditions and critical issues. Participatory reconnaissance surveys and transect walks were conducted to gather primary data on biophysical resources, land management practices (e.g., cropping systems and soil conservation measures) and land degradation processes as described in Sect. 2.4.1.1. Community meetings were important, not only in the identification and listing, but also in the prioritization of land degradation problems. Prioritization of the problems was achieved through pairwise ranking (Table 2).

Development of Participatory Action Plans

After problem identification, listing and prioritization, appropriate SLM solutions and actions were suggested for each land degradation issue, and participatory action plans (PAPs) were developed for each landscape. To complement the PAPs, a range of SLM practices were also reviewed, evaluated, and documented in a participatory manner in technical workshops by multi-stakeholders with expertise in soil and water management, soil fertility, carbon sequestration, crop production and livestock production. The SLM practices were evaluated and ranked on a scale of 1–5 (with 5 being the best) using five indicators; that is, technical feasibility, social acceptability, environmental benefits, carbon benefits, biodiversity benefits and economic viability. The output of this process was a prioritized inventory of potential SLM practices focusing on improving water use efficiency and productivity, improving soil fertility, managing vegetation, conserving soils, and attending to micro-climatic conditions. The highly-ranked SLM practices were selected for scaling and exhibition at the demonstration sites. Finally, the PAPs were translated to comprehensive land use management plans (LUMPs), which blended the indigenous conservation practices and scientifically-selected SLM practices, and represented the entire SLM landscapes.

2.5 *Institutional Frameworks for Scaling SLM*

As mentioned earlier, the delivery of the Project was through community engagement at all stages. However, this was underpinned by a consortium of strategic partners in the agricultural value chain, including local institutions referred to as innovation platforms (IPs) that supported the flow of credit, inputs, and market information towards increased agricultural productivity and environmental sustainability (Fig. 2).

2.5.1 Consortium of Partners

Given the complex nature of the SLM Project, a central consortium of partners and institutional actors with specific roles and responsibilities in the SLM project was constituted (Table 3). The consortium was composed of the United Nations Environment Programme (UNEP), Alliance for a Green Revolution in Africa (AGRA), Kenya Agricultural and Livestock Research Organization (KALRO), County Governments of Kakamega, Vihiga and Nandi, Ministry of Agriculture, Livestock, Fisheries and Cooperatives (MoALFC), Kenya Forest Service (KFS), Kenya Wildlife Service (KWS), Kenya Forestry Research Institute (KEFRI), Masinde Muliro University of Science and Technology (MMUST), Nature Kenya, Anglican Development Services—Western (ADS-W) and Rural Outreach Programme (ROP). There was also a project steering committee chaired by the Ministry of Environment and Forestry, which ensured constructive collaboration between the participating institutions and implementation of the Project as per the agreed framework to achieve its targets. It also provided high-level orientation and guidance on all institutional, political, and operational matters. At the sub-county and county levels, landscape committees mainly consisting of the extension leaders and producer groups, guided and coordinated the delivery of site activities. They also reviewed project progress locally, addressed implementation barriers, harmonized activities, and shared information and experiences.

The consortium was operationalized in three stages; namely, identifying partners with common vision and interests in conservation of the Kakamega-Nandi ecosystem, defining their roles and responsibilities within the components of the Project and signing of the Memorandum of Agreements (MoAs).

Table 3 Project partners and their roles

Institutional partner	Role
1. United nations environment programme	GEF implementing agency; monitoring and evaluation
2. Alliance for a green revolution in Africa	Executing agency; secretariat to project steering committee, with oversight of financial management
3. County Governments of Kakamega, Nandi and Vihiga	Project implementation, in-kind budgetary support, political will
4. Kenya agricultural and livestock research organization	Lead implementing agency on the ground and host to Project Management Unit (PMU); supervision of other implementing partners; technical and financial reporting
5. Ministry of agriculture, livestock, fisheries and cooperatives	Project implementation; linkage with ongoing programmes for up-scaling of SLM
5. Nature Kenya	Development, review and implementation of participatory forest management plans; capacity building of Community Forest Associations on SFM
6. Kenya forest service	
8. Kenya wildlife service	Support to biodiversity conservation, support to eco-tourism activities
9. Kenya forestry research institute	Assessment and product development of non-wood forest products and services; dissemination of SFM technologies
10. Masinde Muliro university of science and technology	Capacity building on agro-biodiversity and social/environmental management framework; support to higher level student trainings
11. Anglican development services—Western	Community mobilization, increase farmers' access to input and output markets
12. Rural outreach programme	

2.5.2 Innovation Platforms

At the SLM landscape level, in collaboration with AGRA, KALRO and other partners, project delivery was coordinated through 10 IPs. The IPs were forums created for relevant and interested actors in the maize, bean and indigenous vegetable value chains to interact and share knowledge and innovations around enhancing food security and livelihoods, while ensuring the sustainability of land resources (Fig. 9). The value chain actors included producer groups, input suppliers, traders (i.e., small and medium-scale agricultural enterprises—SMAEs), researchers, agro-processors, transporters, extension workers, local leaders, creditors (i.e., village savings and loans associations—VSLAs), and community forest associations (CFAs).

The value chain actors (VCAs) performed specific roles in the transfer of SLM technologies across scales, with a view to generating socio-economic and biophysical benefits. For example, the producer groups were charged with, *inter alia*, implementing SLM practices on their farms, managing the 50 SLM demonstration sites

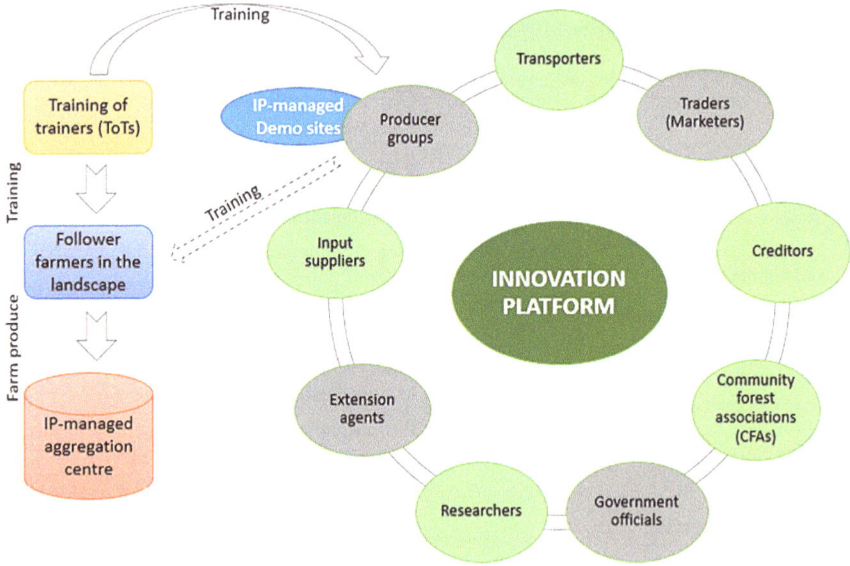

Fig. 9 Conceptual framework of the Innovation Platform

(*demos*) and moving the practices beyond individual farms and *demos* to the wider landscapes. This was achieved by reaching out to other farmers to offer guidance on SLM practices, as well as to share experiences and knowledge. The VCAs were identified and validated through a stakeholder analysis process that was led by the STCs, KALRO, ROP and ADS-W. The identification was based on their potential contribution in the development of functional IPs for scaling SLM innovations (Eneku et al. 2013). As such, their knowledge, strengths, weaknesses, attitudes, practices, and opportunities were all considered. In a nutshell, the IPs provided multi-stakeholder spaces for dialogue, diagnosis of complex land resource constraints and opportunities, and collective prescription of innovative SLM solutions, which promote food security, livelihood, and environmental sustainability (Dusengemungu et al. 2014; Oluwole et al. 2016; Schut et al. 2019). Ultimately, the 10 IPs morphed into functional entities with elected officials and certificates of incorporation.

3 Achievements

Putting the ILM approach to SLM into action in Kakamega-Nandi landscape yielded manifold benefits, including remarkable increments in farm productivity and profitability, land under SLM, and uptake of SLM practices by the targeted beneficiaries.

3.1 Farm Productivity and Profitability

Quantitative data gathered by tracking 396 early adopters of SLM technologies showed an increase in the yields of maize and beans from an average of 0.93 and 0.23 tons ha^{-1} in July 2018 to 3.20 and 0.44 tons ha^{-1} in December 2021, respectively (Figs. 10 and 11). This reflected a rise in farmers' income from an average of USD 105 (KES 11,498) in 2018 to about USD 240 (KES 26,280) in 2021. The increase in yields and incomes were ascribed to improvements in the members' skill set, knowledge and other transferable assets through capacity building conducted by the 100 ToTs who had been trained by the SLM Project. For example, 16,013 farmers were trained on post-harvest management practices, and 50 farmer groups on SLM. Also, between October 2019 and January 2020, 14 farmer groups benefitted economically by aggregating and selling 2169 metric tons of maize and beans valued at USD 9,800 (KES 1.07 million). Overall, this implies that ILM approach can catalyze the progress towards the achievement of food security and reduction of hunger and poverty levels, which form part of the Agenda 2030 on sustainable development.

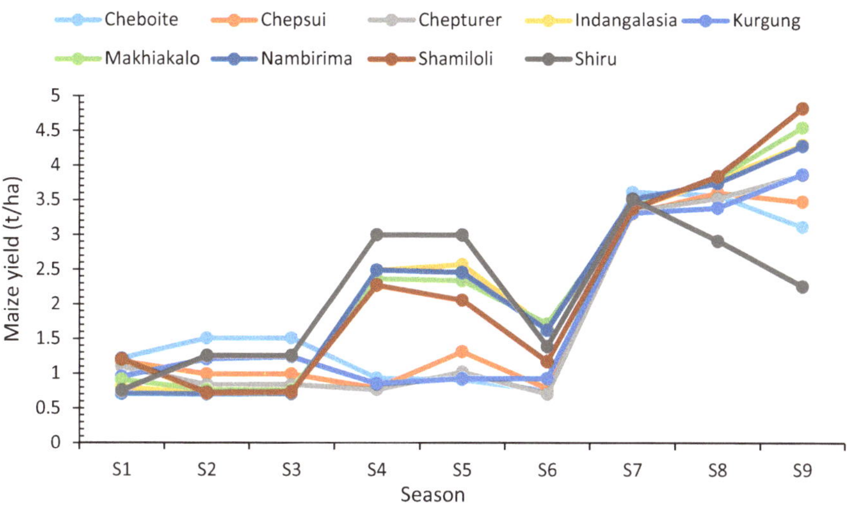

Fig. 10 Seasonal maize yields on farmers' fields in the 10 SLM landscapes. *Notes* S1 = July–Sep 2018; S2 = Oct–Dec 2018; S3 = Jan–March 2019; S4 = April–June 2019; S5 = July–Sep 2019; S6 = Oct–Dec 2019; S7 = Jan–March 2020; S8 = July–Sep 2021; S9 = Oct–Dec 2021

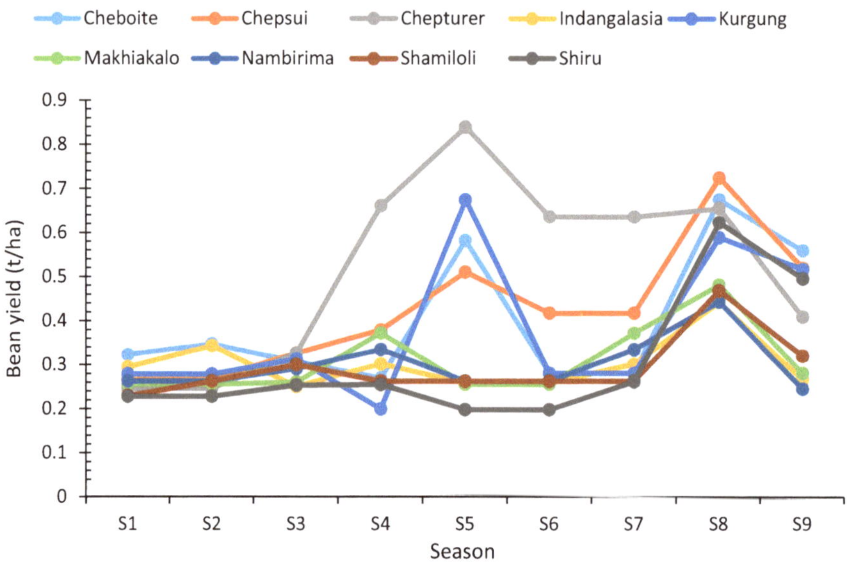

Fig. 11 Seasonal bean yields on farmers' fields in the 10 SLM landscapes. *Notes* S1 = July–Sep 2018; S2 = Oct–Dec 2018; S3 = Jan–March 2019; S4 = April–June 2019; S5 = July–Sep 2019; S6 = Oct–Dec 2019; S7 = Jan–March 2020; S8 = July–Sep 2021; S9 = Oct–Dec 2021

3.2 Area of Land Under SLM and Adoption of SLM Practices

Besides farm productivity and incomes, about 76% of the target farmers benefited by adopting at least one SLM technology (Table 4; Fig. 12). A survey of 184 households, which was also triangulated with Focus Group Discussions (FGDs), revealed that conservation agriculture was the most preferred technology by the respondents followed by agroforestry and soil conservation (Fig. 12). As a result of the uptake of these technologies, over 4000 ha of agricultural land and 7000 a of forest land were put under SLM and 282,758 tree seedlings planted on the farms and hotspots of land degradation. This reduced pressure on forest resources and mitigated land degradation across the Kakamega-Nandi landscape.

Table 4 Land under SLM and adoption of agricultural technologies

Outcome indicator	Project life target	Actual in 2022	Achievement (%)
1. Area (Ha) of land under SLM	3913	4440	113
2. Area (Ha) of land under participatory forest management (PFM)	7772	7628	98
3. Percentage (%) of farmers using at least one SLM technology	91,000	69,160	76

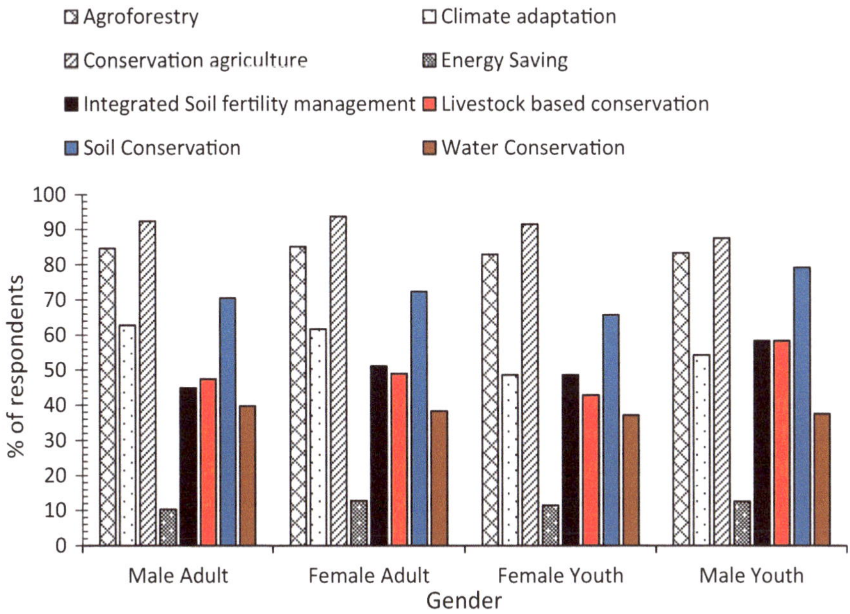

Fig. 12 Adoption levels by smallholder farmers on SLM Preferred technologies

4 Lessons Learned

Experiences with the ILM approach in western Kenya showed that bottom-up engagement, collaborative and hands-on leadership, participatory planning, implementation, and monitoring, as well as regular, open, and honest communication are crucial strategies for the success of multi-institutional, multi-disciplinary and multi-stakeholder agricultural and environmental projects that require positive synergistic endeavours. These strategies foster local ownership and peer-to-peer knowledge dissemination, leading to collective community uptake of SLM and convergence of vision and effort.

5 Challenges

Most of the challenges encountered with the ILM approach were economic and socio-cultural; for example, community behaviours were entrenched and resistance to the new approach owing to the prevailing socio-economic circumstances (e.g., poverty and insecure land rights) was strong. Building trust and convincing farmers to fully embrace the new ideas and participate in their implementation required ample time for major engagements, education, and capacity building. In addition, implementation of ILM approach also placed high demands on the available resources, especially when the different interest groups were being mobilized to participate. Lastly, conflicting interests, expectations and demands on the part of the stakeholders occasionally caused disagreements and increased the complexity of the multi-stakeholder processes. The MoAs however greatly minimized such conflicts.

6 Summary, Conclusions and Recommendations

The innovative ILM approach that was put into practice in western Kenya by the SLM Project has been presented. Specifically, the conceptual framework, strategies, and processes, as well as the achievements, challenges and lessons learnt have been systematically and exhaustively described. At the core of the ILM approach is community participation at all stages, supported by a consortium of partners and institutional actors. Implementation of the ILM approach to SLM in Kakamega-Nandi ecosystem yielded substantial increases in incomes, crop yields, adoption levels and area of land under SLM. Based on these results and evidence, we argue that the new ILM approach is an effective model that has succeeded in deterring further encroachments into the forests, curbing land degradation, reducing poverty levels and improving climate risk management in western Kenya. Corollary to this, the approach can be taken forward and replicated elsewhere to tackle landscape degradation and restore ecosystem services and contribute to the realization of the global sustainable development goals.

Acknowledgements The authors appreciate the Global Environment Facility for funding the SLM Project, as well as the UNEP, AGRA, KALRO, County Governments of Kakamega, Nandi and Vihiga and farmers for their support.

References

Dallimer M, Stringer LC, Orchard SE, Osano P, Njoroge G, Wen C (2016) Economics of Land Degradation (ELD) Kenya study. Costs and benefits of sustainable soil fertility management in Western Kenya. Report for the Economics of Land Degradation Initiative, p 40

Dusengemungu L, Kibwika P, Kiazze FB (2014) Enhancing food security through innovation platforms: experiences from research into use project in Rwanda. Glob Discl Econ Bus 3(1)

Eneku GA, Wagoire WW, Nakanwagi J, Tukahirwa JMB (2013) Innovation platforms: a tool for scaling up sustainable land management innovations in the highlands of Eastern Uganda. Afr Crop Sci J 21(s3):751–760

FAO and ITPS (2015) Status of the world's soil resources—main report. Food and Agriculture Organization of the United Nations and Intergovernmental Technical Panel on Soils, Rome, Italy

Jaetzold R, Schmidt H, Hornetz B, Shisanya C (2005) Farm management handbook of Kenya, Vol. II. Natural conditions and farm management information, 2nd edn. Part A West Kenya, Subpart A1 Western Province. Ministry of Agriculture, Kenya and German Agency for Technical Cooperation (GTZ), Nairobi

Jaetzold R, Schmidt H, Hornetz B, Shisanya C (2010) Farm management handbook of Kenya, Vol. II. Natural conditions and farm management information, 2nd edn. Part B Central Kenya, Subpart B1a Southern Rift Valley Province. Ministry of Agriculture, Kenya and German Agency for Technical Cooperation (GTZ), Nairobi

Kizito F, Tamene L, Koech N, Pondi B, Nganga K (2018) Land degradation assessments using multiscale hierarchical approaches for agroecosystem restoration and improved food security: the case for Kenya and Burkina Faso. CIAT Publication, International Center for Tropical Agriculture (CIAT) and TMG-Think Tank for Sustainability. Nairobi, Kenya, p 56. Available online at: https://cgspace.cgiar.org/handle/10568/97165. Accessed 14 June 2022

KNBS (2019) 2019 Kenya population and housing census. Volume I: population by county and sub-county. Available online at: https://www.knbs.or.ke. Accessed 14 June 2022

Kogo BK, Kumar L, Koech R (2020) Impact of land use/cover changes on soil erosion in Western Kenya. Sustainability 12:9740. https://doi.org/10.3390/su12229740

Mansourian S, Berrahmouni N (2021) Review of forest and landscape restoration in Africa. FAO and AUDA-NEPAD, Accra. Available online at: https://doi.org/10.4060/cb6111en. Accessed 14 June 2022

Mbow C (2020) Use it sustainably or lose it! The land stakes in SDGs for sub-Saharan Africa. Land 9:63. https://doi.org/10.3390/land9030063

Mulinge W, Gicheru P, Murithi F, Maingi P, Kihiu E, Kirui OK, Mirzabaev A (2016) Economics of land degradation and improvement in Kenya. In: Nkonya E, Mirzabaev A, von Braun J (eds) Economics of land degradation and improvement—a global assessment for sustainable development. Springer, Cham. https://doi.org/10.1007/978-3-319-19168-3_16

Musafiri CM, Kiboi M, Macharia J, Ng'etich OK, Kosgei DK, Mulianga B, Okoti M, Ngetich FK (2022) Adoption of climate-smart agricultural practices among smallholder farmers in Western Kenya: do socioeconomic, institutional, and biophysical factors matter? Heliyon 8:e08677

Oluwole FB, Youdeowei A, Ohiomoba SI, Adekunle AA, Akinbanijo OO (2016) Agricultural innovation platforms: Framework for improving sustainable livelihoods in Africa. Forum for Agricultural Research in Africa (FARA), Accra Ghana

Schut M et al (2019) Innovation platforms in agricultural research for development: ex-ante appraisal of the purposes and conditions under which innovation platforms can contribute to agricultural development outcomes. Expl Agric 55(4):575–596

UN, Department of Economic and Social Affairs, Population Division (2017) World population prospects: the 2017 revision, key findings and advance tables. Working paper no. ESA/P/WP/248. Available online at: https://esa.un.org/unpd/wpp/publications/Files/WPP2017_KeyFindings.pdf. Accessed 14 June 2022

UNCCD (2015) Desertification, land degradation & drought (DLDD): some global facts and figures. United Nations Conventions to Combat Desertification. Available online at: http://www.unccd.int/Lists/SiteDocumentLibrary/WDCD/DLDD%20Facts.pdf. Accessed 14 June 2022

Waswa BS (2012) Assessment of land degradation patterns in Western Kenya: Implications for Restoration and Rehabilitation. PhD thesis, University of Bonn, Bonn. Available online at: https://d-nb.info/1043056505/34. Accessed 14 June 2022

Open Access This chapter is licensed under the terms of the Creative Commons Attribution 4.0 International License (http://creativecommons.org/licenses/by/4.0/), which permits use, sharing, adaptation, distribution and reproduction in any medium or format, as long as you give appropriate credit to the original author(s) and the source, provide a link to the Creative Commons license and indicate if changes were made.

The images or other third party material in this chapter are included in the chapter's Creative Commons license, unless indicated otherwise in a credit line to the material. If material is not included in the chapter's Creative Commons license and your intended use is not permitted by statutory regulation or exceeds the permitted use, you will need to obtain permission directly from the copyright holder.

Leveraging on Indigenous Vegetables for Climate Change Adaptation in Central Malawi: Exploring Opportunities and Challenges

Abel Sefasi, Sibongile Zimba-Chimzinga, Mvuyeni Nyasulu, Charity Chanza, Jacinta Nyaika, Gomezgani Nyasulu, and Rowland Maganizo Kamanga

Abstract Malawi's economy is highly dependent on rain-fed agriculture thereby making it vulnerable to droughts and under-nutrition. Indigenous vegetables (IVs) have remarkable potential in contributing to sustainable food and nutrition security in Malawi. These vegetables are highly nutritious and adaptable to the farming systems and climate of Malawi. Despite that IVs have the potential to contribute to sustainable food and nutrition security in Malawi, there is no empirical evidence that provides the status of the IVs. To address this gap, a participatory evaluation approach was used to identify opportunities and challenges that promote or hinder commercialization of IVs. This study identified various challenges affecting the scaling-up of utilization and commercialization of IVs. From 84 farming households surveyed, 56.1% of the farmers use recycled IV seed. Lack of quality seed was indicated as a major challenge both by farmers and stakeholders. The highly cultivated IV was found to be Pumpkin leaves*(Nkhwani*), which is cultivated by over 88.1% of the farmers and representing 21.5% of all other vegetables. The current data paves way for designing proper interventions to improve IV value chain. It would be best to develop technical messages as guidelines on best agronomic and post-harvest practices for developing IVs in Malawi.

Keywords Indigenous vegetables · Climate change · Food security · Quality seed · Malawi

---

A. Sefasi (✉) · S. Zimba-Chimzinga · M. Nyasulu · J. Nyaika · G. Nyasulu · R. M. Kamanga
Horticulture Department, Lilongwe University of Agriculture and Natural Resources, Lilongwe, Malawi
e-mail: asefasi@luanar.ac.mw

C. Chanza
Extension Department, Lilongwe University of Agriculture and Natural Resources, Lilongwe, Malawi

© The Author(s) 2026

B. R. Singh et al. (eds.), *Sustainable Food Systems in Sub-Saharan Africa*,
https://doi.org/10.1007/978-3-032-11288-0_16

1 Introduction

Malawi's current food systems face a multitude of intricate and interconnected challenges. The national food systems dialogue held in 2021 identified five crucial courses of action to tackle the challenges. They include ensuring access to safe and nutritious food for all, promoting healthy and sustainable consumption patterns, enhancing nature-positive food production, advancing equitable livelihoods for food system participants, and building resilience against vulnerabilities, shocks, and stresses (FAO 2021). Agriculture produces most of the food that people depend on for food security, nutrition and resilience. In Malawi, the agricultural sector employs eighty percent of Malawi's population, and sixty percent of the population relies on subsistence agriculture (USAID 2013). However, high population density and growth, along with frequent droughts and floods, pose a distinct set of challenges that make Malawians chronically susceptible to food and nutrition insecurity (Devereux 2007). The entire country has recently demonstrated vulnerability to drought (World Bank 2013). For example, in 2016, Southern African countries including Malawi experienced El Niño related severe drought. Rains, which typically begin in October/November, were delayed by almost two months in some areas of Malawi, leading to failed crops for many subsistence farmers. Due to the poverty levels, population, and environmental pressures, Malawi is especially vulnerable to weather shocks. Recently, in the 2022/2023 rainy season, Malawi received excessive rains which were associated with floods and widespread destruction of crops and livestock, particularly in the Southern region of the country. Droughts and floods bring a two-fold problem of both food and nutritional insecurity. It is evident that chronic under-nutrition (stunting) and micro-nutrient deficiencies is nearly equally spread across the country with stunting being high in all the three regions: Southern (48%), Central (47%) and Northern (45%) (World Bank 2013). Therefore, there is urgent need to manage the effects of climate change and under-nutrition in Malawi. Indigenous Vegetables have potential to serve as a climate change adaptation measure, and to assist in the attainment of sustainable food and nutrition security. Indigenous crops are central in increasing farmers' resilience and adaptation to climate change. Because indigenous vegetables are uniquely adapted to their local environments, they are often tolerant of challenging surrounding conditions (Schippers and Budd 1997) and boast vast reservoirs of genetic diversity, which is important to a plant's ability to adapt to climate change. Due to this stress-resistance-ability, indigenous crops are a rich source of genetic traits that could help buffer major agricultural plants against the warming trend and precipitation changes associated with a changing climate.

Indigenous vegetables contain significant amounts of vitamins and proteins (Abukutsa-Onyango 2011; Mnzava 1997) which are, in most cases equivalent to or higher than those found in popular exotic vegetables like cabbage (Abukutsa-Onyango 2011; Schippers and Budd 1997). Some of the nutritious and pre-adapted indigenous vegetables that are consumed in Malawi include Amaranths *(Amaranthus species)*, spider plant *(Cleome gynandra)*, African vegetable nightshades (*Solanum* species), cowpeas (*Vigna unguiculata*), African eggplant (*Solanum aethiopicum*),

African kale *(Brassica carinata)* and jews mallow *(Corchorus olitorius)* (Schippers and Budd 1997). Also, they have medicinal properties (Kokwaro 1993; Olembo et al 1995); for instance, spider plant has been reported to relieve constipation. It is suggested that increased cultivation, consumption and commercialization of indigenous vegetables has potential to increase incomes, restore biodiversity, and preserve local cultures (Stone et al 2011; Weinberger and Lumpkin 2007). Therefore, it is crucial to recognize, acknowledge, and collaborate with various cultures and communities that have preserved them for centuries. In addition to climate concerns, global conflicts such as the war in Ukraine threaten imports of wheat and other crops to the global south. In circumstances like these, enhancing and conserving Indigenous crops could help bolster these regions against resulting food shortages.

Despite the obvious importance of Indigenous vegetables, their utilization and consumption is declining. Previous studies have documented that this decline is more pronounced among rural communities, particularly the young generation. In some cases, these Indigenous vegetables have been associated with resource-poor people (Madisa and Tshamekang 2012). These vegetables are increasingly being replaced by exotic vegetables which are perceived as a sign of affluence (Nyasulu et al 2021). This has led to substantial loss in indigenous knowledge on collection, preparation, utilization and preservation of indigenous vegetables (Chawanje et al 2011). Challenges hampering the growth of Indigenous vegetables sector in Southern Africa were recently highlighted (Chagomoka et al. 2014). These included lack of access to high quality seeds, high on-farm production losses, high perishability and post-harvest losses, lack of appropriate market infrastructure for handling perishable produce, weak linkages between supply chain actors, lack of well-structured and organized markets leading to high transaction costs along the supply chain. However, specific locations could have variations regarding their priority challenges. In addition, previous reports have demonstrated that the list of priority vegetables will vary over time and between different locations of the same region (Chagomoka et al. 2014; Kwapata and Maliro 1991). This demonstrates that for a country like Malawi any intervention for the improvement of Indigenous vegetables would require proper understanding of their current status. Previous studies on indigenous vegetables in Malawi have primarily focused on documenting IV species and their place in the Malawian diet. However, there is limited empirical evidence regarding the participatory evaluation of the opportunities and challenges of using indigenous vegetables (IVs) as a climate change adaptation strategy in central Malawi. To address this gap, the study aimed to answer the following research questions:

i. What are the key opportunities presented by indigenous vegetables (IVs) in central Malawi as a climate change adaptation strategy, as identified by farmers and stakeholders?
ii. What specific challenges hinder the utilization and commercialization of IVs within the agricultural value chain in the context of climate change adaptation?
iii. What are the common IVs and the magnitude of production in central Malawi?
iv. What are the stakeholders involved in the promotion of IVs in central Malawi?

2 Methodology

2.1 Characteristics of the Study Area

The survey was conducted in Dedza and Nkhotakota districts of central Malawi. Dedza has an altitude of 1000–1500 m above sea level with moderately deep and well drained soils. It experiences frequent dry spells and droughts. The altitude for Nkhotakota is 478 m above sea level. During the cool season, it experiences temperatures between 7.5 and 12.5 °C and relative humidity in the warm season varies between 40 and 60%, with an annual rainfall of 60–80% (GoM 1999). Dedza has a humid sub-tropical mild summer climate with temperatures below 25 °C, while Nkhotakota lies along the lakeshore and experiences higher sub-tropical to tropical temperatures. Nkhotakota is prone to floods and dry spells. The soils in Dedza are predominantly clay loam while in Nkhotakota, the soil is sandy clay loam or sandy (GoM 2020; Munthali et al 2019). Dedza district was considered as an extreme moist/wet area where Amaranthus can easily grow; on the other hand, Nkhotakota district was also considered as an extremey hot/dry area where Amaranthus can easily grow. Amaranthus is a C_4 plant (Managa and Nemadodzi 2023) hence able to allow the efficient usage of CO_2 in a wide range of temperature and moisture-stress environments. Therefore, this study tries to evaluate the availability, utilization and commercialization of Amaranthus in these two areas.

2.2 Sample Size and Sampling Procedure

The major focus for the survey was to establish the current status of IVs in the survey area. Non probability sampling using exponential discriminative snowball technique was used to identify farmers who grew IVs. Data saturation point was reached at 84 farmers. The sampling unit was a household and their IV farm. With the help of the extension workers, a total of 60 male and female participants were purposively selected to participate in FGDs. Three FGDs were conducted in each district, with each FGD featuring 10 participants who were purposively selected based on their knowledge and experience with indigenous vegetable production.

2.3 Stakeholder Consultations

Stakeholder organizations which had worked on IVs for more than 2 years provided supplementary data which was triangulated with data from the survey and FGDs to enhance the validity and reliability of the results. A checklist was used to collect complimentary information from various stakeholders involved in promotion of IVs in Dedza, Nkhotakota and Lilongwe districts. The consultations were aimed at establishing the level of involvement in IV promotion by stakeholders. Principal challenges and major opportunities were also identified by this exercise. These organizations included government departments and Non-Governmental Organizations (NGOs) such as World Vision International (WVI) Malawi, Concern Universal (CU), Nkhotakota Aids Support organization (NASO), Farm Income Diversification Programme (FIDP), Catholic Development Commission of Malawi (CADECOM), Total Land Care (TLC), Katengeza Horticulture Co-operative, FOCCAD, Society for Professional Agriculturists (SPA), Trustees of Agricultural Promotion Program (TAPP), Nkhotakota and Dedza District Agriculture Development Office. Other organizations and programmes not interviewed, but working closely with some of those listed above are District Health Offices (D.H.O.), Support for Service Delivery Integration (SSDI), scaling up nutrition (S.U.N.), Concern Worldwide, and One Village One Product (OVOP).

2.4 Data Collection Methods and Tools

A household survey and FGDs were conducted where a semi-structured questionnaire and checklist were administered to farmers to collect quantitative and qualitative data respectively. The survey and FGDs were aimed at gathering information on the perceptions towards IVs, the major opportunities, principal challenges and constraints to production, utilization and commercialization of IVs.

2.5 Data Analysis

Quantitative data was coded and analyzed using Statistical Package for Social Sciences (SPSS) version 20 to come up with descriptive statistics. Content analysis was used to analyse qualitative data.

3 Results

3.1 Social Economic Characteristics

A total of 84 households were interviewed during the survey. Since most households in Malawi are headed by males, a large number of respondents were males (77.38%). Majority of the respondents (61.9%) involved in IV cultivation were aged between 25 and 55 years, while 33.33% were aged over 55 and 4.76% were aged between 18 and 24. This study also established that most of the respondents had low literacy levels. The proportion of respondents who had not attended formal education was 28.57%. Only 21.43% of them had attended junior primary school education while 38.1% had senior primary school background. About 2.38% had reached to junior secondary school level, while 8.33% had reached senior secondary school level. The rest (1.19%) of the respondents had attended some adult literacy education.

3.2 Geospatial Characteristics

About half of the sampled population owned at least 0.4 to 0.8 ha while 29.76% possessed more than 0.8 ha but less than 4 ha of the agricultural land. In Malawi, the average holding size for small-scale farms is 0.69 hectares. Further, large landholding size was observed in Nkhotakota relative to Dedza. Interestingly, the proportion of famers growing IVs under mixed cropping system in Nkhotakota was higher (71.5%), while in Dedza up to 63.3% of the farmers practiced mixed cropping. Mixed cropping is usually done with maize or tobacco as the major crops.

3.3 Crop Diversification

The results show that among the sampled population, the major source of income is through farming (85.7%). Based on the proportion of farmers growing a particular crop, 10 major crops were identified. Maize was a major crop grown by almost all the farmers (Fig. 1). Other major crops were cassava and sweet potato from which farmers also derive vegetables, locally referred to as *Chigwada* (Cassava leaves) and *Kholowa* (Sweet potato leaves), respectively.

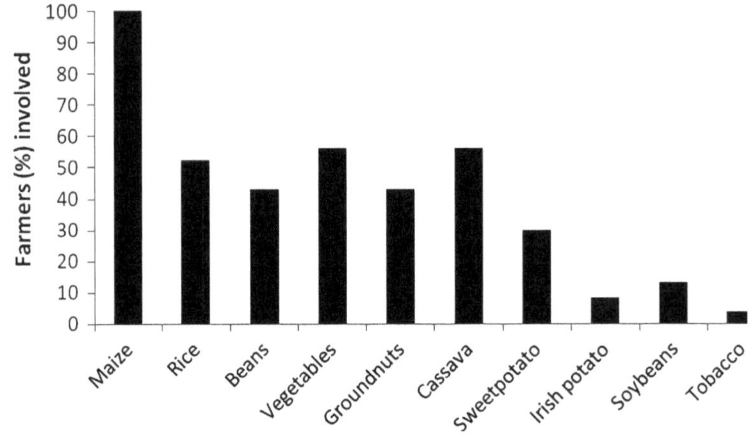

Fig. 1 Major crops grown as determined by per cent of farmers involved in cultivation

3.4 Extent of IV Cultivation

Up to 30 vegetables species were cultivated by farmers. Table 1 shows that Nkhotakota has a large proportion (21.7%) of farmers growing IVs under monocropping, while Dedza has a high proportion (20.2%) for intercropping. Of all the IVs, the most highly cultivated one was pumpkin leaves, locally called *Nkhwani.* It was cultivated by over 88.1% of the farmers, representing 21.5% of all other vegetables (Table 2). Other prominent IVs were cassava leaves*(Manhot esculenta),* Amaranths *(Amaranth sp),* sweet potato leaves *(Ipomoea batatas),* Cowpea leaves *(Vigna unguiculata),* bean leaves *(Phaseolus vulgaris),* Blackjack (*Bidens pilosa*), Jews mallow *(Corchorus olitorius),* Roselle *(Hibiscus sabdariffa),* and Cat's whiskers (*Cleome gynandra*). There were other IVs which were not commonly grown. These represented 6.4% of the total IVs mentioned. Although not commonly cultivated, these vegetables were equally important as traditional medicine and food. Vegetables in this group included *Kazizira, Kandudwa, Pitipiti, Chanusi, Chansungwi, Chikambu, Nkhungudzu, Chimkakala, Nkhanyanga, Kaliputi, Chikhadabura, Katate, Limanda and Njerenjeza.* It was interesting to note in this study that farmers intending to sell vegetables expressed preference for growing exotic species (Fig. 2).

Table 1 Proportion (%) of farmers cultivating indigenous vegetables under a particular cropping system

Cropping system	District	
	Dedza	Nkhotakota
Monocropping	16.5	21.7
Intercropping	20.2	6.8
Mixed cropping	63.3	71.5

Table 2 Proportion (%) of farmers growing major indigenous vegetables

Indigenous vegetable	Farmers (%) growing vegetable		
	Nkhotakota	Dedza	Total
Nkhwani	96.0	76.5	88.1
Chigwada	86.0	0.0	51.2
Bonongwe	44.0	52.9	47.6
Kholowa	76.0	2.9	46.4
Chitambe	46.0	23.5	36.9
Khwanya	18.0	55.9	33.3
Chisoso	18.0	50.0	31.0
Denje	22.0	8.8	16.7
Chithanda	16.0	5.9	11.9
Chidede	12.0	2.9	8.3
Luni	8.0	2.9	6.0
Mwamunaaligone	0.0	14.7	6.0

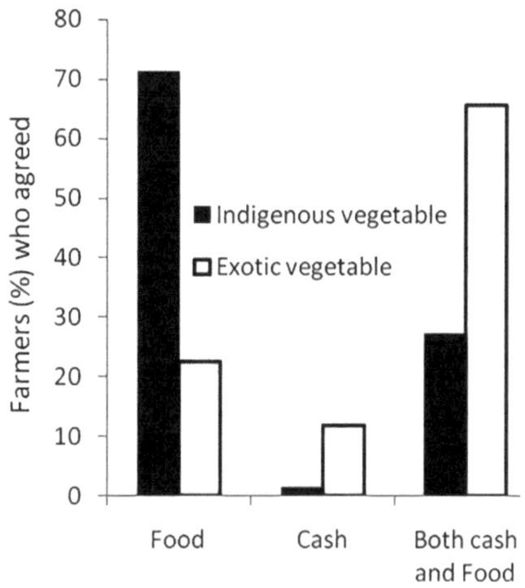

Fig. 2 Farmer's reason for growing Indigenous and exotic vegetables

Stakeholder involvement in promotion of IV

Of the stakeholders interviewed, only ten had demonstrated consistent involvement with IVs at community level for two years or more. Most of the organizations were mainly targeting improvement of nutrition and mitigating effects of HIV/AIDS. The major involvement of most stakeholders was through training of farmers in nutrition aspects, including processing of IVs (Fig. 3). Most organizations trained famers to grow IVs in their backyards with the sole aim of feeding their family. Although commercialization of IVs has emerged as a major challenge, only one organization, namely Farm Income Diversification Programme (FIDP), included development of business skills in the package of its intervention on IVs.

It was found that Amaranths was the dominant IV promoted as a priority by nine of the ten stakeholder organizations. Although Amaranths' seeds have been found to be more nutritious than the leaves, there were no farmers involved in processing such

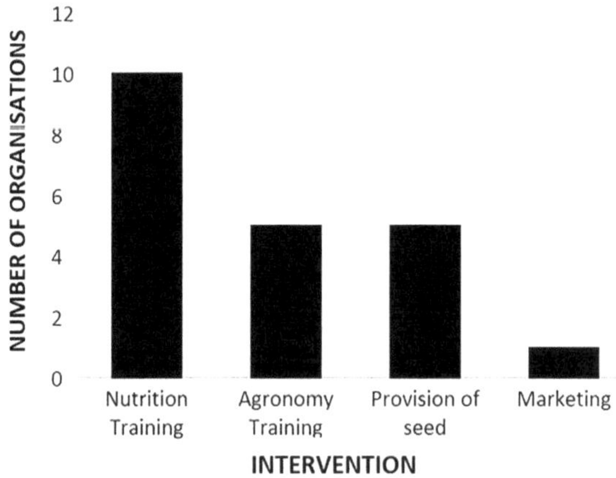

Fig. 3 Major interventions by stakeholders promoting indigenous vegetables

seeds. All the farmers interviewed were only consuming the leaves. The Department of Crop Development of the Ministry of Agriculture had also indicated Amaranths as a priority IV. In the survey with farmers, Amaranths came third after pumpkin leaves and cassava leaves (Table 2). It is likely that this is a result of the disparity in understanding of the definition of IVs by local farmers in comparison with Agricultural officers. Nevertheless, Table 2 still shows that Amaranths is an important IV, considering the high diversity of species grown by local farmers in the study areas.

Other vegetables promoted by the stakeholders include Cats' whiskers (Luni), Pumpkin leaves (Nkhwani), black jack (Chisoso), and sweet potato leaves (Kholowa), cassava leaves, mustard leaf, Moringa, Kamganje (Local rape).

3.5 Challenges Faced in the IV Value Chain

Lack of quality seed was indicated as a major challenge both by farmers and stakeholders. Most farmers (56.1%) use recycled IV seed. Apart from poor access to quality seed, farmers indicated that frequent dry spells, droughts in rainy season is a major challenge. The data showed that 66.6% of farmers grow IVs in rainy season without additional irrigation.

Overall, a large proportion (66.6%) of farmers grows IVs in rainy season while 11.0% grow IVs under irrigation. In Dedza, 24.1% of farmers rely on irrigation while in Nkhotakota only 4.7% cultivate IVs under irrigation. Much of the water for irrigation in Dedza was sourced from wells (68.8%), while in Nkhotakota up to 55.2% of farmers were able to grow IVs under residual moisture in wetlands. Other sources of irrigation water in both districts were rivers and boreholes. It is worth noting that Dedza receives relatively lower rainfall than Nkhotakota. This may explain the large proportion of farmers depending on irrigation in Dedza.

3.6 Current Opportunities in the IV Value Chain

Projects aimed at reducing the impact of HIV/AIDS, such as the one implemented by World Vision International (WVI) in Dedza, have increased awareness in many communities about the importance of including indigenous vegetables (IVs) in their diets. While the connection between HIV and the consumption of IVs has been recognized within these communities, there is limited scientific evidence linking IV consumption to the reduction of the impact of HIV/AIDS. Despite this limitation, there is a growing number of organizations interested in promoting IVs. One possible linkage of indigenous vegetables consumption with HIV/ AIDS is that they may be cheap, nutritious, economically beneficial and their production less labor-intensive. Moreover, IVs may constitute an alternative to food shortages and income problems caused by HIV/AIDS if existing added value technologies are improved and additional appropriate ones developed and made available to farmers for use at low cost.

Accordingly, the government has maintained extension staff at the local community level, making them accessible for linking IV intervention projects to local communities. Furthermore, majority of the respondents indicated that changes in the climate scenario make it essential to produce and utilize IVs, as they are adaptable to adverse environmental conditions. It was also revealed that most consumers are becoming aware of healthy eating habits, which has led to an increased consumption of organically grown foods. Since IVs can be grown without chemical pesticides and with organic fertilizers, they have become an attractive option for most households.

3.7 Marketing Opportunities and Challenges for Horticulture Value Chains

The study found that farmers are involved in various horticultural enterprises including tomato, onion, leafy vegetables and other exotic vegetables. Interestingly, indigenous vegetables were not mentioned during the discussions on horticultural value chains. They also narrated that these vegetables were usually sold at farm gate or they take them to nearby and popular markets closer to them. The respondents demonstrated their knowledge of opportunities that could be used to improve marketing of their product. They mentioned things like good road networks, mobile-phone based platforms like WhatsApp and Facebook. They indicated that they needed to be empowered to utilize such platforms and avenues to their benefit. An interaction with market actors revealed high demand for indigenous vegetables on the market. All the markets visited in the two districts showed a similar trend of high demand. The market stakeholders lamented the limited supply of these indigenous vegetables on the market. Most intermediaries reported that their customers demand these indigenous vegetables but they are always in low supply. Some stakeholders indicated that some wrong attitudes and perceptions limit the production of these indigenous vegetables. For instance, some farmers said these vegetables have been growing naturally hence, they see no need to grow them.

"*These indigenous vegetables are highly demanded on the market. I came with Amaranthus, black jack and pumpkin leaves. I sold them within an hour and now I am just waiting for my colleague to finish her produce then we go home. The problem is that these vegetables grow naturally hence they are in short supply*" one female producer at a local market narrated.

They went further to narrate that they are mostly for women. Market stakeholders also corroborated this by saying these indigenous vegetables are associated with women as quoted below;

"*As you can see here, no man is selling Amaranthus, Cats' whiskers or black jack as they are associated with women. These are sold by women in that shade, for us we sell cabbage, tomato and other vegetables*" One male vendor reported.

4 Discussion

The aim of the present study was to evaluate opportunities and challenges of exploiting IVs as a climate change mitigation measure in central Malawi. A participatory approach was deployed to study the IVs value chain, mainly with farmers and stakeholders as lead informants. The study found poor extension services in relation to promotion of IVs. Although government has the most extensive agricultural extension staff network, it was noted that there were no deliberate efforts by such staff to promote IVs. While there were some trainings done targeting cultivation and postharvest handling of exotic vegetables, there were no such trainings for IVs. This would also explain that the negative perception towards IVs is not only among farmers, but also among agricultural extension staff. It is also possible that the agricultural policy of Malawi needs to be re-focused.

However, an opportunity was noted in that the agricultural extension services in Malawi has recently become pluralistic, both governmental and non-governmental extension staff have direct access to farmers. It is possible therefore for NGO extension staff to reach farmers directly with IV promotion messages or to work with government extension staff to access farmers with technical messages. The growing number of NGOs interested in exploiting vegetables as an HIV/AIDS mitigation measure and a way of improving nutrition is an opportunity for expanding the cultivation and consumption of IVs. While the scientific basis of this observation remains unclear, it is suggested that one possible linkage of indigenous vegetables consumption with HIV/ AIDS is that they may be cheap, nutritious, economically beneficial and their production less labor-intensive. Moreover, IVs may constitute an alternative to food shortages and income problems caused by HIV/AIDS if existing added value technologies are improved and additional appropriate ones developed and made available to farmers for use at low cost.

Another key driving factor for consumption of IVs was found to be the change in eating habits, especially among the urban and nutritionally aware households, which presents a remarkable opportunity for promoting consumption and utilization of IVs. Since IVs are predominantly produced organically, they have become an attractive option, relative to most exotic vegetables that are produced with excessive amounts of dangerous chemicals and inorganic fertilizes. Hence, there is need to address the current drawbacks in the IVs value chain, especially on seed supply, which was reported as a primary setback in production of IVs. Despite the growing demand for, and rising awareness of the healthy benefits of IVs, young people remain less involved in IV production. This study found that there was low participation of young generation in the cultivation, collection, processing and utilization of IVs. This remained a fact although the high proportion of the population is made up of the young generation. The low involvement of the young generation in IV cultivation is not surprising. Previous reports indicate that young people look down on IVs and regard them as weeds (Abukutsa-Onyango 2011; Madisa and Tshamekang 2012) and a primitive food. This perception is highly prevalent in rural areas of Malawi, although anecdotal data shows that some Malawians in urban areas have realized the nutritive value of IVs, and they are now being sold in established markets. Most rural famers grow very small plots of IVs which they almost use exclusively for household consumption. The major contributing factor in the farmers' decision not to include IVs among crops for sale is obviously driven by demand. Hence, there is need to raise the awareness of the importance of IVs, especially among the young people.

This study has also revealed an incredible diversity of IVs utilized in the study areas. Over 30 IVs were reported to be cultivated by the farmers. It must be mentioned that these results could be context specific. Pumpkin leaves were found to be the most predominant in both districts. While cultivation of pumpkin leaves is indeed generally predominant in Malawi, these results may differ in other study areas due to differences in food cultures, climate scenarios, agricultural interventions by government and NGOs, social status among others. Hence, it is important to contextualize these observations in subsequent studies, to elucidate the underlying factors influencing choice of a particular value chain in an area. Furthermore, while the study sites explored are of diverse cultures, future studies are recommended to increase the number of districts studied in order to effectively capture the IV production practices and contexts. In the present case however, the data presented here is only an example and a context-specific opportunity to design interventions, such as technical guidelines, to enhance the value chain, and to address key challenges for sustainable food and nutrition security in Malawi. Additionally, between the two districts, it was revealed that Dedza had a higher dependence on irrigation than Nkhotakota. This is ascribed to low rainfall received in the district. However, this may also be attributed to the fact that although farmers in Nkhotakota are expected to have high access to irrigation water, the district, along the shore of Lake Malawi, has fishing as a major alternative source of livelihood. In Nkhotakota, the proportion of farmers relying solely on rainfall for cultivation of IVs was 70.6%.

IVs are uniquely adapted to their local environments and present a rich source of genetic diversity and traits that could help buffer major agricultural plants against the various abiotic stresses. They are therefore vital for breeding purposes, as they enable plant breeders to produce improved varieties. Because Indigenous crops are uniquely adapted to their local environments, they are often tolerant to challenging surrounding conditions and boast of vast reservoirs of genetic diversity, which is important to a plant's ability to adapt to climate change. Due to this stress-resistance ability, Indigenous crops have recently attracted interest from scientists and plant breeders as a possible source of genetic traits that could help buffer major agricultural plants against the warming trend and precipitation changes associated with a changing climate. However, as these research efforts move forward, it is crucial that when we draw on the promise of Indigenous crops to help address the coming challenges, plant biologists and other experts should recognize, acknowledge, and collaborate with the cultures and countries that have preserved them for thousands of years. Furthermore, in addition to climate concerns, global conflicts such as the war in Ukraine threaten wheat and other imports to the Global South. Enhancing and conserving Indigenous crops could help bolster these regions against the resulting food shortages.

In order to make the most out of IVs, it is imperative to first address the prevailing challenges. Consultations with both the farmers and stakeholders had indicated that access to quality seed of IVs and absence of technical messages on recommended agronomic practices, processing and market access as the major challenges. This is in a sharp contrast with exotic vegetables, which have an established seed system. For example, despite pumpkin leaves being a prominent IV in Malawi, and consumed by over 88.1% of respondents, there are no officially released varieties in Malawi, not to mention availability of quality certified seeds. This gap remains a major impediment, hence calls for urgent attention by various experts including breeders, extension experts and agribusiness specialists.

5 Conclusion

This chapter has explored opportunities and challenges associated with cultivation of indigenous vegetables for climate change adaptation. It was interesting to find that there were more than ten stakeholders involved in the indigenous vegetables value chain. Importantly also, over thirty indigenous vegetables were cultivated by farmers. This illustrates the diversity of IVs, which is vital for breeding purposes as it enables plant breeders to produce new improved varieties. Various IV species have been identified that could be up-scaled in the target areas as climate change mitigation measures in the cultivation of vegetables in the study area. The identification of priority IVs is an important contribution by this study. It would assist future interventions in targeting only those IV species while promotion campaigns for the rest are on-going. Pumpkin leaves, cassava leaves, amaranths and sweet potato leaves have been identified as priority IVs by farmers while stakeholders indicated amaranths as a popular IV. This disparity could be sorely due to the difference in the understanding of the definition of IVs by the two groups. The identification of those IVs only known by vernacular names is on-going in a follow-up study. Most IVs are short-duration annuals able to escape the effects of drought. The data shows that poor access to quality seed and poor extension services targeting IVs were the main challenges. It was however interesting to find that there were more than ten stakeholders, mainly NGOs, involved in promoting IVs at various levels, mainly through nutrition and HIV/AIDS mitigation campaigns.

6 Recommendation

This chapter has highlighted various opportunities and challenges associated with indigenous vegetables regarding the potential for climate resilience. However, it would be best to conduct a similar survey involving a larger section of Malawi in rainy season. It would also be commendable if breeding efforts in Malawi included IVs among their priority crops to improve their water stress tolerance, nutrition (and anti-nutrition status) factors and their shelf life among other traits. This would accelerate adoption of IVs. Farmer trainings through demonstrations would improve capacity of farmers in conducting proper agronomic practices and postharvest management. It would also be best for organizations involved in promotion of IVs to improve on collaboration among themselves, to improve efficient use of resources and reduce duplication of efforts targeting farmers. A stakeholders' workshop involving all stakeholders involved in promotion of IVs in Malawi would be helpful. It would be best to develop sound technical messages and explore ways of developing the IV seed systems which is currently non-existent. The study recommends that more IV studies need to be conducted to identify and establish their common names and scientific names for easy communication and further research interventions.

Acknowledgements The authors would like to acknowledge primarily the funding from Capacity Building for Managing Climate Change (CABMACC) through the project CABMACC 001-2015 LUANAR '*Promotion of indigenous vegetables seed production, conservation and utilization in Malawi*'. Farmers, stakeholders and District agriculture officers are thanked for providing valuable data. The authors would like to thank the Lilongwe University of Agriculture and natural resources (LUANAR) management and staff for the support.

Abbreviations and Acronyms

CABMACC	Capacity Building for Managing Climate Change
CADECOM	Catholic Development Commission of Malawi
CW	Concern Worldwide
CU	Concern Universal
DHO	District Health Offices
FGD	Focus Group Discussion
FIDP	Farm Income Diversification Programme
GoM	Government of Malawi
HIV/AIDS	Human immunodeficiency virus/acquired immunodeficiency syndrome
IVs	Indigenous vegetables
KHC	Katengeza Horticulture Cooperative
LUANAR	Lilongwe University of Agriculture and natural resources
NASO	Nkhotakota Aids Support organization
NGO	Non-Governmental Organization
OVOP	One Village One Product
SPA	Society for Professional Agriculturists
SPSS	Statistical Package for Social Sciences
SSDI	Support for Service Delivery Integration
SUN	Scaling Up Nutrition
TAPP	Trustees of Agricultural Promotion Program
TLC	Total Land Care (TLC)
USAID	United States Agency for International Development
WVI	World Vision International Malawi

References

Abukutsa-Onyango MO (2011) The diversity of cultivated African leafy vegetables in three communities in western Kenya. Afr J Food Agric Nutr Dev (AJFAND) 7(3)

Chagomoka T, Afari-Sefa V, Pitoro R (2014) Value chain analysis of traditional vegetables from Malawi and Mozambique. Int Food Agribus Manage Rev 17(4):59–86

Chawanje CM, Barbeau WE, Gruen I (2011) The importance of indigenous crops in Malawi—a case of *Nymphaea petersiana* (Nyika). University of Malawi, Zomba, Malawi

FAO (2021) United Nations food systems dialogues: the Malawi story so far. In: FAO representative in Malawi. Food and agriculture organization of the United Nations, FAO-Malawi@fao.org, vol 3. Lilongwe, Malawi

Devereux S (2007) The impact of droughts and floods on food security and policy options to alleviate negative effects. AGEC Blue Book J 37:47–58

GoM (1999) Government of Malawi Dedza district Socio-economic profile. Government of Malawi, Lilongwe, Malawi

GoM (2020) Nkhotakota district council socio-economic profile, Malawi. Government of Malawi

Kokwaro JO (1993) Medicinal plants of East Africa, 2nd edn. Kenya Literature Bureau, Nairobi

Kwapata MB, Maliro MF (1991) Indigenous vegetables in Malawi: Germplasm collecting and improvement of production practices. Bunda College of Agriculture, Lilongwe, Malawi

Madisa ME, Tshamekang ME (2012) Conservation and utilization of indigenous vegetables in Botswana. Department of Agricultural Research, Ministry of Agriculture, Gaborone, Botswana

Managa GM Nemadodzi LE (2023) Comparison of agronomic parameters and nutritional composition on red and green amaranth species grown in open field versus greenhouse environment. Agriculture 13:685. https://doi.org/10.3390/agriculture13030685

Mnzava NA (1997) Vegetable crop diversification and the place of traditional species. In: Guarino L (ed) (1997), Traditional African vegetables. Promoting the conservation and use of underutilized and neglected crops. 16 Proceedings of the IPGRI International Workshop on Genetic Resources of Traditional Vegetables in Africa. Conservation and use, 29–31 Aug 1995, ICRAF-HQ, Nairobi, Kenya. Institute of Plant Genetic and Crop Plant Research, Gatersleben/International Plant Genetic Resources Institute, Rome, Italy, pp 1–15

Munthali MG, Davis N, Adeola AM, Botai JO, Kamwi JM, Chisale HLW, Orimoogunje OOI (2019) Local perception of drivers of land-use and land cover change dynamics across Dedza District. Central Malawi Region

Nyasulu M, Sefasi A, Chimzinga S, Maliro M (2021) Agromophological characterisation of amaranth accessions from Malawi. Am J Plant Sci 12:1528–1542. https://doi.org/10.4236/ajps.2021.1210108

Olembo NK, Fedha SS, Ngaira ES (1995) Medicinal and agricultural plants of Ikolomani, Kakamega District

Schippers RR, Budd L (1997) African indigenous vegetables. In: Workshop proceedings, 13–18 Jan 1997, Limbe, Cameroon. NRI-Natural Resources Institute, University of Greenwich, Chatham Maritime. United Kingdom

Stone A, Massey A, Theobald M, Styslinger M, Kane D, Kandy D, Tung A, Adekoya A, Madan J, Davert E (2011) Africa's indigenous vegetables. In: World watch institute (ed), The state of the world 2011: Innovations that nourish the planet. World Watch Institute, Washington, DC

USAID (2013) Feed the future malawi fact sheet 2013. http://feedthefuture.gov/sites/default/files/resource/files/ftf_factsheet_malawi_jan2013.pdf

Weinberger K, Lumpkin TA (2007) Diversification into horticulture and poverty reduction: a research agenda. World Dev 35:1464–1480

World Bank (2013) MVAC report. Malawi Vulnerability Assessment Committee Report

Open Access This chapter is licensed under the terms of the Creative Commons Attribution 4.0 International License (http://creativecommons.org/licenses/by/4.0/), which permits use, sharing, adaptation, distribution and reproduction in any medium or format, as long as you give appropriate credit to the original author(s) and the source, provide a link to the Creative Commons license and indicate if changes were made.

The images or other third party material in this chapter are included in the chapter's Creative Commons license, unless indicated otherwise in a credit line to the material. If material is not included in the chapter's Creative Commons license and your intended use is not permitted by statutory regulation or exceeds the permitted use, you will need to obtain permission directly from the copyright holder.

Smallholder Farmland Degradation in Malawi: Unpacking Restoration Drivers Using a DPSIR Framework

Austin Tibu, Bishal K. Sitaula, Lars Kare Grimsby, Jennifer Joy West, and Moses Limuwa

Abstract Socio-economic drivers are putting pressure on soil fertility and crop productivity in smallholder agriculture in Malawi. Policy responses may be designed to address direct drivers of farmland degradation. However, less tangible indirect drivers may have an equally important role. This chapter reviews literature and policy to analyze interventions and outcomes for land restoration in Malawi, as framed by the DPSIR framework. The framework enables the systematization of societal drivers and responses to environmental issues. We extend it to include values and worldviews, to allow for a deeper understanding of how human agency has been shaped and can shape decision-making in land restoration. Four questions emerge relevant for designing policy and interventions to address land degradation: (1) What values, worldviews, beliefs, and perspectives exist about nature, the environment, and the local food system in the community? (2) How can we explore the underlying reasons and explanations for smallholder farmers' willingness to engage in land restoration practices? (3) How do smallholder attitudes and perceptions concerning the benefits of environmental conservation compare with biophysical characteristics on the ground? (4) How are the current integrated learning approaches to environmental conservation enabling meaning-making, and how are they facilitating the integration of the options of being and becoming stakeholders? To achieve land restoration on degraded smallholder farmland in Malawi, it is necessary to stimulate local participation by integrating local knowledge, capacities, and values as drivers and responses to degraded farmland.

Keywords Land degradation · Land restoration · Drivers-pressures-states-impacts-responses (DPSIR) framework · Smallholders · Malawi · Agricultural production · Motivations · Inner qualities

A. Tibu (✉) · B. K. Sitaula · L. K. Grimsby · J. J. West
Norwegian University of Life Sciences, Ås, Norway
e-mail: austin.tibu@nmbu.no

M. Limuwa
Lilongwe University of Agriculture and Natural Resources, Lilongwe, Malawi

© The Author(s) 2026

B. R. Singh et al. (eds.), *Sustainable Food Systems in Sub-Saharan Africa*,
https://doi.org/10.1007/978-3-032-11288-0_17

1 Introduction

Local smallholder farmers are central to the efforts to restore land globally. Several studies highlight the interplay of local farmers as drivers and victims of land degradation (Agramont et al. 2021). In Malawi, smallholder farmers play a crucial role in the country's agricultural systems amidst several challenges. Cultivating on <5 ha of land per household, they make up a significant proportion of the population, which has increased over the years (Muyanga et al. 2020). Over 80% of the national population lives in rural areas and relies on agriculture for their food and livelihoods, and almost half of the Malawian diet consists of maize (World Bank 2019). However, agricultural production on smallholder farmland continues to face land fragmentation and lack of inputs added to the soil, leading to land degradation (UNEP 2016). Restoration efforts rely on understanding these local actors' interests, motivation, and willingness. Hence decoding smallholder farmers' aspirations, based on understanding their preferences, values, and worldviews, can be a valuable starting point for addressing land degradation. The current approach for addressing land degradation in Malawi focuses on mitigating the impact of population growth on soil erosion, deforestation, and forest degradation (GoM 2017a; UNDP 2021). Several initiatives have been implemented to combat these issues, including subsidized fertilizer programs to replenish soil nutrients, improved land tenure security, and sustainable land management practices to reduce degradation (World Bank 2019). Additionally, efforts are being made to tackle deforestation through tree planting and forest management interventions (GoM 2017a). While recent policies like the National Forest Landscape Restoration Strategy (NFLRS) (2018–2030), the National Soil Conservation and Restoration Strategy (2020–2026) (UNDP 2021) recognize the importance of a holistic approach to land restoration in Malawi, they seem not to consider how environmental systems (e.g. smallholder farmlands) are reproduced and reproduce human values, cultural beliefs, and forms of life.

Strategies to address land degradation have utilized frameworks to comprehend the interconnectedness of social and environmental systems (Agramont et al. 2021). One such framework is the Drivers Pressures State Impact Responses (DPSIR), which has been extensively used in academic studies on environmental degradation, river basin management, watershed management, and soil and water management (Agyemang et al. 2007; Sun et al. 2016; Agramont et al. 2021; Obubu et al. 2022). However, there is limited literature available regarding the application of DPSIR on smallholder farmland in Malawi (Pullanikkatil et al. 2016; Kosamu et al. 2022). Further, some strategies to address land restoration in Malawi appear to implicitly incorporate the DPSIR framework in identifying the drivers of environmental degradation (GoM 2017a; UNDP 2021). This approach could oversimplify the complex socio-ecological system operating at the farmland level. As a result, the derived responses may not match with the drivers in land restoration strategies. Potential mismatches may be prevalent at the smallholder farmland level because local levels tend to be neglected in many DPSIR applications (Agramont et al. 2021). Reviewing the driving forces of the current restoration strategies can unveil the potential to align restoration with local needs and aspirations.

Humans have the ability both to degrade and to restore land. These capacities can stem from individuals' and communities' current and potential values, beliefs, and worldviews by shaping how humans relate and connect to natural resources. These 'internal factors' are crucial in projecting how people view and address sustainability challenges (Wamsler and Brink 2018). Such motivations may be as important as external incentives, for example in promoting sustainable intensification practices (Jambo et al. 2019). Yet these dimensions are often neglected in addressing environmental issues. To improve the connection between human capital and the socio-ecological aspects of any system, such as smallholder farmland, it is essential to consider these inner dimensions and how they are espoused in current frameworks used to unpack drivers of land degradation. Consequently, understanding local capacities and values systematically offered by adapted frameworks can enhance opportunities for targeting deep leverage points for smallholder farmers to engage in land restoration.

This paper therefore reviews drivers of land degradation and their responses to comprehend the interconnectedness of smallholder farmers to their land. The Drivers Pressures State Impact Responses (DPSIR) framework was applied to understand the match between the NFLRS practices and their driving forces, particularly relating to smallholder farmland in Malawi. The review adopts transformation perspectives, highlighting the importance of human agency—values, beliefs, and worldviews—in shaping livelihood strategies such as land restoration. The perspectives can be applied to explore human agency related to their connection, awareness, insights, and purpose, and relate these characteristics to human connectedness to nature (Wamsler et al. 2021). This review's main objectives are: (1) to review the DPSIR as a framework for land degradation assessment and restoration responses; (2) to present a systematic overview of drivers and impacts of land degradation and land restoration responses in Malawi using the DPSIR framework, and to highlight additional considerations for environmental conservation in Malawi; (3) to compare the framing of landscape restoration strategies with the systematic review findings; and finally, (4) to highlight areas for further research and investigation of underlying problems of land degradation and its restoration responses, considering recent perspectives on inner sustainability and societal transformation.

2 Methods

The review analyzed agricultural and environmental policy and research papers in the Malawian context. This review was led by a team of experts from Malawi and Norway in the years 2023 and 2024. It employed a specific string of keywords to filter articles that pertained to Malawi, and topics such as land degradation, conservation, restoration, drivers, response, transformation, and approach. The retrieved articles were then carefully sorted into categories based on their connection to land degradation or restoration, and their relation to smallholder farmland, using an Excel Spreadsheet.

Based on published literature, technical guidelines, tools, strategies, and project documents, the review first described the DPSIR framework (OECD 2003; Tscherning et al. 2012). The resulting framework and indicators are presented in Sect. 3.2, highlighting major drivers, pressures, impacts, and state of land degradation in Malawi. The review also highlights land restoration practices in response to these drivers and pressures of land degradation. Particular attention was focused on smallholder farmers in Malawi because of the importance of the sector to national agricultural development.

Section 3.3 discusses the policies and strategies related to land restoration in Malawi. The NFLRS in Malawi was reviewed to examine the restoration practices in response to the drivers and pressures of land degradation/restoration. This was followed by comparing the framings with the summary of drivers, pressures, states, impacts, and responses to land degradation on smallholder farmlands in Malawi presented in Sect. 3.2.

Section 3.4 illustrates how our 'openness principle to integrating frameworks' can enhance the DPSIR framework for unpacking land restoration planning and implementation in Malawi. The discussion in this section adopts transformation approaches, highlighting the importance of human agency—values, beliefs, and worldviews—in shaping livelihood strategies such as land restoration. Additionally, the section poses thought-provoking questions about the relationship between individuals' lived experiences and worldviews, and the typical top-down approach taken by national strategies and policies aimed at addressing environmental concerns and agrarian change.

3 Results and Discussion

3.1 *The DPSIR Framework*

The review noted that the DPSIR framework emerged in the 1990s as a tool that addresses the root causes of human and natural variability and their impacts on the environment (OECD 2003; Tscherning et al. 2012). It was also able to identify the responses emanating from the impacts of changes on the livelihood assets and outcomes in a society (Carr et al. 2007). The DPSIR framework appears to be a key bridge between researchers and development partners by unifying their varying views to untangle complex socio-ecological systems (Agramont et al. 2021). Along the way, options are being suggested to tackle some of the debates of the frameworks, including the potential of DPSIR to untangle complex socio-ecological systems across spatial scales and social strata. However, the review observed that the application of the DPSIR framework may neglect the local levels and can result in a spatial mismatch between drivers and responses (Agramont et al. 2021).

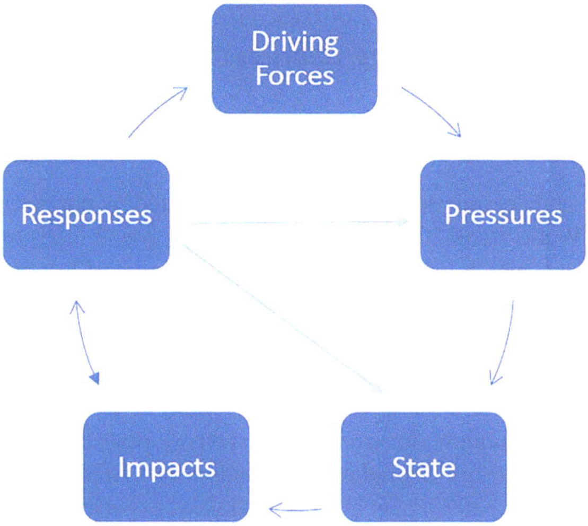

Fig. 1 The drivers-pressures-state-impact-response framework (adapted from OECD 2003)

Figure 1 presents an overview of the framework. The driving forces exert pressure on the environment. These forces can be social and/or economic (Agramont et al. 2021). The environmental impacts are a result of the pressures experienced by the system, which can change its natural state. The changes in the natural state may trigger society to respond to these impacts, pressures, and their driving forces.

The driving forces are the factors that cause changes in the system. These drivers may be direct or indirect. Direct factors refer to social, economic, or ecological processes that drive activities that directly impact the environment. Indirect drivers stimulate indirect causes of degradation (Gessesew 2007).

Pressures are human activities generated by the driving forces, and they directly affect the system. And the state is the condition of the system represented by indicators at a specific time. The nature of the state of the environment changes because of the pressures. The impacts may be socio-economic and ecological factors, which may change because of multiple stressors affecting human health and/or the ecosystem (Kosamu et al. 2022). Responses are activities and programs made by society to reduce the pressures, state, and impacts (Gessesew 2007). Responses may be directed to drivers, pressures, states, and impacts. By connecting the drivers and their responses, the framework has supported development organizations and researchers to respond to practical issues at local, national, and global levels.

While the DPSIR framework is useful in connecting root causes to environmental impacts of land degradation across social strata, it tends to simplify complex realities (Niemeijer and De Groot). As a result, it might miss the importance of human agency in driving responses to tackle relevant issues. This potential gap is the main point of departure for the review.

3.2 Drivers, Pressures, States, Impacts, and Responses to Land Degradation in Malawi

3.2.1 Drivers of Land Degradation

Land degradation in Malawi is aggravated by population growth and climate change (World Bank 2024). Rapid population growth leads to more cultivation on marginal land, which in turn results in soil erosion and loss of fertility (Li et al. 2017). Due to continuous cultivation, smallholder farmers are struggling to implement crop-fallow systems required to restore soil nutrients (World Bank 2019). Monocropping maize and inadequate fertilization have led to significant soil nutrient loss, especially nitrogen. Conservation efforts are being hindered by insecure customary land tenure.

The review discovered that steep slopes and erosive rainfall magnifies existing drivers to land degradation in Malawi. 31% of Malawi's slopes are very steep, ranging from 6 to 30% (Li et al. 2017). Malawi experiences annual rainfall between 725 and 2500 mm from November to April (Chimimba et al. 2023). Areas with Cambisols, comprising 76% of Malawi's soils, are at the highest risk of topsoil loss (Omuto and Vargas 2018). Climate change may worsen the impact of erosive rains on Malawi's degradable soils.

What is emerging from the review is that human decisions and behaviors may contribute to land degradation (Martini et al. 2023). The significance of natural resources influences human actions towards ecosystems (Arias-Arévalo et al. 2017). To comprehend this, one must examine the local agency and identities of smallholders and their link to global threats like land degradation.

The review also noted that historical framing of land degradation may have worsened soil degradation, particularly on smallholder farmland (Green 2009). The transition towards the modern ideas introduced by European settlers in Malawi may have ignored traditional knowledge of land management (Diop et al. 2022). Restrictive measures on soil conservation considerably led to the loss of traditional knowledge. The review observed that some emerging policies in Malawi seem to be driven partly by this perspective. For example, the current Agricultural Input Program (AIP) for increasing access to inorganic fertilizers may be contributing to maize cultivation in unsuitable areas, abandoning sustainable land management practices (World Bank 2024). The World Bank noted that the high cost of AIP limits resources available for other initiatives against unsustainable farming methods. Thus, the review found that the drivers of land degradation are framed primarily on biophysical considerations, overlooking the modified sense of human agency of the population.

3.2.2 Pressures

The review showed that smallholder agriculture contributes to food production in Malawi. However, the increasing smallholder population coupled with limited resources to invest in soil conservation has resulted in soil degradation (Muyanga et al. 2020). Soil degradation will continue to threaten agricultural production as smallholder farm sizes keep on shrinking due to fragmentation. Farmers continue to intensify their use of productivity-enhancing technologies as their farmlands get smaller and the population keeps growing.

At the same time, insufficient measures are taken to protect vegetation and promote sustainable use of resources in small farms. Smallholder farmers often need to have limited capacity to provide labor requirements for soil conservation practices (World Bank 2024). Poor farmers fall short of the resources to benefit from soil conservation. Use of organic manure is limited due to the few livestock on farms, and the erratic availability of organic raw materials for compost-making (Carr et al. 2017). Hence the adoption of land conservation techniques remains low (World Bank 2019; Chinseu et al. 2019). The review also noted that agricultural expansion is replacing forests, grasslands, or wetlands (Missanjo and Kadzuwa 2024). This leads to bare, erodible land. Steep bare land easily erodes.

The review also found that agricultural extension efforts to enhance the adoption of land conservation seem to lack the capacity to upscale land restoration approaches and practices (Chunga et al. 2023). The low rate of extension staff-to-farmer ratio in Malawi hampers access to extension services (Khaila et al. 2015). Where available, the current extension approaches seem to maintain the premise of the belief that smallholders are the major drivers of land degradation.

3.2.3 State

The review shows that renewable natural resources provide subsistence and livelihood income to 80 percent of the population (Yaron et al. 2011). 80% of Malawi's land area is degraded (World Bank 2024), and 75% of the soils are already degraded or threatened by degradation (Omuto and Vargas 2018). Farmland soil fertility has decreased: low total N levels of 0.05% were observed (Njira et al. 2021). Soil organic matter levels in the 0–20 cm depth are very low, hindering plant nutrient retention. As a result, crop yields, like maize, are low at 2.1 metric tons per hectare (GoM 2018).

It was noted that the annual erosion rates have risen to 29 metric tons per hectare per year (Omuto and Vargas 2018). This is a serious threat in Malawi because 31% of the slopes in Malawi are very steep (Li et al. 2017), and 76% of the soils are vulnerable to soil erosion (Omuto and Vargas 2018). 57.4% of land cultivated in Malawi is located in marginal and unsuitable areas (Li et al. 2017). Deforestation rates are estimated at 0.5% per annum (Missanjo and Kadzuwa 2024). The demand for agricultural land and natural resources will continue as the current population is estimated to be close to 20 million (NSO 2019). 86% of the total population lives in rural areas where smallholder agriculture is their major livelihood activity (NSO 2019). Despite understanding these factors, the adoption rate of sustainable agricultural practices remains low (Chinseu et al. 2019). Only 12% of the cultivated land has recommended ridges on contour (World Bank 2019). Agricultural extension services continue to struggle to reach all land restoration actors in Malawi, where only 57% of women and 74% of men are literate (GoM 2018).

From the review above, it can be noted that Malawi's land degradation states (low fertility, yields, high erosion), and deforestation and forest degradation are well documented. However, the current state of human agency for the increasing population is not fully explored. Land degradation can contribute to the loss of human identity (van Noordwijk et al. 2020); and the importance of natural resources to individuals determines human decisions and behaviors toward ecosystem conservation (Arias-Arévalo et al. 2017).

3.2.4 Impacts

The review noted that annual soil erosion rates have increased from 20 metric tons per hectare in the 1990s to 29 metric tons per hectare per year in 2017 (Omuto and Vargas 2018). Projections for future land degradation under different scenarios suggest increasing rates of soil erosion and land degradation (World Bank 2024). This leaves the smallholder farmland with infertile soils that require additional application of inorganic fertilizers to maintain adequate levels of agricultural production. The effects of climate change on rainfall variability are likely to exacerbate the effects of erosive rains on highly degradable soils in Malawi.

The review further observed that the average maize production between 2011 and 2016 was just above 3 metric tons per hectare (GoM 2018). As a result, an increasing number of people require humanitarian assistance during lean periods. For example, during the 2016/2017 season, up to 6.8 million individuals needed such assistance (GoM 2018). A report by NSO (2019)/MVAC 2023 also points out that in 2022, roughly 2.6 million people, or 13% of the country's population lacked adequate food for their households (GoM 2023). In other words, an additional 10% soil loss can lead to more than 270,000 additional individuals descending into poverty (Asfaw et al. 2020).

The review observed that land use change is a key factor in land degradation in Malawi. For example, agricultural land has increased from 47, 053 square kilometers in 1990 to 60, 421 km^2 in 2022 (Missanjo and Kadzuwa 2024). It appears that agricultural expansion in Malawi will continue to take place at the expense of forest, grassland, and wetland conversion (Li et al. 2017). Evidence shows that, cropland increased by 9.6% between 1999 and 2018 (World Bank 2024). This crop expansion can be partly attributed to population increase. Malawi's population is growing at the rate of 2.9% per annum (NSO 2019). Worse still there are limited opportunities outside agriculture in Malawi. Consequently, rural population proportion has remained unchanged from 86% in 1998 to 84% in 2018 (NSO 2019). Hence, some scholars have expressed concerns regarding the possibility of intensifying agricultural production on fragmented pieces of land that have been cultivated for centuries (Muyanga et al. 2020). These trends suggest that land degradation may worsen land use change by encouraging cropping with minimum inputs on the same piece of land and promoting agricultural expansion to fragile land.

While acknowledging that land degradation has far-reaching effects on several productive sectors in Malawi, land degradation may also have impacts on the actors of the restoration drive, including the loss of attention, respect, and ecological knowledge, which may be equally important in designing land restoration responses (van Noordwijk et al. 2020). Changes in how smallholders value natural resources and how natural resource degradation shapes human behavior can contribute to the effective design of NFLRS practices for Malawi.

3.2.5 Responses

The review highlighted efforts to regulate natural resource and agricultural land use in Malawi. Notably, the NFLRS emphasizes land restoration through agricultural technologies, and soil and water conservation (GoM 2017a). Further, the recent land reforms (in 2016) are aimed to improve land tenure security and to promote sustainable land management by smallholder farmers (Calow 2021). The Soil Conservation and Restoration Plan (2020–2026) was launched by the Government for land restoration in Malawi (UNDP 2021). This indicates that the agriculture and forest sectors in Malawi have common goals and objectives for combating land degradation.

The review noted that increased concern exists across the sectors regarding the impacts of land degradation and efforts to address it. Various sectors acknowledge the critical link between natural resource management and agricultural production. The NFLRS provides a detailed plan for land restoration on agricultural land, incorporating natural regeneration, tree planting, agricultural technologies, and conservation practices (GoM 2017a). The Soil Conservation and Restoration Plan highlights sustainable land and water management as a key focus (UNDP 2021). Challenges hindering the implementation of these policies and strategies include limited understanding of how they work across the sectors and at the local levels (GoM 2018; UNDP 2021; Kosamu et al. 2022).

In response, the National Guidelines for Catchment Management and Infrastructure were developed in 2015 to provide detailed integrated approaches for implementing cross-sectoral practices, such as land restoration practices (Government of Malawi 2015). The watershed approach in the guidelines focuses on smaller microcatchment activities that include priority interventions in the NFLRS and the National Restoration Plan (Calow 2021). The guidelines are being adopted for use in several projects on river basin management, food and/or cash transfer programs, public works programs, and other climate resilience projects; but it is not clear whether they are a common reference point for land restoration strategies in Malawi (Calow 2021). The approaches in the guidelines are not explicit on how to change the behavior of nudge smallholder farmers to conserve farmlands. Yet these behavioral interventions are increasingly becoming one of the priorities in recent programs on land restoration (World Bank 2020).

The review found that the National Guidelines for Catchment Management and Infrastructure provide detailed steps for the implementation of the activities, including land restoration practices at the local level (Government of Malawi 2015).

The guidelines considerably address inter-linkages between upstream and downstream communities. These linkages can expand the worldviews of the smallholder farmers and can help them to construct their understanding of the world through experience and reflection (Taylor et al. 2012). However, the guidelines are less mentioned in the recent historical chronology of farmer-centered approaches that would deliver innovations based on the active participation of locals, including smallholder farmers (Tchuwa et al. 2022). The approaches are delivered through a platform that brings all providers to one point at the national and district levels (Masangano and Mthinda 2012).

The review noted that responses can be enhanced by considering human-nature connectedness and how this interplay is shaping human behavior (Bentz et al. 2022). These options have the potential to bundle long-term strategies that may create sustainable pathways towards the restoration of smallholder farmland. Such strategies may contribute to responses that can embed land restoration practices into the activity of the smallholder farmland. The review argues that the application of the DPSIR to unpack response to land degradation may explore the role of these responses as avenues that are conducive to upscaling land restoration practices. A summary of these additional drivers, pressures, states, impacts and responses is presented in Fig. 2. The additional considerations are highlighted in bold for clarity.

3.3 Policy Context and Arguments for Land Restoration in Malawi

The review noted that Malawi has taken significant strides in implementing its NFLRS priority interventions (World Bank 2020; Calow 2021). The strategy effectively employed Restoration Opportunity Assessment Methodologies (ROAM) and Restoration Diagnostic tools and guidelines to pinpoint critical areas that can drive, facilitate, and implement landscape restoration practices (Hanson et al. 2014; GoM 2017b). The strategy seems to be a product of the ROAM, the Diagnostics, and the DPSIR framework. The ROAM framework appears to be quite thorough with concrete steps necessary to achieve restoration objectives (Djenontin et al. 2020).

Djenontin et al. (2020) recently characterized land restoration strategies and projects across Africa to highlight shortcomings that can compromise opportunities for their upscaling. Their study revealed that restoration efforts have multiple socio-ecological goals in both agricultural and forest landscapes. These goals include improving the livelihoods of the local people. However, the authors stress the need for cross-sectoral integration to effectively combat environmental degradation. They suggest that tailored, place-specific approaches that consider local drivers and actors have the potential to improve food security, reduce environmental harm, and facilitate agricultural transformation and climate adaptation at the community level (Djenontin et al. 2021). Additionally, the authors emphasize that assessing the decision-making abilities of smallholder farmers remains a critical factor in the success of restoration

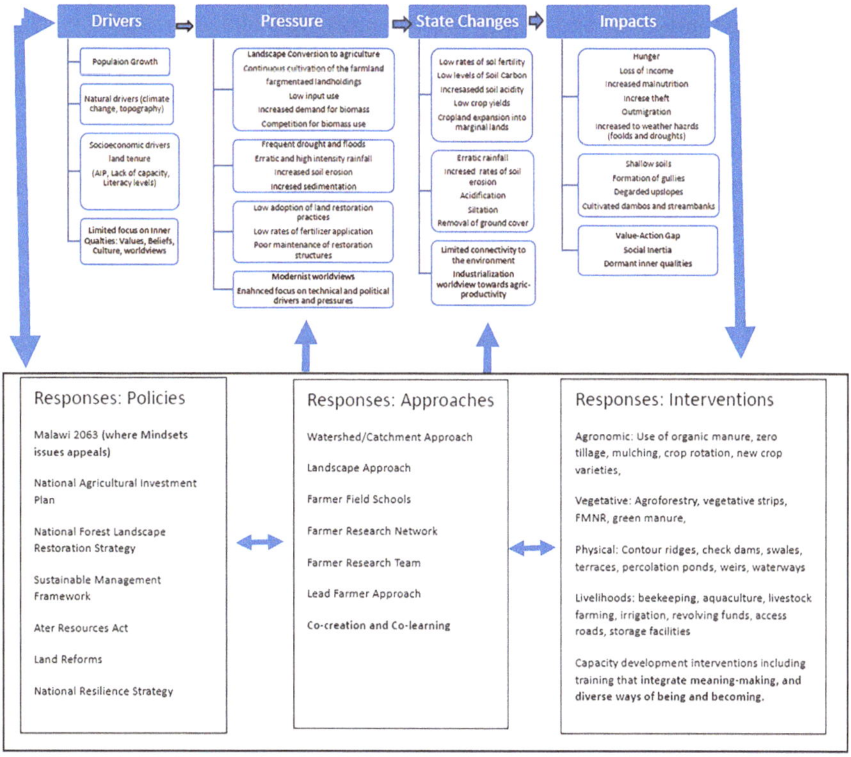

Fig. 2 The modified DPSIR for smallholder farmlands identification of drivers and pressures for upscaling land restoration activities

projects in local contexts (Djenontin et al. 2020). This was particularly important for the Malawi NFLRS, where a number of these shortcomings were noted.

Additionally, there seems to be a lack of clarity on how the ROAM identified the drivers, pressures, impacts, and responses to land degradation in Malawi. It appears that the ROAM identified these drivers and their responses in the NFLRS primarily based on biophysical indicators and qualitative viewpoints (GoM 2017b). To enhance its validity, the ROAM framework was rolled out in the landscape around Liwonde Forest Reserve, but limited evidence exists on its application in other landscapes (GoM 2017b). Yet landscapes can be characterized by heterogeneous farming systems across the globe, including Malawi. These gaps could potentially hinder efforts to tailor differentiated land restoration practices that are suitable and preferred across sectors and contexts, particularly in heterogeneous contexts of smallholders in Malawi.

After reviewing the policy response, specifically the NFLRS against the DPSIR framework, a few key observations can be made.

First, *it is apparent that land degradation is recognized as a critical issue for national development in Malawi.* There is also cross-sectoral agreement that implementing integrated approaches could improve agricultural productivity, build resilience to shocks, and protect and restore environmental services. Secondly, there may be some differences in the desired restoration outcomes sought by various sectors. For instance, the agricultural sector may emphasize sustainable intensification, conservation agriculture, or climate-smart agriculture, while the forestry sector may prioritize biodiversity conservation, and the water sector may focus on the conservation, control, and use of water resources (Calow 2021). Thus, land degradation seems to be regarded as a cross-sectoral issue requiring urgent responses, such as land restoration practices. The review found that the application of the ROAM approach may have emphasized forest resource protection and restoration. This viewpoint is common in restoration ecology. This perspective may disconnect the restoration responses from the agricultural production agenda (van Noordwijk et al. 2020). However, it is important to keep in mind that smallholder farmers may place great emphasis on how land restoration can support agricultural production (Muyanga et al. 2020). To achieve effective restoration and management of farmland, this review argues that involving smallholder farmers in the restoration process can be effective in identifying practical responses to context-driven drivers of land degradation. The review found that the initial test run of the ROAM framework in Liwonde Forest has yielded positive outcomes (GoM 2017b). However, it is essential to consider the socio-ecological diversity of smallholders across other landscapes in Malawi. This consideration can ensure that interventions are suitable and acceptable in other agricultural landscapes across the country.

Secondly, *utilizing the DPSIR framework from an integrated standpoint can assist with comprehending the various factors driving land restoration across the sectors and at a local level.* It is worth noting that putting land users at the center of the land restoration process is critical to its effectiveness, and the DPSIR framework facilitates this (Obubu et al. 2022). It seems that the ROAM's application helped identify NFLRS priorities based on biophysical factors and expert viewpoints. However, the restoration drive seems to be evolving from a technical top-down to an on-the-ground paradigm anchored on local knowledge and practices (Djenontin et al. 2021). This shift calls for a context specific analysis of factors that can provide a concrete starting point for engaging land restoration efforts. However, the ROAM application in Malawi seems to have considered factors that were easy to map, and these provide a high-level starting point for engaging land restoration efforts (GoM 2017b). Connectedness to nature, values, motivation, and other human agency attributes may be difficult to map. Hence, these attributes seem inadequately explored in the current prioritization of NFLRS practices. Tailored frameworks co-created based on a known framework, like the DPSIR framework, can help to comprehend drivers that are complementary to these high-level and easy-to-map drivers and opportunities (van Noordwijk et al. 2020). Figure 2 provides a detailed outline of additional considerations on the drivers, pressures, impacts, responses to land degradation in Malawi.

Thirdly, *the review observed responses from the application of both the ROAM and the DPSIR framework can be more elaborate on how to approach the identified pressures.* For example, the ROAM unpacked five main intervention opportunities focusing on agricultural technologies, soil and water conservation, riverbank restoration, community forestry, and forest management. Some of these interventions may struggle to differentiate themselves from natural resources management activities in Malawi. The review found that some approaches, which are known to be historical to the implementation of these activities, did 'more harm than good' (Kamoto et al. 2013). In some cases, incentives may be perceived as payment for implementing natural resources management efforts, and they thus become a disincentive for sustaining the efforts (Miller et al. 2021).

The limited connectivity to nature can be one of the factors contributing to the weak capacity for implementing natural resources management at the local level (World Bank 2024). To address this problem, farmer-to-farmer settings must be carefully designed to enhance connectedness to natural resources management with sufficient implementation of targeted efforts. The review found that agricultural extension services can further help to bridge this gap by facilitating farmers' behavioral change toward these prioritized agricultural technologies. Land restoration interventions can be bundled into existing agricultural extension platforms, and guidelines that can lead to the desired behavioral change can be developed. Existing platforms and guidelines can be tailored to embrace the sharing of perspectives among smallholder farmers and between local farmers and external sources (Maniraho et al. 2023). The application of DPSIR can provide deep leverage points for targeting context-specific land restoration practices on smallholder farmland. Ultimately, carefully designed integrated extension approaches can potentially enhance the understanding of how land restoration practices can be scaled up on smallholder farmland.

3.4 *Potential Areas for Further Research and Investigation of Underlying Problems Using the DPSIR Framework*

As per the research review in Sects. 3.2 and 3.3, it is apparent that the DPSIR framework is a useful tool for categorizing drivers and responses associated with land restoration. This conclusion aligns with earlier studies that have also employed this framework in comparable scenarios (Agramont et al. 2021; Kosamu et al. 2022; Obubu et al. 2022). The review identified considerable gaps in the drivers and response practices of land degradation, both at the national level and those that are specific to smallholder farmlands. Nevertheless, it is imperative to acknowledge whether the framework adequately accounts for the lived experiences and agency of individuals impacted by land degradation. The review attempted to scrutinize the policies on land restoration in Malawi, and to evaluate them against the observations made on applying the DPSIR framework as a tool to identify the drivers of degradation. The key observation made is that the DPSIR was partly used in the prioritization of land restoration interventions based on the high-level criteria that were mappable. Further, the ROAM application was only piloted in Liwonde landscapes to have a feel of context-specific drivers and opportunities for land restoration (GoM 2017b).

This could undermine the delivery of the NFLRS and compromise its opportunity to scale up. The review noted that the need to define the restoration actors is one of the emerging trends for the successful implementation of forest landscape restoration projects not only in Malawi but across Sub-Saharan Africa (Djenontin et al. 2020). Unfortunately, Malawi NFLRS did not adequately address context-specific analysis of these drivers and responses from sectoral, let alone local actors, including smallholder farmers.

The review also suggests evaluating the way personal values, beliefs, and worldviews can influence the narratives surrounding land degradation at a national or global level. The beliefs, values, and worldviews that individuals hold contribute to the larger discourses and paradigms that ultimately shape how issues are framed, as well as the solutions that are prioritized (O'Brien and Sygna 2013). This calls for a comprehensive approach that takes into account the drivers, pressures, impacts, and responses to environmental issues, from both physical and qualitative perspectives (Bentz et al. 2022). Hence, it is important to understand individuals' views on what constitutes healthy land, soil, and environment, to address global issues at the local level.

Current restoration approaches appear to nudge degradation into restoration by incentivizing communities, including smallholder farmers. For example, the World Bank suggests that providing community environmental conservation funds can be one of the effective tools to support the implementation of forest landscape restoration activities (World Bank 2024). However, other authors have observed that these incentives may translate into payments for externally driven conservation, which may hinder adoption after the end of the incentive (Kamoto et al. 2013). To scale out restoration on smallholder farmers, we need to understand the underlying processes that influence decision-making toward the drivers and responses to degradation that can complement these incentives.

To scale up implementation on smallholder farmland, the NFLRS can benefit from the analysis of the context-specific self-image of these farmers. The review suggests an understanding of the drivers of land degradation to acknowledge farmers' diversity in environmental awareness. Awareness can be measured by environmental beliefs and attitudes (Dunlap 2008). The attitudes are often categorized into (1) the ability to upset the balance of nature, (2) the existence of limits to expand agricultural systems, (3) the rights of these farmers over their agri-ecological systems. Further, the farmers' pattern of factors that can translate external support into action can be understood by assessing five clusters of qualities that influence human awareness, connection, insights, purpose, and agency (Wamsler et al. 2020). These factors can influence beliefs, values, and worldviews that connect humans to their environment. Regarding smallholder farmland, these factors can be seen as connecting farmers to their smallholder pieces of land and will influence the way they might achieve land restoration at the present time and in the future.

Lastly, an enabling environment to support the development of these patterns and qualities can be assessed by considering the existing farmer-to-farmer settings for engaging farmers in land restoration practices. Several farmer-to-farmer settings are used to facilitate farmers' behavioral change towards more productive and sustainable agriculture (Tchuwa et al. 2022). However, one challenge remains—that of providing a safe space for reflection, co-creation and effective learning that can be conducive to scaling up land restoration practices. These farmer-to-farmer settings could be important enabling factors to effectuate decision-making and willingness to scale up land restoration on smallholder farmlands.

Therefore, as researchers and practitioners in land restoration practices, we argue that there is a need to address the following questions in any given context:

1. What values, worldviews, beliefs, and perspectives exist about nature, the environment, and the local food system in the community?
2. How can we explore the underlying reasons and explanations for smallholder farmers' willingness to engage in land restoration practices?
3. How do smallholder attitudes and perceptions concerning the benefits of environmental conservation compare with biophysical characteristics on the ground?
4. How are the current integrated learning and listening approaches to environmental conservation approaches enabling meaning-making and integrating the options of being and becoming stakeholders?

The questions presented above have the potential to spark meaningful change in land restoration practices, particularly on small farms. This review aims to bridge the gaps in current frameworks for addressing environmental degradation in Malawi, as outlined in national strategies and plans (EAD 2010; GoM 2017a, b, 2018; World Bank 2019; Calow 2021; UNDP 2021). Answering these questions could be relevant to agricultural communities in Sub-Saharan Africa facing similar challenges. It is important to recognize that land restoration involves a political process that is shaped by personal dimensions such as values, beliefs, and worldviews. Failure to adequately address these factors can lead to the adoption of poor priority practices, and a failure to achieve the goal of land restoration.

Acknowledgements The review paper is a product of several discussions between the authors. As such the authors extend their gratitude to participating sectors and offices for their time and thoughts. These include the agriculture sector (Land Resources Conservation Department), the forestry sector (Forestry Department), the environmental sector (Environmental Affairs Department), the Local Govermnent sector (Kasungu, Dowa, and Ntcheu), and many other sectors, organizations, and institutions relevant to land restoration in Malawi. The review was conducted with support from FoodMa Programme, a collaborative project between NMBU and LUANAR, with support from the Norwegian Government. Special thanks go to the reviewers for their positive responses to finalize this paper. All errors and omissions in this review are the responsibility of the authors.

References

Agramont A, Van Cauwenbergh N, Van Griesven A, Craps M (2021) Integrating spatial and social characteristics in the DPSIR framework for the sustainable management of river basins: case study of the Katari River Basin. Bolivia. Water Int 47(1):8–29. https://doi.org/10.1080/02508060.2021.1997021

Agyemang I, Mcdonald A, Carver S (2007) Application of the DPSIR framework to environmental degradation assessment in northern Ghana. Nat Resour Forum 31:212–225. https://doi.org/10.1111/j.1477-8947.2007.00152.x
Arias-Arévalo P, Martín-López B, Gómez-Baggethun E (2017) Exploring intrinsic, instrumental, and relational values for sustainable management of social-ecological systems. Ecol Soc 22(4):443
Asfaw S, Pallante G, Palma A (2020) Distributional impacts of soil erosion on agricultural productivity and welfare in Malawi. Ecol Econ 177:106764. https://doi.org/10.1016/J.ECOLECON.2020.106764
Bentz J, O'Brien K, Scoville-Simonds M (2022) Beyond "blah blah blah": exploring the "how" of transformation. Sustain Sci 17(2):497–506. https://doi.org/10.1007/S11625-022-01123-0/METRICS
Calow R (2021) Catchment protection and management in Malawi: a review
Carr ER, Wingard PM, Yorty SC, Thompson MC, Jensen NK, Roberson J (2007) Applying DPSIR to sustainable development. Int J Sustain Develop World Ecol 14(6):543–555. https://doi.org/10.1080/13504500709469753
Carr S, Kool H, Giller KE (2017) Surviving on half a hectare of land: an introduction to the issues surrounding smallholder farming in Malawi
Chimimba E, Ngongondo C, Li C et al (2023) Characterisation of dry spells for agricultural applications in Malawi. Springer, Berlin
Chinseu E, Dougill A, Stringer L (2019) Why do smallholder farmers dis-adopt conservation agriculture? Insights from Malawi. Land Degrad Develop 30(5):533–543
Chunga BA, Graves A, Knox JW (2023) Evaluating barriers to effective rural stakeholder engagement in catchment management in Malawi. Environ Sci Policy 147:138–146. https://doi.org/10.1016/j.envsci.2023.06.006
Diop M, Chirinda N, Beniaich A, El-Gharous M, El-Mejahed K (2022) Soil and water conservation in Africa: state of play and potential role in tackling soil degradation and building soil health in agricultural lands. Sustainability 14(20):13425. https://doi.org/10.3390/SU142013425
Djenontin INS, Zulu LC, Ligmann-Zielinska A (2020) Improving representation of decision rules in LUCC-ABM: an example with an elicitation of farmers' decision making for landscape restoration in central Malawi. Sustainability 12(13):5380. https://doi.org/10.3390/SU12135380
Djenontin INS, Zulu LC, Etongo D (2021) Ultimately, what is forest landscape restoration in practice? Embodiments in sub-Saharan Africa and implications for future design. Environ Manage 68(5):619–641. https://doi.org/10.1007/S00267-020-01360-Y
Dunlap RE (2008) The new environmental paradigm scale: from marginality to worldwide use. J Environ Educ 40(1):3–18. https://doi.org/10.3200/JOEE.40.1.3-18
EAD (2010) Malawi state of environment and outlook report: environment for sustainable economic growth 2010. Lilongwe
Gessesew WS 2007 (2017) Application of DPSIR framework for assessment of land degradation: a review
GoM (2017a) National forest landscape restoration strategy. In: Ministry of natural resources, energy and mining. https://cepa.rmportal.net/Library/government-publications/national-forest-landscape-restoration-strategy/view. Accessed 22 Feb 2022
GoM (2017b) Forest landscape restoration opportunities assessment for Malawi. https://cepa.rmportal.net/Library/inbox/forest-landscape-restoration-opportunities-assessment-for-malawi/view. Accessed 5 Mar 2022
GoM (2018) National agricultural investment plan (NAIP). Ministry of Agriculture, Irrigation and Water Development, Lilongwe
GoM (2023) Malawi high levels of acute food insecurity caused by climate-related shocks, economic decline and high food prices in both rural and urban areas. Lilongwe
Government of Malawi (2015) Malawi national guidelines: integrated catchment management and rural infrastructure. Training workbook. Lilongwe

Green E (2009) A lasting story: conservation and agricultural extension services in colonial Malawi. J Afr Hist 50(2):247–267. https://doi.org/10.1017/S0021853709990028
Hanson C, Buckingham K, Dewitt S, Laestadius L (2014) The restoration diagnostic a method for developing forest landscape restoration strategies by rapidly assessing the status of key success factors
Jambo I, Groot J et al (2019) Motivations for the use of sustainable intensification practices among smallholder farmers in Tanzania and Malawi. Taylor & Francis
Kamoto J, Clarkson G, Dorward P, Shepherd D (2013) Doing more harm than good? Community based natural resource management and the neglect of local institutions in policy development. Land Use Policy 35:293–301. https://doi.org/10.1016/j.landusepol.2013.06.002
Khaila S, Tchuwa F, Franzel S, Simpson B (2015) The farmer-to-farmer extension approach in Malawi: a survey of lead farmers. https://doi.org/10.5716/WP14200.PDF
Kosamu IBM, Makwinja R, Kaonga CC, Mengistou S, Kaunda E, Alamirew T, Njaya F (2022) Application of DPSIR and Tobit models in assessing freshwater ecosystems: the case of lake Malombe, Malawi. Water 14(4):619. https://doi.org/10.3390/w14040619
Li G, Messina JP, Peter BG, Snapp SS (2017) Mapping land suitability for agriculture in Malawi. Land Degrad Dev 28(7):2001–2016. https://doi.org/10.1002/LDR.2723
Maniraho L, Frietsch M, Sieber S, Löhr K (2023) A framework for drivers fostering social-ecological restoration within forest landscape based on people's participation: a systematic literature review. Discov Sustain 4(1):141. https://doi.org/10.1007/s43621-023-00141-x
Martini E, Pagella T, Mollee E, van Noordwijk M (2023) Relational values in locally adaptive farmer-to-farmer extension: how important? Curr Opin Environ Sustain 65:11541
Masangano C, Mthinda C (2012) Pluralistic extension system in Malawi
Miller DC, Mansourian S, Gabay M, Hajjar R, Jagger P, Kamoto JFM, Newton P, Oldekop JA, Razafindratsima OH, Shyamsundar P, Sunderland T, Wildburger C (2021) Forests, trees and poverty alleviation: policy implications of current knowledge. For Policy Econ 131:102566. https://doi.org/10.1016/J.FORPOL.2021.102566
Missanjo E, Kadzuwa H (2024) Activity data and emission factor for forestry and other land use change subsector to enhance carbon market policy and action in Malawi. J Environ Prot 15(4):401–414. https://doi.org/10.4236/JEP.2024.154023
Muyanga M, Nyirenda Z, Lifeyo Y, Burke WJ (2020) The future of smallholder farming in Malawi. MwAPATA Institute Working Paper No 20/03
Niemeijer D, De Groot RS (2006) A conceptual framework for selecting environmental indicator sets. https://doi.org/10.1016/j.ecolind.2006.11.012
Njira KOW, Semu E, Mrema JP, Nalivata PC (2021) Productivity of pigeon pea, cowpea and maize under sole cropping, legume–legume and legume–cereal intercrops on Alfisols in Central Malawi. Agrofor Syst 95(2):279–291. https://doi.org/10.1007/S10457-020-00589-0/FIGURES/3
NSO (2019) 2018 Malawi population and housing census: main report
O'Brien K, Sygna L (2013) Responding to climate change: the three spheres of transformation. In: Proceedings transformation in a changing climate: international conference in Oslo 19–21 June 2013. Oslo
Obubu JP, Odong R, Alamerew T, Fetahi T, Mengistou S (2022) Application of DPSIR model to identify the drivers and impacts of land use and land cover changes and climate change on land, water, and livelihoods in the L. Environ Syst Res 11(1):11. https://doi.org/10.1186/s40068-022-00254-8
OECD (2003) Environmental indicators: development, measurement and use. Reference paper
Omuto CT, Vargas R (2018) Soil nutrient loss assessment in Malawi technical report. Food and Agriculture Organization of the United Nations
Pullanikkatil D, Palamuleni L, Ruhiiga T (2016) Assessment of land use change in Likangala River catchment, Malawi: a remote sensing and DPSIR approach. Appl Geogr 71:9–23. https://doi.org/10.1016/J.APGEOG.2016.04.005

Sun S, Wang Y, Liu J, Cai H, Wu P, Geng Q, Xu L (2016) Sustainability assessment of regional water resources under the DPSIR framework. J Hydrol 532:140–148. https://doi.org/10.1016/J.JHYDROL.2015.11.028

Taylor EW, Duveskog D, Friis-Hansen E (2012) Fostering transformative learning in non-formal settings: farmer-field schools in east Africa. Int J Lifelong Educ 31(6):725–742. https://doi.org/10.1080/02601370.2012.713035

Tchuwa F, Wellard K, Morton J, Kambewa D, Mkweu D, Tchuwa F, Wellard K, Morton J, Kambewa D, Mkweu D, Mhango W (2022) From learning plot to main field: scaling-out soil health innovations in Malawi. https://doi.org/10.3390/su14031532

Tscherning K, Helming K, Krippner B, Sieber S, Paloma SG (2012) Does research applying the DPSIR framework support decision making? Land Use Policy 29(1):102–110

UNDP (2021) National soil conservation and restoration action plan (2021–2026). Lilongwe

UNEP (2016) Global environment outlook: GEO-6: regional assessment for Africa. United Nations Environment Programme, Nairobi

van Noordwijk M, Gitz V, Minang PA, Dewi S, Leimona B, Duguma L, Pingault N, Meybeck A (2020) People-centric nature-based land restoration through agroforestry: a typology. Land 9(8):251. https://doi.org/10.3390/LAND9080251

Wamsler C, Brink E (2018) Mindsets for sustainability: exploring the link between mindfulness and sustainable climate adaptation. Ecol Econ 151:55–61. https://doi.org/10.1016/J.ECOLECON.2018.04.029

Wamsler C, Schäpke N, Fraude C, Stasiak D, Bruhn T, Lawrence M, Schroeder H, Mundaca L (2020) Enabling new mindsets and transformative skills for negotiating and activating climate action: lessons from UNFCCC conferences of the parties. Environ Sci Policy 112:227–235. https://doi.org/10.1016/J.ENVSCI.2020.06.005

Wamsler C, Osberg G, Osika W, Herndersson H, Mundaca L (2021) Linking internal and external transformation for sustainability and climate action: towards a new research and policy agenda. Global Environ Change 71:102373. https://doi.org/10.1016/J.GLOENVCHA.2021.102373

World Bank (2019) Malawi country environmental analysis. Malawi Country Environmental Analysis. https://doi.org/10.1596/31326

World Bank (2020) Malawi watershed services improvement project project appraisal document

World Bank (2024) Malawi WB 2024 country economic monitor

Yaron G, Mangani R, Mlava J, Kambewa P, Makungwa S, Mtethiwa A, Munthali S, Mgoola W, Kazembe J (2011) Economic valuation of sustainable natural resource use in Malawi

Open Access This chapter is licensed under the terms of the Creative Commons Attribution 4.0 International License (http://creativecommons.org/licenses/by/4.0/), which permits use, sharing, adaptation, distribution and reproduction in any medium or format, as long as you give appropriate credit to the original author(s) and the source, provide a link to the Creative Commons license and indicate if changes were made.

The images or other third party material in this chapter are included in the chapter's Creative Commons license, unless indicated otherwise in a credit line to the material. If material is not included in the chapter's Creative Commons license and your intended use is not permitted by statutory regulation or exceeds the permitted use, you will need to obtain permission directly from the copyright holder.

Policy Formulation and Implementation for Smallholders

The Legume Pendulum: The Role of Legumes in Smallholder Farmers' Food and Income Security in Dowa, Malawi

Mayamiko Nathaniel Kakwera, Daimon Kambewa, and Ruth Haug

Abstract Due to various drivers, food and income insecurity in African countries like Malawi hampers Sustainable Development Goals 1 and 2. In attempt to redress this, Malawi's agricultural policy promotes legumes among smallholder farmers for consumption and marketing. We investigated the role of common beans, groundnuts, and soybeans in smallholder men and women farmers' food and income security. We used qualitative and quantitative data from key informants and 515 smallholder farmers in Dowa district, Malawi. Our theoretical framework is rooted in the six dimensions of food security and the farmer typology, which comprises '*moving out*', '*hanging in*', *and* '*stepping up*'. Despite the policy attention, the contributions of the three legumes in various typologies of smallholders' food and income security are minimal and declining. Limited access to productive assets and resources such as land, labour and inputs weakens farmers' agency, as observed among 'moving-out' and 'hanging-in' farmers. With the upcoming marketing potential of the groundnuts and soybeans, men smallholders are increasing their attention in decisions and actions on these legumes, though limited to marketing and revenue use, with women as primary labour providers of all production activities. Enhancing the role of these legumes regarding smallholder farmers' food and income security requires strengthening smallholder farmers, both men's and women's, agency and effective and conducive policy implementation. Overall, we conclude that the role of these legumes in smallholder farmers' food and income security in Dowa is limited and declining.

Keywords Food security · Income · Farmer typology · Legumes · Malawi

M. N. Kakwera (✉) · R. Haug
Department of International Environment and Development Studies, Faculty of Landscape and Society, Norwegian University of Life Sciences, Ås, Norway
e-mail: mayamiko.nathaniel.kakwera@nmbu.no; mkakwera@luanar.ac.mw

M. N. Kakwera · D. Kambewa
Extension Department, Lilongwe University of Agriculture and Natural Resources, Lilongwe, Malawi

© The Author(s) 2026

B. R. Singh et al. (eds.), *Sustainable Food Systems in Sub-Saharan Africa*,
https://doi.org/10.1007/978-3-032-11288-0_18

1 Introduction

The year2030 is the deadline for the United Nations Sustainable Development Goals (SDGs), but global food security is still a major challenge, especially for African countries (FAO et al. 2023; Sokourenko et al. 2022). In Malawi, food insecurity persists, with about 22% (4.4 million people) of the country's total population facing high levels of acute food insecurity in the lean season of 2023, which is the record-highest number in the past six years (GRFC 2024, p. 26) The majority of the households in the acute food insecurity situation are located in rural areas of Malawi, with only about 0.5 million living in cities (GRFC, 2024, p. 39). About 67.8% of smallholder farmers in Malawi reported low food security per the progressive severity scale in 2019 (NSO 2020).

Poverty, low agricultural productivity, and environmental degradation are key drivers of food the insecurity experienced by smallholder farmers in Malawi (Benson 2021; Benson and De Weerdt 2023; Giller 2020). Other drivers include small and fragmented land sizes, which further constrain productivity and production gains (Jayne and Sanchez 2021; Matita et al. 2022). There are the challenges of inadequate rural infrastructure, and limited access to extension and advisory services (Jayne et al. 2023; Ragasa and Mazunda 2018) which in turn affect smallholder farmers productivity and food security (World Bank 2023). In addition, Malawian smallholder farmers are also challenged with expensive agricultural inputs, lack of market incentives for agricultural commodities, gender disparities, and partially implemented or completely unimplemented sector-related policies (Haug et al. 2021a, b). The 44% devaluation of the Malawi Kwacha in 2024, the effects of the COVID-19 pandemic, and related global conflicts, mainly the Russia-Ukraine war, reflected as high energy and fertilizer costs and supply chain disruptions, worsened Malawi's precarious food and income insecurity situation (Chadza et al. 2022; FAO et al. 2021; GRFC 2024; World Bank 2023).

Redress mechanisms for Malawi's food insecurity include developing and implementing policies and interventions to enhance agricultural productivity and trade, prioritisation, and investments in food supply chains such as those for legumes (Government of Malawi 2018, 2022; Hunga et al. 2023). Globally, legumes are becoming increasingly important in food security and transformation as they provide affordable protein sources and help bridge the protein gap for people transitioning from animal-based diets (Jimenez-Lopez et al. 2020). Malawi's trade, agricultural, and food security policies focus on legumes as promising enterprises for addressing multifaceted outcomes within national and household food systems (Government of Malawi 2018, 2022). In addition, soybeans are Malawi's second-largest export and the most advanced value-adding sub-sector, given the notable investments and outputs of soybean food and feed products (World Bank, 2023). At the same time, Malawian groundnuts are increasingly processed for the international markets, building on production arrangements from the tobacco value chain. A contract farming and private sector involvement is already on-going through Pyxus Agriculture Limited, a subsidiary company of Alliance One Tobacco Company (World

Bank 2023). Groundnuts, soybeans, and common beans are earmarked in the policy space to meet Malawians' income and food security needs (Government of Malawi 2022).

The three legumes' attributes that make them a policy tool are their potential to contribute to healthier and cost-effective diets serving as a protein source for consumers (Donovan and Gelli 2019; Gelli et al. 2020). Additionally, legume production offers income opportunities and benefits to various stakeholders (Benson 2021; Benson and De Weerdt 2023; Chimonyo et al. 2019). Legume crops can enhance soil nitrogen without significant greenhouse gas emissions, addressing challenges associated with high inorganic fertilizer costs (Snapp et al. 2019).

It is from this increased attention to legumes that our study is framed. The research question we address is: What role do the three legumes, groundnuts, soybeans, and common beans, play in food and income security for men and women smallholder farmers of different typologies in Dowa, Malawi? We assess this question by drawing on primary data from Dowa. We use the six dimensions of food security (Leach et al. 2020) and the farmer typology, which comprises '*moving out*', '*hanging in*', *and* '*stepping up*' (Dorward et al. 2009; Dorward and Giller 2022; Leach et al. 2020) *as our analytical lens and presentation of results.* We found that the role of these three legumes in smallholder farmers' food and income security in Dowa is limited and declining.

2 Theoretical Framework

The study draws from the six domains of food security (Clapp et al. 2022; FAO et al. 2023) and the farmer typology theoretical frameworks (Dorward et al. 2009; Leach et al. 2020) as detailed in the following subsections.

2.1 Food Security

According to the World Food Summit in 1996, "Food security exists when all people have physical and economic access to sufficient, safe and nutritious food to meet their dietary needs and food preferences for an active and healthy life" (FAO 1996). Evolving from this definition are the six domains of food security, namely food availability, access, stability, utilisation, agency, and sustainability (Clapp et al. 2022; FAO et al. 2023; HLPE 2020).

The availability dimension addresses the actual or potential physical presence of food, including aspects of production, food reserves, markets and transportation, (FAO et al. 2023). The access dimension addresses the question of whether households and individuals have sufficient physical and economic access to food. The stability dimension is the condition that is met when households constantly have food (FAO et al. 2023). Stability addresses short-term or medium to long-term food availability with implications of acute, chronic food insecurity (FAO et al. 2023). The utilisation dimension examines whether households maximise adequate nutrition and energy in the body (Clapp et al. 2022; FAO et al. 2023). The agency dimension is about the capacity of individuals or groups to make their own decisions about what foods they eat, what foods they produce, how that food is produced, processed, and distributed within food systems and their ability to engage in processes that shape food system policies and governance (Clapp et al. 2022; FAO et al. 2023; Galiè et al. 2017). The sustainability dimension looks at the provision of food to the present generation without compromising the capacity and quality of natural resources to provide food for future generations from perspectives of social, economic, and environmental sustainability (Clapp et al. 2022; Moseley and Jordan 2001; United Nations 1987).

2.2 *Typologies of Farmers*

The heterogeneity among farmers calls for alternatives to blanket approaches and recommendations (Alvarez et al. 2018; Guiomar et al. 2018; Marshall et al. 2021). One way to appreciate the farmer's differences is by classifying them into different typologies (Kostrowicki 1977). In their work on poverty alleviation and agricultural commercialisation Dorward et al. (2009) classified rural people into 'hanging-in', 'stepping up', and 'stepping-out'.'Hanging-in' refers to poor people who continue to engage in their current livelihood activities, often in challenging circumstances, to maintain their current level of well-being, while the 'stepping up' typology involves rural people seeking to improve their everyday livelihood activities by increasing productivity, expanding their assets, or accessing better markets activities (Dorward et al. 2009). The 'stepping-out' strategy refers to people transitioning from existing activities and are engaged in accumulating assets, which in time can provide a base for moving into different activities that have initial investment requirements leading to higher or more stable returns (Dorward et al. 2009; Dorward and Giller 2022). Other scholars have used the schema in studying agricultural and food security interventions (Leach et al. 2020; Matita et al. 2022) and impacts on smallholder farmers' commercialising efforts (Chinsinga et al. 2021). Leach et al. (2020) add categories of 'moving out' and 'stepping in' to the typology. The 'moving-out' category typifies households that move out of agricultural activities due to push or pull factors. In contrast, the stepping-in category comprises the urbanised, often medium-scale, investor farmers who come into the countryside to acquire land for commercial agricultural activities (Leach et al. 2020).

Figure 1 shows the adapted analytical framework for this study. We use the typologies of 'hanging-in' and 'stepping up' as conceptualised in the typologies by Dorward et al. (2009) and Leach et al. (2020). We adapt the 'moving-out' typology of Leach et al. (2020) by emphasising that this typology looks at those households who move out of productive agricultural activities due to economic, social and environmental shocks or because of the emergence of alternative livelihood opportunities such as

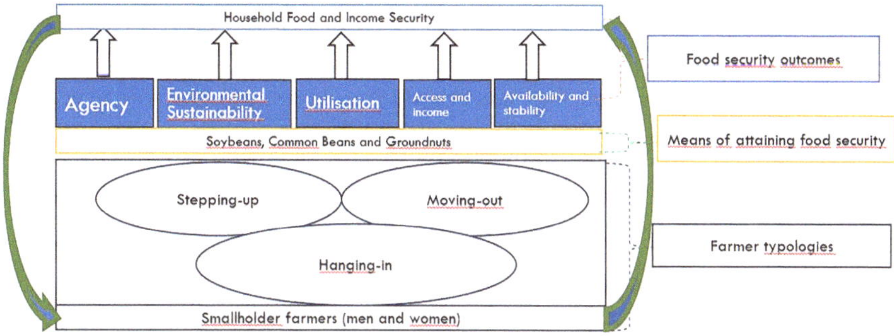

Fig. 1 Analytical framework: food security dimensions and farmer typologies with the three legumes cutting across all cells (based on FAO et al. 2023; Dorward et al 2009)

urban employment and out migration but that does not entirely mean they stopped farming. Therefore, the adapted analytical framework consists of three typologies: 'hanging-in', stepping up, and 'moving-out'.

We applied the adapted farmer typologies across the domains of food security with a focus on the contribution of the three legumes. We also include gender as part of the analytical framework. We assessed food availability and stability by using legume production as a proxy, illuminating how households sustain this production over time. The accessibility dimension was analysed by assessing whether farmers retained enough quantities of each legume and overall food to meet the needs of their households. The dimension of food access overlapped with the income and agency metrics (Alkire 2007). The study analysed income using the proxy of farmers' expenditures on food and non-food items. It assessed the relative contribution of income derived from legumes to the household's total income. While acknowledging the different approaches to assessing agency (Ehlert and Voßemer 2015; Hahn 2012; Kpienbaareh et al. 2023; Zhang 2022) and that measures for food security such as agency are path-dependent (Iversen et al. 2023), the household agency dimension was examined by evaluating men and women farmers' access to land, seeds, extension services, and their decision-making capabilities within the household (Alkire 2007; Galiè et al. 2019).

This study focused on individual men and women farmers' agency instead of collective agency (Clapp et al. 2022). To assess the sustainability dimension, we relied on environmental indicators as other dimensions of sustainability were intertwined with dimensions of food access and agency, for instance, aspects of access to production assets such as seed and utilisation on aspects to do with food diversity (Maluf et al. 2022, p. 187). Utilisation was assessed by exploring how the households retained and used the legumes as food (Mulenga et al. 2021). The typologies of 'hanging-in, stepping up', and 'moving-out' were applied with respect to the farmers activities on the three legumes.

3 Methods

This study employed a mixed-methods approach to investigate the role of legume production on food and income security of smallholder farmers in the Chibvala Extension Planning Area (EPA) within Dowa district, Malawi.

3.1 Selection Study Area

Chibvala Extension Planning Area (EPA) and Dowa district under the Kasungu Agricultural Development Division (ADD) were purposively selected as the study sites. The rationale was that Dowa, on Lilongwe City's outskirts, hosts Malawi's major agricultural market players. The players include mega traders and processors who buy the crop for retailing, export, bulk selling to processors, processing, and packaging for final consumption (Chinsinga and Matita 2021). The district also hosts a refugee camp with a unique host-refugee relationship in agricultural production and markets where the refugees hire land, labour and the commodity trade business (Chinsinga and Matita 2021; Government of Malawi 2017).

3.2 Selection of Informants

Household survey

The data was collected through a survey of 515 households that were sampled from 44 villages across four sections in the Chibvala EPA. Household survey data was collected from May to July 2023. In selecting the households, multistage sampling was used, whereby four sections were chosen purposively from a total of 11, based on their potential for producing soybeans, groundnuts, and common beans. According to records from the EPA, these sections were among the top four producers of the three legumes. Following this arrangement, villages were randomly selected from the four sections based on a proportionate sampling of the farming population resulting in 17 villages in Zidunge, 12 in Mtengowanthenga, six in Chibvala Central, (considered the most significant trading center/market in the EPA), and nine in Chadza. The households were then randomly selected from each village using Excel to assign random numbers to the listed household roster. The number of households per village was also guided by proportionate sampling. In total, 212 farmers were interviewed from the Zidunge section, 133 farmers from the Mtegowanthenga section, 72 from Chibvala Central, and 98 from the Chadza section.

Focus group discussions and field observations

We conducted focus group discussions (FGDs) in 8 villages, adding up to forty-eight FGDs. Each village had 6 FGDs, split evenly between men and women. The discussions were divided into two phases: the first phase occurred before planting and continued through the legumes' vegetative growth stages, October 2022 to February 2023, while the second phase took place during the harvest and marketing season, March to May 2023. In the first phase of FGDs, legume farmers had to establish associations with the three typologies: 'hanging in', 'moving out', and 'stepping up'. In addition to identifying which typology the farmers were in, they also listed the characteristics associated with those particular typologies which were used in the household survey that succeeded the FGDs. We followed the same groups of farmers from phase one into phase two. Two villages were randomly selected from each section, with a minimum of 4 participants and a maximum of 16 participants per FGD. Participants were randomly chosen from a database provided by the EPA, with 18 participants selected from each village to allow for potential absences. Advance communication ensured attendance. Two hundred eighty participants participated in the FGDs, with 162 women. The focus group discussions were employed to mainly solicit in-depth data on gendered dimensions of the three legumes: farmer market interactions, land access and use and access to support services such as agricultural extension services.

While working on the focus group discussions, field observations were made on legume production processes, market arrangements, land access, and use. These field observations were mainly done by the first author and provided a point of reflection in the entire data, analysis and reporting stages.

Key informants

Key informants were selected using purposive sampling and snowballing techniques. The informants included frontline agricultural extension workers, district and national agricultural subject matter specialists from trade, crops, agribusiness departments in the public service, NGO extension/field facilitators and legume researchers at the National Agricultural Research Services. In total, the study included 15 key informants. Through the key informant interviews the study collected data on delivery of support services for legume production and marketing.

3.3 Data Collection

Survey data was collected using a questionnaire. Enumerators, fluent in local languages, administered a structured questionnaire designed for the study. The questionnaire modules covered household demographics, agricultural practices, household food and non-food expenditure, and access to land and seeds. Focus group discussion data was collected using an interview guide and by writing down the notes. Participants were engaged through visual aids, such as figures representing different farm typologies, to initiate discussions and encourage participation.

3.4 Data Analysis

The analytical framework presented in Fig. 1 guided the data analysis. Farmers who participated in the survey and focus group discussions (FGDs) were requested to self-identify with one of the three typologies: 'hanging-in', 'stepping up', and 'moving-out'. The identification criteria for each typology were determined by how farmers perceived changes in their legume farming enterprise and farming resources, such as land, access to seed, and on-farm and off-farm enterprise diversification. Only 23 out of 515 households identified themselves as 'moving-out' while 323 households categorised themselves as 'hanging-in' and 169 as 'stepping-up'. Among the surveyed 515 households, 25% were female headed.

Statistical methods were employed to analyse the quantitative data obtained from the survey to assess the six food security domains, including availability, access plus income, stability, agency, utilisation, and sustainability. Specifically, Multivariate Analysis of variance (MANOVA) was employed to compare means across the three farmer typologies. The variables measuring land, food expenditure, and legume production were transformed using natural logarithm to reduce skewness and achieve normality (Lee 2020).

Additionally, Chi-square tests were run on cross-tabulations. We used a content analysis approach to analyse the qualitative data collected from the focus group discussions and key informant interviews, which were done using Atlas Ti version 7 software. After coding and organising the data, preliminary findings were further analysed in accordance with the food security dimensions, farmer typology and gender aspects, as presented in Fig. 1.

4 Results

The presentation of results were structured following the analytical framework (Fig. 1), which include the six food security domains and the three farmer typologies with the three legumes as the focus.

4.1 Food Availability and Stability

Both yield and production results for all three legumes showed limited availability of the legumes for smallholder farmers' household needs. The results showed significantly different mean production of soybeans among farmers in typologies of 'hanging-in' (71.24 kg) compared to 'stepping-up' (125.41 kg), as indicated in Table 1. The results for groundnuts were not statistically significant; the mean production was 89.53 kg for 'hanging-in' farmers, 117.36 kg for 'stepping up' farmers, and

100.67 kg for 'moving-out' farmers. On the other hand, common beans mean production was significantly different ($p < 0.001$) between 'hanging-in' and 'moving-out' farmers, with mean values of 48.62 kg and 26.38 kg, respectively. Overall, the mean production levels across the three legumes and farmer typologies were suboptimal, although there were some differences across the farmer typologies.

Yield results across the legumes as provided in Table 1 reveal low yields across the three farmer typologies compared to the respective potential yields for the three legumes. Results in Fig. 2 show that legume production has declined over the past five years across all farmer typologies.

Our field observations and focus group discussions indicated that the 2022/23 growing season was characterised by challenges for the legume farmers and farming in general. Using recycled, farmer-saved, and uncertified seeds from the common market was prevalent, driven by the cost of seeds and limited trust in suppliers of certified seeds, which farmers noted as affecting their production. Heavy rains affected the quality of produce across the three legumes. Some produce, like groundnuts, got rotten in the field. Soybean, which had attracted many farmers to grow it because the 2021/22 marketing seasonal prices were better off, was attacked by soybean rust that

Table 1 Results of the three legumes, production, and productivity per the three farmer typologies in 2022/23 rainfed production season

Variable	Farmer typology	N	Mean	Standard deviation	Sig.
Soybean production in kg (log10)	'Hanging-in'	238	1.8231	0.4535	0.0000
	'Stepping up'	131	2.0608	0.48103	
	'Moving-out'	16	1.8378	0.40391	
Groundnut production in kg (log10)	'Hanging-in'	53	1.961	0.42403	0.0400
	'Stepping up'	66	2.1479	0.36036	
	'Moving-out'	8	2.0111	0.52525	
Common beans production kg (log10)	'Hanging-in'	168	1.634	0.47055	0.0000
	'Stepping up'	104	1.9889	0.39097	
	'Moving-out'	19	1.4586	0.50308	
Soybean yield in kg/Ha (log10)	'Hanging-in'	237	2.4512	0.50365	0.0060
	'Stepping up'	131	2.6325	0.54415	
	'Moving-out'	16	2.5385	0.39775	
Groundnuts yield in kg/Ha (log10)	'Hanging-in'	53	2.6612	0.37257	0.0060
	'Stepping up'	66	2.8802	0.3815	
	'Moving-out'	8	2.6679	0.36815	
Common beans yield in kg/Ha (log10)	'Hanging-in'	168	2.1355	0.46303	0.0000
	'Stepping up'	104	2.3929	0.42392	
	'Moving-out'	19	1.9631	0.56694	

Data source Farmer Survey in 2023. Data is in a log-base 10 format, and we provide in the text for the actual values for a more interpretable discussion
Source Farmer survey in 2023

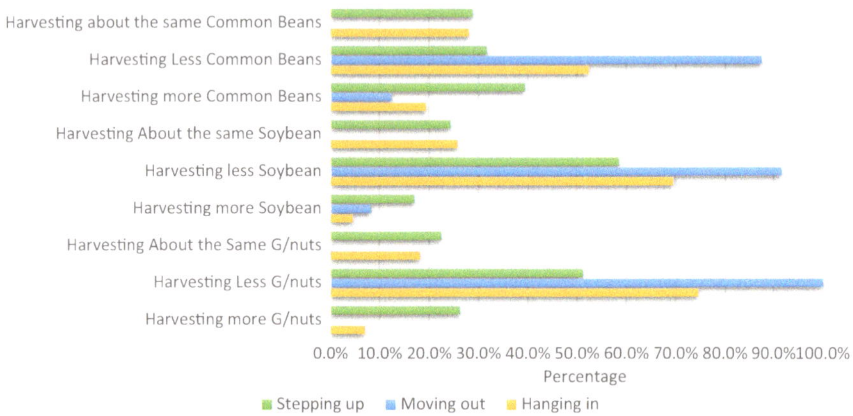

Fig. 2 Farmers' five years reflection on legume production. *Source* Farmer survey in 2023

significantly decimated the yields. Common beans intercropped in maize were the crop farmers reported to have had fewer issues and got better yields and production. Examining the case of soybean rust with key informants and farmers in perspectives of what remedies were available and in use revealed that extension workers lacked the knowledge and skill to address the issue of significant disease in soybeans. The farmers also indicated they did not know how to address the problem. Up to the end of our fieldwork in July 2023 which coincided with harvesting time, farmers and extension workers had yet to figure out a solution to the problem of soybean rust.

Focus group discussions revealed limited access to extension services, insufficient land for expanding production, and the absence of certified seeds for the production and productivity of the three legumes, which posed challenges for the stable access to food in time and space by the farmers. In the FGDs, farmers reported practising intercropping soybean and common bean with maize while noting the lower yields obtained from such practice as a solution to their limited land size.

Our key informants stated that farmers who produce legumes tend to receive less support on extension and advisory services.

> We are clouded out of our time to work with farmers on crops such as legumes as we run activities such as registration of agricultural input support programmes, farmer training on maize production related activities and attend meetings related to our work. [Key informant 1]

One of the key informants also reported that

> Our seniors focus on hearing the food security situation of the communities, which is reduced food situation. This focus on maize in our reporting lines signals to us what to focus on. [Key informant 2]

Farmers observed and reported limited interface with extension workers in the survey and the FGD. We observed and confirmed with the local extension workers the high vacancy rate in the sections of the EPA. Only two of the four sampled sections of the EPA where we did the study had an extension worker servicing them. Apart from the high vacancy rate, there was also the issue of non-domiciled extension workers. The extension workers were operating far from the sections for several reasons. They lacked motorised transport to ease mobility to enable them to provide services to communities living and farming in distant areas. Extension workers primarily contacted the farmers individually and often through a lead farmer, as we witnessed fewer group approaches during our fieldwork. Thus, extension workers were expected to service over 2,000 farming households.

4.2 Food Access and Income

The findings in Table 2 show that 'hanging-in' farmers barely kept enough common beans for household consumption. The 'stepping up' farmers also kept limited beans for household consumption.

> We are failing to meet our food needs throughout the year despite working hard to. The produce we get from our fields is inadequate to take us through. In case where we sell some, the selling prices are not meaningful for us to buy the much-needed food items whose prices are getting higher every time. [Women FGD participant 12]

A finding that corresponds with the household survey results is that a substantial proportion of farmers experience inadequacy of food (see Table 2), both in the short and long term. The 1-month self-reported food security situation revealed that many farmers across all typologies reported that their food situation was "less than adequate for household needs." Specifically, farmers faced challenges in meeting their household food needs, indicating potential vulnerabilities in short-term food security, with 'stepping up' farmers reporting struggles with food adequacy, suggesting that they, too, face food security challenges in the near term. Regarding their food situation for 12 months, 76% of the 'hanging-in', 52% of the 'stepping up', and 70% of 'moving-out' farmers reported having inadequate food.

'Hanging-in' and 'moving-out' farmers reported working for casual labour, commonly called *ganyu*, to supplement their income. Men in some communities engaged in casual jobs such as sand mining and loading the sand into trucks, which usually ferry the sand to construction sites in Lilongwe. Women reportedly picked up piecework to wash the clothes of refugees who resided in the Dzaleka refugee camp for the villages that bordered it. Both men and women also reported having to work on fields for the better-off farmers in their vicinity. The income from the piecework was used in many ways. Most women complained that their husbands spend their piece work income mainly on alcohol and associated entertainment.

Meanwhile, the women used their *ganyu* income to feed the families and sometimes buy inputs such as seeds. The peak of the *ganyu* period was reportedly at the onset of the rainy season, coinciding with the time they needed to work on their own fields. For farmers in the 'hanging-in' and stepping up category, the dilemma of working on their fields first or doing piece work to meet pressing food needs was

Table 2 Legume adequacy, household food security situation and food expenditure share across the farmer typologies

Statement on food situation	Kept enough soybean for household consumption				Kept enough groundnuts for household consumption				Kept enough common beans for household consumption			
Farmer typology	'Hanging-in'	'Stepping up'	'Moving-out'	Total	'Hanging-in'	'Stepping up'	'Moving-out'	Total	'Hanging-in'	'Stepping up'	'Moving-out'	Total
Mean (Likert score)	3.8	3.4	3.3	3.7	3.2	2.8	2.2	2.9	3.8	3.1	3.3	3.5
Std. deviation	1.3	1.4	1.3	1.3	1.4	1.4	0.7	1.47	1.3	1.4	1.1	1.4
N	238	133	16	387	65	75	9	149	190	119	20	329

Food situation/farmer typology	'Hanging-in' (%)	'Stepping up (%)'	'Moving-out' (%)	Chi-square test
1-month self-reported food security situation				
Less than adequate for Hh needs	57	30	44	$P = 0.000$
Just adequate for Hh needs	42	52	52	
More than adequate for Hh needs	1	18	4	
Total	100	100	100	
12 months self-reported food security situation				
Less than adequate for Hh needs	76	52	70	$P = 0.000$
Just adequate for Hh needs	22	29	26	
More than adequate for Hh needs	2	19	4	
Total	100	100	100	

	Total food and non-food expenditure by the household (MK)					Household food expenditure share (%)				
	'Hanging-in'	'Stepping up'	'Moving-out'	Total	Significance	'Hanging-in'	'Stepping up'	'Moving-out'	Total	Significance
Mean	116,662	174,818	90,746	134,589	$P = 0.001$	61	54	68	59	$P = 0.002$

(continued)

Table 2 (continued)

	Total food and non-food expenditure by the household (MK)					Household food expenditure share (%)				
	'Hanging-in'	'Stepping up'	'Moving-out'	Total	Significance	'Hanging-in'	'Stepping up'	'Moving-out'	Total	Significance
Std. dev.	259,705	236,823	66,675	248,182		23	23	17	23	
N	323	169	23	515		323	169	23	515	

Source Survey data 2023

highlighted where they usually had to prioritise work on piece work to address the pressing food needs. Despite indicating the piecework route as a way of diversifying their income sources and enhancing their access to food, the focus group participants had indicated that lately, opportunities to do such work are hard to get by as the number of people seeking piecework is on the rise and the job offers are diminishing.

Legumes contribute less than 10% of household income for all three typologies of farmers. Farmers in the 'stepping up' category get about 9% of their total income from legumes, followed by 'hanging-in' farmers who get about 7% of their income from legumes. Moreover, significant variations in mean food expenditure shares across farmer typologies, presented in Table 2, highlight potential food security implications. Farmers categorised as 'stepping-up' allocate approximately 54% of their income to food items. In comparison, farmers reported higher percentages of 61% and 68% in the 'hanging-in' and 'moving-out' typologies, respectively. This links food expenditure patterns to farmer typologies' financial situation and food security.

Among the 'hanging-in' and 'moving-out' farmers, soybeans are usually sold off early, with income from such sales used to meet pressing food needs, pay medical bills, and support other immediate household needs. 'Hanging-in' and 'moving-out' farmers across men and women supplement their food needs by working as casual labourers. Men reported working on agricultural fields of the 'stepping up' farmers such as refugees, better-off fellow villagers or residents of urban peripheral centres renting land or buying land in the villages. During the agricultural marketing season, some men switch from providing field-related casual labour to working as temporary labourers in tobacco companies, off-taker markets, and sand mining. In the 'hanging-in' and 'moving-out' typologies, women worked for paid chores such as fetching water, washing, and doing on-field activities to meet their household food needs in lean periods. The casual labour supply activities were reported to be conducted throughout the year, with those done off the peak lean hunger season contributing towards income saved in village savings and loan associations by the women. It was reported in the FGD that revenue from the village savings and loan associations was used to buy agricultural inputs, mainly seeds, at the onset of the production season or to pay for household needs.

4.3 Food Utilisation

Results in Fig. 3 show that 44.7% of farmers in the 'stepping up' category consume more beans than five years ago. However, strikingly higher consumption was noted among farmers in the 'moving-out' category, whereby 53% reported consuming more beans now than before. Farmers in all typologies agreed to consume fewer groundnuts and fewer soybeans now than five years ago.

Common beans grown by 'hanging-in' and 'moving-out' farmers provide relish during the peak of the lean hunger period. Some participants of the FGD reported replacing staples with common beans prepared as a smooth porridge by pounding

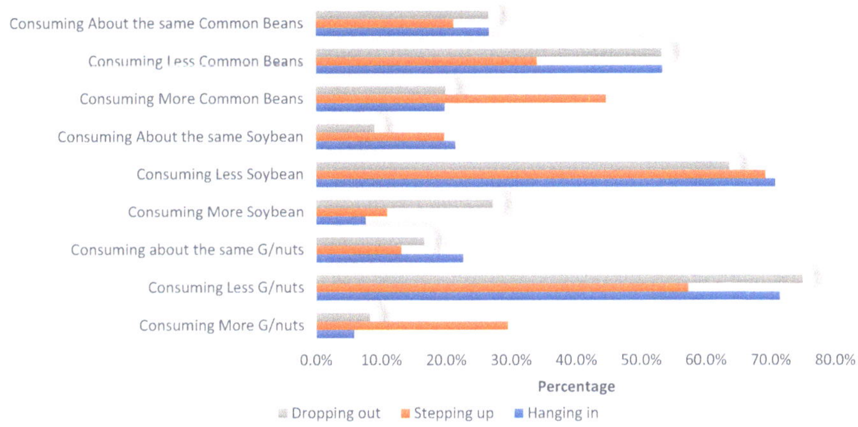

Fig. 3 Farmers' five years reflections on legume consumption. The groundnut Chi-square test was significant at 0.000, while for common beans, it was significant at 0.001. *Source* Farmer survey in 2023

parboiled beans and feeding on them. It is women who play this role. FGDs and KIIs results affirmed the shortages of food for household consumption.

> In the past, production volumes of the three legumes were better than now as we had access to improved and good quality seeds through the farm input subsidy programme, and our soils were fertile. Better yields, which would give us more for household use, is a song of the past since the agricultural input support programme no longer supports legumes. [Female FGD participant 2]

Another FGD with women in the second data collection phase echoed these results. This is what they said:

> Low production to low consumption is the game now. [Female FGD Participant 7]

One of the key informants working on providing farmers with extension services reported that:

> In as much as we try to teach these farmers how to prepare many food recipes from legumes, the consumption of the legumes remains low. They would have to balance what to sell, keep for their home use, and keep for seed from the low volumes attained at harvest. From my observation working as a frontline extension officer in this area, I notice that the legume production volumes are being affected by poor soils as most fertile land is either rented out or used for maize, limited access to good quality seed, and generally poor agricultural practices (such as less use of inoculants). [Key informant 9]

4.4 Environmental Sustainability

The survey showed that some farmers incorporated legume residues to improve soil fertility. However, the scale of operations was limited partly due to land sizes, agricultural practices, labour issues and land tenure issues. Given the small land size, most farmers had challenges practising crop rotation. Intercropping the legumes, common beans, and soybeans with maize was the most common. In cases where legumes are grown, the farmers usually uproot and shred the crop in their homes, leaving limited residues for incorporation into the field. Due to the mixed farming systems of livestock and crops, farmers expressed that when residues of a crop, such as groundnuts, which are usually shredded in the field, are either collected to feed livestock or left for free grazing of the livestock. During the FGD, farmers indicated that much labour is required to return the residues of common beans and soybeans to the field once shredding is completed in their homestead. The few who managed to do this either had sufficient family labour or had oxcarts to ferry the residues to their fields. Land tenure issues also affected the pursuit of legumes among the 'stepping up' farmers. Most grew legumes such as groundnuts and soybeans on rented land, with tenure security usually limited to one growing season.

4.5 Agency

Access to extension services, seed of good quality, affordable input prices and utilisation of land were key limitations by farmers across the farmer categories. Regarding improving smallholder access to land, we learned in the focus group discussions that a few 'stepping-up' smallholders rent land from the poorer smallholders to improve their access to land. This is agency for the better offs, but misery for the poorer ones who are oftentimes being hired back to work on their own land and losing their user-rights to the land.

Like those growing soybean and common beans, their production orientation for either markets or consumption, their capability to retain the produce for household consumption and have excess for sale was not significantly different across the farmers in 'hanging-in,' 'stepping up, and 'moving-out' typologies.

The 'hanging-in' farmers reported that they sometimes undertake piecework to raise money to buy seeds for their farms. Usually, the seed sourced in such cases is not certified. Typically, the seed is grain that is either sold by fellow farmers who are better off and afforded grain from their previous harvest or by traders operating in near-by markets.

Survey results provided in Fig. 4 show that for most farmers, land sizes remained the same. However, some farmers also reported that they had less land now than they did was five years ago. Surprisingly, 'stepping up' farmers led on the perception of owning less land in the past 5 years despite their figures being low at 18.9%; this finding is partly explained by the results in the focus group discussion, which showed that farmers in the 'stepping up' category were always aiming for more of land, production, and productivity.

Results in Table 3 show that 'moving-out' farmers tend to allocate more land to groundnuts, although this difference lacks statistical significance. For soybean cultivation, 'hanging-in' farmers allocate approximately 0.23 ha, 'stepping up' farmers

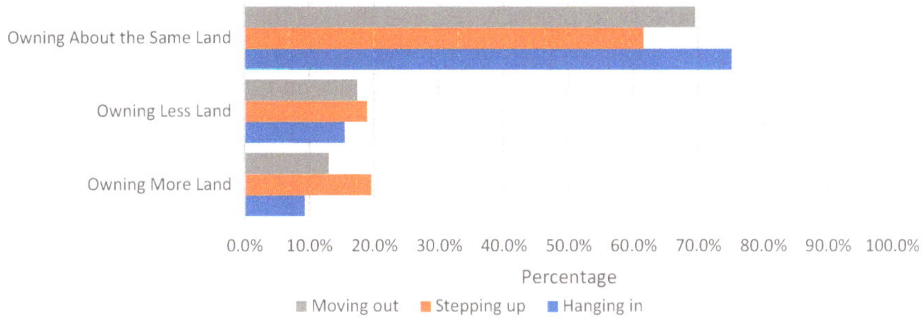

Fig. 4 Farmer's Five Years Reflection on Landownership. *Source* Farmer survey in 2023

allocate around 0.27 ha, and 'moving-out' farmers allocate roughly 0.17 ha, indicating varied strategies among farmers despite lacking statistical significance. Similarly, there is uniformity in groundnut allocation across typologies, with 'hanging-in' farmers allocating an average of 0.17 ha, 'stepping up' farmers approximately 0.18 ha, and 'Moving-out' farmers around 0.20 ha. Farmers allocate limited land to legume production, with common beans averaging 0.32 ha, soybeans at 0.24 ha, and groundnuts at 0.19 ha. Notably, land allocation among common bean farmers varies significantly, with 'hanging-in' farmers allocating an average of 0.31 ha, 'stepping up' farmers around 0.38 ha, and 'moving-out' farmers 0.31 ha, attributed partly to intercropping with maize, a staple crop. Additionally, 'stepping up' farmers consistently allocate more land to soybean and common bean production within the legume category, reflecting their increasing engagement in these crops.

Table 3 Household land allocation to the three legumes

Variable	Farmer typology	N	Mean	Standard deviation	Sig.
Land allocated for soybean production in ha (log10)	'Hanging-in'	237	− 0.6275	0.32874	0.1730
	'Stepping-up'	132	− 0.5704	0.37292	
	'Moving-out'	16	− 0.7007	0.24753	
Land allocated for groundnut production in ha (log10)	'Hanging-in'	64	− 0.7399	0.32433	0.8710
	'Stepping-up'	75	− 0.7197	0.27268	
	'Moving-out'	9	− 0.6944	0.25789	
Land allocated for common beans production ha (log10)	'Hanging-in'	172	− 0.5043	0.3037	0.0670
	'Stepping-up'	109	− 0.4182	0.28087	
	'Moving-out'	19	− 0.5046	0.45698	

Data source Farmer Survey in 2023. Data is in a log-base 10 format, and we provide the text with the actual values for a more interpretable discussion
Source Farmer survey in 2023

Gendering the capability to produce, use and market the legumes revealed that women in 'hanging-in' and 'moving-out' typologies reported taking much control over decisions around field production, harvest, sale, and utilisation of the three legumes. However, soybeans have lately been seen as gaining men's attention. For the 'stepping up' category, the women reported that soybeans and groundnuts are no longer crops under their complete control as men have taken up the key roles and decision-making processes. Men in 'hanging-in' and 'moving-out' typologies expressed that women control legume production and marketing while they concentrate on other crops.

> At the planting stage, we may seem to be doing things jointly, but our wives assume the roles from harvest to the utilisation of soybeans, common beans, and groundnuts. It is taken as an offence when you try to ask for proceeds and utilisation of these legumes. [Men FGD 32]
>
> Lately, we are noticing that our husbands are keen on decisions and actions around soybean production; this is because the crop is now fetching better prices and slowly relegating us to passive decision-makers on this enterprise. [Women FGD 11].

This narration was common among the women FGDs we conducted. Men in the 'hanging-in' and 'moving-out' typologies indicated that groundnuts are usually for women as there is an intense labour enterprise.

A common finding in our female FGDs in this area was the growing feminisation of agriculture. Women presented scenarios where they worked in the fields with limited support from their husbands. The women narrated doing most of the agricultural activities to safeguard their families' livelihoods. For their part, men would drink and avoid agricultural work in their fields.

> Our husbands are lacking responsibility as members of the family. We farm and do most of the work alone. We do not usually confront them to keep peace in our marriages, but if they were involved, our agricultural output would have been better off. [Women FGD participant 18].

The situation of men leaving most work on legume production to their wives usually happened because women had lesser bargaining power in their homes. This is what transpired in the FGD:

> The women expressed limited control over the men's actions given that they mostly farm on land inherited from their husbands' parents, staying, and operating in their husband's patrimonial locations. [Women FGD participant 18].

5 Discussion

Based on the findings presented in the result section, we have identified four themes that we find of particular interest to discuss in addressing our research question. Accordingly, the discussion is structured around the four themes: (a) Agency and declining role of legumes in food and income security in Dowa district; (b) Mismatch of policy aspirations and actions on legumes; (c) 'Stepping up' and 'hanging-in' typologies of farmers; and (c) Intra-household dynamics on decisions and actions.

(a) *Agency and declining role of legumes in food and income security in Dowa district*: Over the past five years, there has been a decrease in per-household production and productivity of soybean, groundnut, and common bean, which has led to limited availability of food from these three legumes. As a result, the three legumes play a minor role in smallholder farmers' food and income security, a finding which concurs with what is reported by Jayne and Sanchez (2021) and Benson and De Weerdt (2023). We further show that farmers have limited agency, which is the ability to change their situation as regards increasing land that can be used for legumes, increasing their production with limited access to quality legume seed, limited access to the market that can offer a decent price for the legumes, limited access to labour to cultivate legumes dependent on women labour, limited access to extension services that can advise on legume production, lack of voice in society to influence policy issues such as the deployment of extension workers and marketing issues. These findings align with growing realisation and attention around agency, which was watered down by the neoliberal consensus on the role of smallholder farmers in food security by focusing on the non-contextualised technocratic aspect of fixing hunger and agricultural development problems (Moseley et al., 2015; Zhang, 2020). As observed by Rahman (2023), food and income insecurity of many smallholder farmers are products of the inability to access formal insurance and credit that are accessible in high-income countries; they have psychological impediments that come from being in the state of poverty, which in turn limit their ability to exercise the agency they might have. UNDP (2024) also allies with our findings by highlighting that addressing global challenges, such as food and income insecurity, is termed as closing agency gaps, requires intentional efforts on economic empowerment, access to information, and social and norm stereotypes, which hold people back from realising their development potential (UNDP 2024).
(b) *Mismatch of policy aspirations and actions on legumes*: Despite government policy aspirations of promoting legumes, the measures taken have not yielded results as far as production, food security or income are concerned. Public extension officers report less priority being given to legumes. The limitation to transition from agricultural public policy goals and aspirations to aspired results is a widely noted problem in scholarship with limited capacity to implement the policies (Haug et al. 2021a, b). There is also the problem of misguided policy development (Haug and Hella 2013; Jayne et al. 2023) and lack of policy ownership at the local level due to heavy influence of external actors in the policy development processes (Jayne et al. 2023). In the case of public extension services in Malawi, maize has remained a key crop due to its role in food security and its related political connotations, as lack of maize is usually interpreted as food insecurity in Malawi (Baulch and Jolex 2021; Sibande et al. 2017; Smale 1995). In our study, access to legume extension services was yet another critical limitation for farmers in achieving their potential production and productivity. The lack of access to extension services was manifested through the lack of basic modern knowledge in the production of legumes, pest and disease management, as well as the absence of operational arrangements like having active farmer clubs which

can pursue issues of markets and marketing. Our study affirms what Chinsinga et al. (2021) noted with the groundnut value chain that extension was hardly accessible for the smallholder farmers. Ragasa and Mazunda (2018) stated that from the nationally representative study of Malawi only 28% of the farming households accessed extension services from government extension workers, while 6% received extension services from non-governmental organisations and only 2% accessed private sector extension services in 2013. Other studies have pointed out that low investment in agricultural research, development, and extension impacted the quality and quantity of service delivered to the farmer and reduced agricultural productivity (Jayne et al. 2023; Jayne and Sanchez 2021).

(c) *'Stepping up' and 'hanging-in' typologies of farmers*: The majority of the farmers were in the *'hanging-in'* category, contrary to the notion of agricultural commercialisation, which has been part of the agricultural policy agenda and discourse in Malawi. Despite legumes playing a minor role, our study shows that few farmers are 'moving-out' of farming. Our findings are similar to Chinsinga and Matita (2021) and Chinsinga et al. (2021) who noted that few farmers are 'moving-out' of farming despite agriculture's dismal role in their livelihoods. Work by Matita et al. (2022) sheds light on the food and income security challenges encountered by farmers categorised as 'hanging-in', where in their study, most farmers were in the typology of 'hanging-in' to livelihoods, regardless of the degree of commercialisation. Matita et al. (2022) further associated the recurrence of 'hanging-in' farmers with the failure of current practices to offer sustainable poverty reduction measures. In line with the work by Chinsinga et al. (2021) and World Bank (2023), we find a narrow difference in farmers food and income security status in the various typologies. However, households in the 'hanging-in' and 'moving-out' typologies rely more heavily on casual labour to support their food and income security during the lean season, when the agricultural sector is more vibrant, and across the year in accordance with World Bank findings (World Bank 2023). Work preceding this study suggests that farming households have more members engaged in *ganyu* employment than non-farming households. However, the *ganyu* employment is affected by low wages due to pay, which is less than one-third of earnings in business or waged employment (World Bank 2023). Apart from the pay, the issue of an unskilled labour force and limited opportunities for *ganyu* employment compromise wages (Shonhe et al. 2022; World Bank, 2023). Of concern to this trend, which our study established and was also reported by Chinsinga et al. (2021) and Giller et al. (2021a, b), is that doing *ganyu* work compromises and conflicts with farmer's production, which affects the food security cycle as employment to earn food or income for work in rural areas is highly seasonal.

(d) *Intra-household dynamics on decisions and actions*: In households with both men and women, our findings indicate that women are overburdened with legume production responsibilities compared to men. However, men decide on sales and how income from legume sales should be used. These findings are in accordance with what Chinsinga et al. (2021) found, where partly in the 'hanging-in' and mainly in the 'stepping up' typologies, men are becoming increasingly interested in legumes such as soya beans and groundnuts despite limiting most of their actions to marketing and using proceeds from such enterprises. The growing interest of men in these enterprises aligns with the observation by World Bank (2023) that for the past decade, soybeans and groundnuts have been demonstrated to garner export markets and thereby drive demand, which somewhat has improved nominal marketing prices of these crops. What we find in this work and what was reported by World Bank (2023) seems to build on what Doss (2002) and Zhang (2022) observed in Ghana and China, respectively, where the gender patterns and roles in agricultural production were skewed to women in production and subsistence agriculture and yet they lost control when it evolved to commercial agriculture. Since women are so important in attaining household food security, their roles in legume production need attention, particularly in striking a balance of economic empowerment that translates to improved well-being without increasing work burden (Haug et al. 2021a, b).

6 Conclusion

The purpose of this paper has been to assess the role of three specific legumes regarding food and income security for smallholder farmers of different typologies in Dowa district, Malawi. We find that despite the policy positions, the contribution of the three legumes in various typologies of smallholders' food and income security is minimal and declining. Limited access to productive assets and resources such as land, labour and inputs weakens farmers' agency, as observed among 'moving-out' and 'hanging-in' farmers. With the upcoming marketing potential of the groundnuts and soybeans, men smallholders are increasing their attention in decisions and actions on these legumes, though they limit themselves to marketing and revenue use, with women serving as primary labour providers of all production activities. Enhancing the role of these legumes regarding smallholder farmers' food and income security requires strengthening both men's and women's smallholders' agency as well as effective and conducive policy implementation. Overall, we conclude that the role of these legumes in smallholder farmers' food and income security in Dowa is limited and declining.

References

Alkire S (2007) Measuring agency: issues and possibilities. Indian J Human Develop 1(1):169–175. https://doi.org/10.1177/0973703020070110

Alvarez S, Timler CJ, Michalscheck M, Paas W, Descheemaeker K, Tittonell P et al (2018) Capturing farm diversity with hypothesis-based typologies: an innovative methodological framework for

farming system typology development. PLoS ONE 13(5). https://doi.org/10.1371/JOURNAL.PONE.0194757
Baulch B, Jolex A (2021) Mixed Fortunes prices paid to soybean farmers have improved in 2021…but not those to maize farmers. https://www.researchgate.net/publication/354328547
Benson T (2021) Disentangling food security from subsistence agriculture in Malawi. International Food Policy Research Institute, Washington, DC. https://doi.org/10.2499/9780896294059
Benson T, De Weerdt J (2023) Employment options and challenges for rural households in Malawi. Washington, D.C. https://ebrary.ifpri.org/utils/getfile/collection/p15738coll2/id/136607/filename/136817.pdf. Accessed 18 Feb 2024
Chadza W, Burke WJ, Jayne TS, Muyanga M (2022) How Russia's invasion of Ukraine will affect the food security situation in Malawi: implications for malawian policy response. Accessed 10 April 2022
Chimonyo VGP, Snapp SS, Chikowo R (2019) Grain legumes increase yield stability in maize based cropping systems. Crop Sci 59(3):1222–1235. https://doi.org/10.2135/CROPSCI2018.09.0532
Chinsinga B, Matita M (2021) Political economy of groundnut value chain in Malawi: Its re-emergence amidst policy chaos, strategic neglect and opportunism (No. 056). Brighton. Accessed 14 Mar 2022
Chinsinga B, Matita M, Chimombo M, Msofi L, Kaiyatsa S, Mazalale J (2021) Agricultural commercialisation and rural livelihoods in Malawi: a historical and contemporary Agrarian Inquiry (No. 75). Brighton. Accessed 3 Mar 2024
Clapp J, Moseley WG, Burlingame B, Termine P (2022) Viewpoint: the case for a six-dimensional food security framework. Food Policy 106:102164. https://doi.org/10.1016/J.FOODPOL.2021.102164
Donovan J, Gelli A (2019) Designing interventions in local value chains for improved health and nutrition: insights from Malawi. World Dev Perspect 16:100149. https://doi.org/10.1016/J.WDP.2019.100149
Dorward A, Giller KE (2022) Change in the climate and other factors affecting agriculture, food or poverty: an opportunity, a threat or both? A personal perspective. Glob Food Sec 33:100623. https://doi.org/10.1016/J.GFS.2022.100623
Dorward A, Anderson S, Bernal YN, Vera ES, Rushton J, Pattison J, Paz R (2009) Hanging in, stepping up and stepping out: Livelihood aspirations and strategies of the poor. Dev Pract 19(2):240–247. https://doi.org/10.1080/09614520802689535
Doss CR (2002) Men's crops? Women's crops? The gender patterns of cropping in Ghana. World Dev 30(11):1987–2000. https://doi.org/10.1016/S0305-750X(02)00109-2
Ehlert, J., & Voßemer, C. (2015). Food sovereignty and conceptualization of agency: a methodological discussion. Austr J South-East Asian Stud 8(1):7–26. https://doi.org/10.14764/10.ASEAS-2015.1-2
FAO (1996) Rome declaration and plan of action. https://www.fao.org/3/w3613e/w3613e00.htm. Accessed 4 Feb 2024
FAO, IFAD, UNICEF, WFP, WHO (2021) The State of Food Security and Nutrition in the World 2021. Transforming food systems for food security, improved nutrition and affordable healthy diets for all. https://doi.org/10.4060/cb4474en
FAO, IFAD, UNICEF, WFP, WHO (2023) The State of Food Security and Nutrition in the World 2023. Urbanization, agrifood systems transformation and healthy diets across the rural–urban continuum. https://doi.org/10.4060/cc3017en
Galiè A, Jiggins J, Struik PC, Grando S, Ceccarelli S (2017) Women's empowerment through seed improvement and seed governance: evidence from participatory barley breeding in pre-war Syria. NJAS Wageningen J Life Sci 81:1–8. https://doi.org/10.1016/j.njas.2017.01.002
Galiè A, Teufel N, Girard AW, Baltenweck I, Dominguez-Salas P, Price MJ et al (2019) Women's empowerment, food security and nutrition of pastoral communities in Tanzania. Glob Food Sec 23:125–134. https://doi.org/10.1016/J.GFS.2019.04.005

Gelli A, Donovan J, Margolies A, Aberman N, Santacroce M, Chirwa E, Henson S, Hawkes C (2020) Value chains to improve diets: diagnostics to support intervention design in Malawi. Glob Food Sec 25:100321. https://doi.org/10.1016/J.GFS.2019.09.006

Giller KE (2020) The food security conundrum of sub-Saharan Africa. Glob Food Sec 26:100431. https://doi.org/10.1016/J.GFS.2020.100431

Giller KE, Delaune T, Silva JV, Descheemaeker K, van de Ven G, Schut AGT, van Wijk M, Hammond J, Hochman Z, Taulya G, Chikowo R, Narayanan S, Kishore A, Bresciani F, Teixeira HM, Andersson JA, van Ittersum MK (2021a) The future of farming: who will produce our food? Food Secur 13(5):1073–1099. https://doi.org/10.1007/s12571-021-01184-6

Giller KE, Delaune T, Silva VJ, Wijk MV, Hammond J, Descheemaeker K, van De Ven G, Schut AGT, Taulya G, Chikowo R, Andersson JA (2021b) Small farms and development in sub-Saharan Africa: farming for food, for income or for lack of better options? Food Secur 1:1431–1454. https://doi.org/10.1007/s12571-021-01209-0

Government of Malawi (2017) Dowa District Council Socio/Economic Profile 2017–2022. https://global-uploads.webflow.com/6061a9d807f5368139d1c52c/610b2a58bacd590cad52ebd6_Dowa-District-Council-Socio-Economic-Profile-2017-2022.pdf. Accessed 20 Mar 2022

Government of Malawi (2018) National Agricultural Investment Plan. Accessed 29 Mar 2022

Government of Malawi (2022) National Export Strategy II

GRFC (2024) FSIN and global network against food crises. https://www.fsinplatform.org/grfc2024

Guiomar N, Godinho S, Pinto-Correia T, Almeida M, Bartolini F, Bezák P et al (2018) Typology and distribution of small farms in Europe: towards a better picture. Land Use Policy 75:784–798. https://doi.org/10.1016/j.landusepol.2018.04.012

Hahn HP (2012) Consumption, identities, and agency in Africa: an overview. Decoding Mod Consum Soc 69–86. https://doi.org/10.1057/9781137013002_5

Haug R, Hella J (2013) The art of balancing food security: securing availability and affordability of food in Tanzania. Food Secur 415–426. https://doi.org/10.1007/s12571-013-0266-8

Haug R, Mwaseba DL, Njarui D, Moeletsi M, Magalasi M, Mutimura M et al (2021a) Feminization of African agriculture and the meaning of decision-making for empowerment and sustainability. Sustainability (Switzerland) 13(16). https://doi.org/10.3390/SU13168993

Haug R, Nchimbi-Msolla S, Murage A, Moeletsi M, Magalasi M, Mutimura M et al (2021b) From policy promises to result through innovation in African agriculture? World 2(2):253–266. https://doi.org/10.3390/world2020016

HLPE (2020) Food security and nutrition building a global narrative Towards 2030. www.kpe-kalamatas.gr. Accessed 26 Mar 2022

Hunga HG, Chiwaula L, Katundu M (2023) Organizational readiness for policy implementation: an assessment of the harmonized seed regulations in Malawi. Environ Develop 45:2211–4645. https://doi.org/10.1016/j.envdev.2023.100828

Iversen TO, Westengen O, Jerven M (2023) Measuring the end of hunger: Knowledge politics in the selection of SDG food security indicators. Agric Hum Values 40(3):1273–1286. https://doi.org/10.1007/s10460-023-10418-6

Jayne, Sanchez PA (2021) Agricultural productivity must improve in sub-Saharan Africa. Science 372(6546):1045–1047. https://doi.org/10.1126/science.abf5413

Jayne, Zingore S, Niang AI, Palm C, Sanchez P (2023) Building twenty-first century agricultural research and extension capacity in Africa. Eur. Rev. Agric. Econ. 50(5):1824–1846. https://doi.org/10.1093/erae/jbad028

Jimenez-Lopez JC, Singh KB, Clemente A, Nelson MN, Ochatt S, Smith PMC (2020) Editorial: legumes for global food security. Front Plant Sci 11. https://doi.org/10.3389/fpls.2020.00926

Kostrowicki J (1977) Agricultural typology concept and method. Agric Syst 2(1):33–45. https://doi.org/10.1016/0308-521X(77)90015-4

Kpienbaareh D, Kansanga M, Luginaah I, Tkaczyk Z, Moseley WG (2023) Dietary Power and self-determination among female farmers in Burkina Faso: a proposal for a food consumption agency metric. Land 2023 12(5):978. https://doi.org/10.3390/LAND12050978

Leach M, Nisbett N, Cabral L, Harris J, Hossain N, Thompson J (2020) Food politics and development. In: World development. Elsevier. https://doi.org/10.1016/j.worlddev.2020.105024
Lee DK (2020) Data transformation: a focus on the interpretation. Korean J Anesthesiol 73(6):503–508. https://doi.org/10.4097/kja.20137
Maluf RS, Burlandy L, Cintrão RP, Jomalinis E, Carvalho TCO, Tribaldos T (2022) Sustainability, justice and equity in food systems: Ideas and proposals in dispute in Brazil. Environ Innov Soc Trans 45:183–199. https://doi.org/10.1016/j.eist.2022.10.005
Marshall Q, Fanzo J, Barrett CB, Jones AD, Herforth A, McLaren R (2021) Building a global food systems typology: a new tool for reducing complexity in food systems analysis. Front Sustain Food Syst 5. https://doi.org/10.3389/fsufs.2021.746512
Matita M, Chirwa EW, Kaiyatsa S, Mazalale J, Chimombo M, Msofi Mgalamadzi L, Chinsinga B (2022) Determinants of smallholder farmers' livelihood trajectories. Evidence from rural Malawi. Agrekon61(4):399–411. https://doi.org/10.1080/03031853.2022.2107030
Moseley WG, Jordan CF (2001) Measuring agricultural sustainability: energy analysis of conventional till and no–till maize in the Georgia Piedmont 41(1):105–116. https://about.jstor.org/terms. Accessed 6 April 2022
Moseley W, Schnurr M, Bezner Kerr R (2015) Interrogating the technocratic (neoliberal) agenda for agricultural development and hunger alleviation in Africa. Afr Geogr Rev 34(1):1–7. https://doi.org/10.1080/19376812.2014.1003308
Mulenga H, Mwangwela AM, Kampanje-Phiri J, Mtimuni B (2021) Influence of gendered roles on legume utilization and improved child dietary intake in Malawi. Afr J Food Agric Nutrition Develop 21(3):17764–17786. https://doi.org/10.18697/ajfand.98.18205
NSO (2020) The fifth integrated household survey (IHS-5) 2020 Report
Ragasa C, Mazunda J (2018) The impact of agricultural extension services in the context of a heavily subsidized input system: the case of Malawi. https://doi.org/10.1016/j.worlddev.2017.12.004
Rahman T (2023) Poverty, agency, and development. Soc Philos Policy 40(1):9–35. https://doi.org/10.1017/S0265052523000353
Shonhe T, Scoones I, Murimbarimba F (2022) Agricultural commercialisation and changing labour regimes in Zimbabwe. J Contemp Afr Stud 40(1):78–96. https://doi.org/10.1080/02589001.2021.1933397
Sibande L, Bailey A, Davidova S (2017) The impact of farm input subsidies on maize marketing in Malawi. Food Policy 69:190–206. https://doi.org/10.1016/j.foodpol.2017.04.001
Smale M (1995) "Maize is life": Malawi's delayed green revolution. World Dev 23(5):819–831. https://doi.org/10.1016/0305-750X(95)00013-3
Snapp SS, Cox CM, Peter BG (2019) Multipurpose legumes for smallholders in sub-Saharan Africa: identification of promising 'scale out' options. Glob Food Sec 23:22–32. https://doi.org/10.1016/J.GFS.2019.03.002
Sokourenko K, Haddad L, Beal T, Mutyasira V, Keizire B (2022) African food systems: a regional data-based snapshot. Geneva. https://doi.org/10.36072/bp.9
UNDP (2024) Human Development Report 2023/2024. Breaking the Gridlock: reimagining cooperation in a polarized world
United Nations (1987) Our common future
World Bank (2023) A narrow path to prosperity. Washington, D.C. www.worldbank.org
Zhang L (2022) From left behind to leader: gender, agency, and food sovereignty in China. In: Social innovation and sustainability transition, pp 177–189. https://doi.org/10.1007/S10460-020-10114-9

Open Access This chapter is licensed under the terms of the Creative Commons Attribution 4.0 International License (http://creativecommons.org/licenses/by/4.0/), which permits use, sharing, adaptation, distribution and reproduction in any medium or format, as long as you give appropriate credit to the original author(s) and the source, provide a link to the Creative Commons license and indicate if changes were made.

The images or other third party material in this chapter are included in the chapter's Creative Commons license, unless indicated otherwise in a credit line to the material. If material is not included in the chapter's Creative Commons license and your intended use is not permitted by statutory regulation or exceeds the permitted use, you will need to obtain permission directly from the copyright holder.

Intrahousehold Gender Relations in Sorghum and Millet Production Resources: Evidence from Male- and Female-Headed Households in Southern Malawi

Charity Chanza, Sibongile Zimba Chimzinga, Andrew Dougill, Christine Bosch, and Stefan Kepinski

Abstract Sorghum and millet are important sources of food and nutrition security among smallholder farmers in Malawi. The majority of farmers engaged in sorghum and millet production in southern Malawi are women, and they are disadvantaged by inequalities in the form of access to, control over and entitlement to, production resources compared to their male counterparts. This chapter addresses intrahousehold gender relations in sorghum and millet production among adult males and adult females in male- and female-headed households in Chikwawa and Nsanje Districts of southern Malawi. The study employed mixed methods approach. A Harvard access and control matrix was employed to learn who had access to and control over production resources. To understand the gender relations dynamics, Chi-Square tabulations and content analysis were used to analyse the association between the gender of the household head and the gender of the adult household member with access to, control

Authors Charity Chanza, Sibongile Zimba Chimzinga contributed equally

C. Chanza (✉)
Extension Department, Lilongwe University of Agriculture and Natural Resources, Lilongwe, Malawi
e-mail: cchanza@luanar.ac.mw

S. Z. Chimzinga
Horticulture Department, Lilongwe University of Agriculture and Natural Resources, Lilongwe, Malawi
e-mail: sibozim@yahoo.co.uk

S. Z. Chimzinga · S. Kepinski
Centre for Plant Sciences, School of Biology, University of Leeds, Leeds, UK
e-mail: S.Kepinski@leeds.ac.uk

A. Dougill
Department of Environment and Geography, University of York, York, UK
e-mail: andy.dougill@york.ac.uk

C. Bosch
School of Food Science and Nutrition, University of Leeds, Leeds, UK
e-mail: C.Bosch@leeds.ac.uk

© The Author(s) 2026

B. R. Singh et al. (eds.), *Sustainable Food Systems in Sub-Saharan Africa*,
https://doi.org/10.1007/978-3-032-11288-0_19

over, and entitlement to production resources. Our findings reveal that there were more adult male than female members in MHHs who had access to all production resources; on the other hand, there were more adult female than male members in FHHs who had access to the same resources. In MHHs, there were more adult male members than female members who controlled and were entitled to land, labour, seed, farm tools, manure, credit, extension services and transportation. Likewise, in FHHs more adult female members than male members controlled and were entitled to seed, credit, extension services and transport. In FHHs, more adult males than females controlled land. Adult male and female members in FHHs equally controlled and were entitled to labour, manure and farm tools. The chapter concludes that the pattern of access to, control over and entitlement to resources in sorghum and millet does not expedite equal outcomes for adult male and female members in MHHs and FHHs, except for the control of and entitlement to labour, manure and farm tools in FHHs. Agricultural development interventions promoting equitable access to institutional support services and productive resources in Sub Saharan Africa must be informed by intrahousehold dynamics affecting men and women to bridge gender gaps in context and reduce the structural disadvantages against women farmers.

Keywords Intrahousehold gender relations · Sorghum · Millet · Access · Control · Malawi

1 Introduction

Sorghum and millet are climate-resilient crops, important nutritious grains, especially in tropical and sub-tropical fragile regions (Chaturvedi et al. 2022). They are among traditional crop species which are often neglected and underutilized, rely on biological functioning of the ecosystem, require low input of synthetic fertilisers, pesticides and irrigation (Modi and Mabhaudhi 2013; Van Rensburg et al. 2007). These crops can substitute the other major staple cereals that are susceptible to drought and heat, while ensuring farm resilience and food diversification (Hadebe et al. 2017). These comparative advantages make sorghum and millet an obvious choice for farmers in an era of climate change and low input farming systems. Despite sorghum and millet being native to sub-Saharan Africa (SSA), their neglect and underuse is a matter of concern given their potential to improve food and nutritional security, coupled with socioeconomic conditions of farmers in this region (Motsi et al. 2022). In terms of production, these crops rank among the major six cereal crops, providing a staple food for approximately 250 million globally (Bhagavatula et al. 2013; Chaturvedi et al. 2022).

In Malawi, sorghum and millet are staple foods mainly grown in the southern region, and their production is dominated by female farmers who experience gender inequalities in agricultural inputs and resources relative to male farmers (Zimba et al. 2023). The extent of agricultural productivity between male and female farmers varies across and within countries in sub-Saharan Africa (Rao et al. 2019). Gender-specific restrictions in access to farm inputs have been blamed for the differences in productivity between men and women, as they result in lower and unequal levels of access to fertiliser, labour, and land use (Chigbu et al. 2019; Khan et al. 2021), as well as to institutional support services, such as extension and advisory services, and credit facilities (Sell et al. 2018). Thus, women and female-headed agricultural households in SSA regularly face production constraints, major among which are discrepancies in access to resources such as land and agricultural inputs that are associated with higher rates of poverty and food insecurity among women farmers than men (Otieno et al. 2021). Differences in economic opportunities and access to productive resources also leave women more exposed to climate change impacts because they are usually poorer, less educated, and are excluded from political, community and household decision-making processes (Phiri et al. 2022). Recent evidence in selected African countries reveals that closing the gender gap will foster poverty reduction, gains in agricultural Gross Domestic Product (GDP), increases in women's agricultural productivity, and a reduction of global hunger (Gebre et al. 2021).

Neglected and underutilized species (NUS) are predominantly grown by smallholder farmers in local communities (Swetha and Muthukumar 2016; Mayes et al. 2011); and these communities have deep-seated gender roles and norms about who has the right to control and benefit (or not) from NUS cultivation and sale (Malkanthi 2016). Most NUS, including sorghum and millet, are closely associated with women (and thus considered women's crops), and can be an entry point for gender equality and empowerment (Conti et al. 2019). It is important to note that while the prevailing narratives around gender and NUS hold some truth, the reality is often much more complex. Gender roles, decision-making processes, power dynamics, and trade-offs can overlap and sometimes even contradict each other. While minor crops like sorghum and millet can be used to promote women's empowerment and improve households' well-being, there is limited research to support this claim. It remains unclear whether the conditions are ripe for women to benefit equally from the opportunities that these crops present.

A review of literature reveals that gender studies on access to, control over, and entitlement to resources arising from sorghum and millet production, and the power relations governing such distributions are not Malawi-specific. Most studies conducted in Malawi on sorghum and millet have focused on the agronomical, plant breeding (Chintu et al. 1996; Ngwira et al. 2019; Munkombwe et al. 2020; Nkongolo et al. 2008), agrobiodiversity and climate change resilience aspects of the crops (Kerr 2014; Vanya 2012). Recent work in the area focused on how gender and climate risk awareness influence sorghum and pearl millet farmers' varietal trait and cropping system choices in Malawi (Zimba et al. 2023), and not gender relations per se. A gender analysis study of sorghum and millet is therefore necessary for the generation of pertinent sex-disaggregated data on gender benefits, access to, control over, and entitlement to resources arising from sorghum and millet production, and the power relations governing such distributions. These would guide the formulation of interventions that would narrow down gender gaps by addressing the inequalities uncovered and improve lives of women and men in the sorghum and millet farming communities.

In this chapter, we examine the intrahousehold gender differences in access to, control over and entitlement to resources among adult household members in male and female-headed households involved in sorghum and millet production in southern Malawi. The chapter addresses the following research question: What are the differences between adult male and adult female household members from male and female-headed households in their access to, control over and entitlement to resources in sorghum and millet production and why? An understanding of gender relations in sorghum and millet production is essential to build empowerment pathways within the household and close gender gaps between male and female farmers at society level.

1.1 Theoretical Framework

Quantitative and qualitative gender research is increasingly focusing on analysing detailed information from individuals within the household, rather than treating the household as a unified decision-making unit. Different intrahousehold models are being developed to better understand how the drivers and impacts of gender inequality considering the various preferences, needs, and abilities of different household members (Lambrecht 2016). Households can be both spaces of cooperation and conflict, and decision-making power is often influenced by the bargaining power of the individual household members (Sen 1990).

This study is informed by the Feminist Political Ecology (FPE). The integration of FPE within the realm of political ecology has been instrumental in unraveling the intricate relationships between humans and their natural environment, particularly in understanding access to, and control over, productive resources. FPE, pioneered by Rocheleau et al. (1996) introduced a gendered perspective to the political economy framework, emphasizing the significance of gender in assessing access to and control over resources. This lens does not solely focus on gender; it also recognizes its interplay with class, race, ethnicity, and culture, offering a more holistic understanding.

Luna (2018) and Elmhirst (2015) recommended FPE as a framework that scrutinizes intra-household dynamics, specifically the gendered politics influencing resource allocation and decision-making within households. Ribot and Peluso (2003) expanded on this, highlighting "access" to include gendered politics of resource control and distribution within households. Their theory described what they called "access" using social identity, which offered a deeper understanding into differential power ('bundles of powers'). They expounded that in a household, differential power held by individuals essentially dictated access to, control over and mediation of, the access of others.

This chapter underscores access to, control over and entitlement to, productive resources in sorghum and millet production by farming households. FPE posits that gender roles and family hierarchies grant men and women distinct rights to resources (Elmhirst 2015). It also views households as complex production and consumption units rather than simplistic cooperative entities, aligning with feminist post-structural viewpoints that acknowledge the nuanced nature of households (Gengenbach et al. 2018). Moreover, FPE considers women as a heterogenous group with varied interests regarding resource access and control (Nyantakyi-Frimpong 2019). This framework critically examines existing inequalities between men and women within households from an intersectional perspective, recognizing that gender is not a binary concept but intersects with other dimensions like class, ethnicity, and others (Gengenbach et al. 2018).

'Access' in this study means user rights, and 'control' is defined as *power* over decision-making on resources, while 'entitlement' refers to ownership of resources. The 'household' in this study is conceived as an institution of decision making, an area of joint consumption, production and investment in which both labour and resource allocation decisions are made (Agarwal 2011). In this study a household is defined as female headed if it meets the following criteria: de jure female headed household (FHH) (if it is managed by a widowed, divorced, separated or single woman) and de facto FHH (where the woman is responsible for maintenance of the household because the husband is mostly or permanently not physically present on the farm (Takane 2009).

2 Methodology

2.1 Study Site

The study was carried out in Nyachirenda and Mikalango Extension Planning Areas (EPAs) in Nsanje and Chikwawa districts respectively. Nyachirenda EPA has 12 sections with a total of 16,277 farming households, while Mikalango EPA has 29 sections with a total of 26,531 farming households.

2.2 Sampling Technique

Mankhokwe and Ulalo Sections, and Madyadya and Thunyeni Sections were purposively selected in Nyachirenda EPA and Mikalango EPA respectively, because most of sorghum and millet production take place in these sections. One village from each Section was randomly sampled, resulting in the selection of Bitirinyu Village in Ulalo Section, Kalumbi Village in Mankhokwe Section, Jasi 1 Village in Madyadya Section and Fodya Village in Thunyeni Section. A list of households' names from the selected villages was obtained from the Agricultural Extension Development Coordinator (AEDC). Proportionate random sampling was employed to select a sample size of 152 households using Cochran's (1977) formula. With the help of the Agricultural Extension Development Officers who staff the sections, purposive sampling was done to select 44 FGD participants based on their knowledge of the subject matter.

2.3 Data Collection Methods

A mixed methods approach was used to collect qualitative and quantitative data. Quantitative data was collected in household surveys, using a semi-structured questionnaire and Harvard access and control kits. These interviews generated quantitative data concerning the socio-economic characteristics of the households, access to, control over and entitlement to resources in sorghum and millet production. Qualitative data was collected using participatory techniques: interviews with key informants (KII) and focus group discussions (FGDs) using a flexible checklist which enabled the collection of data on who has access to what resources and why, who has control over which resources and why, and who is entitled to what resources and why, in the production process of sorghum/ millet. A total of 44 participants were involved in focus group discussions (FGDs). In each EPA, two FGDs were conducted differentiated along gender lines, comprising of 10 participants per discussion group. The groups were divided based on gender because of the sensitive nature of some topics for discussion. In female participant groups, care was taken to balance the representation of married women and female household heads. The study deployed a mixture of these data collection tools to triangulate the information collected.

2.4 Data Analysis Methods

To assess the gender differences in access to, control over, entitlement to resources in sorghum and millet production, Chi-Square (χ^2) tests, Harvard access and control matrix and content analysis were used. Quantitative data was analysed using descriptive statistics. Pearson Chi-Square (χ^2) tests of independence were used to test for significant association between gender of the household head and gender of the household members (adult male, adult female, and both adult male and adult female) as regards access to, control over, entitlement to resources. In this study, the resources that farmers accessed, had control over, and were entitled to were grouped into natural and human resources, raw materials, and institutional resources. The Harvard analysis of access and control matrix was deployed to learn who had access to natural and human resources, raw materials, and institutional resources required for sorghum and millet production. Qualitative data were analyzed using thematic framework analysis (Ritchie and Spencer 1994). This involved analyzing the content of the transcribed interviews by examining the underlying themes in the text material that contain information related to the particular themes of the research.

3 Results and Discussion

3.1 *Socio-economics and Demographic Characteristics of the Respondents*

The sample size for the study was 152 households, comprising of 54% female respondents and 46% male respondents. Sixty two percent were male headed households (MHHs) while 38% were FHHs. The majority of the respondents were married (72%) while the rest were widowed (16%), divorced (7%), separated (4%) and single (1%). The majority of the respondents were primary school dropouts (55%). Also, most of them perceived sorghum (88%) and millet (82%) as their main staple crops, while 49% reported maize as their main crop. Out of 152 respondents only 30% belonged to a farming club. Out of those who belonged to a farming club, 52% were males while 48% were females. The mean age for the respondents was 44 years with 20 years and 88 years being minimum and maximum ages respectively. Average landholding size for the households was 1 ha which was supplemented by an average of 0.40 ha of rented land. Out of the available land, 0.4 ha was allocated to sorghum production, while 0.30 ha and 0.34 were allocated to millet and maize respectively. The respondents reported harvesting an average of 209 kg of sorghum, 132 kg of millet and 266 kg of maize in the cropping season that the study was carried out.

3.2 *Access to, Control Over and Entitlement to Sorghum and Millet Production Resources Among Household Members*

Gender inequalities exists in access to, control over and entitlement to, valuable production resources such as land, credit, seed, fertilisers and extension services that would enhance production capacity. Findings are presented in tables that show Chi-square tabulations, which test for significant association between the gender of the household head and the gender of a household member with access to, control over, and entitlement to resources. This was to understand gender dynamics in access to, control over, and entitlement to resources within households in relation to the gender of the household head. Those with significant P-values provide statistical evidence that the difference in the gender of a household member (adult male, adult female, and both adult male and adult female) with access to, control over, and entitlement to resources is significantly associated with the gender of the household head.

3.2.1 Gendered Access to, Control Over and Entitlement to Natural and Human Resources in Sorghum and Millet Production

The natural and human resources required for sorghum and millet production included land and labour. Tables 1 and 2 present findings on access to, control over and entitlement to natural and human resources in sorghum and millet production respectively.

Access to, control over and entitlement to land

The gender of a household member with access to land was significantly associated with gender of the household head among sorghum (p-value = 0.000) and millet (p-value = 0.002) producing households. Further, the gender of those who had access to land was influenced by the gender of the household head. Male headed sorghum- and millet-producing households had more adult males (sorghum 7%, millet 8.5%) than adult females (sorghum 1%, millet 1%) having access to land, and there were more adult females (sorghum 25%, millet 18.4%) than adult males (sorghum 1%, millet 5%) having access to land in FHHs. Even so, the proportions of adult female members in FHHs having access to land were way bigger than those of adult males having access to land for both crops.

In many households, land access and use are not determined by an individual acting alone, but rather in conjunction with other household members (Mishra and Sam 2016). For instance, Lambrecht (2016) found that in Ghana, men's responsibility as household heads affords them the grounds for implicit and explicit rules that favour male access to land and labour within the household. This helps to explain why more adult male members had access to land in MHHs and vice versa. Access to and use

Table 1 Access to, control over and entitlement to natural and human resources among household members in sorghum production

Resources and gender MHH(n)=94 FHH(n)=58	Sorghum								
	Access %			Control %			Entitlement %		
	MHH	FHH	*P*-value	MHH	FHH	*P*-Value	MHH	FHH	*P*-value
Land									
Adult male	7.0	1.4	0.000	52.3	50.0	0.000	54.7	48.6	0.000
Adult female	1.2	25.0		3.5	36.1		3.5	37.5	
Both	65.1	65.3		16.3	5.6		12.8	4.2	
None	26.7	8.3		27.9	8.3		29.1	9.7	
Labour									
Adult male	14.0	2.8	0.000	47.7	37.5	0.000	44.2	33.3	0.000
Adult female	9.3	25.0		9.3	34.7		9.3	36.1	
Both	46.5	61.1		14.0	16.7		17.4	20.8	
None	30.2	11.1		29.1	11.1		29.1	9.8	

Note p-values are based on the Chi statistic

Table 2 Access to, control over and entitlement to natural and human resources among household members in millet production

Resources and gender MHH(n)=94 FHH(n)=58	Millet								
	Access %			Control %			Entitlement %		
	MHH	FHH	*P*-value	MHH	FHH	*P*-value	MHH	FHH	*P*-value
Land									
Adult male	8.5	5.3	0.002	53.5	42.5	0.000	57.7	43.4	0.000
Adult female	1.4	18.4		7.1	39.0		7.0	36.8	
Both	66.2	65.8		15.5	5.3		9.9	5.3	
None	23.9	10.5		23.9	13.2		25.4	14.5	
Labour									
Adult male	16.9	4.0	0.009	46.5	38.9	0.002	44.3	32.9	0.008
Adult female	12.7	24.0		14.1	40.8		15.7	39.5	
Both	46.5	58.7		15.5	17.1		15.7	15.8	
None	23.9	13.3		23.9	13.2		24.3	11.8	

Note *p*-values are based on the Chi statistic

of land by one household member affects other members in terms of access to land, labour and other resources (Kevane and Gray 2010). As a result, male and female household members may need to negotiate with other household members to gain access to land within the household or seek approval from them to acquire land outside of it (Lambrecht 2016). In this study, it was noted that household land was supplemented by land rentals to improve household access to land (see Sect. 3.1). Despite that land rental markets have become increasingly active in SSA (Holden and Otsuka 2014), women are disadvantaged when it comes to participating in land markets transactions because they have less access to resources than men, and positive and negative gender biases stemming from cultural norms, legal frameworks and other societal factors may work against their ability to secure access to land (Deere and Leon 2003). Conversely, studies have found that women may experience positive bias in rental markets due to their credibility for paying rent or inability to claim ownership of the land (Giovarelli 2006). It is important to note that these biases are not universal and may vary depending on the specific context and cultural norms.

Regarding control over land, findings showed that the gender of a household member with control over land was significantly associated with the gender of the household head among sorghum (p-value = 0.000) and millet (p-value = 0.000) producing households. Findings showed that MHHs had more adult male members (sorghum 52%, millet 54%) controlling land for sorghum and millet production with very few adult female members (sorghum, 4%, millet 7%) having control. Similarly, FHHs had more adult male members (sorghum 52%, 43%) dominating control over land unlike adult female members (sorghum 36%, millet 39%). This demonstrates that regardless of the gender of the household head, adult male members had an upper hand on land control.

Similar findings were observed on entitlement to land where the gender of a household member with entitlement to land was significantly associated with the gender of the household head among sorghum (p-value = 0.000) and millet (p-value = 0.000) producing households. In MHHs, more adult male members (sorghum 55%, millet 58%) were entitled to land than their female counterparts (sorghum 4% millet 7%). Likewise, in FHHs more adult male members (sorghum 49%, millet 43%) controlled land than their female counterparts (sorghum 38%, millet 37%). This implies that adult male members from both household types owned the land. Control over resources is a precondition for distribution of resources to different farming enterprises (van Eerdewijk et al. 2017), and it is useful in predicting which household members are likely to benefit from a particular farming enterprise. Men and women have different kinds of control and ownership of resources happening both nationally and within communities and households (Hausmann et al. 2010). Although women may have substantial ownership of certain types of assets as compared to men, a growing collection of factual evidence below demonstrates that in different environments worldwide, women usually control fewer overall assets than men.

Under patrilineal traditions in Malawi, women traditionally do not own, control or inherit land and acquire access to land through male family members (Slavchevska et al. 2017). Paul (2018) and Yadav & Sharma (2017) also reported findings that women unlike men have lesser control over land as a result of restrictive patriarchal traditional rules that appoint men as decision-makers over land. Women, thus, have little say when it comes to controlling and making decisions on the usage of the household land. Control over land unlike access is important for decisions to do with land allocation such as allocating land for sorghum and millet production over other land uses (Kusakabe 2003; Ndanga et al. 2013). In SSA, men frequently obtain use and certain management rights over land through bequeathal or allocation by their family or descent, while, for women, marriage is the most customary means to secure access to land (Doss et al. 2015: Meinzen-Dick et al. 2014; World Bank 2009; FAO 2011).

Access to, control over and entitlement to labour

Just like access to land, the gender of a household member with access to labour was significantly associated with the gender of the household head among sorghum (p-value = 0.000) and millet (p-value = 0.009) producing households. Moreover, the gender of those who had access to labour was influenced by the gender of the household head, implying that male-headed sorghum and millet producing households had more adult males (sorghum 14%, millet 9%) than adult females (sorghum 17%, millet 13%) with access to labour. On the other hand, there were more adult females (sorghum 25%, millet 24%) than adult males (sorghum 3%, millet 4%) in FHHs having access to labour.

The gender of a household member with control over labour was significantly associated with the gender of the household head among sorghum (p-value = 0.000) and millet (p-value = 0.002) producing households. MHHs had more adult males (sorghum 48%, millet 47%) controlling labour for sorghum and millet production than adult females (sorghum 9%, millet 7%). In FHHs, there are slight differences in the proportions of adult females (sorghum 35%, millet 42%) and adult males (sorghum 37% millet 40%) controlling labour, implying that both genders more or less controlled labour to produce the crops. The findings further show that the gender of a household member with entitlement to labour was significantly associated with the gender of the household head among sorghum (p-value $= 0.000$) and millet (p-value $= 0.008$) producing households. More adult males (sorghum 44%, millet 44%) were entitled to labour than their female counterparts (sorghum 9%, millet 16%) in MHHs, while in FHHs, as before, we see small differences in proportions of adult females (sorghum 36%, millet 39%) and adult males (sorghum 33%, millet 33%) reporting that they were entitled to labour. This implies that both genders were generally entitled to labour for production of the crops. In FHHs, there were slight disparities among adult males and adult females in their control of, and entitlement to, labour for the production of the crops. This indicated that women, just like men, had autonomy and power over household labour. A finding by Palacios-Lopez et al. (2017) indicated that women contribute 52% of the agricultural labour in Malawi. On the contrary, Kabeer (2003) explained that women have less control over family labor due to protracted differences regarding women's paid work. As a result, women tend to provide a lot of unpaid work on farms to maintain their marriages, with existing intra-household patriarchal bias restricting their decision-making regarding the use of family labour. Chant and Brickell (2013) attributed this phenomenon to unequal intra-household power relations, with men having greater authority over the use of household members on their farms. This may help to explain women's lack of power to control labour in MHH in this study.

In line with the findings of this study, Ankrah et al. (2020) noted that in households where women are the main breadwinners, they tend to have more autonomy and decision-making power when it comes to using family labour. This may be due to a shift in social dynamics and gender roles within the household. Other studies demonstrate that gender roles in many societies make it difficult for women to control labour within the household (Hargreaves 2010). However, in situations where agricultural labour is scarce and women can provide a significant amount of agricultural labour, their contribution to the household is widely recognized. This often brings higher decision-making power, higher wages for women and greater access to land (Yngstrom 2002).

3.2.2 Gendered Access to, Control Over and Entitlement to Raw Materials in Sorghum and Millet Production

Raw materials that were needed to produce sorghum and millet were seeds, fertilisers, farm tools and manure. Table 3 present findings on access to, control over and entitlement to raw materials among household members in sorghum production while Table 4 provide the same on millet production.

Access to, control over and entitlement to seeds

Table 3 Access to, control over and entitlement to raw materials among household members in sorghum production

Resources and gender MHH(n)=94 FHH(n)=58	Sorghum								
	Access %			Control %			Entitlement %		
	MHH	FHH	*P*-value	MHH	FHH	*P*-value	MHH	FHH	*P*-value
Seed									
Adult male	16.3	4.2	0.001	35.3	31.9	0.000	33.7	27.8	0.000
Adult female	3.5	32.4		16.5	45.8		12.8	44.4	
Both	53.5	54.9		21.5	13.9		26.7	18.1	
None	26.7	8.5		27.1	8.3		26.7	9.7	
Fertiliser									
Adult male	0.0	1.4	0.167	8.1	2.8	0.380	7.0	2.8	0.348
Adult female	0.0	2.8		1.2	2.8		1.2	4.2	
Both	9.3	4.2		1.2	2.8		1.2	2.8	
None	90.7	91.7		89.5	91.7		90.6	90.2	
Farm tools									
Adult male	17.4	2.8	0.000	43.5	31.9	0.000	43.0	30.6	0.000
Adult female	2.3	27.8		4.7	36.1		5.8	37.5	
Both	52.3	61.1		23.5	22.2		22.1	16.7	
None	27.9	8.3		28.2	9.7		29.1	15.3	
Manure									
Adult male	7.0	0.0	0.022	20.9	11.1	0.076	18.6	8.4	0.026
Adult female	1.2	8.3		3.5	12.5		3.5	11.1	
Both	23.3	22.2		7.0	9.7		8.1	12.7	
None	68.6	69.4		68.6	66.7		69.8	67.8	

Note p-values are based on the Chi statistic

Results show that the gender of a household member with access to sorghum and millet seed was significantly associated with the gender of the household head among sorghum (*p*-value = 0.001) and millet (*p*-value = 0.000) producing households. Findings reveal that MHHs had more adult male members (sorghum 16%, millet 16%) having access to the seed than adult female members (sorghum 3%, millet 4%). FHHs have more adult female members (sorghum 32%, millet 29%) having access to the seed than adult male members (sorghum 4%, millet 5%). MHHs had more members without access to seed, unlike FHHs who maintained very few members without access to seed. We see a similar pattern of results regarding seed control in that the gender of the household member with control over seed was significantly associated with the gender of the household head among sorghum (*p*-value = 0.000) and millet (*p*-value = 0.006) producing households. Yet again, more adult male members (sorghum 35%, millet 37%) had control over the seed than their female counterparts (sorghum 17%, millet 15%) in MHHs; on the other hand, more

Table 4 Access to, control over and entitlement to raw materials among household members in millet production

Resources and gender MHH(n)=94 FHH(n)=58	Millet								
	Access %			Control %			Entitlement %		
	MHH	FHH	*P*-value	MHH	FHH	*P*-value	MHH	FHH	*P*-value
Seed									
Adult male	15.5	5.3	0.000	37.1	32.9	0.006	36.6	23.7	0.002
Adult female	4.2	25.0		15.7	40.8		15.5	44.7	
Both	56.3	57.9		22.9	13.2		23.9	15.8	
None	23.9	11.8		24.3	13.2		23.9	15.8	
Fertiliser									
Adult male	1.4	0.0	0.141	8.5	1.3	0.060	7.0	1.3	0.065
Adult female	0.0	5.3		1.4	5.3		1.4	6.6	
Both	7.0	3.9		0.0	2.6		0.0	2.6	
None	91.5	90.8		90.1	90.8		91.5	89.5	
Farm tools									
Adult male	14.1	5.3	0.001	38.0	36.8	0.001	45.1	30.3	0.000
Adult female	2.8	19.7		11.3	38.2		7.0	38.2	
Both	57.7	63.2		25.4	13.2		22.5	10.5	
None	25.5	11.8		25.4	11.8		25.4	21.1	
Manure									
Adult male	5.6	0.0	0.057	14.1	7.9	0.404	15.5	5.3	0.125
Adult female	1.4	7.9		5.6	11.8		5.6	10.5	
Both	21.1	21.1		8.5	9.2		7.0	11.8	
None	71.8	71.1		71.8	71.1		71.8	72.4	

Note p-values are based on the Chi statistic

adult female members (sorghum 46%, millet 41%) had control over the seed than adult male members (sorghum 32%, millet 33%) in FHHs. FHHs had more female members control the seed unlike male members, but MHHs maintained the large number of adult male members controlling the type of seed to be grown compared to adult female members. Regarding entitlement to seed, the study shows that the gender of the household member with entitlement to sorghum and millet seed was significantly associated with the gender of the household head among sorghum (p-value $= 0.000$) and millet (p-value $= 0.002$) producing households. In MHHs there were more adult male members (sorghum 44%, millet 37%) who were entitled to the seed than adult female members (sorghum 13%, millet 15%). In FHHs, more adult female members (sorghum 44%, millet 45%) were entitled to the seed than their male counterparts (sorghum 28%, millet 24%). In our previous study in the same area, a significant association was observed between seed accessibility and gender, which influenced decision-making on the variety type to grow. Male farmers had a 62%

chance of accessing seeds from seed distributors or local markets, compared to only 37% of female farmers (Zimba et al. 2023). Unlike this study, Ankrah et al. (2020) found no significant relationship between gender and access to improved maize seeds or between gender and access to fertilizer. In their study, this finding was explained by a programme which advanced improved seeds and fertilizers to farmers.

Access to, control over and entitlement to inorganic fertilisers

Results show that there were no significant associations between the gender of the household member and the gender of the household head in access to (sorghum *p*-value = 0.167, millet *p*-value = 0.141), control over (sorghum *p*-value = 0.380, millet *p*-value = 0.060), and entitlement to (sorghum *p*-value = 0.348, millet *p*-value = 0.065) inorganic fertilisers in both sorghum and millet production. Neither adult male members nor adult female members in MHHs accessed inorganic fertiliser in sorghum production; and in millet production, 1% of adult male members and none of the adult female members accessed inorganic fertilisers. Similarly negligible proportions of adult females (sorghum 3%, millet 5%) and adult males (sorghum 1%, millet 0) in FHHs accessed the fertilizer. As far as inorganic fertilizers were concerned, there was equal control among adult male (3%) and adult female (3%) members in FHHs sorghum production, unlike in millet production where adult females (5%) had an upper hand than adult males (1%). In MHHs, more adult male members (sorghum 8%, millet 9%) had control of inorganic fertilisers than female members (sorghum 1%, millet 1%) in both sorghum and millet production. For both sorghum and millet, more adult male members (sorghum 7%, millet 7%) were entitled to inorganic fertilisers in MHHs than their female counterparts (sorghum 1%, millet 1%), while more adult female members (sorghum 4%, millet 7%) were more entitled to inorganic fertilisers than adult male members (sorghum 3%, millet1%) in FHHs.

The lack of significance of the findings on inorganic fertilisers could be the result of the fact that most of sorghum and millet farmers do not usually use inorganic fertilisers in their fields, as evidenced by a narration from the discussions below:

> Sorghum and millet are part of our traditional crops, and we choose to grow them because they are tolerant to heat, drought ….and can grow in poor soil fertility without the need for fertilisers or pesticides as other cereals.
>
> (Female participant no. 4 in women's FGD group, Thunyeni Section, Mikalango EPA)

While Freeman and Omiti (2003) and Horrell and Krishnan (2007) found that household head gender had no significant effect on the household's access to inorganic fertilizer in Kenya and Zimbabwe, Chirwa (2005) in Malawi, reported that FHHs were less likely to access inorganic fertilizer. Even though this study has found that many adult female members generally had access to land for sorghum and millet production, Whitehead and Tsikata (2003) found that many women had to limit the size of their farmland because of poor access to the basic inputs needed to cultivate land such as labour and fertilisers, or to make investments to improve the land. The lack of the required inputs indirectly reduces women's access to land. Kang et al. (2020) also reported outstanding differences in inorganic fertilizer use across plot ownership structures, with the amount of inorganic fertilizer used in female owned plots being smaller than in male owned and jointly owned farm plots.

Access to, control over and entitlement to farm tools

The study found that there was a significant association between the gender of the household member with access to farm tools and the gender of the household head in sorghum (p-value = 0.000) and millet (p-value = 0.001) producing households. MHHs had more adult male members (sorghum 17%, millet 14%) having access to farm tools than adult female members (sorghum 2%, millet 3%). We found that FHHs had more adult female members (sorghum 28%, millet 20%) that had access to the resource than adult male members (sorghum 2.8%, millet 5%). We see here that MHHs had more household members without access to farm tools, unlike FHHs who maintained very few members without access to farm tools. Just like access to seed, there was a significant association between the gender of the household member with control over farm tools and the gender of the household head in sorghum (p-value = 0.000) and millet (p-value = 0.001) producing households. In MHHs, adult male members (sorghum 44%, millet 38%) highly controlled use of farm tools, with very little control by adult female members (sorghum 5%, millet 11%). FHHs postulate almost equal chance on the control of tools having 32% adult males and 36% adult females (sorghum) and 38% adult males and 37% adult females (millet). Likewise, there was a significant association between the gender of the household members with entitlement to farm tools and the gender of the household head in sorghum (p-value = 0.000) and millet (p-value = 0.000). Findings similar to household members' control over farm tools were obtained in household members' entitlement to farm tools where more adult males (sorghum 43%, millet 45%) than adult females (sorghum 6%, millet 7%) were entitled to farm tools in MHHs. FHHs assume almost equal prospect of entitlement to farm tools having 31% adult males and 37% adult females (sorghum) and 30% adult males and 38% adult females (millet).

Access to, control over and entitlement to manure

Findings show insignificant association between the gender of a household member and the gender a household head regarding access to, control over and entitlement to manure in millet production. In sorghum production, however, there was significant association between the gender of a household member with the gender of household head with access to (p-value = 0.022) and entitlement (p-value = 0.026) to manure while there was insignificant association between the variables on control over manure. We note that more adult male members (7%) than adult female members (1%) accessed manure under MHHs, and only adult female members (8%) accessed the manure under FHHs. More adult males (19%) than adult females (3%) were entitled to manure under MHHs. Different from MHHs, FHHs present almost equal proportions of adult female members (11%) and adult male members (8%) who were entitled to manure. Control of manure was dominated by adult male members (21%) unlike adult female members (3%) in MHHs yet in FHH it is equally controlled by adult male members (11%) and adult female members (12%).

Various studies which assessed gender-based differences in access to and/or ownership of draft animals reported about the necessity of the animals for the functioning of manual plows and provision of manure, which are among the crucial factors for increasing agricultural productivity among the poor farming households (Smith 2008). Oladele and Monkhei (2008) and Oluka et al. (2005) observed significant variations in the populations of animals owned by men and women in Botswana and Uganda respectively, where men were significantly more likely to own cattle, donkeys, and horses, whereas women were significantly more likely to own small stocks such as goats and chickens. Freeman et al. (2008) reported that livestock farming was the second most important livelihood activity after crop production in Chikwawa and Nsanje districts of Malawi. Most of the livestock in the districts were owned by households that fell in food secure category dominated by MHH. The food insecure and extremely food insecure categories mostly comprised of FHH, who mostly kept goats and poultry, and poultry respectively (ibid). Diverse and larger herds of animals kept in MHHs may explain why more adult male members-controlled manure and were more entitled to manure than adult female members in this study, thereby indicating a further gender-based disadvantage in productive use of soil fertility inputs.

3.2.3 Gendered Access to, Control Over and Entitlement to Institutional Resources in Sorghum and Millet Production

The institutional services that the study considered for productive engagement in sorghum and millet farming included credit, extension services and transport. Tables 5 and 6 present findings on access to, control over and entitlement to institutional resources among household members in sorghum production and millet production respectively.

Access to, control over and entitlement to credit

The gender of the household member with access to credit was significantly associated with the gender of the household head in sorghum (p-value = 0.035) and millet (p-value = 0.026) producing households. Study results show that access to credit in MHHs was almost equal among adult male members (sorghum 2%, millet 3%) and their female counterparts (sorghum 1%, millet 3%). In FHHs access to credit was diverted towards the adult female members (7%) with adult male members not accessing any credit. The proportion of adult female members in FHHs having access to credit was way higher than the proportion of adult males having access to credit in MHHs. Generally, more adult females than adult males accessed credit for sorghum and millet production. There are two probable explanations for this. First, FGDs revealed that most female participants belonged to women's informal savings groupings called '*Banki nkhonde*' from where they obtained their financial credit, which could have been used to support sorghum and millet production. Men participants stated that formal banks in the area do not have initiatives that target smallholder farmers, and they do not get credit from them because of high interest rates and lack

Table 5 Access to, control over and entitlement to institutional services among household members in sorghum production

Resources and gender MHH(n)=94 FHH(n)=58	Millet								
	Access %			Control %			Entitlement %		
	MHH	FHH	*P*-value	MHH	FHH	*P*-value	MHH	FHH	*P*-value
Credit									
Adult male	2.3	0.0	0.035	9.3	2.8	0.009	10.5	2.8	0.004
Adult female	1.2	6.9		1.2	8.3		1.2	8.3	
Both	15.1	5.6		5.8	0.0		7.0	0.0	
None	81.4	87.5		83.7	88.9		81.4	88.9	
Extension									
Adult male	25.6	14.1	0.000	41.9	23.6	0.000			
Adult female	7.0	43.7		4.7	43.1				
Both	30.2	18.1	0.000	15.1	6.9				
None	37.2	23.9		38.4	26.4				
Transport									
Adult male	26.7	5.6	0.000	44.2	23.6	0.000	43.0	23.6	0.000
Adult female	1.2	20.8		1.2	29.2		1.2	31.9	
Both	26.7	38.9		9.3	12.2		11.6	13.9	
None	45.3	34.7		45.3	34.7		44.2	30.6	

Note *p*-values are based on the Chi statistic

of collateral. Secondly, sorghum and millet are considered 'women's crops' and any initiative that aims to provide farm inputs in support of the two crops is likely to target female farmers. Male participants narrated the following during FGDs when they were asked 'Why is it that it is mostly women who grow millet and sorghum?'.

> Hhhmmm…. You know… women are responsible for making sure that there is food in the house. Millet and sorghum are grown for household consumption- we do not sell these crops so it is their responsibility to grow food crops.
>
> (Male participant no. 2 in men's FDG group, Mankhokwe Section, Nyachirenda EPA)
>
> In a household men and women jointly decide that they will grow millet and sorghum in any given year. But women take a leading role in deciding which variety of sorghum and millet to grow- this is so because it is their "department" since they are the ones who mostly grow the crops in this area.
>
> (Male participant no. 3 in men's FDG group, Mankhokwe Section, Nyachirenda EPA)
>
> Yes… this is what happens. Women are responsible for growing sorghum and millet in the uplands while men go to the marshy land to grow crops like maize, beans, pumpkins, tomatoes…. we mainly grow these crops for sale.
>
> (Male participant no. 4 in men's FDG group, Mankhokwe Section, Nyachirenda EPA)

Findings showed that the gender of the household member with control over credit was significantly associated with the gender of the household head in sorghum (*p*-value = 0.009) and millet (*p*-value = 0.003) producing households. The gender of

Table 6 Access to, control over and entitlement to institutional services among household members in millet production

Resources and gender MHH(n)=94 FHH(n)=58	Millet								
	Access %			Control %			Entitlement %		
	MHH	FHH	*P*-value	MHH	FHH	*P*-value	MHH	FHH	*P*-value
Credit									
Adult male	2.8	1.3	0.026	11.3	1.3	0.003	14.1	1.3	0.002
Adult female	2.8	8.0		1.4	8.0		0.0	6.7	
Both	15.5	2.7		5.6	0.0		5.6	1.3	
None	78.9	88.0		81.7	90.7		80.3	90.7	
Extension									
Adult male	26.8	14.7	0.000	42.3	17.1	0.000			
Adult female	9.9	38.7		7.0	40.8				
Both	28.2	10.6		15.5	5.3				
None	35.2	36.0		35.2	36.8				
Transport									
Adult male	29.6	7.9	0.000	46.5	22.4	0.000	45.1	21.3	0.000
Adult female	2.8	18.4		4.2	28.9		4.2	30.7	
Both	26.8	34.2		7.0	10.5		9.9	13.3	
None	40.8	39.5		42.3	38.2		40.8	34.7	

Note *p*-values are based on the Chi statistic

the household member with entitlement to credit was significantly associated with the gender of the household head in sorghum (*p*-value = 0.004) and millet (*p*-value = 0.002) producing households. In this study, major control over and entitlement to credit was observed from adult male members as opposed to adult female members in MHH while in FHH, more adult female members exerted control over and entitlement to the same than their male counterparts. It is important to note that despite the fact that sorghum and millet are considered women's crop in the study area, and that more adult female members than adult male members accessed credit for the production of the crops, in MHHs, adult male members controlled and were entitled to the credit which was accessed by the adult females. This concurs with Ankrah et al. (2020) who reported that women's access to credit is blocked by traditional rules and norms, placing them in the background and granting men space to act as guarantors. As a result, in cases where women's credit application is approved, men tend to assert control over the obtained funds.

Access to, and demand driven extension services

Extension contact plays a critical role of availing information to farmers, which is expected to encourage adoption of technologies and hence enhance farm productivity (Danso-Abbeam et al. 2018). This study sought to find out if there were any disparities in access to and demand for (control) extension services among adult male and female members in MHHs and FHHs in the production of sorghum and millet. It was found that the gender of the household member with access to extension services was significantly associated with the gender of the household head in sorghum (*p*-value = 0.000) and millet (*p*-value = 0.000) producing households. The study learnt that more adult male members (sorghum 26%, millet 27%) had access to extension services than adult female members (sorghum 7%, millet 10%) in MHHs while more adult female members (sorghum 44%, millet 39%) had access to extension services than adult male members (sorghum 14%, millet 15%) in FHHs. This demonstrates unequal access to extension services between the genders in both household types. Considering that sorghum and millet are generally grown by women in the study area, lack of or inadequate access to extension services by adult females in MHHs may result in poor farming practices affecting sorghum and millet yields. This is turn may affect the income and well-being of the households. With regard to demand for extension services, the study found that there was a significant association between the gender of the household member who demanded extension services and the gender of the household head in sorghum (*p*-value = 0.000) and millet (*p*-value = 0.001) producing households. Apart from the planned extension visits by the extension agents (typically once a month), more adult males (sorghum 42%, millet 42%) than adult females (sorghum 5%, millet 7%) in MHHs and more adult females (sorghum 40%, millet 43%) than adult males (sorghum 23.6%, millet 13%) in FHHs demanded extension services. KIIs revealed that generally, there were very few female extension agents working in the study area. During discussions, female participants expressed preference for agricultural extension agents of the same gender to deliver extension services. This is corroborated by Britwum and Akorsu (2016) who presented same sex preference in extension services. A married female participant in a women's group narrated that:

> Our district has very few female extension workers… a male extension worker visits our farms and homes…. Having a male extension worker frequently visit you, in your home, when your husband is present can be inconvenient and uncomfortable…. sometimes our husbands are suspicious for no reason at all although they (husbands) know that they (male extension workers) come to deliver extension services. We would have loved if the government sent us more female extension workers to work with us (female farmers).
>
> (Female participant no. 5 in women's FGD group, Thunyeni Section, Mikalango EPA)

Most unmarried female participants, on the other hand, did not mind about male extension workers visiting their houses to deliver extension services. One unmarried female participant indicated that:

> It does not matter whether a female or male extension worker visits me at home…what do I have to be afraid of? It is my house, anyone can visit me anytime… even if I want to visit the extension worker at his office, I am free to do so…
>
> (Female participant no. 8 in women's FGD group, Thunyeni Section, Mikalango EPA)

These FGD findings help to explain why more adult male members than adult female members in MHHs, and more adult females than male members in FHHs, accessed and demanded extension services. This may mean that the mere presence of a male extension worker discouraged adult females in MHHs from participating in extension activities targeting sorghum and millet production. This implies that men in MHHs received extension services on behalf of women who are responsible for producing the crops. This is unlike unmarried women who, in addition to accessing the extension services at home or farm, can even take the initiative to visit the extension agent's office to demand for additional extension services. Gender inequalities undervalue women's roles and contributions, often leading to the perception that men have the right to make decisions over women's work and well-being without considering their needs. This highlights how men's decisions can hinder women's access to resources and opportunities.

Apart from the explanations of married women participants above, married men participants mostly assigned the blame to women, citing reasons that women were less self-confident, or not intelligent enough or not serious enough—to justify why men sometimes received extension training on behalf of women. This finding is supported by Mudege et al. (2017) in their study of the role of gender norms in accessing agricultural training in sweet potato production in Malawi, who reported that in certain households, the decision of who could attend training was based on the perceived ability to understand it. The authors observed that husbands, worried about their wives' lower levels of literacy, were apprehensive of women's struggle to comprehend the training. Furthermore, some men attending agricultural training still regarded it as their duty and prerogative to learn on behalf of their wives, even though the sweet potato was clearly regarded as a woman's crop (ibid).

Norms that designate men as household heads were identified as potential barriers to women's access to extension services in MHHs. During discussions, some men suggested that women may choose not to participate in extension training because they prefer men to take the lead, given their status as household heads. One married male participant mentioned that it was his responsibility to attend trainings while his wife managed household chores and took care of children. Little and Panelli (2003) suggested that traditional meanings of femininity that confine women to caregiving and motherhood may actually constrain women's opportunities. In Malawi, the prevailing perception in communities and among extension agents is that men are household heads and therefore represent the family (Mudege et al 2017). In both developing and developed countries, the perception that men are the primary managers of farms and agricultural organizations still prevails (Brandth and Haugen 2010). This can perpetuate gender-based power imbalances and hinder the implementation of progressive government policies aimed at promoting full women's participation in agricultural production.

Access to, control over and entitlement to transportation

Transportation in rural areas often involves walking or using intermediate means of transport, as noted by the World Bank in 2010 and Tamene and Megento in 2017. The most available traditional or local means of transport, including animal (cattle) drawn carts, bicycles, motorcycles or on foot through either head-loading or back-loading, play an important role in filling the transport gap as conventional motorised transport services are poor in the study area. Public transport is particularly difficult due to poor road networks. This lack of infrastructure can have a negative impact on women, who face difficulty in accessing essential services such as markets.

The study sought to find out if there were any gender differences in access to, control over and entitlement to transport among adult household members in MHHs and FHHs. The gender of the household member with access to, control over and entitlement to transportation was significantly associated with the gender of the household head in sorghum and millet production at (p-value = 0.000). Findings showed that more adult male members (sorghum 27%, millet 30%) than adult female members (sorghum 1%, millet 3%) in MHHs had access to transportation while in FHHs more adult female members (sorghum 21%, millet 18%,) had access to transportation than their male counterparts (sorghum 6%, millet 8%). Control over transportation was very skewed towards adult male members (sorghum 44%, millet 47%) as opposed to adult female members (sorghum 1%, millet 4%) in MHHs while in FHHs, there were almost balanced proportions of adult females (sorghum 29%, 29% millet) and adult males (sorghum 24%, millet 22) controlling transport. The picture is similar with entitlement to transport, where more adult males (sorghum 43%, millet 45%) than their female counterparts (sorghum 1%, millet 4%) were entitled to transportation in MHHs. In FHHs, the differences in entitlement to transportation among adult females (sorghum 32%, millet 31%) and adult males (sorghum 24%, millet 21%) are minimal. Our results show that gendered differences in access to, control over and entitlement to transportation in sorghum and millet production are more pronounced in MHHs than in FHHs.

Discussions revealed that while most of sorghum and millet is consumed at a household level, some is kept for sale either at the farmgate or transported to the local or main district market. Women participants from both districts stated that the ownership and use of traditional or local means of transport is generally male dominated due to economic and/or socio-cultural factors. In Chikwawa women participants narrated several limitations that women face in accessing and using local transport means. These included the fact that women's income was limited, making them unable to buy local transport means such as bicycles, and that they could not access such means without the consent of their male counterparts. There was also the perception that women lack the physical strength to handle draught animals or push animal drawn carts. Most of the male participants stated that they owned bicycles, which they used to transport produce to the local market. Most female participants stated that they did not own bicycles, as such they either walked to the market carrying their load on head or paid bicycle owners to transport them to the local markets. Worst still, most women participants seemed to concur that their multiple reproductive roles tended to determine their mobility in terms of how far and how long they can travel from the home, while men constantly traveled to distant places using bicycles, cattle drawn carts or motorcycles. Even though most of the women participants did not have means of transport, they perceived that they had access to, control over and entitlement to through paying for the bicycles. It can be inferred that this is the premise for the finding in the household survey that adult female members in FHHs had access to, control over and entitlement to transportation.

4 Conclusion

The study established that in MHHs, there were more significant adult male than female members who had access to land, labour, seed, farm tools, manure, credit, extension services and transportation. In FHH, there were more adult female than male members who had access to the same resources. The study hence concludes that there were gender differences in favour of adult male members in MHHs in accessing land labour, seed, farm tools, manure, credit, extension services and transport in sorghum and millet production. The study further concludes that there were gender differences in favour of adult female members in FHHs in accessing land, labour, seed, farm tools, manure, credit and extension services and transportation in sorghum and millet production.

The study showed that in MHHs, more adult male than female members controlled and were entitled to land, labour, seed, farm tools, manure, credit, extension services and transportation. The study hence concludes that there were gender disparities in favour of adult male members in MHHs regarding control over and entitlement to land, labour, seed, farm tools, manure, credit, extension services and transportation in sorghum and millet production. The study revealed that in FHHs, more adult females than male members controlled and were entitled to seed, credit, extension services and transportation. It can therefore be concluded that there were gender differences in favour of adult female members in FHHs in the control over and entitlement to seed, credit, extension services, and transportation in sorghum and millet production. The study found that more adult males than female members controlled land in FHHs. The study hence determines there were gender differences in favour of adult male members in FHHs in the control of land for sorghum and millet production.

The study reported that adult male and female members in FHHs equally controlled and were entitled to labour, manure and farm tools. The study thus concludes that there was gender equality in the control over and entitlement to labour, manure and farm tools in FHHs in sorghum and millet production. The chapter thus advances that the pattern of access to, control over and entitlement to resources in sorghum and millet does not expedite equal outcomes for adult male and female members in MHHs and FHHs except for the control of and entitlement to labour, manure and farm tools in FHHs. Even though sorghum and millet production are the domain of women, their lack of control over natural and human resources, other key resources, and institutions place them in a disadvantaged position to equally benefit from sorghum and millet production. On the other hand, men appear to have more control due to cultural norms that position them over women. This therefore implies that males have greater chances to invest, expand and benefit from sorghum and millet as opposed to females. Agricultural development interventions promoting equitable access to institutional support services and productive resources in Sub Saharan Africa must be informed by intrahousehold dynamics affecting men and women to bridge gender gaps in context and achieve desirable outcomes.

References

Agarwal B (2011) Bargaining and gender relations: within and beyond the household. Feminist Econ 3(1):1–51. https://doi.org/10.1080/135457097338799

Ankrah D, Freeman Y, Afful A (2020) Gendered access to productive resources—evidence from smallholder farmers in Awutu Senya West District of Ghana. Sci Afr 10:e00604. https://doi.org/10.1016/j.sciaf.2020.e00604

Bhagavatula S, Rao PP, Basavaraj G, Nagaraj N (2013) Sorghum and millet economies in Asia "facts, trends and outlook". International Crops Research Institute for the Semi-Arid Tropics

Brandth B, Haugen MS (2010) Doing farm tourism: the intertwining practices of gender and work. J Woman Culture Soc 35(2):425–446. https://doi.org/10.1086/605480

Britwum AO, Akorsu AD (2016) Qualitative gender evaluation of agricultural intensification practices in northern Ghana. International Institute for Tropical Agriculture, Ibadan, Nigeria

Chant S, Brickell K (2013) Domesticating (and de-patriarchalising) the development agenda. New Front Feminist Polit Econ 12:85. https://doi.org/10.4324/9781315884745

Chaturvedi P, Govindaraj M, Govindan V, Weckwerth W (2022) Editorial: sorghum and pearl millet as climate resilient crops for food and nutrition security. Front Plant Sci 13:851970. https://doi.org/10.3389/fpls.2022.851970

Chigbu U, Paradza G, Dachaga W (2019) Differentiations in women's land tenure experiences: implications for women's land access and tenure security in sub-Saharan Africa. Land 8(2):1–22. https://doi.org/10.3390/land8020022

Chintu EM, Chigwe CFB, Obilana AB, Chirwa RW, Msiska FS (1996) Sorghum variety release in Malawi: the case of Pilira 1 and Pilira 2. In: Leuschner K, Manthe CS (eds) Drought tolerant crops for Southern Africa: Proceeding from SADC/ICRISAT/Regional sorghum and pearl millet workshop, 25–29 July 1994, Gaborone Botswana, Patancheru, Andhra Pradesh, India. International Crops Research Institute for the Semi-Arid Tropics, pp 19–25

Chirwa EW (2005) Adoption of fertilizer and hybrid seeds by smallholder maize farmers in southern Malawi. Develop S Afr 22:1–12. https://doi.org/10.1080/03768350500044065

Cochran WG (1977) Sampling techniques, 3rd edn. Wiley, New York

Conti MV, Campanaro A, Coccetti P, De Giuseppe R, Galimberti A, Labra M, Cena H (2019) Potential role of neglected and underutilized plant species in improving women's empowerment and nutrition in areas of Sub-Saharan Africa. Nut Rev 77(11):817–828. https://doi.org/10.1093/nutrit/nuz038

Danso-Abbeam G, Ehiakpor DS, Aidoo R (2018) Agricultural extension and its effects on farm productivity and income: insight from northern Ghana. Agric Food Secur 7:74. https://doi.org/10.1186/s40066-018-0225-x

Deere CD, Leon M (2003) The gender asset gap: land in Latin America. World Develop 31(6):925–947. https://doi.org/10.1016/S0305-750X(03)00046-9

Doss C, Kovarik C, Peterman A, Quisumbing A, Bold A (2015) Gender inequalities in ownership and control of land in Africa. Agric Econ 46(3):403–434. https://doi.org/10.1111/agec.12171

Elmhirst R (2015) Feminist political ecology. In: The Routledge handbook of gender and development. Routledge, pp 82–90

FAO (2011) State of food and agriculture. Women and agriculture: closing the gender gap for development. FAO, Rome

Freeman HA, Omiti JM (2003) Fertilizer use in semi-arid areas of Kenya: analysis of smallholder farmers' adoption behavior under liberalized markets. Nutr Cycl Agroecosyst 66(1):23–31. https://doi.org/10.1023/A:1023355011400

Freeman HA, Kaitibie S, Moyo S, Perry BD (2008) Livestock, livelihoods and vulnerability in Lesotho, Malawi and Zambia. In: Designing livestock interventions for emergency situations. ILRI Research Report 8. ILRI Nairobi, Kenya

Gebre GG, Isoda H, Rahut DB, Amekawa Y, Hisako Nomura H (2021) Gender differences in agricultural productivity: evidence from maize farm households in southern Ethiopia. GeoJournal 86(2):843–864. https://doi.org/10.1007/s10708-019-10098-y

Gengenbach H, Schurman RA, Bassett TJ, Munro WA, Moseley WG (2018) Limits of the new green revolution for Africa: reconceptualising gendered agricultural value chains. GeoJournal 184(2):208–214. https://doi.org/10.1111/geoj.12233
Giovarelli R (2006) Overcoming gender biases in established and transitional property rights systems. In: Quisumbing AR, Pandolfelli L (eds) (2010). Promising approaches to address the needs of poor female farmers: resources, constraints, and interventions. World Develop 38(4):581–592
Hadebe ST, Modi AT, Mabhaudhi T (2017) Drought tolerance and water use of cereal crops: a focus on sorghum as a food security crop in Sub-Saharan Africa. J Agron Crop Sci 203(3):177–191. https://doi.org/10.1111/jac.12191
Hargreaves S (2010) As mining steals the basis for survival, peasant women show a way out of the crisis. In: Women and land rights: questions of access, ownership and control. Perspectives political analysis and commentary from Africa Heinrich Böll Foundation Africa No. 213, pp 19–22.
Hausmann R, Tyson LD, Zahidi S (2010) The global gender gap report. World Economic Forum, Geneva. https://www.cig.gov.pt/siic/pdf/2014/Gender_GAP_report_2010.pdf
Holden ST, Otsuka K (2014) The roles of land tenure reforms and land markets in the context of population growth and land use intensification in Africa. Food Policy 48:88–97. https://doi.org/10.1016/j.foodpol.2014.03.005
Horrell S, Krishnan P (2007) Poverty and productivity in female-headed households in Zimbabwe. J Develop Stud 43(8):1351–1380. https://doi.org/10.1080/00220380701611477
Kabeer N (2003) Gender equality, poverty eradication and the millennium development goals: promoting women's capabilities and participation. Gender & Development Discussion Paper Series No. 13
Kang M, Schwab B, Yu J (2020) Gender differences in the relationship between land ownership and managerial rights: implications for intrahousehold farm labor allocation. World Develop 125:104669. https://doi.org/10.1016/j.worlddev.2019.104669
Kerr BR (2014) Lost and found crops: agrobiodiversity, indigenous knowledge, and a feminist political ecology of sorghum and finger millet in northern Malawi. Ann Assoc Am Geogr 104(3):577–593. https://doi.org/10.1080/00045608.2014.892346
Kevane M, Gray LC (2010) A woman's field is made at night: gendered land rights and norms in Burkina Faso. Feminist Econ 5(3):1–26. https://doi.org/10.1080/135457099337789
Khan MDR, Haque MI, Zeeshan NK, Kaushik I, Shree K (2021) Caste, land ownership and agricultural productivity in India: evidence from a large-scale survey of farm households. Develop Practice 31(4):421–431. https://doi.org/10.1080/09614524.2020.1853679
Kusakabe K (2003) Women's involvement in small-scale aquaculture in northeast Thailand. Develop Practice 13(4):333–345. https://doi.org/10.1080/0961452032000112392
Lambrecht IB (2016) "As a husband I will love, lead, and provide". Gendered access to land in Ghana. World Develop 88:188–200. https://doi.org/10.1016/j.worlddev.2016.07.018
Little J, Panelli R (2003) Gender research in rural geography. Gender Place Culture 10(3):281–289. https://doi.org/10.1080/0966369032000114046
Luna F (2018) Globalisation, gender and research. In: Globalising feminist bioethics. Routledge, pp 254–265
Malkanthi, SHP (2016) Gender contribution to cultivation and use of underutilised crops: case of Moneragala district in Sri Lanka. Int J Agric Res Govern Ecol 12(1):77–92. https://doi.org/10.1504/IJARGE.2016.074680.
Mayes S, Massawe FJ, Alderson PG, Roberts JA, Azam-Ali SN, Hermann M (2011) The potential for underutilized crops to improve security of food production. Exp Bot 63(3):1075–1079. https://doi.org/10.1093/jxb/err396
Meinzen-Dick R, Johnson N, Quisumbing AR, Njuki J, Behrman JA, Rubin D, Waithanji E (2014) The gender asset gap and its implications for agricultural and rural development. In: Quisumbing RA, Meinzen-Dick R, Raney LT, Croppenstedt A, Behrman AJ, Peterman A (eds) Gender in

agriculture: closing the knowledge gap. Springer, Dordrecht, pp 91–115. https://doi.org/10.1007/978-94-017-8616-4_5
Mishra K, Sam AB (2016) Does women's land ownership promote their empowerment? Empirical evidence from Nepal. World Develop 78:360–371. https://doi.org/10.1016/j.worlddev.2015.10.003
Modi AT, Mabhaudhi T (2013) Water use and drought tolerance of selected traditional and indigenous crops. Water Research Commission; Pretoria, South. Africa: 2013. Final Report of Water Research Commission Project K5/1771//4; WRC Report No. 1771/1/13
Motsi H, Molapo M, Phiri EM (2022) A review of the adaptive capacity of sweet sorghum to improve food security and poverty alleviation in Sub-Saharan Africa. S Afr J Bot 150:323–329. https://doi.org/10.1016/j.sajb.2022.07.040
Mudege NN, Mdege N, Abidin PE, Bhatasara S (2017) The role of gender norms in access to agricultural training in Chikwawa and Phalombe, Malawi. Gend Place Cult 24:1689–1710. https://doi.org/10.1080/0966369X.2017.1383363
Munkombwe G, Nthewa C, Mutaliano JA (2020) Strengthening seed delivery system for enhanced adoption of improved sorghum varieties among smallholder farmers in Malawi, Mozambique and Zambia. Earth Environ Sci 482 (1):012008. https://doi.org/10.1088/1755-1315/482/1/012008
Ndanga L, Kwamena Q, Dennis J (2013) Economically feasible options for increased women participation in Kenyan aquaculture value chain. Aquaculture 414–415:183–190. https://doi.org/10.1016/j.aquaculture.2013.08.012
Ngwira A, Madzonga O, Chunga P, Siyeni D, Chintu J, Simwaka P, Yohane E (2019) Guide for groundnut, pigeon pea, sorghum and finger millet production in Malawi. Available at https://wwwresearchgate.net/publication/349352279. Accessed 10 Jan 2024
Nkongolo KK, Chinthu K, Malusi M, Vokhiwa Z (2008) Participatory variety selection and characterization of sorghum (*Sorghum bicolor* (L.) Moench) elite accessions from Malawian gene pool using farmer and breeder knowledge. Afr J Agric Res 3:273–283. https://doi.org/10.5897/AJAR.9000527
Nyantakyi-Frimpong H (2019) Combining feminist political ecology and participatory diagramming to study climate information service delivery and knowledge flows among smallholder farmers in northern Ghana. Appl Geogr 112:102079. https://doi.org/10.1016/j.apgeog.2019.102079
Oladele OI, Monkhei M (2008) Gender ownership patterns of livestock in Botswana. Volume 20, Article #156. Retrieved 17 Jan 2024. http://www.lrrd.org/lrrd20/10/olad20156.htm
Oluka J, Owoyesigire B, Esenu B, Ssewannyana E (2005) Small stock and women in livestock production in the Teso farming system region of Uganda. In: Small stock in development: proceedings of a workshop on enhancing the contribution of small livestock to the livelihoods of resource-poor communities. Masaka: Hotel Brovad, 15–19 Nov 2004
Otieno G, Zebrowski WM, Recha J, Reynolds TW (2021) Gender and social seed networks for climate change adaptation: evidence from bean, finger millet, and sorghum seed systems in East Africa. Sustainability 13:2074. https://doi.org/10.3390/su13042074
Palacios-Lopez A, Christiaensen L, Kilic T (2017) How much of the labor in African agriculture is provided by women? Food Policy 67:52–63. https://doi.org/10.1016/j.foodpol.2016.09.017
Paul MM (2018) A Review on gender differences in access to and control over agricultural resources in farmer families. Sci Park Res J 1(4):2321–8045. https://doi.org/10.9780/23218045/1202013/49
Phiri AT, Toure HM, Kipkogei O, Traore R, Afokpe PM, Lamore AA (2022) A review of gender inclusivity in agriculture and natural resources management under the changing climate in sub-Saharan Africa. Cogent Soc Sci 8:2024674. https://doi.org/10.1080/23311886.2021.2024674
Rao N, Gazdar H, Chanchani D, Ibrahim M (2019) Women's agricultural work and nutrition in south Asia: from pathways to a cross-disciplinary, grounded analytical framework. Leveraging Agric Nutrition South Asia 82:50–62. https://doi.org/10.1016/j.foodpol.2018.10.014

Ribot, JC, Peluso, NL (2003) A theory of access. Rur Soc 68(2):153–181. https://doi.org/10.1111/j.1549-0831.2003.tb00133.x

Ritchie J, Spencer L (1994) Qualitative data analysis for applied policy research. In: Bryman A, Burgess RG (eds) Analysing qualitative data. Routledge, London, pp 173–194

Rocheleau D, Thomas-Slayter B, Wangari E (1996) Feminist political ecology: global issues and local experiences. CPC Press, London, UK, p 164. ISBN 0415120268

Sell M, Bäckman S, TettehAnang B, Niemi JK (2018) The unequal efficiency gap: key factors influencing women farmer's efficiency in Uganda. Cogent Food Agric 4(1):1–18. https://doi.org/10.1080/23311932.2018.1551750

Sen A (1990) Gender and cooperative conflicts. In: Tinker I (ed) Persistent inequalities: women and world development. Oxford University Press, Oxford, pp 12–149

Slavchevska V, De la O Campos AP, Brunelli C, Doss C (2017) Beyond ownership: tracking progress on women's land rights in Sub-Saharan Africa. In: Paper presented at the World Bank land and poverty conference, Washington DC, 20–24 Mar 2017

Smith RE (2008) Cattle's effect on land and labour productivity: evidence from Zambia. J Int Develop 20(7):905–919. https://doi.org/10.1002/jid.1467

Swetha MP, Muthukumar SP (2016) Characterization of nutrients, amino acids, poly-phenols and antioxidant activity of ridge gourd (*Luffa acutangula*) peel. J Food Sci Techn 53:3122–3128. https://doi.org/10.1007/s13197-016-2285-x

Takane T (2009) Disparities and diversities among female-headed households in rural Malawi after 20 years of economic liberalization. Singapore J Trop Geogr 30:358–372. https://doi.org/10.1111/j.1467-9493.2009.00379.x

Tamene S, Megento TL (2017) The effect of rural road transport infrastructure on smallholder farmers' agricultural productivity in Horro Guduru Wollega Zone, Western Ethiopia. AUC Geographica 52(1):89–99. https://doi.org/10.14712/23361980.2017

van Eerdewijk A, Wong F, Vaast C, Newton J, Tyszler M, Pennington A (2017) White paper: a conceptual model of women and girls' empowerment. Royal Tropical Institute, KIT, Amsterdam. https://www.researchgate.net/publication/315516870_White_Paper_a_Conceptual_Model_on_Women_and_Girls%27_Empowerment. Accessed 13 Jan 2024

Van Rensburg WJ, Van Averbeke W, Slabbert R, Faber M, Van Jaarsveld P, Van Heerden I, Wenhold F, Oelofse A (2007) African leafy vegetables in South Africa. J Water SA 33:3. https://doi.org/10.4314/wsa.v33i3.49110

Vanya CL (2012) The impact of climate variability and change on sorghum production in the lower shire valley in Malawi. M. Sc. Thesis, University of Nairobi, Nairobi

Whitehead A, Tsikata D (2003) Policy discourses on women's land rights in Sub-Saharan Africa: the implications of the return to the customary. J Agrarian Change 3(1–2):67–112. https://doi.org/10.1111/1471-0366.00051

World Bank (2009) Transport Week. Working session for transport task teams on gender mainstreaming, S Holste, Washington DC (Gender and the Rural Access Program in Afghanistan)

Yadav B, Sharma A (2017) Gender roles analysis of ornamental fish enterprises in Maharashtra State, India. Asian Fish Sci (Special issue) 30S:333–342. https://doi.org/10.33997/j.afs.2017.30.S1.020

Yngstrom I (2002) Women, wives and land rights in Africa: situating gender beyond the household in the debate over land policy and changing tenure systems. Oxford Develop Stud 30(1):21–40. https://doi.org/10.1080/136008101200114886

Zimba S, Dougill A, Chanza C, Boesch C, Kepinski S (2023) Gender differential in choices of crop variety traits and climate-smart cropping systems: insights from sorghum and millet farmers in drought-prone areas of Malawi. Plants People Planet 1–15. https://doi.org/10.1002/ppp3.10467

Open Access This chapter is licensed under the terms of the Creative Commons Attribution 4.0 International License (http://creativecommons.org/licenses/by/4.0/), which permits use, sharing, adaptation, distribution and reproduction in any medium or format, as long as you give appropriate credit to the original author(s) and the source, provide a link to the Creative Commons license and indicate if changes were made.

The images or other third party material in this chapter are included in the chapter's Creative Commons license, unless indicated otherwise in a credit line to the material. If material is not included in the chapter's Creative Commons license and your intended use is not permitted by statutory regulation or exceeds the permitted use, you will need to obtain permission directly from the copyright holder.

Can Farmer-Centred Approaches Deliver Impact at Scale? Evidence from Interventions Supporting Soil Health Innovation in Rural Malawi

Tchuwa Frank, Wellard Kate, Morton John, Kambewa Daimon, Mhango Wezi, Lamboll Richard, and Mkweu Daniso

Abstract Increasingly, agricultural research and development interventions are embracing farmer-centred approaches. The approaches are touted as effective in building farmers' capacity to address their farming needs and problems, and considered by many as a way to strengthening platforms for co-creation of knowledge and contributing to sustainable agri-food systems. But can these bottom-up approaches deliver their promises at scale in developing countries like Malawi? This chapter describes how four farmer-centred approaches, namely: Lead Farmer, Farmer Field School; Farmer Research Team; and Farmer Research Networks were operationalised to support soil health innovation in rural Malawi (Mzimba, Kasungu and Ntcheu districts). In these areas resource-poor farmers face food and nutrition security challenges that are associated with low crop productivity in the context of degrading soils and unpredictable rainfall pattern among others. The chapter also presents an analysis of how social network processes created in farmer-centred approaches influence the flow of information from farmers participating in soil health interventions to the broader farming community at scale. Evidence presented in this chapter reveals that whilst vertical networks link the participating farmers to diverse and novel ideas on options for managing soil health, the homophilic tendency in horizontal networks restricts the flow of this information to the broader farming community.

Keywords Participatory approaches · Social networks · Sustainable agriculture · Agri-food systems

T. Frank (✉) · K. Daimon · M. Wezi · M. Daniso
Bunda College, Lilongwe University of Agriculture and Natural Resources, Lilongwe, Malawi
e-mail: ftchuwa@luanar.ac.mw

W. Kate · M. John · L. Richard
Natural Resources Institute, University of Greenwich, London, UK

© The Author(s) 2026

B. R. Singh et al. (eds.), *Sustainable Food Systems in Sub-Saharan Africa*,
https://doi.org/10.1007/978-3-032-11288-0_20

1 Introduction

Strengthening networks among actors in agriculture and food systems is highlighted as one of the solutions to enhancing sustainable and inclusive agricultural development whilst feeding the growing global population with healthy and nutritious food (Food and Agriculture Organisation (FAO) 2021; Spielman et al. 2011). Proponents posit that the nature, quality and extent of linkages among diverse food system actors has a major influence on farmers' access to agricultural services and how the overall system functions. The actors, particularly smallholder farmers in developing countries, get linked to markets (Kaaria et al. 2009) as well as to innovation platforms where social learning and co-creation of knowledge occurs (Sanginga et al. 2009). They provide spaces where farmers collectively engage with agricultural scientists and extension workers in large-scale experimentation to generate knowledge and adapt innovations suitable to farmers' contexts (Nelson et al. 2016). Apart from the function of generating novel ideas, Hermans et al. (2012) and Spielman et al. (2011) stress the crucial role that the networks play in scaling innovations (i.e., the spread of innovations among farmers). The linkages essentially facilitate the flow of new knowledge and ideas from innovators to other stakeholders in the agriculture or food system.

Since the 1980s, the participatory paradigm has informed the debate and sometimes the design and implementation of agricultural research and development interventions. Early proponents of participatory approaches argued that the involvement of farmers would put more power in the hands of farmers and facilitate the generation of sustainable innovations that are suitable to resource-poor farmers and their diverse farming systems (Chambers and Jiggins 1987). However, Pretty (1995) observed different typologies of participation that ranged from those aiming at achieving efficiency by making farmers agree with and support new developments, services or innovations, to those that view engagement of farmers as a fundamental right. The latter looked at farmer engagement as instrumental to the initiation of mobilization for collective action, empowerment and institution building. The question is: what type of farmer-centred approaches are needed to create social networks that strengthen innovation in agriculture and food systems of resource-poor communities?

Studies have reported evidence on strengthened social linkages as a result of farmer-centred approaches (Scoones and Thompson 2009). For instance, social ties increased among members and between groups participating in Local Agricultural Research Committees in Latin America and Farmer Field Schools (FFS) in West Africa (Davis et al. 2012; David and Asamoah 2011) and Kenya (Bourne et al. 2021). The Lead Farmer concept in Nigeria and FFSs in Kenya, Tanzania and Uganda are reported to have helped create strong ties and trust among community members (Friis-Hansen and Duveskog 2012; Franke et al. 2006). Similarly, there was wide recognition of researcher farmers as innovation brokers of communities in Honduras (Classen et al. 2008). Braun et al. (2000) also reported on FFS members and communities as becoming linked to agricultural professionals from universities, researchers, extension workers and other development organisations in Asia.

The success stories associated with farmer participation has over the years prompted development practitioners to look to various farmer-centred approaches as instruments for disseminating information and technologies to the wider farming community, especially in contexts where the extension worker to farmer ratio is high (e.g., 1:3000 in Malawi) (GoM 2010; Khaila et al 2015). This focus on technology diffusion emanates from the belief that agricultural and food systems will improve if farmers access and adopt the disseminated technologies. However, the problems affecting agriculture and food systems are not only about technologies; they are also social in nature (Hall 2006). The emphasis on technical problems has obscured social processes that influence the generation/creation, sharing and use of the novel knowledge and innovations by farmers participating in farmer-centred approaches and the wider community.

The network analysis approach, which involves mapping actors (called nodes) in a system and the connections (called ties) of a specific actor to other actors in the system, has been recognised as instrumental to analysing and understanding social relations and interactions important for strengthening agriculture and food systems (Rocker et al 2022; Christensen and O'Sullivan 2015). The approach enables scholars to investigate social ties such as kinship (e.g., brother and sister relationship), cognitive (e.g., who does the actor know?), affective (e.g., who does the actor like or dislike?), interaction (e.g., to whom does the actor give advice?) and transaction (e.g., to whom does the actor sell items?). Recently, some studies have used the network approach to establish adoption of supply-driven technologies. For example, adoption of plant health recommendations in China and Zambia (Danielsen et al. 2020), common bean technologies in Tanzania (Kansiime et al. 2018), pit planting technology in Malawi (Beaman et al. 2021), improved groundnut varieties in Kenya and Uganda (Thuo et al. 2013), as well as plantain technologies and agroforestry practices in Ghana (Weyori et al. 2018; Isaac 2012).

This chapter applies the network approach to address the question of how vertical and horizontal social networks created in interventions that apply farmer-centred approaches facilitate the generation and flow of information for improving farm soils in agrarian communities in Malawi. The vertical networks involve interactions among actors operating at different levels and are often characterised by unequal power (e.g., agricultural research and extension workers from national and district offices interacting with farmers in the community). The horizontal networks involve the interactions among actors operating at the same level and tend to have more equal power (e.g., farmers interacting with fellow farmers within the community) (Rose 1999; Putnam 1993). In theory, whilst vertical networks (bridging ties) and horizontal networks (bonding ties) could be instrumental in linking farmers to novel information and innovation (Bourne et al 2021), processes related to culture, gender and socio-economic status could constrain the flow of information among farmers in the broader farming community. These barriers in the networks could exclude marginalised groups, including women, from accessing vital information and innovations for addressing challenges such as degrading farm soils.

Land management and good soil health are critical components for transforming agriculture and food systems towards sustainability and resilience (Critchley et al.

2023; HLPE, 2019). Moreso, the need for healthy soils is paramount for agrarian communities in countries like Malawi that are facing frequent negative impacts of climate change including prolonged dry spells and floods caused by cyclones (e.g., Ana and Gombe in 2022 and Freddy in 2023) leading to accelerated land degradation and disruption of agricultural and food production systems. The term soil health in this chapter refers to the continued capacity of soil to function as a living system, within ecosystem and land-use boundaries, to sustain biological productivity, maintain the quality of air and water environments, and promote plant, animal and human health (Stewart et al. 2018). Soils are not healthy when the interaction of chemical (e.g., soil nutrients), physical (e.g., soil texture and structure) and biological (e.g., plants and animals) properties is unbalanced.

Alongside concerns about soil health, soil loss in Malawi is estimated at 29 ton/ha/year (GoM 2015). The leading causes of soil loss in the country include cultivation of fragile soils on steep slopes because of population increase and pro-maize policies, markets and institutions that have pushed farmers into practicing continuous maize monoculture (World Bank Group 2019). The loss of soils has enormous effects on crop productivity, especially maize, a dominant staple crop that is allocated to over 70% of arable land. For instance, Asfaw et al. (2018) reported 32–61% losses in maize productivity in scenarios where the rate of soil loss reached 22 ton/ha/year, and hence accelerating the loss of soil nutrients, particularly nitrogen (N) and potassium (K). Moreover, the study revealed that the negative impacts of soil loss on maize productivity mainly affected marginalised groups such as the female headed farm households.

This chapter describes different farmer-centred approaches that have been used to address soil health and loss issues in Malawi, particularly those approaches that are implemented in administratively and agroecologically demarcated Extension Planning Areas (EPAs), namely Kandeu (Ntcheu district), Mkanakhoti (Kasungu district) and Zombwe (Mzimba district). The chapter also delves into the social processes that lead to creation of vertical networks in farmer-centred approaches, and how horizontal networks may support or hinder sharing of soil heath information to farmers in the broader community.

2 Farmer-Centred Approaches for Healthy Soils and Resilient Food Systems in Malawi

Various approaches are employed to engage farming communities in implementing soil health improving interventions. They include lead farmers, famer field schools and farmer research networks. Some of them prioritise modernization of local agricultural systems through promotion of technologies and soil amendments tested and approved by formal research organizations. Others, advocate for the involvement of farmers in creating solutions for improving the health of farm soils sustainably. In the case of Malawi, the approaches used to work with farmers in agricultural

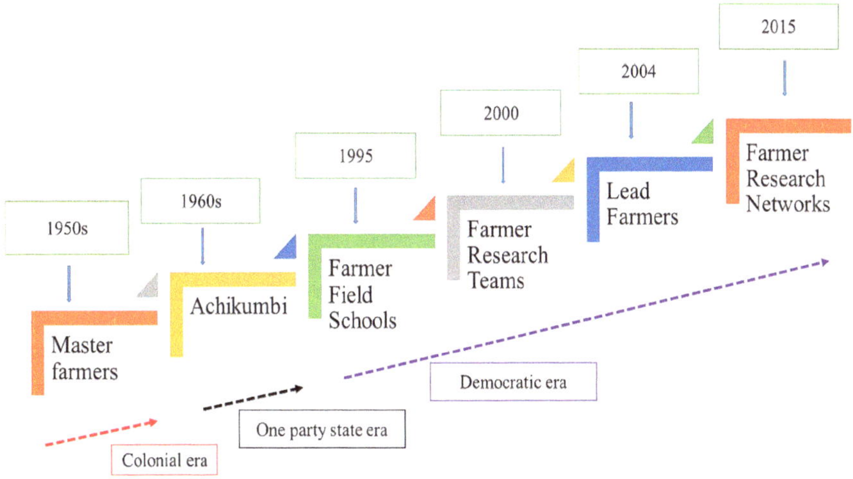

Fig. 1 Introduction of selected farmer-based research and extension approaches in Malawi. *Source* Authors' own construction

activities have evolved from top-down to participatory approaches. Figure 1 shows the different approaches for working with farmers introduced in Malawi (before and after independence in 1964). The notable ones include the master farmers in the colonial era, progressive farmers during the one-party state, and the lead farmers, FFS, Farmer Research Team (FRT) and Farmer Research Network (FRN).

Generally, the approaches were introduced and applied to facilitate diffusion of agricultural information and technologies and or support local institutions in innovation processes. For example, the lead farmers continue to be useful in spreading soil health information and technologies to the broader farming communities through on-farm demonstrations (Khaila et al. 2015). These fields managed by lead farmers are used to demonstrate the superiority of technologies recommended by "expert researchers and extension workers" from mainstream agricultural organisations over those practised by farmers. The approach gives limited space to farmers to make changes to "approved" technologies promoted by external agents, and has no space for farmers to experiment with and develop local innovations. In contrast, there are cases when approaches such as the FFS, FRT and FRN are employed to support farmer-led experimentation whose goal is to identify, test and adopt soil health innovations that suit local contexts. This section describes the farmer-centred approaches applied to support soil health interventions, with a focus on those that were introduced during the democratic[1] era. The strengths and weaknesses of each approach are also discussed in the section.

[1] The agriculture sector in Malawi has transitioned through three political eras. The first is the pre-independence era when the country was known as Nyasaland and its agriculture sector managed by the British administration (1891–1964). The second era starts in 1964 when the country gained its independence. During this era, a one-party political system was adopted, and all agricultural

a. **Lead farmers**

A lead farmer is an individual expert farmer who delivers extension services to other farmers in a community (10 to 30 follower farmers) (GoM 2010). This approach of using model farmers in the diffusion of innovations evolved from the early approaches enforced by pre- and post-independence administrations in the country. The "master farmer approach", for instance, was implemented during the colonial administration in the 1950s. It was succeeded by the post-colonial "progressive farmers" locally known as "*Achikumbi*", implemented during the one-party government led by Dr. Hastings Kamuzu Banda of the Malawi Congress Party. The master and progressive farmer approaches were both top-down, as farmers were expected to practise agricultural recommendations dictated by the Government. During the colonial administration, for example, failure to apply recommended land resources conservation practices was punishable by law. Likewise, the one-party state under Dr Banda rewarded only the "*Achikumbi*[2]" who were practising government-recommended technologies and farm practices. The winning *Achikumbi* benefited from frequent visits from field extension agents, and were also given modern farm inputs (e.g., hybrid seed and inorganic fertilizers) and tools (e.g., hand hoes) (Chowa et al. 2013).

The use of model farmers resurfaced in the present multi-party and democratic government system mainly due to the shortage of trained field extension agents in rural areas. In 2004, the "Lead Farmer Project" implemented by Self-Help Africa introduced the Lead Farmer approach in northern Malawi. Since then, there has been a rapid increase in the number of interventions using lead farmers in the country. Unlike master and progressive farmers who were supervised by government officials (during the colonial and the one-party state administrations), lead farmers are promoted and receive support from both non-state actors including NGOs and the Government. Lead farmers can be found in all the districts of the country (Ragasa, 2020; Wellard et al. 2013).

A case study on lead farmers in Kandeu EPA of central Malawi established that the main reason for applying the lead farmer model was to cover for the shortage of field extension staff. For instance, in the 2015/16 farming season, the EPA had only four extension workers serving 24,385 farm families. Thus, lead farmers were given the responsibility of training fellow farmers (mainly through demonstrations) on recommended farming practices and technologies (including soil fertility improving technologies) that are recognised and promoted by the Ministry of Agriculture and NGOs. However, the key informants who were implementing the approach in the area reported several challenges including costs of training and supervising lead

institutions were nationalised. In the third era, Malawi embraced democracy and constituted a multi-party-political system (1994–present). This time pluralism in the agriculture sector was adopted, with non-state actors allowed to deliver services to farmers.

[2] Achikumbi were the best farmers selected in a competition supported by the Government. Selection of the farmers was done by government officials based on criteria that focused on recommended technologies and practices promoted by the Ministry of Agriculture. However, no matter how well the farmer performed, the position of *mchikumbi* number 1 (first position) was reserved for the Head of State (Dr Hastings Kamuzu Banda). Farmers would compete only for *mchikumbi* number 2 and below.

farmers, as well as inputs for setting up demonstrations. Also, the selection criteria that emphasised literacy, communication and personal skills and ownership of land favoured rural elites who were often resented by fellow farmers, especially when they were perceived to be the ones benefiting from the resources provided by the external organisations. By focusing on "recommended practices and technologies" the approach failed to tap into existing local knowledge.

b. **Farmer Field Schools**

A Farmer Field School is a group of farmers (15–30 farmers) who get together to collectively explore a particular topic or issue. The group is led by a community-based facilitator who is trained in facilitation and technical agricultural skills. The facilitator is selected from the community and is equivalent to a model/lead farmer. The approach provides farmers with a learning opportunity through which they acquire agricultural and management skills that make them experts on their farms. It draws principles from non-formal education as well experiential, conscientisation and emancipatory learning. As such, farmers participating in the FFS drive the learning process while working with scientists to co-produce knowledge and solutions suitable to their local context (FAO 2016). The approach was originally conceived in the late 1980s to support a continued process of learning, action and empowerment by its participants (van den Berg et al. 2020).

In Malawi, the FFS approach was piloted in Mzuzu, Lilongwe, Salima and Machinga Agricultural Development Divisions (ADDs) by FAO in the mid to late 1990s (around 1995–1999). The early FFSs in the country focused on promoting Integrated Pest Management (IPM), particularly in the field of horticulture. By the year 2007, there were different agricultural institutions using the approach in the country. Some of them included research institutions (Lunyangwa Agriculture Research Station) and NGOs (Concern Universal, World Vision Malawi and Care Malawi). Also, the private sector (Smallholder Coffee Farmers Trust) adopted the approach. Recently, the approach has been introduced in all the districts, following the implementation of the KULIMA programme (2018–2022) with support from the European Union.

Discussions with extension staff operating at community, district and national levels revealed different views about the purpose for implementing the FFS approach in Malawi. For example, the field level extension workers in Mkanakhoti EPA of Kasungu district considered the approach as an improved method of training and visiting farmers (from field demonstration plots to school plots). Unlike demonstration plots, school plots provide the opportunity for farmers to have an experience with the technology and, therefore, help them to make informed decisions on whether to adopt or not adopt the technology. The district and national extension experts considered the FFS as addressing the low extension worker–farmer ratio and dwindling quality of extension services delivered through the lead farmers, as well as allowing farmers to test and adapt technologies to suit their environment. Issues of gender, access to credit (village saving and loans) and climate change were also integrated in the FFS activities. However, field officers noted challenges related to costs of supervising the FFSs and inputs for setting up school plots. The limited number of members in the FFS groups (30) was viewed as a barrier to inclusion of several other farmers, including women and the youths, who expressed interest to participate. The approach also demanded technical support in designing the school plots where technologies would be tested and validated systematically. In many cases, the required technical support, especially research services, was lacking.

c. **Farmer Research Teams**

A Farmer Research Team is a group of farmers selected by the community to conduct research (Bezner-Kerr and Chirwa 2004). FRTs were introduced in Malawi in 2000 through the Soil Food and Health Community (SFHC) project implemented in Mzimba District in northern Malawi. Later, they were introduced in Dedza district in Central Malawi. The FRT approach aims at supporting a farmer-managed agricultural research service that is accountable to the community. The FRT creates a platform where a selected team of farmers takes charge of the agricultural research process that benefits both themselves and their community. Each FRT member works with 20 farmers (project participants) who also test one to two technologies learned from the FRT member. The farmers drive the process of evaluating and adapting soil fertility technologies, which mainly involves conducting field experiments to compare the established practices with the unproven and locally known practices. The findings from the experimentation are reported to the community where suitable recommendations are identified and disseminated to the broader farming community.

According to the SFHC project staff in Zombwe EPA in Mzimba district, the FRT was introduced to allow farmers to evaluate the soil fertility technologies in their farms. The approach allows farmers to adapt the technologies to their situation. FRT members collect information related to strengths and weaknesses of the tested soil fertility technologies and then provide feedback to the project office, as well as disseminate technologies from experimental plots to other farmers in the villages. Gender and good nutrition practices were also key areas addressed in the FRT approach. However, sustainability of the approach is constrained by input costs (e.g., seed) required for setting up the trial plots for many FRT members, as well as conflicts and resentment towards the FRT members who are perceived by community members as the only ones benefiting from inputs and training provided by external organisations. For example, a feeling of resentment emerges among community members when only a few FRT members are repeatedly selected for exchange visits to other districts.

d. **Farmer Research Networks**

A Farmer Research Network is as an association of farmer groups working together with research and development organisations to facilitate sharing of information and data as well as access to technical, institutional and financial support. The members of the network collaborate to conduct high-quality and credible research on mutually agreed topics to address agreed-on constraints collectively. Farmers engage as equal partners with scientists, development agents and other relevant stakeholders in generating, adapting and scaling out innovations (Nelson and Coe 2014). Unlike the approaches that engage farmers in testing, evaluating and disseminating (to other farmers in the community) technologies generated elsewhere, the FRN provides farmers space to create knowledge for improving soil health. The approach is guided by three key principles (1) engaging farmers that represent the social and biophysical diversity of their communities participate in the whole research process, (2) ensuring that research is rigorous, democratic, and focused on understanding biophysical and social variations, and (3) strengthening networks that foster collaboration and opportunities for learning and knowledge sharing (Richardson et al, 2021).

FRNs were introduced in central (Ntcheu and Kasungu Districts) and northern Malawi (Mzimba District) by the Lilongwe University of Agriculture and Natural Resources (LUANAR). According to field extension workers and researchers from the university, the FRN was introduced to address the shortfalls of researcher-led soil management solutions. The researcher-led soil fertility technologies promoted in the areas often failed to match the complex farming systems of farmers. A case in point was a decade (2005–2015) of promoting maize-legume intercropping and rotation technologies in Mkanakhoti EPA, Kasungu district. There was limited integration of legumes (especially pigeon peas) into smallholder farmers' maize-based farming system because of a lack of quality seed, pest damage and livestock feeding on the crops in the field. The FRN approach provided a network that linked individual farmers and farmer groups (10–20 members) at the local level to each other, as well as to agricultural organisations operating in the area and at district, national and international levels. Such a network would facilitate engagement of farmers in experimentation (informed by indigenous and scientific knowledge) to other farmers in the community, with the aim of identifying, testing, adapting and scaling suitable soil health innovations within their farming community, including those farmers from marginalised groups (women, children and the disabled).

Researchers involved in the approach observed several issues related to its resource and knowledge intensive characteristics, which may in the long run compromise the sustainability of soil health-improving activities. First, there is the issue of costs of inputs and supervision of plots managed by individual farmers. Secondly, the approach requires sufficient technical support when designing the farmer-led trials to ensure robustness of the data and innovations generated and tested with farmers. Thirdly, the individual farmer researchers recruited in the approach may not all have time for collecting data or may not be in position to allocate land to trials in addition to other planned farm enterprises. Finally, there is a risk of rural elite capture and exclusion of farmers who do not have resources, knowledge and experience in systematic research processes.

Table 1 Provides a summary of the main differences between the four approaches presented in this section.

Table 1 Differences between the four farmer-centred approaches in Malawi

Feature	LF	FFS	FRT	FRN
Organization at community level	1 model farmer leading a group of 10–30 follower farmers	1 community-based facilitator leading a group of 15–30 farmers	1 farmer researcher leading a group of 20 farmers	A group of 10–20 farmer researchers (no one leads)
Type of learning plots	Demonstration plot to showcase recommended technologies and practices	Demonstration/ trial plot to test and validate technologies and practices	Trial plot to test and validate technologies and practices	Trial plot to test and validate technologies and practices
Location of learning plots	1 plot in the community used to train follower farmers about a new recommended technology and practice	1 plot in the community used as a classroom where farmers observe and reflect on performance of a technology and practice	1 Mother plot in the community (with 3–5 technologies) and 20 baby plots spread in the community (with 1–2 technologies from mother plot)	10–20 plots randomly spread in the community (all testing 3–6 similar technologies plus 1 innovation suggested by a farmer)
Potential for integrating local knowledge	Low (emphasis on showcasing approved technologies)	Moderate (facilitator tend to prefer validation of recommended technologies)	High (farmer researchers allowed to collect data on local solutions)	High (farmer researchers allowed to collect data on local solutions including their innovations)

Source Authors' own construction from field data

3 How Vertical Networks Help Farmers Access Soil Health Information

Bourne et al. (2021) observed that farmer-centred approaches such as the FFS create bridging ties (vertical networks) between farmers and actors from outside the community. The vertical networks enable farmers to acquire novel ideas about how to improve soil health, which is important for achieving sustainable food systems. In the case of Ntcheu, Kasungu and Mzimba districts in Malawi, individuals participating in farmer-centred approaches accessed information about different soil health options by interacting with various stakeholders. Farmers, researchers and extension workers shared information on integration of soil management practices such as maize-legume intercropping, crop rotation, crop residue incorporation, manure application and agroforestry, whilst minimising the application of inorganic soil fertility inputs. These soil health management options are aligned to some of the agroecological elements, namely diversity (different crop species), synergies (intercropping and rotation), recycling (utilizing crop residues and animal wastes to improve soil quality) and efficiency (reducing use of external inputs such as inorganic fertilizers) (Barrios et al. 2020).

Data on actor interactions provided by field officers implementing the approaches (key informant interviews, n = 10) and farmers participating in the approaches (4 focus group discussions, n = 80) revealed that the vertical networks were created during village meetings, field supervision, training, meetings and field days conducted within on-farm demonstration plots, as well as during data collection and sharing of results obtained from on-farm experimental plots. These spaces allowed farmers participating in the approaches to share local knowledge and acquire soil management ideas from influential extension workers and researchers who represented government, NGOs, universities, private companies and development partners. The network data was further processed in the Netdraw software to visualize and determine the degree of centrality, which shows the number of connections that an actor has in a network (Borgatti 2002). Figures 2, 3, 4 and 5 show the results on the linkage of farmers to external actors supporting soil health management. In the figures one can observe that various actors of different status were interacting through the vertical networks. A focus on farmers (red nodes) shows that except for FRT, lead farmers and members of FFSs and FRN had a high degree of centrality, meaning that they interacted with different external actors. The interactions with various stakeholders increase the likelihood of farmers accessing diverse and novel ideas for improving soil health.

Follow up discussions with line managers of extension services at district and national level, with regards to the relevance of farmers interacting with external actors revealed that despite the existing vertical networks helping farmers to access novel information for improving soil health, they tend to perpetuate the transfer of technology model where farmers are seen as receivers of wisdom from researchers and extension workers. Farmers also tend to look at these interaction spaces as opportunities for accessing handouts in form of farm inputs and equipment. Moreover, the vertical networks are mainly accessible to a few elite farmers who are selected to participate in soil health interventions. Often, these farmers reside in easy-to-reach communities with good roads, have resources such as land and are more educated than the average farmer in the broader community. A point in case is the selection criteria for lead farmers that emphasise abilities to read and write, as well as being early adopters of technologies promoted by extension workers. Perhaps, such shortfalls of the vertical networks necessitate the creation of horizontal networks, which involve interaction and sharing of soil health information among farmers with equal status and hence more likely to enhance sharing of soil health information in the broader community.

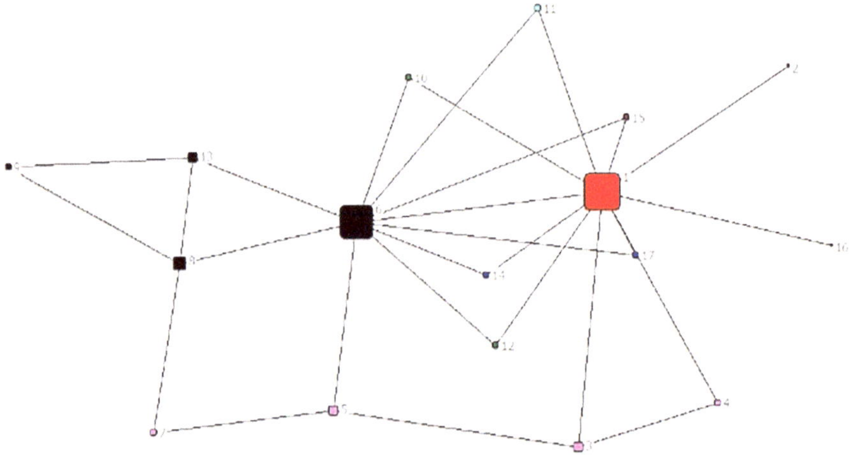

Fig. 2 Vertical networks for sharing soil health information in the Lead Farmer approach. *Source* Authors' own construction from field data. *Notes* Size of the node represent the degree centrality measure or the number of connections that each actor has in the network. The bigger the node, the more connected the actor is. *Key for colours* Red = Farmers, Pink = Local institutions, Black = Public extension, Green = Non-Governmental Organisation (NGO), Dark blue = University, Light blue = Research organisation, Yellow = Development partners, Brown = Private companies. *Actors in the Lead Farmer* 1 = Lead farmer, 2 = Follower farmers, 3 = Village Agricultural Committee (VAC)/Village Development Committee (VDC), 4 = Village head, 5 = Area Development Committee (ADC), 6 = Field officers at Extension Planning Area (EPA), 7 = District Stakeholder Panel (DSP), 8 = District Agricultural Development Office (DADO), 9 = Agricultural Development Division (ADD), 10 = National Smallholder Farmers Association of Malawi (NASFAM), 11 = Africa RISING, 12 = Catholic Development Commission in Malawi (CADECOM), 13 = Department of Agricultural Extension Services (DAES), 14 = Lilongwe University of Agriculture and Natural Resources (LUANAR), 15 = Seed co., 16 = Parnnar seed, 17 = Michigan State University

4 Social Processes and Soil Health Information Flow in Horizontal Networks

The horizontal networks involve the sharing of information about different soil health options between the farmers participating in different farmer-centred approaches and non-participants in the broader community. These farmer-to-farmer knowledge exchanges occur in different spaces, but the discussions in the fields and at home are popular spaces where knowledge is shared. The study conducted in Ntcheu, Kasungu and Mzimba districts revealed that knowledge exchange was common among community members who interacted in farmer-managed demonstration and experimental fields designed to showcase, validate or test soil health management options, often mounted in fields around dwelling compounds (houses) and along the road (Fig. 6). In these spaces, community members easily interacted with farmers managing soil health related demonstration and experimental fields mounted in different sites across the community (e.g., individual FRN members managed 20

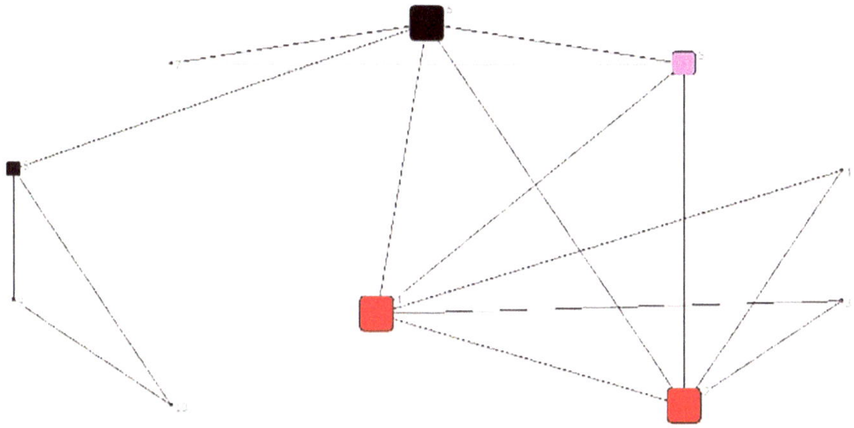

Fig. 3 Vertical networks for sharing soil health information in FFS approach. *Source* Authors' own construction from field data. *Actors in the FFS* 1 = Facilitator, 2 = FFS members, 3 = Field committee, 4 = Village Savings and loans (VSL) committee, 5 = Village head, 6 = Field officers at Extension Planning Area, 7 = Group village head, 8 = District Agricultural Development Office, 9 = Agricultural Development Division, 10 = Department of Agricultural Extension Services

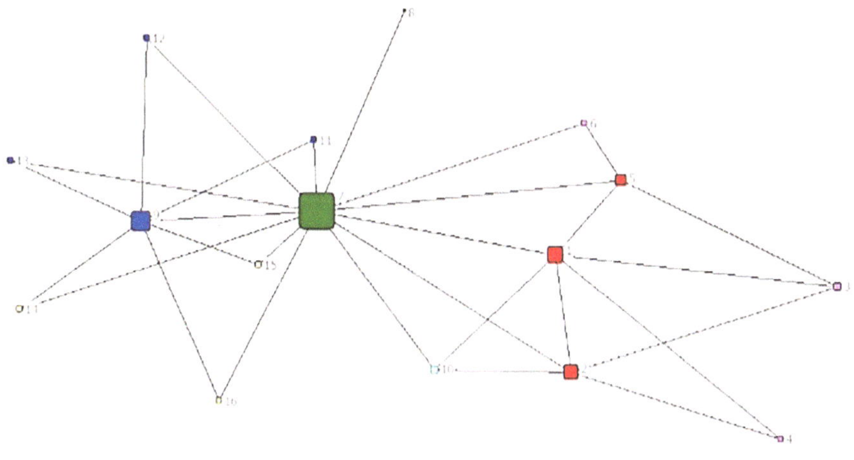

Fig. 4 Vertical networks for sharing soil health information in the FRT approach. *Source* Authors' own construction from field data. *Actors in the FRT* 1 = FRT member, 2 = Follower farmers, 3 = Village committee, 4 = Village head, 5 = Promoters, 6 = Area committee, 7 = Soil Food and Health Communities (SFHC), 8 = Field officers at Extension Planning, 9 = Chancellor College of the University of Malawi, 10 = International Crops Research Institute for the Semi-Arid Tropics (ICRISAT), 11 = Western University, 12 = University of Manitoba, 13 = Cornell University, 14 = Global Affairs Canada, 15 = Presbyterian world services for development, 16 = Canadian Food grains Bank

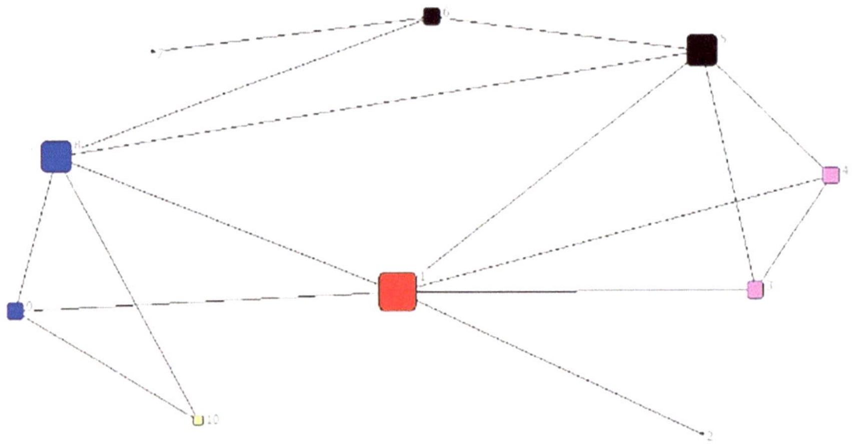

Fig. 5 Vertical networks for sharing soil health information in the FRN approach. *Source* Authors' own construction from field data. *Actors in the FRN* 1 = Farmer Research Group (FRG), 2 = Group committee, 3 = VAC/VDC, 4 = Village head, 5 = Field officers at Extension Planning Area, 6 = DADO, 7 = ADD, 8 = LUANAR, 9 = Natural Resources Institute (NRI) of the University of Greenwich, 10 = McKnight foundation

experimental plots of about 800 m^2 size that were spread within a community). On the contrary, non-participating farmers in the broader community had limited access to a single FFS plot that was established to support learning for only 20 members.

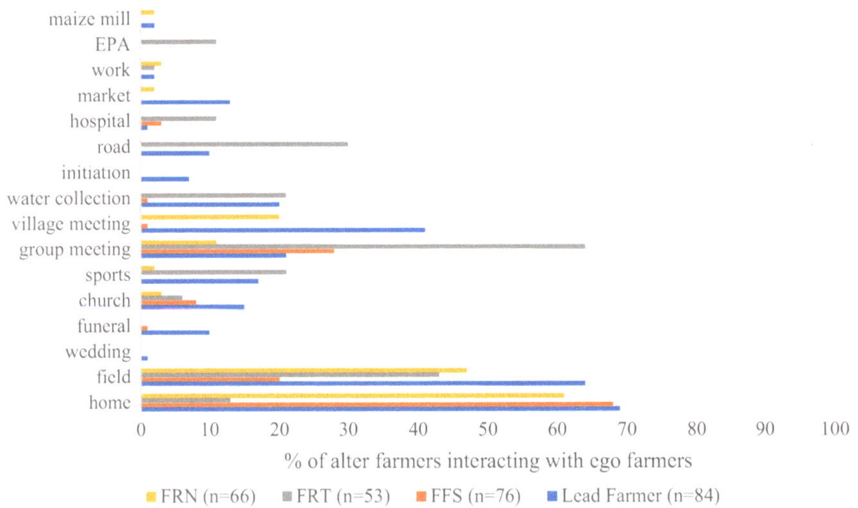

Fig. 6 Spaces where horizontal networks were used to share soil health information to community members (%). *Source* Authors' own construction from field data

The horizontal networks created in farmer-centred approaches are often shaped by the homophilic behaviour (Kadushin 2012) where farmers choose to share soil health information with individuals; they are socially tied to each other by kinship, friendship or neighbourhood. Whilst the homophilic tendency is considered as essential to strengthening bonds among actors with similar characteristics, it restricts the flow of ideas into and out of a network (Granovetter 2005). These restrictions have implications for scaling out soil health options to farmers in the broader community. The findings from the Malawi ego network study that investigated flow of information between 197 randomly selected farmers participating in farmer-centred approaches (i.e., egos) and 282 non-participants (i.e., alters) suggested that while the homophily facilitated the flow of soil health information among farmers with close ties, it restricted the sharing of information to farmers of similar gender and living in the same villages.

The manifestation of homophilic behaviour in horizontal networks was influenced by socio-cultural contexts in which the farmer-centred approaches were implemented. For example, Fig. 7 shows that female farmers in all the approaches demonstrated a strong homophilic tendency compared to their male counterparts. The homophilic behaviour among the females may be explained by the patriarchal system that facilitates the construction and reinforcement of norms responsible for hindering female farmers from interacting with male farmers. Studies in Malawi have highlighted the norms that discourage adult females from interacting with adult males. In addition, female farmers receive less recognition from male farmers as being credible sources of information (Nyantakyi-Frimpong 2017; Lambrecht 2016).

Figure 8 reveal homophilic behaviour among female farmers in Ntcheu district (Kandeu EPA) where they mainly shared soil health information to fellow females within their villages. This strong homophilic tendency may be attributed to the uxorilocal system practiced in Central and Southern Regions of Malawi. In uxorilocal system married females live in their natal villages and hence have limited ties to individuals from other villages but strong ties with relatives and friends within their villages. On the contrary, the virilocal system, where a woman leaves her villages to live with her husband, practiced in parts of the Central region (Mkanakhoti EPA) and Northen region (Zombwe EPA) influenced female farmers to share soil health information to other villages. In the virilocal system, many of the married female farmers migrate to their husbands' villages but continue to interact with relatives and friends from their original villages. The importance of lineage systems to rural development processes in Malawi have been acknowledged in other studies (for example, see Djurfeldt et al. 2018; Berge et al. 2014).

5 Conclusions

This chapter has discussed the application of farmer-centred approaches to address soil health problems, which have significant implications for achieving sustainable food systems in developing countries like Malawi. The results presented in the chapter

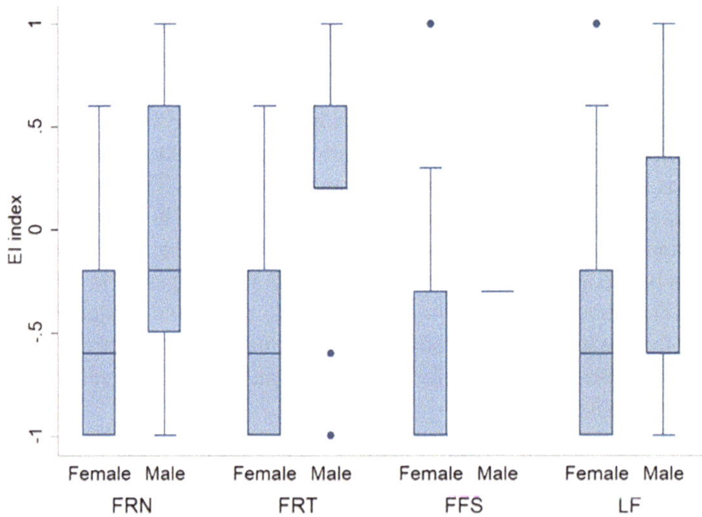

Fig. 7 Which gender do farmers share soil health information to? n: FRN = 50, FRT = 33, FFS = 53, LF = 53: U test p-value for comparing female and males: FRN = 0.0002***: FRT = 0.0008***, FFS = 0.3853, LF = 0.2720. *Source* Authors' own construction from field data. *Notes* The External-Internal (EI) index measures the similarity of individuals sharing information in a social network. It ranges from -1 (perfect homophily) to + 1 (perfect heterophily). A perfect homophily implies that farmers shared soil health information to only those farmers who are members of their gender group or village. A perfect heterophily means that they shared information to only those who belonged to a different gender group or village (Crossley et al 2015). The formula for calculating the EI index is expressed as: $EI = \frac{E-I}{E+I}$, where E is the external ties (e.g., the number of ties of a female ego farmer to alters from a different gender group (male farmers). I is the internal ties (e.g., the number of ties of a female ego farmer to alters from the same gender group (female farmers)

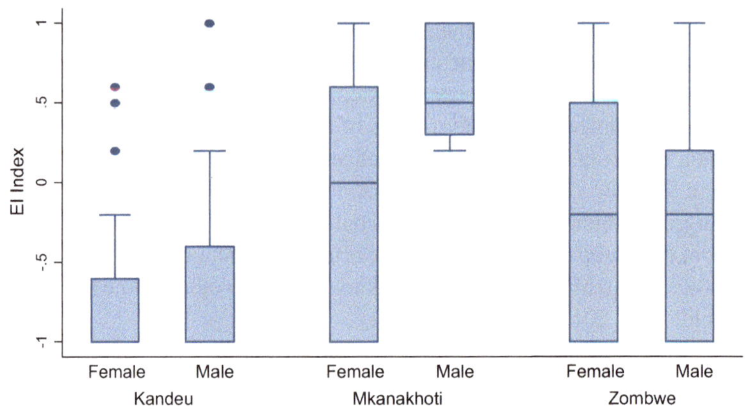

Fig. 8 Do farmers share soil health information outside their villages? n: Kandeu = 68, Mkanakhoti = 71, Zombwe = 50: Kruskal–Wallis p-value for comparing study sites = 0.0001***. n: Female = 131, Male = 58: U test p-value for comparing females and males: Kandeu = 0.485, Mkanakhoti = 0.026*, Zombwe = 0.7939. *Source* Authors' own construction from field data

demonstrate how different farmer-centred approaches (Lead farmer, FFS, FRT and FRN) create vertical networks that enable farmers participating in the approaches to interact with various actors in the agriculture system, including agricultural research and extension workers. Evidence shows that the vertical networks are created in the learning plots where soil health options are demonstrated or evaluated by farmers. Through these vertical networks, farmers acquire information on diverse soil health management options, including integration of legume crops and trees in maize fields, crop residue incorporation, manure application and micro-dosing of external soil inputs. However, when poorly facilitated the vertical networks may perpetuate top-down flow of information and hinder co-learning in farmer-centred approaches. These networks also risk being captured by rural elites at the expense of marginalised farmers living in hard-to-reach areas.

The horizontal networks that mainly occur in farm fields and at home facilitate the flow of soil health information from farmers participating in the approaches to other farmers in the broader community. However, the homophily tendency restricts the flow of information to farmers with shared spatial (villages) and social characteristics, including gender, kinship, friendship and neighbourhood. This homophilic tendency is reinforced by socio-cultural contexts in which the approaches are implemented. These socio-cultural factors are related to gender, patriarchal and marriage systems. The evidence presented in this chapter has implications for scholars and practitioners who are calling for transformation of agri-food systems by using participatory approaches in the development and adaptation of soil health innovations at scale. The assumption that horizontal networks will naturally help the broader agrarian community access novel soil health information may not hold in certain social-cultural contexts. Soil health interventions that use farmer-centred approaches must invest in processes that support co-learning and engagement of farmers who represent diverse socio-economic categories. The design of the interventions needs to integrate strategies for mainstreaming gender network weaving within and across communities, and use of Information and Communication Technologies (ICT) to support interactions and collaborative learning among agricultural research and development professionals, farmer innovators and the broader community.

References

Asfaw S, Orecchia C, Pallante G, Palma A (2018) Soil and nutrients loss in Malawi: an economic assessment. Food and Agriculture Organization of the United Nations, UNDP-UNEP Poverty-Environment Initiative and Ministry of Agriculture, Irrigation and Water Development, Malawi

Barrios E, Gemmill-Herren B, Bicksler A, Siliprandi E, Brathwaite R, Moller S, Batello C, Tittonell P (2020) The 10 elements of agroecology: Enabling transitions towards sustainable agriculture and food systems through visual narratives. Ecosyst People 16(1):230–247. https://doi.org/10.1080/26395916.2020.1808705

Beaman L, BenYishay A, Magruder J, Mushfiq AM (2021) Can network theory-based targeting increase technology adoption? Am Econ Rev 111(6):1918–1943. https://doi.org/10.1257/aer.20200295

Berge E, Kambewa D, Munthali A, Wiig H (2014) Lineage and land reforms in Malawi: do matrilineal and patrilineal landholding systems represent a problem for land reforms in Malawi? Land Use Policy 41(November):61–69. https://doi.org/10.1016/j.landusepol.2014.05.003

Bezner Kerr R, Chirwa, M (2004) Participatory research approaches and social dynamics that influence agricultural practices to improve child nutrition in Malawi. EcoHealth 1(S2):SU109-19. https://doi.org/10.1007/s10393-004-0038-1

Bourne M, de Bruyn LL, Prior J (2021) Participatory versus traditional agricultural advisory models for training farmers in conservation agriculture: a comparative analysis from Kenya. J Agric Educ Ext 27(2):153–174. https://doi.org/10.1080/1389224X.2020.1828113

Borgatti SP (2002) NetDraw software for network visualization. Analytic Technologies, Lexington, KY. https://sites.google.com/site/netdrawsoftware/home

Braun AR, Thiele G, Fernández M (2000) Farmer field schools and local agricultural research committees: complementary platforms for integrated decision-making in sustainable agriculture. Overseas Development Institute

Chambers R, Jiggins J (1987) Agricultural research for resource-poor farmers part I: transfer-of-technology and farming systems research. Agric Adm Ext 1(27):35–52. https://doi.org/10.1016/0269-7475(87)90008-0

Classen L, Humphries S, FitzSimons J, Kaaria S, Jiménez J, Sierra F, Gallardo O (2008) Opening participatory spaces for the most marginal: learning from collective action in the Honduran Hillsides. World Develop 36(11):2402–2420. https://doi.org/10.1016/j.worlddev.2008.04.007

Chowa C, Garforth C, Cardey S (2013) Farmer experience of pluralistic agricultural extension, Malawi. J Agric Educ Extension 19:147–166. https://doi.org/10.1080/1389224X.2012.735620

Christensen L, O'Sullivan R (2015) Using social networking analysis to measure changes in regional food systems collaboration: a methodological framework. JAFSCD 113–129. https://doi.org/10.5304/jafscd.2015.053.013

Critchley, W, Harari, N, Mollee, E, Mekdaschi-Studer, R, Eichenberger, J (2023) Sustainable land management and climate change adaptation for small-scale land users in Sub-Saharan Africa. Land 1206(12). https://doi.org/10.3390/land12061206

Crossley N, Bellotti E, Edwards E, Everett MG, Koskinen J, Tranmer M (2015) Social network analysis for ego-nets: social network analysis for actor-centred networks. Sage

Danielsen S, Mur R, Kleijn W, Wan M, Zhang Y, Phiri N, Chulu B, Zhang T, Posthumus H (2020) Assessing information sharing from plant clinics in China and Zambia through social network analysis. J Agric Educ Ext 26(3):269–289. https://doi.org/10.1080/1389224X.2019.1699125

David S, Asamoah C (2011) The impact of farmer field schools on human and social capital: a case study from Ghana. J Agric Educ Extension 17(3):239–252. https://doi.org/10.1080/1389224X.2011.559076

Davis K, Nkonya E, Kato E, Mekonnen DA, Odendo M, Miiro R, Nkuba J (2012) Impact of farmer field schools on agricultural productivity and poverty in East Africa. World Develop 40(2):402–413. https://doi.org/10.1016/j.worlddev.2011.05.019

Djurfeldt AA, Hillbom E, Mulwafu WO, Mvula P, Djurfeldt G (2018) The family farms together, the decisions, however are made by the man: matrilineal land tenure systems, welfare and decision making in rural Malawi. Land Use Policy 70(January):601–610. https://doi.org/10.1016/j.landusepol.2017.10.048

Food and Agriculture Organisation (FAO) (2021) The State of food security and nutrition in the World: transforming food systems for food security, improved nutrition and affordable healthy diets for all. https://docs.wfp.org/api/documents/WFP-0000130141

Food and Agriculture Organisation (FAO) (2016) Farmer field school guidance document: planning for quality programmes. Food and Agriculture Organisation, Rome. https://www.fao.org/3/i5296e/i5296e.pdf

Franke AC, Ellis-Jones J, Tarawali G, Schulz S, Hussaini MA, Kureh I, White R (2006) Evaluating and scaling-up integrated striga hermonthica control technologies among farmers in Northern Nigeria. Crop Prot 25(8):868–878. https://doi.org/10.1016/j.cropro.2005.11.014

Friis-Hansen E, Duveskog D (2012) The empowerment route to well-being: an analysis of farmer field schools in East Africa. World Develop 40(2):414–427. https://doi.org/10.1016/j.worlddev.2011.05.005
Government of Malawi (GoM) (2015) Soil loss assessment in Malawi. Ministry of Agriculture, Irrigation and Water Development (MAIWD), Lilongwe, Malawi
Government of Malawi (GoM) (2010) Lead farmer concept guidelines. Ministry of Agriculture and Food Security (MoAFS). Department of Agricultural Extension Services (DAES). Lilongwe, Malawi
Granovetter M (2005) The impact of social structure on economic outcomes. J Econ Perspect 19(1):33–50. https://doi.org/10.1257/0895330053147958
Hall, A (2006) Embedding agricultural research in a system of innovation. Report of the Science Forum on CGIAR Priorities: Science for the Poor. Science Council Secretariat, Rome
Hermans F, Stuiver M, Beers PJ, Kok K (2012) The distribution of roles and functions for upscaling and outscaling innovations in agricultural innovation systems. Agric Syst 115:117–128. https://doi.org/10.1016/j.agsy.2012.09.006
HLPE (2019) Agroecological and other innovative approaches for sustainable agriculture and food systems that enhance food security and nutrition. A report by the High-Level Panel of Experts on Food Security and Nutrition of the Committee on World Food Security, Rome. https://www.fao.org/fileadmin/user_upload/hlpe/hlpe_documents/HLPE_Reports/HLPE-Report-14_EN.pdf
Isaac ME (2012) Agricultural information exchange and organizational ties: the effect of network topology on managing agrodiversity. Agric Syst 109(June):9–15. https://doi.org/10.1016/j.agsy.2012.01.011
Kaaria S, Njuki J, Abenakyo A, Delve R, Sanginga P (2009) Enabling rural innovation: Empowering farmers to take advantage of market opportunities and improve livelihoods. In: Innovation Africa: enriching farmers' Livelihoods. Earthscan, pp 167–185
Kadushin C (2012) Understanding social networks: theories, concepts, and findings, 1st ed. Oxford University Press, Oxford
Kansiime MK, Watiti J, Mchana A, Jumah R, Musebe R, Rware H (2018) Achieving scale of farmer reach with improved common bean technologies: the role of village-based advisors. J Agric Educ Ext 24(3):215–232. https://doi.org/10.1080/1389224X.2018.1432495
Khaila S, Tchuwa F, Franzel S, Simpson S (2015) The farmer-to-farmer extension approach in Malawi: a survey of lead farmers. World Agroforestry Centre (ICRAF). http://www.worldagroforestry.org/downloads/publications/pdfs/WP14200.PDF.
Lambrecht IB (2016) As a husband I will love, lead, and provide. Gendered access to land in Ghana. World Develop 88:188–200. https://doi.org/10.1016/j.worlddev.2016.07.018
Nelson R, Coe R (2014) Transforming research and development practice to support agroecological intensification of smallholder farming. J Int Aff 67(2):107
Nelson R, Coe R, Haussmann BIG (2016) Farmer research networks as a strategy for matching diverse options and contexts in smallholder agriculture. Experim Agric 1–20. https://doi.org/10.1017/S0014479716000454
Nyantakyi-Frimpong H (2017) Agricultural diversification and dietary diversity: a feminist political ecology of the everyday experiences of landless and smallholder households in Northern Ghana. Geoforum 86:63–75. https://doi.org/10.1016/j.geoforum.2017.09.003
Pretty JN (1995) Participatory learning for sustainable agriculture. World Develop 23:1247–1263. https://doi.org/10.1016/0305-750X(95)00046-F
Putnam R (1993) Making democracy work. Princeton University Press, Princeton
Ragasa C (2020) Effectiveness of the lead farmer approach in agricultural extension service provision: nationally representative panel data analysis in Malawi. Land Use Policy 99:104966. https://doi.org/10.1016/j.landusepol.2020.104966
Richardson M, Coe R, Descheemaeker K, Haussmann B, Wellard K, Moore M, Maland Cady J, Gubbels P, Tchuwa F, Paz YR, Nelson R (2021) Farmer research networks in principle and practice. Int J Agric Sustain 1–18. https://doi.org/10.1080/14735903.2021.1930954

Rocker S, Kropczynski J, Hinrichs C (2022) Using social network analysis to understand and enhance local and regional food systems. In: Peters C, Thilmany D (eds) Food systems modelling. Academic, pp 231–256. https://doi.org/10.1016/B978-0-12-822112-9.00015-1
Rose R (1999) What does social capital add to individual welfare? An empirical analysis of Russia. Centre for the Study of Public Policy, University of Strathclyde
Sanginga P, Waters-Bayer A, Kaaria K, Njuki JWC (2009) Innovation Africa: beyond Rhetoric to Praxis. In: Innovation Africa. Enriching Farmers' Livelihoods. Earthscan, pp 374–386
Scoones I, Thompson J (2009) Innovation for agricultural research and development. In: Farmer first revisited. Practical Action Publishing, pp 3–38
Spielman DJ, Davis K, Negash M, Ayele G (2011) Rural innovation systems and networks: findings from a study of Ethiopian smallholders. Agric Human Values 28(2):195–212. https://doi.org/10.1007/s10460-010-9273-y
Stewart RD, Jian J, Gyawali AJ, Thomason WE, Badgley BD, Reiter MS, Strickland MS (2018) What we talk about when we talk about soil health. Agric Environ Lett 3:180033. https://doi.org/10.2134/ael2018.06.0033
Thuo M, Alexandra A, Bell BE, Bravo-Ureta DK, Okello ENO, Kidula NL, Deom MC, Puppala N (2013) Social network structures among groundnut farmers. J Agric Educ Extension 19(4):339–359. https://doi.org/10.1080/1389224X.2012.757244
van den Berg H, Phillips S, Dicke M, Fredrix M (2020) Impacts of farmer field schools in the human, social, natural and financial domain: a qualitative review. Food Sec 12:1443–1459. https://doi.org/10.1007/s12571-020-01046-7
Wellard K, Rafanomezana J, Nyirenda M, Okotel M, Subbey V (2013) A review of community extension approaches to innovation for improved livelihoods in Ghana, Uganda and Malawi. J Agric Educ Extension 19(1):21–35. https://doi.org/10.1080/1389224X.2012.714712
Weyori AE, Amare M, Garming H, Waibel H (2018) Agricultural innovation systems and farm technology adoption: findings from a study of the Ghanaian plantain sector. J Agric Educ Ext 24(1):65–87. https://doi.org/10.1080/1389224X.2017.1386115
World Bank Group (2019) Malawi country environmental analysis. World Bank. https://openknowledge.worldbank.org/handle/10986/31326. License: CC BY 3.0 IGO

Open Access This chapter is licensed under the terms of the Creative Commons Attribution 4.0 International License (http://creativecommons.org/licenses/by/4.0/), which permits use, sharing, adaptation, distribution and reproduction in any medium or format, as long as you give appropriate credit to the original author(s) and the source, provide a link to the Creative Commons license and indicate if changes were made.

The images or other third party material in this chapter are included in the chapter's Creative Commons license, unless indicated otherwise in a credit line to the material. If material is not included in the chapter's Creative Commons license and your intended use is not permitted by statutory regulation or exceeds the permitted use, you will need to obtain permission directly from the copyright holder.

The Dynamics of Orange-Fleshed Sweetpotato Production, Consumption, and Marketing Among Malawi's Smallholder Farmers: Case Study of Mulanje and Kasungu Districts

Sarah Chilungo, Monica Chimbaza, Grace T. Munthali, Admire Katunga, and Smith G. Nkhata

Abstract Vitamin A deficiency (VAD) remains a pressing health concern, particularly in impoverished developing-country households, notably in Africa. Current strategies to address VAD emphasize a food-based approach, focusing on breeding crops rich in beta-carotene, a Vitamin A precursor. Malawi, through the Department of Agricultural Research Services, developed and released biofortified orange-fleshed sweetpotato (OFSP) varieties rich in pro-vitamin A carotenoids over a decade ago. Despite this, factors influencing the adoption of OFSP varieties in Malawi are inadequately documented. This study was conducted in Kasungu and Mulanje districts with 273 randomly sampled farmers, with the aim of elucidating the impact of farmers' knowledge on OFSP production, assessing promotional channels that influence adoption, and evaluating smallholder farmer participation in the OFSP market. The data was collected through individual interviews, focus group discussions, and market surveys. The findings reveal that farmers are motivated to cultivate OFSP because of its nutritional benefits. The additional factors that influence their decision include: Vitamin A content, taste, yield, nutrition, maturity, and market demand. The study identifies extension agents as key information channels, despite their infrequent visits. Challenges in marketing OFSP include long distances, poor pricing, market instability, and exploitation by middlemen. Positive influences on market participation include group membership, higher education levels, access to extension services, and attendance at sweet potato demonstration events. The study recommends that sweetpotato breeders should consider farmer-identified factors

S. Chilungo · G. T. Munthali · A. Katunga
Department of Agricultural Research Services, Chitedze Research Station, Lilongwe, Malawi

M. Chimbaza (✉) · S. G. Nkhata
Department of Agriculture and Food Systems, Natural Resources College, Lilongwe University of Agriculture and Natural Resources, Lilongwe, Malawi
e-mail: mchimbaza@luanar.ac.mw

S. G. Nkhata
e-mail: snkhata@luanar.ac.mw

© The Author(s) 2026

B. R. Singh et al. (eds.), *Sustainable Food Systems in Sub-Saharan Africa*,
https://doi.org/10.1007/978-3-032-11288-0_21

when developing OFSP varieties to enhance adoption, while government should prioritise the enhancement of road networks and marketing interventions.

Keywords OFSP · Production · Adoption · Marketing · Consumption · Malawi

1 Introduction

Vitamin A deficiency (VAD) is a leading form of micronutrient deficiency, especially among pregnant mothers, lactating mothers, and children under five years. This condition manifests through impaired night vision, and in its more severe forms, it has been associated with cases of blindness and heightened mortality in young children (World Health Organization (WHO) 2009). VAD affects nearly 127 million-school-aged children worldwide and accounts for 6% of all deaths among children under 5 years of age (Fawzi et al. 1994; Villamor and Fawzi 2002; Black et al. 2008). The current VAD data for Malawi shows that about 4.5% of children are vitamin A deficient, and the prevalence for other vulnerable groups is almost insignificant (NSO 2016). The low prevalence rate of VAD among children under five and other vulnerable groups is partially attributed to high consumption of Orange-fleshed sweetpotato (OFSP), a biofortified crop rich in provitamin A carotenoid. VAD is a highly preventable condition that can be mitigated through vitamin A supplementation and consumption of food rich in preformed vitamin A, in the form of retinol, or beta-carotene, a precursor to vitamin A (Low et al. 2000; Bai et al. 2011). Biofortification, which involves breeding staple food crops including sweetpotatoes for substantially improved micronutrient content, is a recent and promising strategy to alleviate micronutrient malnutrition, including VAD. It is a feasible and cost-effective complementary strategy of delivering micronutrients to rural populations that may have limited access to diverse diets and other micronutrient interventions, such as supplementation and food fortification (Bouis and Saltzman 2017).

β-carotene, a provitamin A carotenoid found in biofortified sweet potato, is the most widely studied carotenoid, and is regarded as the main source of Vitamin A in the human diet and is important for vision, immunity and reproduction (Haskell 2012; Grune et al. 2010). It has the highest vitamin A activity among other provitamin A carotenoids such as α-carotenes and β-cryptoxanthins (Donhowe et al. 2014); it also has highest conversion ratio to vitamin A (Yeum and Russell 2002). The deficiency of β-carotene can result in xerophthalmia, blindness, and premature death (Mayne 1996). β-carotene has been considered nearly non-toxic to humans because they can tolerate high dietary intake of the nutrient without harm (Bendich 1988; Hathcock et al. 1990). Humans lack the ability to synthesize vitamin A in vivo (Haskell 2012); hence, they must get adequate amounts of it from the dietary sources rich in β-carotene, like dark green leafy vegetables and fruits (Gul et al. 2015). Awareness creation and capacity building on β-carotene, the health benefits, its sources, food preparation and consumption are all important in the fight against VAD among vulnerable groups. The increased awareness about β-carotene and its health benefits has led to the development of biofortified crops such as OFSP and orange maize.

For a very long-time smallholder farmers in Malawi have been producing white-fleshed sweet potatoes (WFSP) for household consumption but also income generation. Over a decade ago, Malawi developed and released the biofortified orange-fleshed sweetpotato (OFSP) variety that is rich in pro-vitamin A to help fight VAD in the country. Following the release of the first OFSP variety, several other OFSP varieties have been developed and released for widespread adoption by the farming community. Consequently, the Malawian government, in collaboration with CGIARS and NGOs, implemented several activities like vine multiplication and distribution as well as dissemination of extension messages. It also invested in country wide farmer training in production and utilization of OFSP varieties, with anticipation of widespread adoption by the farming community, in the hope of subsequently reducing the impact of VAD. This approach contributed to a significant decline in VAD among vulnerable groups as documented by NSO 2016.

To date, OFSP is considered the single most successful example of biofortification intervention worldwide (Bouis and Saltzman 2017). Past studies have reported increased pro-vitamin A intake resulting from increased consumption of OFSP among rural women and children (Hotz et al. 2012). Evidence of reduced incidence of VAD among these vulnerable groups was reported by van Jaarsveld et al. (2005). Consumption of just 100 g of boiled OFSP roots regularly can meet the vitamin A requirement of a young child; and with community level nutrition education campaign, a significant impact was made on vitamin intakes and status (Hotz et al. 2012; Low et al. 2007).

While the goals of the OFSP program are commendable, there exists an ongoing debate on whether or not consumers would accept OFSP varieties. Some researchers argue that consumers might not adopt these varieties because they are either too watery and/or too sweet to meet local taste preferences (Harvest Plus 2012). Elsewhere, significant studies on determinants of OFSP adoption have been conducted and results are variable. Women farmers were likely to adopt the OFSP varieties if the clones were sufficiently high in starch, low in fiber, and if they were introduced through community-level education programs that focused on the health of young children (Jenkins et al. 2018). Furthermore, farmer adoption of technology is an economic decision based upon expected marginal benefits and costs (Kassie et al. 2011).

Few studies have specifically looked at the adoption of OFSP in Malawi. Recently, due to increased interface of production and utilization interventions, studies on adoption have been on "impact of knowledge of production and consumption on the adoption" (Caeiro and Vicente 2020; Gatto et al. 2021). A study by Gatto et al. 2021, tried to look at the impact of training on production and consumption of OFSP on smallholder farmer adoption but did not go in-depth due to methodological issues. Our paper adds to this literature by analyzing the causal impacts of an intervention promoting OFSP, which emphasized education for nutrition and production as well as farmer market participation. The study assessed production and consumption of

OFSP and effect on adoption, evaluated the effect of institutional factors on adoption of OFSP, and assessed the smallholder farmer OFSP market participation.

2 Materials and Methods

2.1 Setting

Prior to the survey, an exploratory study was done in Mulanje, Thyolo, Lilongwe and Kasungu districts. The aim of the study was to get insights into the interventions which were implemented to popularize the OFSP; to understand the dissemination channels used; and to take note of the stakeholders engaged in disseminating OFSP. Some modifications were done to the research design based on the exploratory study's findings. For instance, Thyolo district was dropped as a study area because most of its information was similar to that of Mulanje district. In addition, Lilongwe district was dropped since not much work had been done there to promote OFSP. As a result of the exploratory study, a mixed survey approach was adopted (quantitative and qualitative approaches), and the objectives were revised to incorporate the exploratory findings.

To kick start the study, enumerators were trained on data collection tools. The training was done to acquaint enumerators with the data collection tools. The data collection tools included individual questionnaire, checklists for focus group discussions and market survey questionnaire. Modifications to the data collection tools were done depending on the challenges experienced during pre-testing in the field.

2.2 Study Area

The study was conducted in rural areas of Kasungu and Mulanje districts among sweetpotato smallholder farm households in Malawi. These districts were purposively selected since they are some of the largest sweetpotato growing areas in the Southern and Central regions of Malawi, respectively, and OFSP interventions had been implemented there for a number of years.

2.3 Sampling Technique and Sample Size

The study employed a quasi-experimental research design to assess the impact of an agriculture-nutrition intervention aimed at promoting OFSP, which emphasized education for nutrition and production as well as farmer market participation. Since 2009, the Malawi government, through the departments of research and extension, has been popularizing the OFSP through various interventions including the following: vine dissemination, demonstrations, agronomic trainings, nutrition trainings, cooking demonstrations, nutrition sensitization through radio and print media, and marketing trainings. These interventions were administered in specific Extension Planning Areas (EPAs) within a district. Survey data was collected in four EPAs in the two survey districts, with two EPAs forming intervention while the other two were counterfactual.

At EPA level, simple random selection was applied to come up with a representative sample of 273 households. To break down the sample size per district, 124 respondents were sampled from Kasungu district, out of whom 79 were adopters and 45 were non-adopters. About 149 respondents were sampled from Mulanje district, 98 of whom were adopters while 51 were non-adopters. It is important to note that the adopters who were interviewed for this study were sweet potato farmers who had planted the different varieties of OFSP for at least one year. The non-adopters were sweet potato farmers who were aware of the technology but did not plant the OFSP variety.

2.4 *Model Specification*

On the basis of the binary nature of adoption, which is the dependent variable, a binary regression technique was adopted to analyze how each of the explanatory variables (age, gender, marital status, and educational level) influence the probability of an individual taking up OFSP variety. The training in the production and consumption of OFSP, extension approaches etc., were also important considerations. Based on previous research on youth employment (Okicic et al. 2020), which is similar to this current study, the probit model was adopted as a primary methodological approach where the dichotomous dependent variable was currently adopt vs. currently non-adopt. To estimate the model, probit regression analyses procedures using STATA version 13.

The probit model assumed that the probability of the individual to have adopted or not depends on an observable utility index I_i that is determined by one or more explanatory variables X_i, such that the larger the value of the index I_i, the greater the probability of an individual taking up OFSP. The index I_i is expressed as

$$\text{Model 1}: \; I_i = \beta_1 + \beta_1 \, X_i$$

where X_i are explanatory variables and $I_i = E\,(Y_i = 1|X_i)$ which is non-adoption. The model assumes that there is a threshold level of the index, I_i^*, such that if I_i exceeds I_i^*, the farmer has adopted, otherwise not. Given normality assumption, the probability that I_i^* is less than or equal to I_i can be computed from the standardized normal cumulative distribution function (CDF) as;

$$P_i = P(Y = 1|X) = P\left(I_i^* \leq I_i\right) = P(Z_i \leq \beta_1 + \beta_2 X_i) = F(\beta_1 + \beta_2 X_i)$$

where $Z_i = \beta_1 + \beta_2 X_i$

$$F(I_i) = 1/\sqrt{2\pi} \int_{(-\infty)}^{(I_i)} \left[\left[e^{([\![-z]\!]^{2/2})dz}\right]\right]$$

$$= 1/\sqrt{2\pi} \int_{(-\infty)}^{(\beta_1+\beta_2 X_i)} \left[\left[e^{([\![-z]\!]^{2/2})dz}\right]\right]$$

$$I_i = F^{(-1)}(I_i) = F^{(-1)}(P_i)$$
$$= \beta_1 + \beta_2 X_i$$

Marginal effect was adopted as a secondary methodology which was run after the probit analysis to calculate the probability for the dependent variables (X_k). The marginal effect shows how P(Y = 1) changes if X changes from 0 to 1. Therefore, ME for variable X_k would be;

$$\text{Model 2 : Marginal effect for } X_k = \Phi(XB) * b_k$$

2.5 *Analytical Tools*

Statistical tools used for analysis of the data was STATA version 13. Descriptive and econometric analysis were employed to meet the main objective of the study. In the case of descriptive analysis tables, t-tests and chi-square tests were computed. The econometric analysis used binary regression model to identify where knowledge through training is an important factor in OFSP adoption. Furthermore, the study used marginal effect to measure the probability of a household adopting OFSP varieties.

3 Results and Discussions

3.1 *Demographic and Socio-economic Characteristics of Respondents*

The results showed that the majority of household heads for both adopters and non-adopters were males 71.43% and 78.57% respectively (Table 1). The average age of household head for adopters was 43 and 41 years for non-adopters. The average household size for both adopters and non-adopters was 5 persons, and chi-square indicated that there is relationship between household size and adoption of OFSP at 10% significant level. The results also showed that the majority of household heads attended primary education—67.72% (adopters) and 61.90% (non adopters). However, the districts results indicated that majority of the adopters were from Mulanje (74.07%), and the majority of non-adopters were from Kasungu (89.29%). The relationship between the district and the adoption of OFSP was statistically significant at 1% level. On average, the farm size of both adopters and non-adopters was 3 acres. The average estimated annual off-farm income of adopters was MK138, 317.80, and that of non-adopters was MK155, 574.50. On the other hand, the average annual on-farm income for adopters was MK134, 614.10, and that of non-adopters was MK127.032.90.

3.2 *Knowledge on Health Benefits of OFSP*

OFSP varieties were developed to help fight VAD among the vulnerable groups. It was therefore important to ask farmers about their knowledge on the importance of OFSP consumption. Analysis of the results revealed that 92% of adopters and 66.6% of non-adopters are aware of vitamin A (Table 2). About 45.7% of adopters were aware that vitamin A enhances eye sight compared to 17.8% of non-adopters. About 28.5% of non-adopters and 24% of adopters mentioned protection from diseases as the importance of vitamin A (Table 2). OFSP is the main source of pro-vitamin A carotenoid among the African diets. Inadequate or excessive intake of vitamins A can lead to various disorders. For instance, it has been reported that VAD is considered to be the main cause of childhood blindness (Roodhooft 2002). The most important nutritional benefit of vitamin A includes proper immune function, epithelial growth and repair, bone growth, reproduction, and normal embryonic and fetal development (West 2006). Although excessive intake of preformed vitamin A is associated with physiological disorders, carotenoids in foods, even when consumed over long periods and in large amounts, are not known to produce adverse health effects.

Although no causal assessment was done, review of literature revealed that farmers with knowledge about the nutritional content of OFSP were about three times more likely to adopt OFSP than those who did not have this knowledge (Kaguongo et al. 2011). In a separate study, Okello et al (2017) reported that household food security and knowledge of OFSP benefits affect the decision to adopt OFSP varieties. The current results suggest that efforts to promote biofortified crops should be accompanied by nutrition trainings to increase farmer knowledge and enhance adoption.

3.3 *OFSP Production and Adoption*

As expected, Mulanje district being the largest producer of sweetpotato registered the highest number of OFSP adopters (65.8%) while Kasungu district registered on

Table 1 Descriptive statistics of the household

Variables	Adopters (%)	Non-adopters (%)	X^2
Sex			
Male	71.43	78.57	1.5280
Female	28.57	21.43	
Age			
19–35	28.57	41.67	4.6043
36–50	48.15	38.10	
51 +	23.28	20.24	
Mean	42.656	41.214	
Household size			
Less than 5	31.22	45.24	5.5331*
5–8	63.49	52.38	
9+	5.29	2.38	
Mean	5.534	5.048	
Education			
No education	5.82	10.71	5.9768
Primary	67.72	61.90	
Secondary	25.93	23.81	
Tertiary	0.53	3.57	
Mean	7.037	6.821	
District			
1. Mulanje	74.07	10.71	94.1723***
2. Kasungu	25.93	89.29	
Farm size (acres)			
Mean	2.606	3.0717	T = 0.1558
Estimated annual income			
Off-farm	138,317.8	155,574.5	T = -0.4251
On-farm	134,614.1	127,032.9	T = 0.3297

* Significant at 10% level, ** Significant at 5% level; *** Significant at 1% level

63.7% (Table 3). Overall, 64.5% of farmers from the two districts adopted OFSP. This is in line with the findings of Gatto et al (2021) who reported that half of the interviewed respondents grew OFSP. The lower number of OFSP adopters in Kasungu might be attributed to the commercial importance of the crop. Kasungu district is the largest producer of tobacco in the country which is a Malawi's biggest cash crop, so the farmers tended to grow more tobacco than sweetpotato. The current results agree with findings by Strauss et al., who reported that adoption of a technology is influenced by the economic benefits and marginal returns. In areas where big cash

Table 2 Knowledge of consumption of OFSP and VAD

Knowledge measure	Adopter		Non-adopter	
	N	%	N	%
Heard of Vit A				
0. No	14	7.41	28	33.33
1. Yes	175	92.59	56	66.67
Importance of Vit A				
1. Protection from diseases	42	24.00	16	28.57
2. Enhances eye sight	80	45.71	10	17.86
3. Energy giving	6	3.43	8	14.29
4. 1 and 2	10	5.71	4	7.14
5. All of the above	35	20.00	11	19.64
6. Don't know	2	1.14	7	12.50

crops are in dominance, adoption of OFSP might be influenced by emphasizing on value addition and commercial benefits of the crop.

Disaggregating further by treatment, about 55% of respondents in the intervention group and 95% in the control group adopted OFSP compared to 45% in the intervention group and 5% in the control who did not adopt the intervention and these results were significant at 1% level (Table 4). The results confirm the impact of the various interventions, including vine multiplication and distribution implemented in the control areas, which facilitated the adoption of OFSP varieties. In Mozambique, the adoption of the introduced OFSP variety was positively associated with participation in promotion activities and vine distribution, among other factors (Gilligan et al. 2014). The adoption rate was reduced in subsequent seasons after OFSP introduction, mainly because of unavailability of vines, limited propagation capacity, and lower drought tolerance of OFSP (Gilligan et al. 2014; Mazuze 2005). The low adoption rate of OFSP among the intervention group, particularly in Mulanje district, might be attributed to the economic importance of the crop. The white-fleshed sweetpotato variety, is a cash crop in Msika wa Najala EPA where the intervention group was selected, and farmers have been making profits from selling it. It is possible that farmers feared that adoption of OFSP varieties might compromise the market as it is a new crop and consumers are not used to it.

Table 3 Percentage of OFSP adopters and non-adopter

District	Adoption (%)				Total (%)	
	No %	n	Yes %	n	%	n
Mulanje 149	34.2	51	65.8	98	100	149
Kasungu 124	36.3	45	63.7	79	100	124
Total	35.2	96	64.5	177	100	273

Table 4 Percentage of OFSP adopters and non-adopters among intervention and control groups

	Intervention		Counterfactual		X^2
	N	%	N	%	
Adaptors	96	55	93	95	47.28***
Non-adopters	79	45	5	5	
Total	175	100	98	100	

* Significant at 10% level, ** Significant at 5% level; *** Significant at 1% level

Figure 1 presents the proportion of farmers growing different varieties of orange-fleshed and white-fleshed sweetpotato varieties. According to the results, about 30% of the respondents planted Kadyaubwerere, Kenya and Sakananthaka, while about 20% planted Kaphulira and Anakwanire and nearly 10% planted Zondeni and Chipika. Kenya and Sakanathaka are white-fleshed sweetpotao varieties while the rest are orange-fleshed. During focus group discussions, we further probed to understand reasons for farmer preference of the sweetpotato varieties (results not shown). Among the OFSP varieties Kadyaubwerere is widely grown by farmers owing to its sweet taste, good aroma, nutrition and high market demand. Kaphulira and Anakwanire are preferred due to the vitamin A content, early maturing and high yield. The low adoption of Zondeni and Chipika is mainly attributed to low yield and susceptibility to sweetpotato weevil. Kenya variety is popular among the sweetpotato growers due to the high dry matter content, good taste and adaptability to the local environment, while Sakananthaka is preferred due to high yield, early maturing and good taste.

The results on farmers' preference for OFSP indicate that the reasons influencing adoption are diverse, and they include taste, yield, nutrition, maturity among others. Similarly, Jenkins et al. (2018) reported that diverse factors influenced the adoption and retention of OFSP, including organoleptic qualities, taste preferences, access to planting material, agronomic traits, environmental conditions, lack of capital for

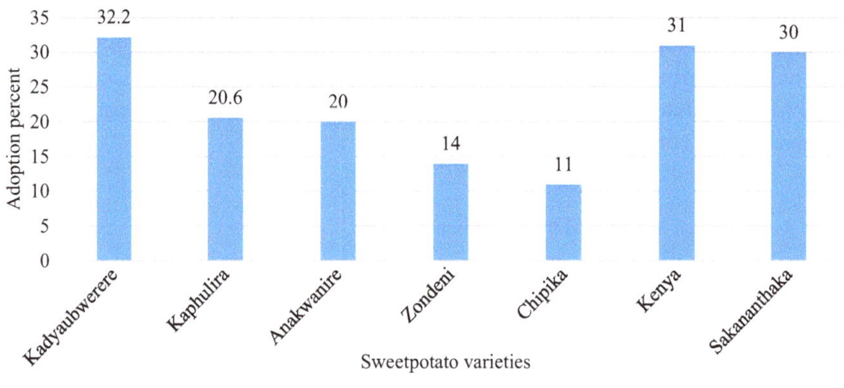

Fig. 1 Proportion of farmers growing different varieties of sweet potato: % (N = 273)

inputs and labor, unstable markets, and limited sharing of information and planting material across farmer networks. Sweetness, taste and dry matter have joint significant effects on adoption of an OFSP variety (Adekambi et al. 2020). Those farmers who grew WFSP varieties emphasized drought tolerance, flavor, and high dry matter content as reasons for adoption and retention (Jenkins et al. 2018). The production attributes, including yield and pest and disease tolerance mentioned by farmers, form an integral part of the sweetpotato breeding program in Malawi.

Although the current study did not probe more to get details on OFSP adoption and retention, Kaguongo et al (2011) reported that knowledge of value addition and nutritional benefits, availability of vines were the key factors for adoption and participation in a value chain extension programme that enhanced the probability of adoption. Level of commercialization and having a child of up to five years were also mentioned as factors influencing intensity of adoption. Unstable markets were reported as another factor that affects adoption and retention of OFSP varieties (Jenkins et al. 2018). Where markets are not stable, OFSP adoption is likely to be low. Malawi is doing a commendable job on value addition of OFSP and inclusion of mothers with children under-five in OFSP interventions to influence adoption. In some districts OFSP was included in school feeding program. With the current interventions, the adoption of OFSP is expected to increase in the near future.

3.4 Institutional Factors

Farmers were interviewed to understand how institutional factors influence adoption of OFSP and results are presented in Table 5. The results revealed that a little proportion of farmers had access to credit (36% for OFSP adopters and 7% for non-adopters). For those farmers that had, access to credit a large proportion of them accessed it from village banks—37 and 19% for adopters and non-adopters respectively.

The pivotal role of education in shaping the adoption and resolution of diverse nutritional challenges within communities cannot be overstated. To comprehend farmers' involvement in OFSP production training, interviews were conducted. As evidenced by the data presented in Table 5, a significant 56% of adopters and 8% of non-adopters underwent training in OFSP production. Notably, the study did not directly assess farmers' participation in OFSP nutrition training. Nevertheless, existing literature, such as the work of Kapur (2019), underscores the positive impact of nutrition education in cultivating balanced feeding behaviors. Furthermore, research by McNulty (2013) underscores the effectiveness of nutritional education in transforming dietary patterns, particularly those influencing chronic diseases. Consequently, incorporating nutritional education into crop production training emerges as a crucial component. Recognizing the transformative potential of education, especially in nutritional aspects, becomes imperative for addressing challenges and fostering positive change within agricultural communities.

Table 5 Institutional characteristics of the household

Variables	Adopters (%)	Non-adopters (%)	X^2
Access to credit			
Yes	36	7	24.4731***
No	64	93	
Main sources of credit			
1. None	31.75	48.81	22.3159***
2. Commercial bank	1.06	2.38	
3. NGO	12.70	2.38	
4. Village bank	37.57	19.05	
5. Relatives/friends	7.41	14.29	
6. Microfinance	8.99	13.10	
7. Contract farming	0.53	0.00	
Access to production training			
Yes	56.08	8.33	54.6608***
No	43.92	91.67	
Extension contacts			
Yes	87.3	66.7	16.0588***
No	12.7	33.3	
Number of times visited by extension agent			
Less than 5	40.1	53.6	5.5339*
5–10	29.7	32.1	
More than 10 time	29.7	14.3	
Mean			
Membership to social organizations			
Yes	**84.7**	**56**	26.1376***
No	15.3	**44**	
Type of social organization			
Care group	5.63	4.26	6.5641
Farmer club	50.00	59.57	
Community Development Group	11.88	10.64	
Village Savings and Loan (VSL)	20.00	19.15	
Cooperative	5.63	2.13	
Association	6.88	4.26	

* Significant at 10% level, ** Significant at 5% level; *** Significant at 1% level

In terms of extension contact, about 87% of adopters and 66% of non-adopters indicated that they have contact with extension services. Usually, the majority of the respondents have less than 5 contacts with extension services as shown by 40% adopters and 53% non-adopters. This indicates that farmers are not frequently visited by extension workers.

Membership to (a) social organization/s is another variable of interest in the institutional characteristics of the household. The analysis of results showed that 84% of adopters were members of social organizations compared to 55% of adopters. Farmers were further interviewed to get names of the social organizations they are affiliated to. According to Table 5, 59% of non-adopters and 50% of adopters belonged to farmers club. The next popular type of social organization was village savings bank. The other types of social organization included care groups, community development groups, as well as cooperative groups and other associations. Adekambi et al (2020) reported that participation in farmer organizations has a strong positive effect on the adoption of OFSP varieties. Membership to farmers' organizations facilitates the sharing of information, knowledge learning, and access to improved seeds, and consequently influence the adoption of new technologies (Nagarajan and Smale 2006).

According to the results in Table 5, 86.7% of adopters get agricultural information from extension agents, 10% from extension circulars, and 59% from radio. A similar trend was observed for the non-adopters, whereby 55.9% indicated extension agents, 36.9% radio, and 8.3% extension circulars. Farmers were further asked to rank the preferred source of information; and according to the results in Table 5, both adopters and non-adopters, the most preferred source of information was extension agents, followed by radio; and extension circulars was the least preferred. The results emphasize the importance of extension agents in disseminating agricultural information.

Distance to markets is also an important institutional household-related factor that can determine adoption of OFSP. According to Table 5, the non-adopters were located closer to markets, within a distance of 6 km; on the other hand, the adopters were located further away from the market (12 km). This makes it harder for the adopters to sale their produce to the market; as a result, vendors come and buy their produce from their gardens, which reduces their bargaining power. The proximity to the market of non-adopters motivates them to adopt and grow OFSP because it has a ready market.

3.5 OFSP Marketing and Factors Affecting Farmers' Participation in Marketing

In Malawi marketing of agricultural commodities is disorganized and characterized with high transport costs, low prices, exploitation by middle men and inadequate market information. Individual interviews were done to understand the marketing system of OFSP. Table 6 shows descriptive statistics for Heckman Model. According to the results, the average quantity of OFSP sold was 2096 kg, the average distance to the produce market was 10 km, and 78% of the respondents participated in the market.

Analysis of market participation revealed that farmer participation in farmer groups increased the probability of market participation by 15%. An increase in education level of household head by 1 year increased the probability of market participation by 0.2%. A farmer who accessed extension services increased the probability of market participation by 0.4%. If the farmer attended sweetpotato demonstration market participation also increased. Access to credit did not increase market participation, indicating that credit is not a big factor influencing farmer participation in OFSP marketing. Further analysis showed that factors that affected market intensity included; age of household head, group participation, as well as access to training opportunities. If a farmer participated in a farmer group, this increased the quantity of sweetpotato sold by1438 kg. Also, participation in training in sweetpotato production increased the quantity of sweetpotato sold by 1312 kg (Table 7).

The current findings indicate that marketing of OFSP is facing big challenges, especially the long market distance which limits farmer participation. It is also clear

Table 6 Smallholder farmer OFSP market participation

Variables	Mean Std. Dev		Min	Max
Quantity of sweet potato sold (kg)	2096.129	3108.097	8	21,600
Age of household head (years)	42.212	13.557	19	81
Distance to produce market (Kilometers)	10.341	42.596	0.1	500
Household size (persons)	5.385	2.041	1	15
Farmer group participation	0.758	0.429	0	1
Training on sweet potato production	0.414	0.493	0	1
Market participation	0.777	0.417	0	1
Farmer group participation	0.758	0.429	0	1
Education level of household head (years)	6.971	3.452	0	17
Sweet potato demonstration participation	0.392	0.489	0	1
Access to nutrition information	0.498	0.501	0	1
Training on sweet potato consumption	0.410	0.493	0	1
Credit accesses	0.271	0.445	0	1

Table 7 Factors affecting farmers' participation in marketing

Market participation	Marginal effects (dydx)	Std. Err	Z
Sex of household head	0.054	0.099	0.550
Farmer group participation	0.155***	0.059	2.620
Training on sweet potato production	0.121	0.102	1.180
District	− 0.002	0.009	0.000
Education level of household head	0.002***	0.000	14.040
Access to extension services	− 0.004*	0.002	− 1.930
Sweet potato demonstration participation	0.049**	0.021	2.330
Access to nutrition information	0.007	0.009	0.800
Training on sweet potato consumption	0.001	0.142	0.000
Credit accesses	− 0.006	0.142	− 0.040
/athrho	17.450***	191.187	0.090
/lnsigma	8.098	0.049	164.970
Rho	1.000	0.000	
Sigma	3288.497	161.426	
lambda	3288.497	161.426	

* Significant at 10% level, ** Significant at 5% level; *** Significant at 1% level

from the results that age, education and membership to farmers' groups and access to extension services increases farmer participation in OFSP marketing. Educated farmers are more knowledgeable on the nutrition benefits of OFSP, and they are more capable of persuading consumers to buy their crop. In addition farmers share information about marketing during group meetings and they are more likely to access better market as a group. Analysis of factors influencing farmers' participation in collective marketing initiatives showed that collective marketing was a better way of marketing commodities; farmer were more able to negotiate prices and to reduce marketing costs. Also, training in collective action positively influenced the participation of farmer groups in collective marketing (Mate 2016).

Comparison with WFSP marketing was not done, but market visits revealed more WFSP being sold on the market than OFSP. However, sellers claimed that OFSP sells faster than WFSP because of the nutrition value of OFSP. In Uganda, HarvestPlus established a strategy to promote marketing of OFSP in 2016. Marketing points of OFSP were created across the country. The points were located in strategic market areas, painted orange and located where there was clear visibility along the roads. Market awareness campaigns targeted consumers through radio talk shows, agricultural and tradeshows, and there were branded packing materials for various outlets. In general, increase in OFSP sales were reported as a result of this approach. Branding significantly improved visibility of the product and enhanced consumer awareness, attracting them to buy more (http://www.sweetpotatoknowledge.org/).

The main marketing challenges identified during focus group discussions were lack of big and stable markets, as well as poor pricing and marketing skills of producers. Farmers loudly grumbled about the exploitation by middlemen who buy at very low prices. Another big challenge highlighted by farmers was the high transportation costs, which greatly reduced their profits. This could be attributed to long distances to the markets and poor road networks. Most sweetpotato producers travel about 10 km to market their produce. In a similar study, the main marketing challenges were found to be the lack of transportation, lack of clear pricing decisions and marketing skills (Belete et al. 2016). Government needs to take note of the transport constraint and improve road networks to ease transportation and marketing of crops.

When asked about marketing, two marketing channels were identified; namely—straight/direct and hawking. In the first marketing channel, farmers sell their produce directly to consumers, while in the second channel, sweetpotatoes tubers are sold in bulk to middlemen, who then sell to the consumers. The first channel was reported to be more profitable than the second one which is characterized by poor pricing and exploitation by hawkers/middlemen. Market prices for both straight and hawking are highly variable, as they depend on season and availability of the crop, but farmers reported high profits during offseason.

4 Conclusions and Recommendations

The study assesses production, consumption and marketing of OFSP among stallholder farmers. It found that a significant number of farmers from both the intervention and control sites adopted OFSP. The popular varieties mentioned by farmers included Kadyaubwerere, Anakwanire, and Kaphulira. The reasons for production of OFSP varieties varied from VA content to taste and market demand. Farmers are still growing WFSP varieties like Kenya and Sakananthaka, due to their yield, taste and tolerance to the sweet potato weevil. The study also established that farmers are aware of the nutritional benefits of OFSP consumption. The results imply that sweet potato breeders need to consider the factors mentioned by farmers in sweet potato breeding to develop varieties that will meet farmer needs.

Another interesting finding is that farmers consider extension agents as the most appropriate channel to access information pertaining to OFSP, although they are not frequently visited by extension workers. The finding suggests the need to intensify extension services. Apart from relying on extension agents, we also can explore other means of information dissemination like media and farmer field schools. We can also explore the lead farmer approach as it has been reported to be quite effective in information dissemination. Farmers learn better when trained by their fellow farmers.

Another major lesson from this study relates to the finding on distance to market, which has an implication on transaction costs and market participation. The finding implies the need to reduce the distances that farmers travel to access the market since it has implications on the marketing of OFSP. The lesson here is that local governments need to invest in road and market development to reduce distance travelled by farmers to access markets and increase their market participation. Farmers should also be encouraged to form cooperative societies to improve their bargaining power and improve their capacity to sale to distance markets.

Declaration of Conflicting Interests The author(s) declared no potential conflicts of interest with respect to the research, authorship, and/or publication of this article.

References

Adekambi SA, Okello JJ, Rajendran S, Acheremu K, Carey EE, Low J, Abidin PE (2020) Effect of varietal attributes on the adoption of an orange-fleshed sweetpotato variety in Upper East and Northern Ghana. Outlook Agric 49(4):311–320

Bai C, Twyman R, Farré G, Sanahuja G, Christou P, Capell T (2011) A golden era—pro-vitamin A enhancement in diverse crops. In Vitro Cell Develop Biolo Plant 47(2):205–221

Belete A, Setumo MP, Laurie SM, Senyolo MP (2016) A stochastic frontier approach to technical efficiency and marketing of orange fleshed sweet potato (OFSP) at farm level: a case study of Kwazulu-Natal Province, South Africa. J Hum Ecol 53(3):257–265

Bendich A (1988) The safety of β-carotene. Nutr Cancer 11:207–214. https://doi.org/10.1080/01635588809513989

Black RE, Allen LH, Bhutta ZA, Caulfield LE, De Onis M, Ezzati M (2008) Maternal and child undernutrition: global and regional exposures and health consequences. Lancet 371:243–260. https://doi.org/10.1016/S0140-6736(07)61690-0

Bouis HE, Saltzman A (2017) Improving nutrition through biofortification: a review of evidence from HarvestPlus, 2003 through 2016. Glob Food Sec 12:49–58

Caeiro RM, Vicente PC (2020) Knowledge of vitamin A deficiency and crop adoption: evidence from a field experiment in Mozambique Agric Econ 51(2):175–190

CIP. Marketing, Processing and Utilization Community of Practice. Accelerating Orange-fleshed Sweetpotato (OFSP) Value Chain Development for Nutrition and Livelihoods in Africa. Proceedings of the Fourth Annual Meeting held in Sovereign Hotel, Kisumu, Kenya. http://www.sweetpotatoknowledge.org/

Donhowe EG, Flores FP, Kerr WL, Wicker L, Kong F (2014) Characterization and in vitro bioavailability of β-carotene: effects of microencapsulation method and food matrix. LWT—Food Sci Technol 57:42–48. https://doi.org/10.1016/j.lwt.2013.12.037

Fawzi WW, Herrera MG, Willet WC, Nestel P, el Amin A, Mohamed KA (1994) Dietary vitamin A intake and the risk of mortality among children. Am J Clin Nutr 59:401–408

Gatto M, Okello J, Pradel W, Kwikiriza N, Mgomezulu W, Nyirenda Z, Okuku H, Suarez V, van Vugt D, Chipungu F, Kapalasa E, Kumwenda W, Hareau G, Heck S (2021) Adoption and effects of orange-fleshed sweetpotato varieties in Malawi. International Potato Center, Lima, Peru

Gilligan DO, Kumar N, McNiven S et al (2014) IFPRI discussion paper 01353: bargaining power and biofortification: the role of gender in adoption of orange sweet potato in Uganda. International Food Policy and Research Institute (IFPRI), Washington, DC

Grune T, Lietz G, Palou A, Ross AC, Stahl W, Tang G, Bielsalski HK (2010) β-carotene is an important Vitamin A source for humans. J Nutr 140:2268S-2285S. https://doi.org/10.3945/jn.109.119024

Gul K, Singh AK, Jabeen R (2015) Nutraceuticals and functional foods: the foods for future world. Crit Rev Food Sci Nutr. Retrieved from https://doi.org/10.1080/10408398.2014.90338

Haskell MJ (2012) The challenge to reach nutritional adequacy for vitamin A: β-carotene bioavailability and conversion-evidence in humans. Am J Clin Nutr 96:1193S-1203S. https://doi.org/10.3945/ajcn.112.034850
Harold MM (2016) Analysis of factors influencing farmers' participation in collective marketing initiatives in Maara, Meru South and Mbeere South Sub-Counties, Kenya. M. Sc. Thesis, Kenyatta University
Hathcock JN, Hattan DG, Jenkins MY, McDonald JT, Sundaresan PR, Wilkening VL (1990) Evaluation of vitamin A toxicity. Am J Clin Nutr 2:183–202
HarvestPlus (2012) Disseminating orange-fleshed sweetpotato: findings from a HarvestPlus project in Mozambique and Uganda. H Plus—HarvestPlus, Washington, DC
Hotz C, Loechl C, Brauw A et al (2012) A large-scale intervention to introduce orange sweet potato in rural Mozambique increases vitamin A intakes among children and women. Br J Nutr 108(1):163–176
Jenkins M, Shanks CB, Brouwer R, Houghtaling B (2018) Factors affecting farmers' willingness and ability to adopt and retain vitamin A-rich varieties of orange-fleshed sweet potato in Mozambique. Food Secur 10(6):1501–1519. https://doi.org/10.1007/s12571-018-0845-9
Kaguongo W, Ortmann G, Wale E, Darroch M, Low J (2011) Factors influencing adoption and intensity of adoption of orange flesh sweet potato varieties: evidence from an extension intervention in Nyanza and Western provinces, Kenya. Afr J Agric Res 7(3):493–503
Kapur R (2019) The significance of nutrition education. ACTA Scientific Nutritional Health. ISSN: 2582-1423
Kassie M, Shiferaw B, sMuricho G (2011) Agricultural technology, crop income, and poverty alleviation in Uganda. World Dev 39(10):1784–1795
Low J, Uaiene R, Andrade MI, Howard J (2000) Orange-flesh sweet potato: promising partnerships for assuring the integration of nutritional concerns into agricultural research and extension in Mozambique. Ministry of Agriculture and Rural Development (Mozambique), Directorate of Economics, Maputo
Low JW, Arimond M, Osman N, Cunguara B, Zano F, Tschirley D (2007) A food-based approach introducing orange-fleshed sweet potatoes increased vitamin A intake and serum retinol concentrations in young children in rural Mozambique. J Nutr 137(5):1320–1327
Mayne ST (1996) Beta-carotene, carotenoids and disease prevention in humans. FASEB J 10:690–701
Mazuze FM (2005) Analysis of adoption and production of orange-fleshed sweetpotatoes: the case study of Gaza province in Mozambique. Unpublished M.Sc. thesis, Michigan State University
McNulty J. "Challenges and Issues in Nutrition Education" (2013).
Nagarajan L, Smale M (2006) Community seed systems and the biodiversity of millet crops in Southern India. In: Smale M (ed) Valuing crop biodiversity: on farm genetic resources and economic change. CABI Publishing, Wallingford, pp 211–232
National Statistical Office (2016) Malawi Micronutrient Survey Report 2015–2016. http://wwwnso malawi.mw/
Okello JJ, Sindi K, Shikuku K, McEwan M, Low J (2017) A study of household food security and adoption of biofortified crop varieties in Tanzania: the case of orange-fleshed sweetpotato. https://www.researchgate.net/publication/317150890
Okicic J, Jukan MK, Pilav-Velic A, Jahic H (2020) Impact of individual and household characteristics on the employment probability among youth from Bosnia and Herzegovina. https://doi.org/10.1080/1331677X.2020.1776138
Roodhooft JM (2002) Leading causes of blindness worldwide. Bull Soc Belge Ophthalmol. 283:19–25
van Jaarsveld PJ, Faber M, Tanumihardjo SA, Nestel P, Lombard CJ, Benadé (2005) β-Carotene-rich orange-fleshed sweet potato improves the vitamin A status of primary school children assessed with the modified-relative-dose–response test. Am J Clin Nutr 81:1080–1087
Villamor E, Fawzi WW (2000) Vitamin A supplementation: implications for morbidity and mortality in Children. J Infect Dis 182(Suppl 1):S122–S133

Villamor E, Mbise R, Spiegelman D, Hertzmark E, Fataki M, Peterson KE, Ndossi G, Fawzi WW (2002) Vitamin A supplements ameliorate the adverse effect of HIV-1, malaria, and diarrheal infections on child growth. Pediatrics 109(1):e6–e6
West KP Jr (2006) Vitamin A: deficiency and interventions. In: Caballero B, Allen L, Prentice A (eds) Encyclopedia of human nutrition, 2nd edn. Elsevier, Amsterdam, pp 348–359
World Health Organization (WHO) (2009) Global prevalence of vitamin A deficiency in populations at risk 1995–2005. WHO Global Database on Vitamin A Deficiency, Geneva, Switzerland
Yeum KJ, Russell RM (2002) Carotenoid bioavailability and bioconversion. Ann Rev Nutr 22(1):483–504

Open Access This chapter is licensed under the terms of the Creative Commons Attribution 4.0 International License (http://creativecommons.org/licenses/by/4.0/), which permits use, sharing, adaptation, distribution and reproduction in any medium or format, as long as you give appropriate credit to the original author(s) and the source, provide a link to the Creative Commons license and indicate if changes were made.

The images or other third party material in this chapter are included in the chapter's Creative Commons license, unless indicated otherwise in a credit line to the material. If material is not included in the chapter's Creative Commons license and your intended use is not permitted by statutory regulation or exceeds the permitted use, you will need to obtain permission directly from the copyright holder.

Preferred Communication Channels for Accessing Climate-Smart Agriculture Messages by Smallholder Farmers in Phalombe, Nkhotakota, and Dowa Districts in Malawi

Alexander Mussa, Charity Chanza, and Tasokwa Kakota Chibowa

Abstract Poor identification of proper communication channels, coupled with farmers' limited involvement in content creation and selection of communication methods, significantly hampers their access to essential agricultural information. This challenge becomes even more critical in situations where farmers have restricted access to established communication channels or lack the technical expertise to utilize the channels effectively. This study was conducted to determine preferred communication channels for accessing Climate-Smart Agriculture (CSA) messages by smallholder farmers in Dowa, Nkhotakota and, Phalombe districts of Malawi. Specifically, the study sought to identify communication channels for disseminating CSA messages, determine farmer's access to communication channels for disseminating CSA messages, ascertain farmer's preference for communication channels and analyze the relationship between farm and farmers' characteristics and preference for communication channels for accessing CSA messages. The study was grounded on two theories: the communication theory and the Uses and gratification theory. The study used qualitative and quantitative research methods and involved 320 smallholder farmers. It was established that primary communication channels for disseminating CSA messages were extension workers (26%), radio (25%), and lead farmers (25%). In terms of access, farmers had more access to the radio (6.83 days per month) followed by lead farmers (4.84 days per month). The study also revealed that farmers preferred accessing CSA messages through extension workers (29%), followed by lead farmers (28%), and radio (25%). Chi-square analysis showed a significant relationship between farm and farmer characteristics such as gender and choice of radio, extension worker, and lead farmer; farm size was associated with the choice of radio, extension worker, and lead farmer; education was associated with

A. Mussa (✉) · C. Chanza
Faculty of Development Studies, Extension Department, Lilongwe University of Agriculture and Natural Resources, Lilongwe, Malawi
e-mail: allexmussah@gmail.com

T. K. Chibowa
Faculty of Agriculture, Department of Basic Sciences, Lilongwe University of Agriculture and Natural Resources, Lilongwe, Malawi

© The Author(s) 2026

B. R. Singh et al. (eds.), *Sustainable Food Systems in Sub-Saharan Africa*,
https://doi.org/10.1007/978-3-032-11288-0_22

the choice of posters, while age was associated with the choice of lead farmer. The study concluded that preference towards communication channels was influenced by farm and farmer characteristics. Consequently, it is advised that prior to disseminating CSA messages, a comprehensive analysis of socio-demographic and farmer characteristics be conducted. This approach ensures effective communication of the messages to the targeted farmers.

Keywords Communication · Communication channel · Climate change · Climate smart agriculture · Smallholder farmer · Lead farmer

1 Introduction

Agriculture is an important sector in Malawi in achieving food security, economic growth and wealth creation, yet production has remained low. More than 64% of the country's population is directly and indirectly employed in the agricultural sector, which also accounts for 90% of foreign exchange earnings and 30% of Gross Domestic Product (Government of Malawi, 2016; Japan International Cooperation Agency, Malawi, 2022). Agriculture occupies about 56% of the land area covering 5.3 million hectares of the country's 9.4 million hectares, and supplies at least 65% of the manufacturing sector's raw material requirements (Suvedi et al. 2023). Despite its importance, the agricultural sector is facing several challenges, including overdependence on rain-fed farming, limited absorption of improved technologies, weak private sector engagement, and lack of investment capacity in mechanization. Other challenges include weaknesses in the current public extension system, shrinking public sector resources, HIV/AIDS pandemic which has reduced the labor force for farming activities, low irrigation development and climate change (Masangano and Mthinda 2012; Muyanga et al. 2020). Recently, climate change has emerged as the biggest challenge to the agricultural sector in Malawi.

The main impacts of climate change experienced in Malawi include severe impacts of droughts, dry spells, floods and late-onset of rainfall (Risk-informed Early Action Partnership [REAP], 2021). These impacts of climate change have resulted in crops drying before maturity, crop damage due to floods, soil degradation (soil erosion, loss of soil fertility), shortage of water, loss of land, reduction in yields or total crop failure and consequently food insecurity (Laurentiu 2015; TroCare 2017). Given the challenges resulting from climate change and climate variations, the Government of Malawi and other stakeholders such as farmer's organizations and Civil Society Organizations (CSOs) are promoting the adoption and practice of Climate Smart Agriculture to enhance the resilience of smallholder farmers to climate shocks.

CSA is an approach which involves the direct incorporation of climate change adaptation and mitigation measures into agricultural development, planning and investment strategies to increase the technical, policy and investment inputs in the environment to achieve sustainable agricultural growth for food production under climate change (Palombi and Sessa 2013). CSA aims at three main objectives: To sustainably increase agricultural productivity; to support equitable increases in farm incomes, food security and development, adapting and building the resilience of agricultural and food security systems to climate change at multiple levels; and to reduce greenhouse gas emissions from agriculture (Mizik 2021; FAO 2013).

The Ministry of Agriculture Irrigation and Water Development, through the Department of Land Resource and Conservation, is the key player in promoting CSA messages. Subject matter specialists at the ministry develop the messages and communicate the information to farmers through the Agriculture Communications Branch (ACB), which is in the Department of Agriculture Extension Services (DAES). Through DAES, the ministry disseminates the messages using varied methods and channels including print media such as leaflets, posters, magazines; electronic media such as radio, cinema, films; as well as interpersonal channels such as lead famers and extension agents.

However, the department has not yet fully ventured into online messaging because majority of farmers in Malawi are rural based and mostly low income earners who can hardly afford a computer or a smart phone to access internet services (Ndilowe 2013). Information is key to the development of individuals and communities, and farmers need appropriate and suitable information for enhanced agricultural production (Bonephace et al. 2022).

The above goal can only be attained by propagating up to date technologies and practices for enhanced production to farmers using suitable and appropriate communication channels (Agbamu 2006). Experience shows that any of the various forms of communication channels are applicable, as long as one clearly defines their audience well enough (Dingwall et al. 2019). To enhance communication effectiveness, it is essential to ensure a suitable alignment between the audience, the message, and the chosen medium. Previous studies have reported challenges in attempts to communicate with farmers resulting from poor understanding of the target audience for agricultural messages and the channels they use to access messages (Ndilowe, 2013; Mbukwa 2015).

Other studies also reported gaps in farmers' participation in content creation and channel selection for disseminating messages, despite the country's commitment to promoting farmer-led agricultural transformation as indicated in the National Agricultural Policy (Masambuka-Kanchewa 2019). Poor identification of communication channels, compounded by lack of farmer's participation in content creation and selection of communication channels affects farmers access to information. This is especially true in cases where farmers have limited access to the existing communication channels, or lack the technical understanding to use them. Increased use of communication channels preferred by and accessible to communicating institutions has also led to the use of channels that do not meet or address farmers' needs and preferences (Masambuka-Kanchewa 2013).

2 Methodology

2.1 Study Areas

The study was conducted in three districts; Phalombe, Nkhotakota and Dowa. Specifically, the study was conducted in Kasongo EPA of Phalombe, Linga EPA of Nkhotakota and Chibvala EPA of Dowa district. The first two districts, Phalombe and Nkhotakota were purposively selected because they were among the 16 districts experiencing severe impacts of climate change in Malawi, such as prolonged dry spells and late onset of rainfall and because they were among the six priority areas from previous studies on vulnerability to climate change by the Climate Adaptation for Rural Livelihoods and Agriculture (CARLA) project. On the other hand, Dowa district was selected because it was the implementation site of the Lead Farmer model by the Development Fund of Norway (CABMACC Baseline Report 2014; Kakota et al. 2017). A lead farmer is defined as a locally based and partially trained farmer who voluntarily facilitates the delivery of at most three good agricultural practices (Ragasa 2020) (Fig. 1).

2.2 Study Population

The study population comprised smallholder farmers who were practicing CSA technologies in Linga, Kasongo and Chibvala Extension Planning Areas. Smallholder farmer refers to those farmers with less than 2.0 hectares of land (IFAD 2013), or farmers owning small plots of land on which they grow subsistence or cash crops using family labour (Aaron 2012). At the time of the study, the number of families practicing CSA in Kasongo EPA were 1044, while Linga and Chibvala had 1507 and 4195 farming families respectively. From these farming families, the researcher selected a sample of 320-smallholder farmers, of which 52 were drawn from Kasongo, 85 form from Linga EPA in Nkhotakota, 185 from Chibvala EPA of Dowa district. The study population also comprised 8 Key informants from District Agriculture Offices, Department of Agriculture and Extension Services and the Department of Land Resources Conservation.

2.3 Sampling Procedures

In the present study, the researcher combined systematic random sampling and purposive sampling techniques. Systematic sampling technique was used to select individual farmers to participate in the study. Prior to the study, the researcher consolidated lists of all households practicing CSA technologies in selected villages of each particular EPA. After compiling the list of households, the researcher randomly

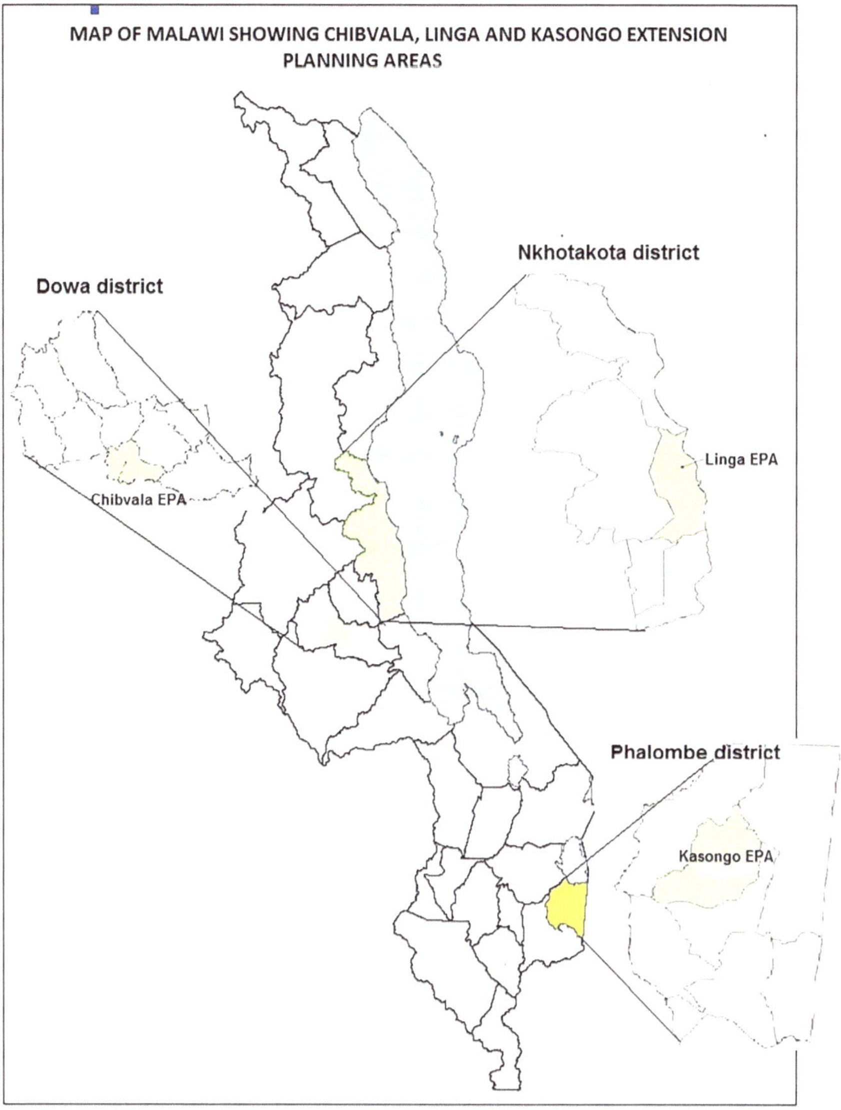

Fig. 1 Map of Malawi showing Linga, Chibvala and Kasongo EPAs

selected a first number and the remaining households were selected at a regular interval to determine the actual households to participate in the study from a given EPA. With regards to selection of Key Informants, the researcher used purposive sampling technique. A total of 8 key informants were involved in the study and these were: 3 Agricultural Extension Development Officers, 3 Agricultural Extension Development Coordinators, 1 official at the Department of Land Resources Conservation and 1 official at the Department of Agriculture and Extension Services.

2.4 Sampling Frame

Before undertaking the study, a reconnaissance survey was conducted to establish the number of villages and households practicing various climate smart agriculture technologies in the selected EPAs. Household listing was conducted in each village and the names of the households were assigned serial numbers. Households to be included in the study were selected using systematic random sampling technique. In this study, a household was defined as a person or group of persons, related or unrelated, who lived together in the same dwelling unit, and made common provisions for food and normally take their food from the same pot or shared the same grain store, or who pooled their income for the purpose of purchasing food.

2.5 Sample Size Formula

The following formula was used to determine the sample size for the study.

$$n = \frac{\left[Z^2(1-p)p\right]}{e^2}$$

where: Z = 1.96 (2 tailed test), e = 0.05 (margin of error), p = proportion of farmers practicing CSA technologies. The value of p as captured during the reconnaissance survey was 0.035 for Kasongo EPA and for Linga EPA was 0.06 while Chibvala EPA had 0.14. The study used 95% (Z = 1.96) level of confidence and a 5% or ($e = 0.05$) margin of error. Using the above formula, the total calculated sample for the study was 322 (Phalombe 52, Nkhotakota 85, and Dowa 185).

2.6 Data Collection Methods and Tools

This study was conducted between the months of January and March 2016. Data was collected using semi-structured questionnaires, interview guides and focus group discussion (FGD) guides. A total of 320 semi-structured questionnaires were administered to 320 smallholder farmers, of which 207 were female. Semi-structured questionnaires were adopted because of their flexibility in striking a balance between structured and unstructured formats, allowing for a mix of closed-ended and open-ended questions (Emde 2014). This flexibility permits respondents to elaborate on their answers, providing richer qualitative data. Key informant guides were also used to solicit information from 8 key informants. Three focus group discussions were conducted in each extension planning area, making a total of 9 focus group discussions for the whole study. Each focus group discussion comprised 10 smallholder farmers, making a total of 90 participants for the focus group discussions. Out of the three FGD done at each extension planning area, one group comprised male respondents only, the second group comprised female participants only while the third group combined both men and women. FGDs were used to collect information because they are a socially oriented research method for capturing real-life data. In addition, focus group discussions have flexibility, high face validity, provide speedy results and they are relatively cheap (Hennink et al. 2020). Besides the above, Focus Group Discussions were also used because they assist to gather in-depth information on the phenomena under study (Gundumogula 2020).

2.7 *Validity and Reliability of Data Collection Tools and Instruments*

To ensure that data collected met recommended quality, validity and reliability tests were applied. Ary et al. (2018) defined validity as the degree to which an instrument accurately measures what it is expected to measure. On the other hand, reliability was defined as the degree of consistency with which an instrument measures whatever it is measuring. The validity of the data collection instruments/tools (questionnaires and interview guides) was achieved by sharing the tools with extension experts who carefully reviewed and approved them for the study, while reliability was achieved by ensuring that the questionnaires were pre-tested before the actual study. In addition, the researchers applied the half split reliability coefficient of test–retest to ensure that some of the participants who were interviewed using questionnaires were also included in the FGDs. This procedure was followed to ensure internal consistency.

2.8 *Research Questions*

The study sought to answer four main questions: (1). What are the communication channels used in disseminating CSA messages to smallholder farmers in the Linga, Kasongo and Chibvala EPAs? (2). How often do farmers access available communication channels to get CSA messages? (3). What are the farmers preferred communication channels for accessing CSA messages in the Linga, Kasongo and Chibvala EPAs? And, (4). What is the relationship between farmers' attributes and their choice of communication channels for accessing CSA messages? To address these questions, a survey approach was adopted whereby a questionnaire was administered. To answer question 1, respondents were asked to list the communication channels being used by the Ministry of Agriculture, through the Department of Agriculture and Extension services and other organizations in the dissemination of CSA messages in their areas. On farmer's access to available communication channels, the respondents were asked to indicate the number of days in a month that they accessed a given communication channel to get CSA messages. Similarly, to answer the question on farmers preferred communication channels, the respondents were asked to list their preferred communication channels for accessing CSA messages, indicate reasons for their preference and rank the channels from the most to least preferred channel. In addition to determining available communication channels and farmers' preference for communication channels, the study went further to understand the relationships between farm and farmers' characteristics and preference for channels of communication. In this regard, the study used Chi-squire test.

2.9 Ethical Considerations

Ethical considerations were observed by ensuring informed consent and willingness to participate in the study. All the participants were briefed on the purpose and benefits of the study and on how the information they provided would be used. The researcher also ensured that the privacy and confidentiality of participants were respected by assigning numbers to respondents instead of using names. In the presentation of the results, the researcher also ensured that no information relating to the identities of the participants was included. This was done to conceal the identities of the participants.

2.10 Theoretical Framework

The present study adopted two theories: The Communication theory by Shannon and Weaver (1949) and the Uses and Gratification theory by Bumler and Katz (1973). The communication theory was adopted to help explain the impact of communication channel on the quality of communication. The communication theory assumes that the channel is the most important component in the communication process because it determines the quality and success of any communication process (Keyton 2010). On the other hand, the Uses and Gratifications theory was used to help explain why farmers utilized certain communication channels and not others. The Uses and Gratification theory is mainly concerned with identifying how people utilize methods of communication to satisfy needs and achieve goals. (Katz et al. 1973). The U&G theory was adopted to help explain the motivations for consuming particular media content. The theory has also been selected because it takes into account the recipients' motives, thereby discarding the premise that media recipients are passive and that they are all influenced by media in the same manner. Despite their criticisms, the two theories remain relevant in the Malawian context because they clearly explain the barriers to effective communication, and give insights into the motivations for farmers to consume particular media content.

3 Results and Discussion

This section presents the results and discussion of the results. The section commences with an exploration and analysis of results on communication channels for disseminating CSA messages, and continues to discuss farmers' access to communication channels and preferred communication channels for accessing CSA messages. The section concludes by addressing the correlation between farmers' characteristics and attributes, and their selection of communication channels.

3.1 Communication Channels for Disseminating CSA Messages

The most common communication channels for disseminating CSA messages in the study sites were extension workers (26%), lead farmers (25%) and radio (25%), while the least available communication channels were printed materials such as leaflets (1%) and newspapers (0%). The results were similar to findings by Kakota et al. (2017) who also reported that extension workers, lead farmers, radios, village meetings, field days and demonstration plots were the most common channels for disseminating CSA messages in Malawi. The current findings concur with a study by Mbukwa's (2015) in Bolero, Rumphi district, Malawi, where extension workers emerged as the most frequently utilized communication channel, followed by lead farmers and radio.

Elsewhere in Nigeria, Okwu and Daudu (2011) reported that the most common communication channels used were interpersonal communication channels such as friends/neighbors/relations, followed by extension service providers. Just like in Malawi, the availability of agricultural information through television and printed media was limited. Disaggregation of data based on study sites revealed distinct variations in the distribution of communication channels across the three study areas. Nevertheless, across all three study sites, extension workers, lead farmers, and radio emerged as predominant communication channels, whereas printed materials were consistently ranked as the least frequently utilized channels (Fig. 2).

Key informants from District Agricultural Offices and Agricultural Extension Planning Areas reported that extension workers and lead farmers were readily available in the communities and that there were several radio stations broadcasting CSA messages, including community radio stations such as Nkhotakota Community radio in Nkhotakota and Chanco Radio in Phalombe. This was also confirmed during FGDs where participants indicated that printed materials were used on special occasions such as field days and not as an everyday communication channel. Similarly, Farooq et al. (2007) and Pathak and Patel (2022) both noted that while print media is a key source of agricultural information, it is not the most utilized, with fellow farmers being the primary source.

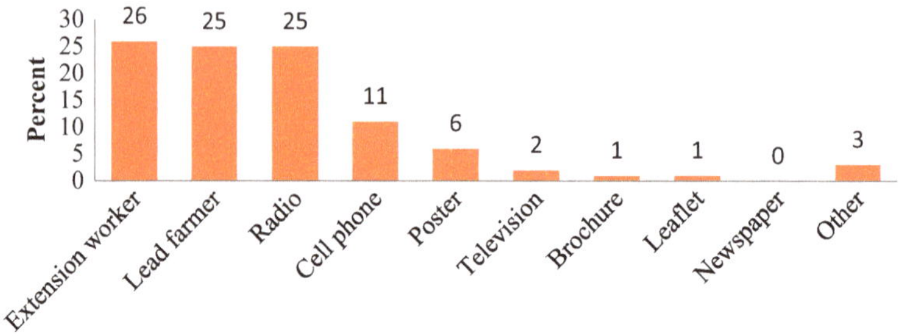

Fig. 2 Communication channels for disseminating CSA messages (N = 320)

The study established that printed materials such as posters were used to disseminate CSA messages during special events only such as Field days. During another Focus Group Discussion in Kasongo AP in Phalombe district, one respondent reported that: *"Printed materials are brought by extension workers from the town and shared with lead farmers who in turn distribute them to us, farmers, or paste them in strategic places such as shops and along busy roads"*. The findings were also consistent with the results reported by Rehman et al. (2011) who reported limited usage of printed materials in Pakistan. The cost of printed materials also affected the choice and use of printed materials in disseminating and accessing agricultural messages.

In Cambodia, Norin (2008) also established that radio was a widely available and used communication channel, while printed materials were an inaccessible communication channel. However, these findings are in contrast to the results reported by Hassan et al. (2010) during a study on Agriculture communication in Malaysia, whereby printed materials, especially newspapers were among the most dominant channels for disseminating agricultural messages. The differences in the results established during the current study and the one conducted in Malaysia might be because of the differences in the economic context of the two countries, whereby Malawi has 3 major newspapers while Hassan et al. (2010) reported that Malaysia had 50 newspapers, of which 13 were in local languages, hence they were widely used.

3.2 Farmer's Access to Communication Channels for Disseminating CSA Messages

To establish farmers access to communication channels, respondents were asked to indicate the number of days they accessed a given channel to access agricultural messages. The results showed that farmers had more access to radio and lead farmers, with an average of 6.83 (± 8.14) and 4.84 (± 5.47) days in a month respectively. Printed materials such as posters, brochures and newspapers were the least accessed communication channels, with an average of 0.23, 0.06 and 00 respectively.

3.2.1 Access to Radio

Radio was a widely accessed communication channel, confirming results of the study which indicated that 51.3% of the farmers owned radio sets. During focus group discussions, one male participant reported that: "We access most of the agricultural messages through radio because majority of us have radios sets. For farmers that do not have, they still listen from their neighbours' radios. Sometimes, we listen to the agricultural programmes on the radio at places like the market place or other places where we meet to pass time" (Male participant, Chibvala EPA). The focus group discussions also revealed that men had more access to radio than women because men controlled access to the radio. It was reported that men could sometimes carry the radio to the market place, garden or other places where they went to pass time while women usually stayed at home.

Besides men controlling access to radio, it was also reported that some women did not know how to operate the radio sets; as such, they could only listen to the radio when their husbands or older children were around. These findings concur with Zossou et al. (2015) who studied the link between local rice processors' access to rural radio, gender, and livelihoods in Benin. It was also reported that whereas radio was a common source of agricultural messages, access to radio and listening habits between men and women were different. More men (63.3%) listened to radio than women because men had more access and control over radio than women at the household level.

3.2.2 Access to Lead Farmer

Farmers access to lead farmers on an average of 4.84 ± 5.47 days in a month. High access to lead farmers was attributed to the fact that they lived within the same locality with the farmers, unlike extension agents who lived outside the communities and managed relatively bigger areas. Focus group discussions and key informant interviews revealed that the majority of the villages in the sampled EPAs had a lead farmer. During a focus group discussion in Dowa, one participant reported that: "We have easy access to lead farmers because they live among us in the communities, compared to extension workers who normally live in town and only visit the village when there is a need. As such, we visit lead farmers more often and anytime to ask for information on the different CSA technologies. Sometimes they visit us in our homes or gardens to appreciate the technologies that we are implementing" (Female participant, Chibvala EPA, Dowa district). However, these findings were contrary to the results established during a study on farmer-to-farmer extension approach in Malawi by Khaila et al. (2015), which reported that lead farmers had challenges reaching out to farmers due to mobility challenges. The study further reported that 84% of the lead farmers had no means of transport, and that only 16% had push bicycles, making it difficult for them to deliver their services.

3.2.3 Access to Extension Workers

Farmers had less access to extension workers (2.07 $\pm$ 2.15 days in a month). Limited access to extension workers was attributed to few extension workers available in the study areas. The problem of inadequate extension staff was compounded by a lack of mobility and the desire of most extension workers to stay in urban areas. These findings were consistent with the results reported in a study on Pluralistic Agricultural Extension in Malawi: Evidence and Lessons from 3 years of Research and Assessments by Ragasa et al. (2019), which reported that the ratio of extension worker to farmer was between 1: 2240 and 1: 3316, as opposed to the recommended ratio of 1:1750 (Masangano and Mthinda, 2012).

Andersen established that relatively more educated extension agents refused to stay in the rural areas due to unfavorable conditions such as lack of electricity. Inadequate extension staff, lack of mobility and the desire of extension workers to stay in urban areas had a huge impact on the delivery of extension services because it affected the frequency at which extension agents interacted with farmers.

3.2.4 Access to Cell Phones and Printed Media

The study indicated that farmers had little access to cell phones and printed materials. The study found that only 40.5% of the respondents had cell phones. The respondents cited lack of money to purchase the phones, maintenance costs and the need to charge phones as the main prohibitive factors to owning cellphones. "Phones are good for communication because they are fast and give you the opportunity to engage with the other person directly. However, the challenge is that phones are expensive to buy and they require charging" (Male participant, Linga EPA).

Although 40.5% of the respondents had cell phones, Dowa was the only district where respondents reported that they had ever received agricultural messages through phone. According to the AEDC for Chibvala EPA in the district, there was a project in the past in the area that was sending agricultural messages to farmers, and one had to register to access the messages. As such, even though some farmers had cell phones, they could not get the messages because they had not registered their numbers. With regards to printed materials, it was established that printed materials such as posters, leaflets and brochures were rarely used. During focus group discussions, participants reported that printed materials were normally used during special events such as field days and not as everyday communication channels.

3.3 Farmers' Preference for Communication Channels Used to Access CSA Messages

Figure 3 presents farmers' preference for communication channels for accessing CSA messages in Chibvala, Linga and Kasongo EPAs.

The results above indicate that extension workers, lead farmers and radio were the most preferred communication channels in all study sites, while cell phones and printed materials such as posters and brochures were the least preferred channels. The data was further disaggregated according to the study locations to appreciate differences in preferences for the communication channels based on study locations.

Despite the study being conducted in three separate locations, the most preferred communication channels remained extension workers, lead farmers and radio. However, there were minor variations among the locations, whereby extension workers were the most preferred channel in Kasongo and Chibvala at 37% and 29% respectively, followed by lead farmers in Chibvala and Linga at 27% and 29% respectively. Radio ranked similar to lead farmers in Linga at 29%, followed by Kasongo at 28% and Chibvala at 22%. The reasons for preference for the three communication channels were similar in all study locations. Extension workers and lead farmers were preferred due to continued presence in the communities, which made them accessible to farmers, as well as instant feedback and individualized presentations. Lead farmers were also preferred because of easy communication, since they shared the same cultural beliefs and language as the community. On the other hand, radio was preferred because of speed and wide coverage. Radio was also preferred because of its flexibility in that one could be listening to an agricultural program while working in their gardens or performing household chores.

Although farmers indicated preference for Extension workers, they also reported challenges associated with the channels. For instance, farmers indicated inadequate extension workers and tendency of some extension workers to reside outside their catchment areas, which resulted in limited contact with farmers. Other key challenges included the use of technical language, or language not familiar with the

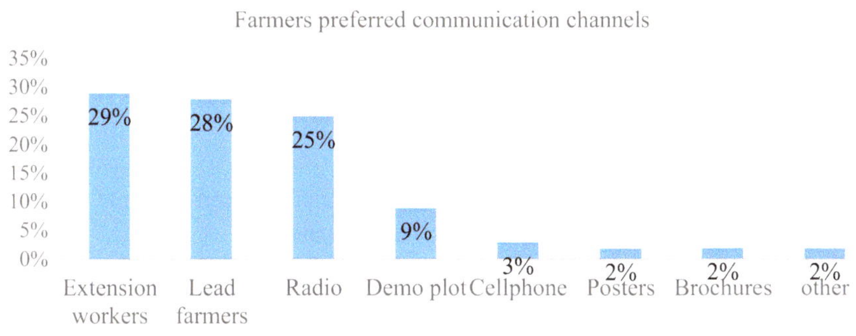

Fig. 3 Preferred communication channels across the three study site (Chibvala, Linga and Kasongo EPAs)

farmers which prevented farmers from understanding the messages. In communication theory, the use of technical language could be regarded as noise which may prevent effective communication. Using various channels to disseminate messages to farmers is not a guarantee that the information will be understood by the farmers. There is need to ensure that appropriate channels are used to deliver the messages to the farmers as intended.

Challenges associated with lead farmers were that they were not conversant with many technologies as they had not undertaken comprehensive training (Malawi Ministry of Agriculture and Water Development, Department of Agriculture Extension Services [DAES], 2015; Khaila et al. 2015; Kakota et al. 2017). With regards, to radio, participants described it as an inflexible medium and not good for disseminating complex messages. The participants explained that radio provided a lot of information at once, and that it did not provide an opportunity for the respondents to pause the program to ask questions or seek clarification. Despite incorporation of other ICT based platforms such as cell phones, whereby listeners had the opportunity to call in during live programs, participants complained that most of the programs were short while in other instances, the programs were pre-recorded thereby reducing the farmer to a mere listener. In addition, the programs were aired at specific times of the day, and if one missed the program, they had no opportunity of tuning in to the program.

Application of the U&G theory to the findings on farmers' preference for communication channels corroborates well with the first and second assumptions, which assert that people use media to gratify specific wants and needs, in this case information. The audience has free will of choice and can select the media that most satisfies them (Katz et al. 1974). The study established that farmers were goal-driven and accessed specific channels based on the benefits they get from them. For example, farmers could choose extension workers or lead farmers because they provided instant feedback or could use radio because of immediacy of information. On the other hand, the farmers could not use certain channels such as printed materials because they lacked feedback mechanisms.

3.4 Relationship Between Farm and Farmers' Attributes and Preference for Channels of Communication for Accessing CSA Messages

The study further sought to understand the relationships between farm and farmers' characteristics and preference for communication channels.

Table 1 presents cross tabulation of age, education level, gender, farm size and occupation against preferred communication channels. The results show that in terms of age, the majority of older farmers preferred radio and extension workers, while the majority of the younger farmers preferred extension workers. The results also show that farmers that attended primary school preferred extension workers, those that attended secondary school preferred radio and extension workers, and those that attended tertiary education preferred radio. The results also show that male small holder farmers preferred radio and extension workers while female small holder farmers preferred extension workers. Furthermore, the study established that farmers with small landholding size preferred extension workers while those with larger land holding size preferred extension workers and radio. Similarly, those respondents with formal employment preferred extension and radio while those involved in petty trading preferred radio and extension workers. Lastly, the study sought to determine the relationship between the main occupation of the farmers against preference for communication channels. The table shows that respondents whose main occupation was agriculture preferred radio and extension workers. These findings correlate with the Uses and Gratification theory which assumes an active audience that makes particular choices, and that the audience has free will of choice and can select the media that most satisfies them. In this case, despite the farmers having access to a wide range of communication channels, they deliberately selected specific channels which could satisfy their information needs (Table 2).

In the present study, the researcher used: age, education levels, gender, farm size and main occupation. The findings are presented in Table 1.

The study findings revealed a significant relationship between choice of radio as a communication channel for accessing CSA messages and the following variables; education level ($P = 0.00$), gender ($P = 0.01$) and farm size ($P = 0.01$). Out of the three variables, education level had a relatively higher Cramer's of 0.2 exhibiting a comparatively stronger relationship. Although age is not significantly associated with choice of radio as a communication channel, the Phi of 0.42 shows a stronger relationship. Farm size show a significant ($P = 0.00$), and a stronger relationship with choice of extension worker as a communication channel. Gender does not show a significant relationship with choice of extension worker as a communication channel, however, it exhibits Phi of 0.32. The results further show that farm size ($P = 0.03$), gender ($P = 0.04$) and age ($P = 0.06$) are significantly associated with choice of lead farmer as a communication channel. Cramer's V/Phi of 0.15, 0.14 and 0.13 respectively shows a relatively stronger relationship. Although education level is not significantly associated with choice of posters as a communication channel for CSA messages, however, the Cramer's V of 0.13 suggests a relatively strong relationship.

Further analysis revealed that farmers with more farm land preferred radios, extension worker and lead farmer as the main sources of CSA messages. These finding concur with Kimaru-Muchai et al. (2013) in Kenya who found a positive relationship between land size and preference for radio as a communication channel among the smallholder farmers. Analysis also showed that male farmers showed preference towards radio and lead farmers. This result differed with Mapiye et al. in South Africa where he found that male farmers had a higher likelihood of using extension agents than female farmers. The difference might be due to social cultural perceptions associated with the role and functions of extension agents. Results also show that educated farmers opted for radio, and older farmers preferred lead farmers. The above findings concur with Okwu and Daudu (2011) in Benue State, Nigeria who

Table 1 Frequency distribution of preference for communication channels against farm and farmer characteristics

		Age			Education level				
Communication channel		Youth	Older	Total	Never	Primary	Secondary	Tertiary	Total
Radio	Least preferred	84	44	128	29	82	17	0	128
	Moderately preferred	36	23	59	3	44	12	0	59
	Most preferred	84	52	136	12	80	39	2	133
Extension	Least preferred	36	22	58	9	39	9	1	58
	Moderately preferred	101	69	170	27	99	43	1	170
	Most preferred	67	28	95	8	68	16	0	92
Lead farmer	Least preferred	92	53	145	24	95	25	1	145
	Moderately preferred	63	25	88	11	56	20	1	88
	Most preferred	49	41	90	9	55	23	0	87
Poster	Least preferred	187	112	299	38	195	64	2	299
	Moderately preferred	15	4	19	6	9	4	0	19
	Most preferred	0	0	0	0	0	0	0	0

		Gender			Farm size			Occupation			
Communication channel		Male	Female	Total	Small	Big	Total	Agriculture	Formal Employment	Petty trading	Total
Radio	Least preferred	93	35	128	87	41	128	123	2	2	127
	Moderately preferred	50	9	59	32	27	59	54	3	2	59
	Most preferred	115	18	188	69	67	136	125	4	4	133

(continued)

Table 1 (continued)

Communication channel		Gender			Farm size			Occupation			
		Male	Female	Total	Small	Big	Total	Agriculture	Formal Employment	Petty trading	Total
Extension	Least preferred	46	12	58	38	20	58	57	1	0	58
	Moderately preferred	133	37	170	108	62	170	157	6	6	169
	Most preferred	79	13	92	42	53	95	88	2	2	92
Lead farmer	Least preferred	109	36	145	89	56	145	135	5	4	144
	Moderately preferred	72	16	88	57	31	88	84	3	1	88
	Most preferred	77	10	87	42	48	90	83	1	3	87
Poster	Least preferred	240	59	299	178	121	299	282	8	8	298
	Moderately preferred	16	3	19	10	9	19	18	1	0	19
	Most preferred	0	0	0	0	0	0	0	0	0	0

Table 2 Relationship between farmers' characteristics and preference for communication channels for accessing CSA messages

Variable	Radio		Extension worker				Lead farmer			Poster		
	Chi-square	*P* value	Phi/ Cramer's V	Chi-square	*P* value	Phi/ Cramer's V	Chi-square	*P* value	Phi/ Cramer's V	Chi-square	*P* value	Phi/ Cramer's V
Age	0.56	0.75	0.42	3.27	0.19	0.10	5.63	0.06	0.13	2.08	0.15	0.08
Education level	24.83	0.00	0.20	9.32	0.16	0.16	4.93	0.55	0.09	5.61	0.13	0.13
Gender	8.75	0.01	0.17	2.31	0.32	0.32	6.30	0.04	0.14	0.18	0.67	0.02
Farm size	8.52	0.01	0.16	10.90	0.00	0.18	7.09	0.03	0.15	0.35	0.55	0.03
Main Occupation	2.68	0.61	0.09	3.08	0.55	0.07	2.22	0.70	0.06	0.93	0.63	0.05

found a significant relationship between communication channels used by farmer's and their education level and farm size.

4 Conclusion

This section presents a summary of the key findings. Based on the study results, radio, lead farmers and extension agents were the most accessible means for acquiring CSA messages in the three study sites. Lead farmers and extension agents were more accessible because they operated from within their localities while radio messages were more accessible because many people owned radio sets. The study further established that the main preferred communication channels were extension agents, lead farmers and radios. The main reasons for preferring extension workers and lead farmers were the ability to provide individualized messages and face to face interaction, which offered instant feedback. On the other hand, radio was preferred because of speed and coverage.

In summary, farmer's preference for a given communication channel is driven by the desire to acquire more elaborate and contextual information that is suitable and usable in their respective contexts. The study established significant relationships between farm and farmer characteristics such as farm size, gender, age and education level and choice of communication channels. The study therefore shows that younger male farmers who have had access to basic education were the main group of farmers that opted to acquire CSA messages through extension workers, lead farmers and radio.

References

Aaron J (2012) A framework for the development of smallholder farmers through cooperative development. Department of Agriculture, Forestry and Fisheries. Republic of South Africa, pp 1–8. Available at https://www.google.com/url?sa=i&rct=j&q=&esrc=s&source=web&cd=&cad=rja&uact=8&ved=0CAMQw7AJahcKEwiYra343On4AhUAAAAAHQAAAAAQAg&url=https%3A%2F%2Fwww.nda.agric.za%2FdoaDev%2FsideMenu%2Fcooperativeandenterprisedevelopment%2Fdocs%2FFRAMEWORK-%2520OF%2520SMALL%2520FARMERS%2520(2).pdf&psig=AOvVaw3totpnC0dxH__hxwFAmLXn&ust=1657384533730757. Accessed on 10 Mar 2022

Agbamu JU (2006) Essentials of agricultural communication in Nigeria. Malthouse Press Ltd

Ary D, Jacobs LC, Irvine CKS, Walker D (2018) Introduction to research in education. Cengage Learning

Bonephace EM, Gui X, Makawia PJ (2022) Assessment of awareness of Farmers on agricultural information services in Mara Region, Tanzania. Library Philosophy & Practice

Blumler JG, Katz E (1973) Uses and gratifications research. Public Opin Q 37(4):509–523. https://doi.org/10.1086/268109

Dingwall JR, Labrie C, McLennon T, Underwood L (2019) Professional communications

Emde M (2014) Open-ended questions in Web surveys—using visual and adaptive questionnaire design to improve narrative responses

FAO (2013) Sourcebook on climate smart agriculture, forestry and fisheries. Available at https://www.fao.org/docrep/018/i3325e/i3325e.pdf. Accessed 29 May 2020
Farooq S, Muhammad S, Chauhdary KM, Ashraf I (2007) Role of print media in the dissemination of agricultural information among farmers
Government of Malawi (2016, September) National Agriculture Policy Ministry of Agriculture, Irrigation, and Water Development. Available at https://www.canr.msu.edu/fsp/countries/malawi/malawi_national_agriculture_policy_25.11.16.pdf
Gundumogula M (2020) Importance of focus groups in qualitative research. Int J Human Social Stud 8(11). https://doi.org/10.24940/theijhss/2020/v8/i11/HS2011-082
Hassan MS, Shaffril HAM, Samah BA, Ali MSS, Ramli NS (2010) Agriculture Communication in Malaysia: the Current Situation. Am J Agric Biol Sci 5(3): 389–396. https://doi.org/10.3844/ajabssp.2010.389.396
Hennink M, Hutter I, Bailey A (2020) Qualitative research methods. Sage
IFAD, U (2013) Smallholders, food security and the environment. Rome: International Fund for Agricultural Development. In: 29 international conference on information and communication technologies. Available at https://www.ifad.org/documents/38714170/39135645/smallholders_report.pdf/133e8903-0204-4e7d-a780-bca847933f2e. Accessed on 27 June 2028
Japan International Cooperation Agency, Malawi Office (2022, May) Sector position paper: Agriculture. Available at https://www.jica.go.jp/malawi/english/activities/c8h0vm00004bpzlh-att/agriculture.pdf.
Kakota TV, Maonga BB, Synnevag G, Chonde C, Mainje M (2017) Harmonization of extension messages on climate smart agriculture in Malawi: Do we speak with one voice, and to whom? J Agric Extension Rural Develop 9(11):255–261. https://doi.org/10.5897/jaerd2017.0905
Kakota T V, Maonga B, Synnevåg G, Chonde C, Mainje M (2017) Harmonization of extension messages on climate smart agriculture in Malawi: Do we speak with one voice, and to whom? J Agric Ext Rural Dev 9(11):255–261 https://doi.org/10.5897/jaerd2017.0905
Khaila S, Tchuwa F, Franzel S, Simpson S (2015) The farmer-to-farmer extension approach in Malawi: a survey of lead farmers. World Agroforestry Centre (ICRAF), Nairobi, Kenya.
Katz E, Blumler JG, Gurevitch M (1973) Uses and gratifications research. Public Opin Quart 37(4):509–523. https://doi.org/10.1086/268109
Katz E, Blumler JG, Gurevitch M (1974) Utilization of mass communication by the individual. In: JG Blumler, E Katz (eds) The uses of mass communications: Current perspectives on gratifications research (pp. 19–32) Sage Publications
Keyton, J (2010) Communication and organizational culture: a key to understanding work experiences. Sage Publications
Kimaru Muchai SW, Mucheru Muna M, Mugwe JN, Mairura FS, Mugendi DN (2013) Gender disparities in uptake of information on soil fertility management in the central highlands of Kenya. International J Agric Sci Res 3(1):197–206
Laurentiu PI (2015) Strategic directions for food security by balancing the report of demand and supply on global agricultural market. In: The Research Institute for Agriculture Economy and Rural Development. International Symposium. Agrarian Economy and Rural Development: Realities and Perspectives for Romania. Proceedings. The Research Institute for Agriculture Economy and Rural Development, p 41
Malawi (2015, May) Ministry of Agriculture, Irrigation and Water Development Department of Agricultural Extension Services (DAES), Nutrition Unit & Agricultural Communication Branch. Nutrition Handbook for Farmer Field Schools. Lilongwe: Government Printer
Masambuka-Kanchewa F (2013) Agricultural information perceptions and behaviors of smallholder farmers in the central region of Malawi (Doctoral dissertation, Purdue University). Available at https://ag.purdue.edu/department/asec/_docs/graduate/thesis/Fally-Masambuka.pdf. Accessed on 21 Dec 2018
Masangano C, Mthinda C (2012) Pluralistic extension system in Malawi. Available at https://www.ifpri.org/publication/pluralistic-extension-system-malawi. Accessed on 13 May 2018

Masambuka Kanchewa F (2019) Agricultural information perceptions and behaviors of smallholder farmers in the Central Region of Malawi (Master's thesis). Purdue University. Available at https://docs.lib.purdue.edu/dissertations/AAI1544423/
Mbukwa P (2015) Investigating the influence of communication channels on the adoption of CSA technologies in Bolero, Rumphi district. Unpublished M.Sc. thesis. Mzuzu University. Available at http://repository.mzuni.ac.mw. Accessed on 27 July 2019
Mizik T (2021) Climate-smart agriculture on small-scale farms: a systematic literature review. Agronomy 11(6):1096
Muyanga M, Nyirenda Z, Lifeyo Y, Burke WJ (2020) The future of smallholder farming in Malawi (No. 20/03). MwAPATA Institute Working Paper. https://doi.org/10.22004/ag.econ.319855
Ndilowe (2013) An investigation of the role of communication in the Malawi agriculture sector-wide approach-special project (ASWAP- SP), a closer look at conservation agriculture: the case of Mitundu extension planning area (EPA) in Chisamba village. Master in the Nordic Media University of Oslo. Henry Holt and Company, New York
Norin P (2008) Communication channels for agriculture information to farmers; a case of Kampong Speu. Available at http://uc.edu.kh. Accessed on 21 Oct 2017
Okwu OJ, Daudu S (2011) Extension communication channels' usage and preference by farmers in Benue State, Nigeria. J Agric Ext Rural Dev 3(5):88– 94
Palombi L, Sessa R (eds) (2013) Climate smart agriculture: Sourcebook Food and Agriculture Organization of the United Nations. Available at https://www.fao.org/4/i3325e/i3325e.pdf
Pathak S, Patel MB (2022) Farmers perception about print media in agricultural information dissemination: a review. Bhartiya Krishi Anusandhan Patrika
Ragasa C, Mthinda C, Chowa C, Mzungu D, Kalagho K, Kazembe C (2019) Assessing and strengthening Malawi's pluralistic agricultural extension system: evidence and lessons from a three-year research study. Intl Food Policy Res Inst
Ragasa C (2020) Effectiveness of the lead farmer approach in agricultural extension service provision: nationally representative panel data analysis in Malawi. Land Use Policy 99:104966 https://doi.org/10.1016/j.landusepol.2020.104966
Rehman F, Muhammad SH, Ashraf I, Hassan S (2011) Factors affecting the effectiveness of print media in the dissemination of agricultural information. Sarhad J Agric 27(1):119–124
Risk-informed Early Action Partnership (REAP) (2021) REAP country case study: Malawi. https://www.early-action-reap.org/sites/default/files/2021-12/REAP%20Case%20study%20-%20Malawi.pdf
Shannon CE, Weaver W (1949) The mathematical theory of communication. University of Illinois Press
Shannon W (1949) Shannon and Weaver model of communication. Communication Theory
Suvedi M, Sasidhar PVK, Agwu AE, Chanza C, Ukamaka MD, Liverpool-Tasie S, Elapata MS (2023) Strengthening agricultural extension training in Nigeria, Malawi, South Africa, Uganda, and Kenya
TroCare (2017) Feeling the heat: how climate change is driving extreme weather in the developing world
Zossou E, Vodouhe DS, Van Mele P, Agboh-Noameshie AR, Lebailly P (2015) Linking local rice processors' access to rural radio, gender, and livelihoods in Benin. Develop Practice 25(7):1057–1066. https://doi.org/10.1080/09614524.2015.1069261

Open Access This chapter is licensed under the terms of the Creative Commons Attribution 4.0 International License (http://creativecommons.org/licenses/by/4.0/), which permits use, sharing, adaptation, distribution and reproduction in any medium or format, as long as you give appropriate credit to the original author(s) and the source, provide a link to the Creative Commons license and indicate if changes were made.

The images or other third party material in this chapter are included in the chapter's Creative Commons license, unless indicated otherwise in a credit line to the material. If material is not included in the chapter's Creative Commons license and your intended use is not permitted by statutory regulation or exceeds the permitted use, you will need to obtain permission directly from the copyright holder.

Impacts of Extension and Advisory Services on Commercial Agriculture and Livelihoods Among Smallholder Farmers in Rural Malawi

Loveness Msofi Mgalamadzi and Charity Chanza

Abstract This chapter delves into the significance of extension and advisory services in promoting agricultural commercialisation and improving livelihood of smallholder farmers in rural Malawi. The underlying premise is that access to such services can enhance farmers' efficiency, encourage the adoption of advanced technologies, and expand their knowledge and skills, ultimately leading to increased production and surplus for sale. Moreover, these services play a crucial role in shaping farmers decision-making processes, enabling them to make informed choices regarding enterprises and adopt market-based farming approaches, resulting in a transformation of their attitudes and perceptions. Numerous studies have showcased how access to extension and advisory services positively impacts food security, boosts household incomes, and enhances overall welfare. To investigate this further, the chapter adopts a mixed-method research design, combining qualitative and quantitative data. Qualitative data was collected through focus group discussions (6) and key informant interviews (11), and quantitative data involving 126 households was collected using household survey. Qualitative data was analysed using thematic content analysis and quantitative data was analysed using descriptive statistics, employing statistical measures such as means, percentages, cross-tabulation, correlations, chi-square tests, and Analysis of Variance. Findings show that despite the anticipated contributions of extension and advisory services, their actual impact in driving commercial agriculture is limited. Other critical factors such as access to land, labour, and capital, as well as the policy and marketing environment, and weather conditions, play significant roles in influencing commercial agriculture and determining livelihoods. These factors remain beyond the scope of extension and advisory services.

L. M. Mgalamadzi (✉) · C. Chanza
Extension Department, Faculty of Development Studies, Bunda College, Lilongwe University of Agriculture and Natural Resources, Lilongwe, Malawi
e-mail: lmgalamadzi@luanar.ac.mw

C. Chanza
e-mail: cchanza@luanar.ac.mw

© The Author(s) 2026

B. R. Singh et al. (eds.), *Sustainable Food Systems in Sub-Saharan Africa*,
https://doi.org/10.1007/978-3-032-11288-0_23

Keywords Extension and advisory services · Commercial agriculture · Livelihoods · Smallholder farmer · Rural malawi

1 Introduction

Agriculture remains the most critical sector in many developing countries (Fuglie et al. 2020). It continues to contribute significantly to GDP and employment (Djurfeldt et al. 2018), serving as a primary source of income for the world's poor (Barrett et al. 2017) and a key livelihood for disadvantaged communities (Dorward et al. 2004). In Malawi, agriculture is the backbone of the economy, accounting for more than a third of the GDP. It is the main source of foreign earnings, employing the majority of the population, most of whom live in rural areas (Phiri et al. 2012). But, despite its importance, the sector faces numerous challenges, including unpredictable weather patterns, increasing pressure on land, poor crop management practices, and low adoption of modern technologies (Phiri et al. 2012). Additionally, heavy reliance on rain-fed agriculture exposes production to climate risks, while high transaction and transportation costs—partly due to poor road infrastructure—further hinder growth. Unstructured produce markets allow for unfair trading and market inefficiencies, while delays in input supply disrupt production. Moreover, high levels of poverty and food insecurity persist among farming communities (Chirwa et al. 2006).

Given agriculture's critical role in Malawi's economy, various initiatives have been implemented to enhance productivity and address these challenges. These include the formulation of policy frameworks aligned with national development goals. For instance, the Malawi Growth and Development Strategy (MGDS) I (2006–2011), II (2011–2016), and III (2017–2022) emphasize agriculture's role in driving economic growth, with a key objective of achieving sustainable development. Similarly, the National Agriculture Policy (NAP) (2016–2022) outlines a vision for Malawi's agricultural sector, focusing on increasing productivity and farm incomes while aiming to reduce poverty, enhance economic growth, and improve food security (GoM 2016; Lindsjö et al. 2021). More recently, the Malawi Agenda 2063 was introduced, envisioning a transition to a self-reliant, industrialized, upper-middle-income country by 2063. One of its three key pillars is agricultural productivity and commercialization (NPC 2020). Additionally, Malawi aligns its development efforts with global and regional frameworks such as the Sustainable Development Goals (SDGs), where agriculture plays a vital role in achieving SDG 1 (ending poverty), SDG 2 (ending hunger and promoting sustainable agriculture), and SDG 12 (ensuring sustainable consumption and production). The country is also committed to the African Union's Agenda 2063.

The drive to commercialize agriculture in Malawi dates to the colonial era when the colonial government promoted cash crop cultivation on estates and developed transport systems to improve market access (de Haas 2017). Von Braun and Kennedy (1994) define agricultural commercialization as the transition from subsistence farming—where production is primarily for household consumption—to commercial farming, where crops are grown for sale. Some scholars (Carletto et al. 2017) view this shift as a positive step toward agricultural sector development, economic growth, and improved livelihoods. The push for agricultural commercialization is rooted in the modernization paradigm, which regards subsistence farming as outdated and inefficient. Given agriculture's importance to Malawi's economy, significant efforts have been made to develop the sector and promote commercial farming. Key initiatives include input subsidies such as the Targeted Input Programme (TIP), the Farm Input Subsidy Programme (FISP) (Chirwa and Dorward 2013), and more recently, the Affordable Input Programme (AIP). These programs aim to increase agricultural productivity while also supporting low-income households by balancing economic and social welfare objectives (De Weerdt and Duchoslav 2022).

Several national strategies have been introduced to support agricultural growth and commercialization. These focus on market liberalization, rural marketing infrastructure development, agricultural market information systems, agro-processing, value addition, and the establishment of commodity exchanges. Key policies include the Malawi Growth and Development Strategy (MGDS), the National Export Strategy (NES), the Agricultural Sector-Wide Approach (ASWAp), the National Agriculture Policy (NAP) (2016), the National Agribusiness Strategy (NAS), and the Malawi Agenda 2063 (2020). Additionally, the Contract Farming Strategy aims to create an enabling environment for efficient, competitive, and fair contract farming, while the National Agricultural Extension and Advisory Services Strategy (NAEASS) (2020) focuses on building the capacity of farmers and other value chain players to meet the demands of both domestic and international markets. Private sector investments also play a role in agricultural commercialization, with opportunities arising through commodity exchanges and warehousing.

One of the key sub-sectors receiving focused efforts within agriculture is agricultural extension and advisory services. According to the Global Forum for Rural Advisory Services (GFRAS), these services play a vital role in enhancing agricultural productivity and reducing rural poverty in Malawi (Cai and Davis 2017). Additionally, research by Van den Ban and Samanta (2006) and Gebremedhin et al. (2006a, b) highlights the need for agricultural extension systems to evolve alongside the changing agricultural landscape, particularly in response to the goal of agricultural commercialization. Agricultural extension services have been defined in various ways, but for the purposes of this chapter, they refer to the provision of information, knowledge, and skill development to promote the adoption of improved agricultural technologies. They also facilitate connections with other institutional support services such as input supply, output marketing, and credit access (Berhanu et al. 2006).

Several studies have examined the current state of agricultural extension services in Malawi, with many identifying significant challenges that hinder effective implementation and limit desired outcomes. Researchers (Masangano and Mthinda 2012; Ragasa and Kaima 2017; Phiri et al. 2012; Chowa et al. 2013; Knorr et al. 2007; Beintema et al. 2016) have documented issues such as inadequate funding, the top-down nature of extension approaches, a lack of specialized skills among extension workers to address diverse farmer needs, insufficient human capacity, and coordination difficulties due to the pluralistic extension system outlined in the current agricultural extension policy. Despite ongoing efforts to strengthen extension services, some argue that these services have remained largely production oriented. Swanson and Rajalahti (2010), for example, note that for an extended period, extension programs have focused primarily on increasing production rather than supporting other activities across agricultural value chains. This narrow focus has led to surplus production that farmers often struggle to store or sell effectively. This chapter explores the role of agricultural extension in driving commercialization and its impact on livelihoods by addressing the following research questions:

1. Who has access to which extension services?
2. What challenges exist in delivering these services?
3. How does agricultural extension contribute to commercial farming?
4. What are the livelihood impacts of access to extension services and participation in commercial agriculture?

2 Theoretical and Conceptual Frameworks

This chapter examines the role of agricultural extension in helping farmers transition to commercial farming, highlighting key challenges, shortcomings, and gaps. The chapter contributes to debates about the role and significance of agricultural extension commercial based agriculture, with some scholars arguing that exposure to these extension services equips farmers with the necessary knowledge, skills, attitudes, and mindset towards commercialisation (Daane et al. 2012; Van den Ban and Samanta 2006; Msofi 2023). Enhanced knowledge and skills help farmers to make informed production decisions, improve efficiency, and increase productivity for both consumption and sale. Furthermore, access to agricultural information influences farmers' attitudes and perceptions, encouraging greater market participation. The evolving mandate of agricultural extension services is to facilitate this shift from subsistence to commercial farming by providing support beyond production, including marketing activities across the agricultural value chain (Gebremedhin et al. 2006b, 2012; GFRAS 2012; Lemma et al. 2014; Van den Ban and Samanta 2006). However, some scholars (Anderson and Feder 2003; Zhou et al. 2013) argue that while extension services create an enabling environment, they are not the sole determinant of commercial orientation among farmers. Other factors, such as access to production resources and market opportunities, often have a more direct impact on commercialization, even when farmers receive extension support.

This chapter is framed within the Diffusion of Innovations Theory (Alene and Manyong 2006; Islam and Madsen 2018; Padel 2001; Rogers 1983) to assess how agricultural extension contributes to commercialization. The diffusion of innovation theory explains how new ideas, innovations and practices spread across a social system overtime focusing on the characteristics of the innovation, the communication channels used to disseminate the innovation, and the social context of the potential adopters (Rogers 1983). The Diffusion of Innovations Theory is applied to understand how access to extension services and the social characteristics of farmers influences farmers' decision to adopt commercial farming. Previous studies (Birkhaeuser et al. 1991; Cawley et al. 2018; Dercon et al. 2018; Hamilton and Hudson 2017; Jin and Huffman 2016; Nakano et al. 2018; Olagunju and Adesiji 2013; Ragasa and Mazunda 2018; Waddington et al. 2010; Wossen et al. 2017) have highlighted various roles of agricultural extension, including: facilitating the adoption of new technologies and innovations; improving productivity and efficiency in farming; enhancing household income and livelihoods; reducing poverty levels; strengthening food security. According to Mur et al. (2016), the overarching goal of agricultural extension is to empower farmers to make informed decisions that enhance their farming practices and overall well-being. This implies that access to extension services plays a crucial role in helping farmers achieve their agricultural and economic goals.

The chapter also draws on Modernization Theory to examine the rationale behind agricultural commercialization. The study defines agricultural commercialisation as the transition from subsistence farming (primarily for consumption) to commercial farming (production for market sale) (Von Braun and Kennedy 1994). The modernization paradigm assumes that subsistence farming is outdated and inefficient, advocating for the adoption of market-oriented agricultural practices (Msofi 2023). This perspective has largely been promoted by institutions like the World Bank across many African countries (Bernstein 2008). Within this modernization framework, agricultural extension is expected to play a critical role in encouraging farmers to adopt modern technologies (Scott 1998), emphasizing the need to analyse how extension services contribute to commercialization.

This chapter also explores disparities in access to extension services and their subsequent impacts, particularly across different genders and social classes. To achieve this, the chapter is grounded in the Social Differentiation Perspective (Dancer and Hossain 2018), which postulates that households are not homogenous, and these intrinsic differences can lead to imbalanced access to and benefits from agricultural extension services. Differences in access to extension services may shape farmers' market participation, and unequal participation in commercialization may further reinforce existing gender and class disparities. Previous research has documented gender and class based disparities in access to agricultural extension services (Gwandu et al. 2013; Loki et al. 2021; Mbo'o-Tchauawou and Colverson 2014; Mudege et al. 2016; Umeta et al. 2011a). Additionally, studies have shown that crop commercialization itself can exacerbate gender and class inequalities (Mgalamadzi et al. 2021; Mgalamadzi et al. 2024). This chapter seeks to provide a deeper understanding of how agricultural extension shapes and is shaped by commercialization and their implications for social equity and livelihoods in general.

Conceptually, the assumption is that differentiated access to agricultural extension services—which also vary in content and how they are delivered—result in differentiated outcomes in livelihood activities which include commercialisation. Furthermore, the assumption is that the differential engagement in commercial farming result in different livelihood outcomes at household level. Figure 1 presents the conceptual framework.

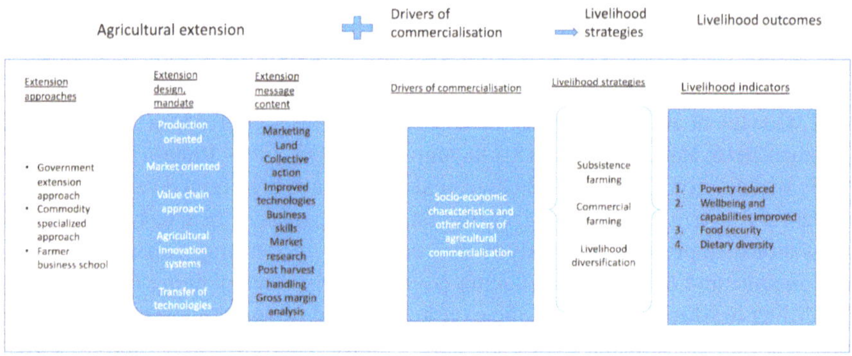

Fig. 1 Conceptual framework (Msofi 2023)

3 Methodology

3.1 Study Approach

The study used mixed methods research design employing both qualitative and quantitative methods to collect data both sequentially and simultaneously with the aim of drawing on the strengths of both methods and offsetting the weaknesses of both methods by using them in combination. The quantitative methods used was household survey using household questionnaire, and the qualitative methods used were focus group discussion, and key informant interviews. The choice of the methods was based on the type of data to be collected to answer the research questions.

3.2 Study Sites

The study was done in Lilongwe district, Mitundu Extension Planning Area (EPA) focusing on three villages that are accessing extension services from specific extension service providers based on the crops they grow. The villages include Chimera, Chinkhowe and Kachono. The choice of the district was based on the presence of the groups of farmers that were accessing extension services from specific extension service providers using specific extension approaches. Another district could have been chosen that had similar characteristics, but Lilongwe was preferred because it is a capital city which is also vibrant in terms of cropping activities. The EPA and villages were chosen based on recommendation from the extension service providers.

3.3 Sampling

The study focused on farming households engaged in the cultivation of groundnuts, tobacco, and maize, all of which are key value chains receiving extension services. The extension service providers included: government agencies, representing public extension services; National Smallholder Farmers Association of Malawi (NASFAM), representing farmer-based extension providers; Agricultural Research and Extension Trust (ARET), representing private extension providers. These providers were selected based on their extensive coverage and their specific support for the chosen value chains, which are vital for both food security and income generation in Malawi. A total of 126 farming households participated in the study, with 42 households sampled from each of the three villages. The sample size was determined using the 'rule of thumb' for statistical analysis, which suggests that a minimum sample of 30 is sufficient for making statistically sound inferences about a population (Israel 2013). For focus group discussions (FGDs), 43 farmers participated, comprising 18 men and 25 women from all three villages. Each FGD consisted of 6–9 participants, aligning with the recommended range of 6–12 participants to ensure adequate representation while maintaining a manageable discussion size (Onwuegbuzie 2009). For key informant interviews, 11 individuals were purposively sampled, including: front-line extension workers, subject matter specialists, lead farmers working with the study participants. Key informants were selected based on their expertise and direct involvement in extension service delivery. Sampling for household surveys was done using systematic random sampling; for focus group discussions (FGDs) sampling was done using stratified random sampling, where men and women were grouped into different strata before selection. Table 1 presents the demographic and socioeconomic characteristics of the study participants.

3.4 Data Collection Tools and Data Collection

Both primary and secondary data were collected for this study. Secondary data were obtained through a desk review of relevant literature, while primary data were gathered using household surveys, focus group discussions (FGDs), and key informant interviews (KIIs). Data collection methods include a household questionnaire which was used to collect survey data; an FGD guide that was used for focus group discussions; and the KII guide was used for key informant interviews. The study received ethical approval from the University of the Western Cape's Humanities and Social Science Research Ethics Committee (HSSREC). Information sheets and consent forms were developed in the participants' local language, and informed consent was obtained from all participants before their involvement in the study.

Table 1 Characteristics of study participants

Characteristics	Description	Percentage
Household size	1 person	0.79
	2–3 people	19.05
	4–5 people	44.44
	6 and above	37.71
Age of household head	15–24	0.79
	25–54	70.16
	55–64	11.29
	65 and over	14.52
Sex of household head	Male	86.51
	Female	13.49
Sex of respondent	Male	55.56
	Female	44.44
Marital status	Married	85.71
	Divorced	2.38
	Separated	3.97
	Widowed	7.94
Education	None	11.90
	Primary	75.40
	Secondary	11.90
	Adult literacy	0.79
Sources of income	Nonfarm	2.38
	Farm	8.73
	Farm and nonfarm	88.89

3.5 *Data Analysis*

Qualitative data was transcribed and organised in ATLAS.ti where coding per different themes was done. Data was analysed using thematic content analysis. quantitative data was entered and cleaned in SPSS and Stata and was analysed using descriptive statistics to derive means, frequencies, crosstabulations and correlations. Further analysis was done using Chi-square tests and Analysis of Variance (ANOVA). Study produced other composite indices in the analysis of the data, and these are: Crop diversification index which was created using Principal Component Analysis, which considered the type and number of crops the household grew. The higher the index, the higher the diversification and vice versa. Household commercialisation Index which is the proportion of the produce sold and was computed suing the following formula:

$$\text{HCI} = (\text{Gross value of crop sales/gross value of crop production}) * 100$$

The study also computed the food consumption scores (FCS) which measures how often the household consumes certain types of food and household dietary diversity score (HDDS) which measures the diversity of household's diets as indicators for food and nutrition situation.

4 Results and Discussion

This section presents and discusses results to answer the research questions. It first presents results of participants' access to extension services, the impact of extension services and its contribution in commercial farming, and the livelihood outcomes of commercial farming.

4.1 Access to Extension and Advisory Services

The study finds that majority of participants (87%) access extension services from different providers, while a few (13%) indicated they do not participate in any extension activities. However, with extension normally having spill over effects, it is likely that even those that indicated that they do not formally access extension services, may benefit from observing what others are doing (Varshney et al. 2019), or they could be those that have other means of learning such us through internet sources. Access to agricultural extension services is regarded as crucial in helping farmers make informed decisions to improve their farming and their livelihoods (Tefera 2015). Although (Davidson 2007) argues that despite the enormous work that extension is doing, farmers are still struggling to maintain their livelihoods. It has been argued by (Adesina and Baidu-Forson 1995) that extension workers are the most important source of information, whether agricultural or others, that farmers, especially in rural areas, have. However, over time, extension approaches have changed from coercive approaches during colonial times and early years after independence to more participatory approaches in contemporary times (Farrington 2002). Furthermore, as adult learners, farmers are autonomous, self-directed; and goal-oriented, as such, they partake in activities that will help them achieve their immediate goals (Knowles 1980). Participation is voluntary and depends on the availability of extension service providers but also the interests and needs of farmers (Msofi 2023).

The study further found that farmers access extension services from various providers depending on their objectives and needs. Figure 2 shows sources of extension services that for farmers. It can also be observed that despite pluralism in extension service delivery (Kelly 2013; Masangano and Mthinda 2012), the government remains the dominant provider of extension services (Anderson and Feder 2003; Ragasa and Kaima 2017; Jensen et al. 2019; Ragasa and Mazunda 2018; Gwandu et al. 2013b). There are several possible reasons why the government is still the major provider of extension services: (1) Coverage—government extension workers are stationed countrywide positioned to reach out to households even in remote areas unlike other providers who concentrate on specific geographic area and their work is time-bound (Muyanga and Jayne 2006). (2) Inclusivity—government extension system provides services on a variety of enterprises unlike other providers with limited focus in areas of concentration, leaving out those that are not involved in those activities or enterprises. (3) Resource requirements—unlike other providers, government extension activities demand less requirements for participants to access services, for example, to be a member of a farmer group, farmers must pay group membership fees or contribute land and inputs for demonstrations. (4) Communication channels—government has a wide range of communicating extension information to farmers including extension workers, lead farmers, radio, print media which makes it likely for farmers to have access to at least one of them. The implications of this for commercial farming are that there is growing demand for extension services to assist farmers as they increasingly becoming engaged in commercialisation or selling their commodities for survival. But, with the heavy burden largely borne by the government extension services, which are also facing challenges, farmers are less likely to benefit from extension services.

In terms of age, those participating in extension activities were relatively younger (43 years on average) than those that were not (51 years), consistent with what (Suvedi et al. 2017) reported, and from technology adoption view, that socioeconomic factors such as age, influence adoption of technologies. Although contrary to what (Loki et al. 2021) observed that those with access to extension services were older, less educated, and owned livestock. More male headed households and men than female headed households and women had access to extension services. These results agree with

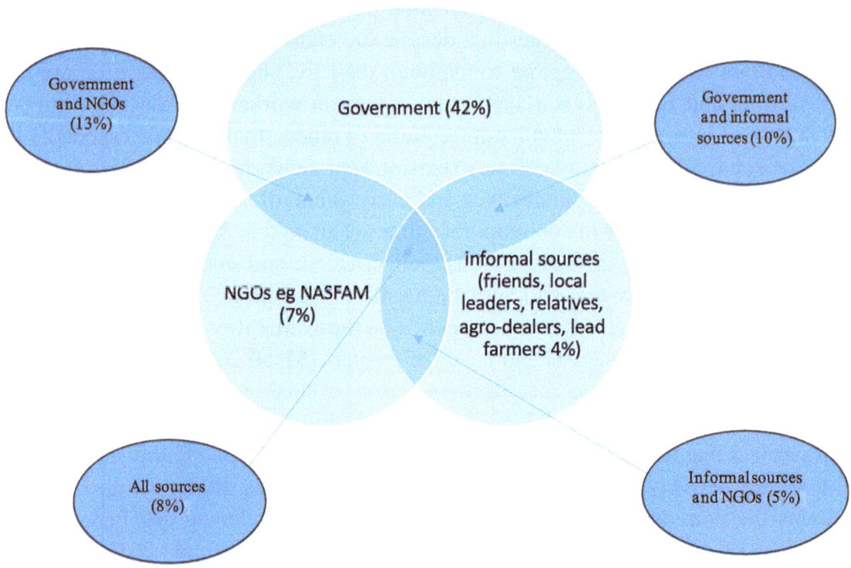

Fig. 2 Sources of extension services (Msofi 2023)

what was reported in Ethiopia by (Ragasa et al. 2012), and (Mudege et al. 2016, 2017). More richer households than poorer households, and those who grow tobacco, were more likely to participate in extension activities than poorer households who are more concerned about their immediate needs than long term solutions which extension can offer (Davidson 2007). The implication for these findings is that extension service providers should be able to recognise the needs and interests including challenges of different groups of farmers and come up with programmes and interventions that are tailored to these needs and interest but also counter the challenges experienced by the different groups.

Majority (average of 70%) accessed messages on recommended husbandry practices, crop diversification, certified seeds, which is mostly on production compared to those that accessed messages on marketing (average of 60%) including value addition, savings and investment, business planning, gross margin analysis and market research. This study looked at extension messages along the value chains ranging from production to marketing of produce. This shows that majority access extension services that are production oriented compared to those that are market oriented. This was also noted by Van den Ban and Samanta (2006) that despite the changing needs of farmers, extension services remain production oriented. Yet there is need for extension services to adapt to the changing needs of farmers and help them produce for the market (Kahan 2014), but also the changing focus of the Ministry of Agriculture which is promoting agricultural commercialisation through various initiatives including at policy level. The implications of these findings are that extension service delivery should tailor the extension messages to the objectives of farmers, thus those that are commercial oriented versus those that are subsistence oriented and avoid one size fits all solutions (Msofi 2023).

The Majority (65%) of the households that had access to these messages rated them highly in terms of usefulness for their farming activities and for commercial farming. This is because most farmers still view extension services key to their farming activities and for most farmers especially those in rural areas, extension services are their main source of information be it agricultural or other general information. However, qualitative data showed farmers' dissatisfaction with the current state of services. For instance, some were of the views that transferring information to them is not enough if farmers do not have access to affordable agricultural inputs and if they do not have access to lucrative markets for their produce. However, some commended the work that extension workers are doing. For instance, the following show their sentiments. "*The advice that extension workers are giving us is helpful. We can see changes in our farms and our homes, especially advice on winter cropping.*" FGD with males in Chimera village. "*In the past the government would ensure availability of cheaper inputs. So, with the availability of inputs and extension workers advice, the farmer was able to see the effectiveness of extension services. But now things are different even if we get the messages, with no inputs it is useless and futile.*" FGD with males in Chinkhowe.

These findings illuminate the challenges that the agricultural extension system continue to face which affects the effectiveness of these services which has been echoed by others and so there is need deal with these challenges for farmers to benefit from agricultural extension (Msofi 2023).

4.2 Impacts of Extension and Advisory Services

4.2.1 Impact of Agricultural Extension on Production

The study found that extension participants are more likely to have higher production levels than non-participants. Higher productivity among extension participants have been linked to increased technical efficiency on the farm because of participating in extension. This was also established in Greece by (Dinar et al. 2007), where farms that had access to both public and private extension services had higher technical efficiency than those with access to either public or private extension services and then those with no access to extension services at all. The results are also consistent with what (Ragasa and Mazunda 2018) found in Malawi, that those farmers who received extension services and rated them very useful, had higher productivity than those that received services that they rated less useful and those that did not receive any extension services at all. Although (Ragasa et al. 2012) found contradicting results that extension contact did not result in any productivity differences among farmers in Ethiopia. Furthermore, a study by (Olagunju and Adesiji 2013) in Nigeria found that non-participants also had better production levels owing to the 'trickle down' effect of extension information. Extension services play an important role in helping farmers achieve their objectives (Anderson and Feder 2003), since they provide the knowledge and skills which is one of the determinants of farmers' performance. Despite the envisioned importance of agricultural extension, evidence suggest that there are other structural factors that affect the work of extension, such as market distortions and infrastructural bottlenecks (Anderson and Feder 2003), which often threaten the effectiveness of extension services.

4.2.2 Agricultural Extension and Crop Diversification

A higher crop diversification index (0.057) was observed among extension participants than non-participants (− 0.04), suggesting a positive relationship between crop diversification and extension participation. The high diversification among the extension participants could be because of, either those that are participating in extension services gain expertise in diversification and have improved their technical efficiency to diversify, or it could be that it is the crop diversification behaviour that pushes households to seek additional technical advice to improve their output and benefits.

4.2.3 Agricultural Extension and Commercialisation

Both extension participants and non-participants are actively involved in both input and output markets, and no significant differences exist in market participation. this shows that market participation is driven by other factors other than agricultural extension. But among extension participants, output market participation could be triggered because of linkages to markets created by participating in extension services (Kaaria et al. 2004). The study further finds that the relationship between agricultural extension and crop commercialisation is not straightforward such that even those that do not participate in extension activities had a higher (51%) HCI compared to those who participate (46%) HCI. What is mostly observed is distress driven commercialisation where farmers are engaged in output markets for survival, to be able to provide for immediate needs, and hence commercialising because even those that do no harvest much, sell most of their produce. These 2 case studies illustrate the situation where a highly commercialised household is not having access to extension services and a household that has access to extension services does not commercialise much.

Case 1

Siyani Mimbulu (not his real name) is a 40-year-old farmer from Chimera village, married to Dinesi (not her real name), who is 32 years old. Neither of them has formal education. They have a household of six members and are engaged in farming, though they do not participate in any agricultural extension activities. They cultivate maize, soybeans, and tobacco on their two-acre plot. During the 2019–2020 growing season, they harvested 20 kg of maize, which they kept for household consumption, and produced 100 kg of soybeans and 100 kg of tobacco, both of which they sold. Their total earnings from crop sales amounted to MK72,000 ($88) out of a potential MK108,000 ($132), resulting in a Household Commercialization Index (HCI) of 67%. Siyani and Dinesi actively participate in input markets, purchasing seeds and fertilizers. Their labor strategy varies by crop: they hire labor for maize production, rely on family labor for soybeans, and use both family and hired labor for tobacco. They sell soybeans to local vendors immediately after harvest in April to meet household needs, while tobacco is sold to private companies in August, aligning with the designated tobacco auction period. In addition to farming, they generate income from small businesses and casual labor (ganyu). Their business activities brought in MK540,000 ($659), while ganyu provided an additional MK30,000 ($37), bringing their total household income to MK670,000 ($817). Their annual household expenditure stands at MK415,400 ($597). The family owns basic farming tools, including hoes and simple irrigation equipment. Their food supply from their own production lasts for about five months, after which they rely on purchased food until the next harvest. Their Food Consumption Score (FCS) is 25, indicating borderline food security, and their dietary diversity score is 3 out of a possible 10.

Case 2

Henele Gonapanja (not his real name) is a 43-year-old farmer from Chimera village. He is married and has a household of four members. He attained a primary level of education. Henele and his family cultivate maize and raise pigs on a total of four acres—two acres of their own land and an additional two acres they rent. During the 2019–2020 growing season, they harvested 350 kg of maize, which they kept entirely for household consumption, resulting in a Household Commercialization Index (HCI) of 0%. Although they participate in input markets, they used recycled maize seeds but purchased fertilizer. They rely solely on family labor for their farming activities. Apart from agriculture, they earn additional income through casual labor (ganyu), which brought in MK64,000 ($78). The family owns basic farming tools, including hoes, an axe, simple irrigation equipment, and a bicycle. Unlike some farmers, they actively engage with extension services, primarily receiving guidance from government extension workers. They have access to comprehensive agricultural advisory messages along the value chain, which they find highly useful. Their food supply lasted for eight months, and they recorded a Food Consumption Score (FCS) of 39.5, indicating good food security. Additionally, they had access to credit from friends, which they used to purchase food.

These two case studies highlight that while agricultural extension services are valuable for equipping farmers with technical knowledge, their direct impact on crop commercialization appears limited. Other factors play a more significant role in determining commercialization outcomes. In the first case, despite having no access to extension services, the household actively engaged in selling its produce, hiring labour, and purchasing inputs. However, the majority of its income came from off-farm sources, which allowed the household to invest in farming and supplement food production to sustain its livelihood. Conversely, the second household actively participated in extension programs but continued to use recycled seeds, did not sell any crops, and relied entirely on family labour. In fact, they supplemented their income by selling family labour. However, their decision to rent additional land indicates some level of commercial intent. These cases suggest that while extension services can provide important knowledge, commercialization is driven more by financial capacity, labour decisions, and market access rather than just technical training. Table 2 presents some of the sentiments from study participants on the role of extension in commercial farming.

The sentiments are clear that extension services are crucial in helping farmers commercialise although not much is happening. It was also clear that much of the extension advice is on production and less on marketing. This is what a group of women from Chimera village had to say; "*My only wish is that extension workers should resume their work because their presence mean that we will get helpful agricultural advice such as on making manure since fertiliser is becoming expensive. We also wish that extension workers should visit us frequently, help us find good markets for our produce, and help us access cheap inputs. In fact, the little advice that we get from extension workers is not helping us commercialise. Extension workers should*

Table 2 Participants sentiments on the role of agricultural extension

Roles of agricultural extension	Participants' sentiments
Linking farmers to markets	"*As much as they are not able to visit each farmer individually, extension workers do help in commercialisation by linking farmer clubs to organisations to buy their produce.*" Key informant interview with lead farmer in Kachono, March 2020 "*Extension services help farmers to find market for their produce by linking them to potential buyers, although this is not really happening and most of the times buyers do not come, or they come very late.*" FGD with females in Kachono, March 2020 "*Things could have been different if extension workers were helping us more, things can change for the better.*" FGD with males in Kachono, March 2020
Promote recommended agricultural practices	"*Extension workers from ARET do not provide loans but they help with markets as they encourage farmers to produce quality tobacco which could be sold at a better price and reduce rejection rate.*" Male farmers during FGD in Chimera, March 2020
Promoting collective marketing through farmer groups	"*Those that do not join groups they do not access extension services, so they usually follow traditional ways of farming, that is why their farming does not improve.*" FGD with females in Kachono, March 2020
Linking farmers to input markets	"*These extension workers do not help us at all in finding agricultural inputs because farmers buy on their own, and they look for the input markets themselves hence they end up buying expensive fertiliser and fake seed. In addition, after getting the money from sales, they do not know how to spend wisely to invest it back into farming because they lack advice.*" FGD with females in Chimera, March 2020 "*Farmers are failing to advance in their farming because of lack of farm input which renders extension advice useless.*" FGD with males and females in Chinkhowe, March 2020

be helping us to find markets for our produce apart from just advising us on what to grow and what methods to use." FGD with males in Chimera, March 2020. Figure 3 presents the relationship between commercialisation and extension participation.

Despite these findings, other studies have shown a positive link between extension services and agricultural commercialization. For instance, research by Muricho (2015), Ayele et al. (2021), and Andaregie et al. (2021) in Ethiopia found that the frequency of extension contact significantly influenced smallholder farmers' market participation decisions. Several arguments can be made regarding the role of agricultural extension in commercial farming. First, the impact of extension services depends on how they are perceived. This is because other factors—such as access to credit, market information, and farmer group membership—are also considered part of extension support. Second, structural and institutional challenges, such as limited access to inputs and poor market infrastructure, can hinder the effectiveness of extension services in promoting commercialization. Another key argument is that

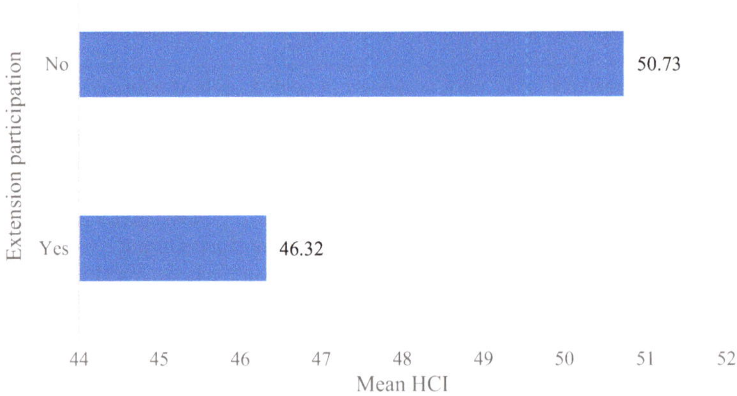

Fig. 3 Commercialisation and extension participation (Msofi 2023)

while extension services facilitate the commercialization process, they do not determine or guarantee outcomes. Extension provides knowledge and skills, but adoption and implementation depend on farmers' financial and resource capabilities. Similarly, extension services can recommend suitable inputs and market linkages, but they cannot ensure input availability, fair pricing, or guaranteed sales. Many of these aspects fall outside the direct mandate of extension services, a point also emphasized by Anderson and Feder (2003). Additionally, methodological limitations in assessing the role of extension services present challenges. Extension involvement varies in intensity and type, and merely measuring participation in extension activities may overlook other valuable aspects of extension support that contribute to commercialization. Similarly, measuring commercialization itself poses challenges, as highlighted by Poulton (2017) and Gebreselassie and Sharp (2008). The commercialization index, which focuses on the proportion of output sold, may not fully capture a household's overall economic situation. To address these complexities, this study incorporated qualitative data to provide a clearer understanding of extension's role. Findings confirm that while farmers recognize the benefits of extension services, there is still room for improvement. Furthermore, the broader impact of extension is often overshadowed by persistent structural and economic challenges within the agricultural sector (Msofi 2023).

4.2.4 Impact of Agricultural Extension on Income

The average income from crop sales is significantly higher among farmers who participate in extension services compared to non-participants. This difference can be attributed to higher production levels among active participants, leading to surplus crops available for sale and, consequently, increased income.

These findings align with previous studies. For instance, Loki et al. (2021) examined the impact of extension access on production and income in the Eastern Cape, South Africa, and found that farmers with access to extension services earned higher incomes than those without. Similarly, Nordin and Höjgård (2017) in Sweden reported positive financial benefits from extension services on farm incomes.

Additionally, expenditure patterns differ between participants and non-participants, with those engaged in extension services having relatively higher annual expenditures than their counterparts.

4.2.5 Impacts of Agricultural Extension on Food Security

Table 3 presents results of food and nutrition security among extension participants and non-participants.

Farmers who participate in extension services tend to retain more food from their own production, reflecting their higher production levels compared to non-participants. While the difference in the number of months that food lasted is not statistically significant, there are notable and significant differences in food consumption scores and household dietary diversity scores between participants and non-participants.

These findings suggest a positive relationship between extension participation and food security, aligning with the results of Ragasa and Mazunda (2018) in Malawi. This relationship can be explained by previously noted factors, such as higher production levels, increased income, and greater crop diversification among participants.

Conversely, food insecurity may also contribute to lower extension participation. Households struggling with food shortages often prioritize immediate survival, spending most of their time working to secure food rather than engaging in extension activities. Additionally, some extension programs require not only time but also financial resources, such as membership fees, which financially constrained households may be unable to afford. This further limits their ability to participate in extension services.

Table 3 Food security situation and extension participation (Msofi 2023)

Variable	Yes	No	*p* values
Mean produce kept for food (kgs)	451.68	290	0.012***
Mean number of months food last	6.7	6.2	0.521
Mean food consumption score	39.95	32.97	0.039**
Mean dietary diversity score	5.7	4.5	0.007***

5 Conclusions

This chapter aimed to enhance the understanding of the role of agricultural extension and advisory services in promoting agricultural commercialization. The underlying assumption was that extension services support farmers in producing for sale by strengthening their technical capacity and disseminating appropriate technologies. If adopted, these technologies can increase production and generate surplus for the market. However, the study found that the contribution of extension services to agricultural commercialization is minimal. This is because commercialization is driven by multiple factors beyond the control of extension services. Key factors include household resource endowments, which determine the ability to produce surplus, market conditions that influence input and output prices, and broader market failures (Msofi 2023).

Additionally, there are fundamental limitations to what agricultural extension can achieve. While extension can provide farmers with knowledge and skills, commercialization depends on other enabling conditions. For instance, extension services can connect farmers with buyers but cannot guarantee fair prices or ensure that produce will be purchased. This highlights the need for a supportive policy environment to maximize the effectiveness of extension services in achieving commercialization goals. Another potential explanation for the limited impact of extension services on commercialization lies in the measurement of commercialization itself. The current measure does not account for the scale of production, meaning that even farmers who sell their entire, albeit small, harvest due to financial distress may be classified as highly commercialized. This skews the relationship between extension access and commercialization, particularly among non-participants who engage in distress sales (Msofi 2023).

The study identified multiple extension service providers in the study sites, each with different approaches and areas of focus, confirming the pluralistic nature of extension services. However, government (public) extension services remain the dominant provider. The findings suggest that extension services are primarily production-oriented rather than market-oriented. As a result, they are less effective in helping farmers achieve commercialization objectives. Moreover, government extension services face significant challenges, such as inadequate funding, making them less likely to support market-driven farming.

From a social differentiation perspective, the study confirms disparities in access to extension services and levels of commercialization across socioeconomic classes and gender groups. Wealthier households are more likely to benefit from extension services, enabling them to improve their farming practices and commercialize more. Additionally, because they are more active in farming, they are also more likely to be targeted by extension providers, further reinforcing their advantages over poorer households. Gender disparities are also evident. Male-headed households and men are more likely to participate in extension activities than women. This is largely due to traditional gender roles, where men are designated as household heads and are prioritized in extension programs, while women's participation is constrained by their multiple responsibilities. In many cases, men also control household decision-making, including whether women can engage in extension activities. Female-headed households are generally less involved in extension services, as many of these women have limited engagement in farming due to challenges in accessing inputs and labour (Msofi 2023). As a result, male farmers and male-headed households are better positioned to leverage extension services to enhance their farming practices and increase commercialization.

List of Abbreviations

AU	African Union
ASWAp	Agriculture Sector Wide Approach
DAES	Department of Agricultural Extension Services
FCS	Food Consumption Score
FGD	Focus Group Discussion
FISP	Farm Input Subsidy Programme
GDP	Gross Domestic Product
GFRAS	Global Forum for Rural Advisory Services
GoM	Government of Malawi
HCI	Household Commercialisation Index
MGDS	Malawi Growth and Development Strategy
NAP	National Agriculture Policy
NAS	National Agribusiness Strategy
NES	National Export Strategy
NPC	National Planning Commission
SDG	Sustainable Development Goals
TIP	Targeted Input Programme

References

Adesina A, Baidu-Forson J (1995) Farmers' perceptions and adoption of new agricultural technology. Agric Econ 13:1–9

Alene AD, Manyong VM (2006) Farmer-to-farmer technology diffusion and yield variation among adopters: the case of improved cowpea in northern Nigeria. Agric Econ 35:203–211

Andaregie A, Astatkie T, Teshome F (2021) Determinants of market participation decision by smallholder haricot bean (phaseolus vulgaris l.) farmers in Northwest Ethiopia. Cogent Food Agric 7(1). https://doi.org/10.1080/23311932.2021.1879715

Anderson JR, Feder G (2003) Rural extension services. World Bank working paper

Ayele T, Goshme D, Tamiru H (2021) Determinants of cereal crops commercialization among smallholder farmers in Guji Zone, Ethiopia. Cogent Food Agric 7(1). https://doi.org/10.1080/23311932.2021.1948249

Barrett CB, Christian P, Shiferaw BA (2017) The structural transformation of African agriculture and rural spaces: introduction to a special section. Agric Econ 48:5–10. https://doi.org/10.1111/agec.12382

Berhanu GDH, Tegegne A (2006) Commercialization of Ethiopian agriculture: extension service from input supplier to knowledge broker and facilitator. IPMS (improving productivity and market success) of Ethiopian Farmers Project Working Paper 1. ILRI (International Livestock Research I

Birkhaeuser D, Evenson RE, Feder G (1991) The economic impact of agricultural extension: a review. Econ Dev Cult Change 39(3):607–650

Cai T, Davis K (2017) Malawi: desk study of extension and advisory services, June 6–36

Carletto C, Corral P, Guelfi A (2017) Agricultural commercialization and nutrition revisited: empirical evidence from three African countries. Food Pol 67:106–118. https://doi.org/10.1016/j.foodpol.2016.09.020

Cawley A, O'Donoghue C, Heanue K, Hilliard R, Sheehan M (2018) The impact of extension services on farm-level income: an instrumental variable approach to combat endogeneity concerns. Appl Econ Perspect Policy 40(4):585–612. https://doi.org/10.1093/AEPP/PPX062

Chirwa EW, Dorward A (2013) Agriculture Input Subsides; the recent Malawi experience, 1st edn. Oxford University Press

Chirwa EW, Kydd J, Dorward A (2006) Future scenarios for agriculture in Malawi: challenges and dilemmas. www.future-agricultures.org

Chowa C, Garforth C, Cardey S (2013) Farmer experience of pluralistic agricultural extension in Malawi. J Agric Edu Ext 19(2):147–166. https://doi.org/10.1080/1389224X.2012.735620

Daane J, Blum M, Rajalahti R, Dolly D, Ruz E (2012) The "new extensionist": roles, strategies, and capacities to strengthen extension and advisory services

Dancer H, Hossain N (2018) Social difference and women's empowerment in the context of the commercialisation of African agriculture, 43

Davidson AP (2007) Participation, education, and pluralism: towards a new extension ethic. Dev Pract 17(1):39–50

de Haas MA (2017) Rural livelihoods and agricultural commercialization in colonial Uganda: conjunctures of external influences and local realities (Issue September). http://edepot.wur.nl/419566C1 - NN08202,6752

De Weerdt J, Duchoslav J (2022) Guiding principles of AIP reform. In policy brief. https://massp.ifpri.info/files/2022/01/Guiding-principles-of-AIP-reform.-Final.pdf

Dercon SO, Gilligan D, Hoddinott J, Woldehanna T (2018) The impact of agricultural extension and roads on poverty and consumption growth in fifteen Ethiopian villages 91(November 2009):1007–1021. https://doi.org/10.1111/j.1467-8276.2009.01325.x.

Dinar A, Karagiannis G, Tzouvelekas V (2007) Evaluating the impact of agricultural extension on farms' performance in Crete: a nonneutral stochastic frontier approach. Agric Econ 36(2):135–146. https://doi.org/10.1111/J.1574-0862.2007.00193.X

Djurfeldt AA, Dzanku FM, Isinika AC (2018) Agriculture, diversification, and gender in rural Africa

Dorward A, Kydd J, Morrison J, Urey IAN (2004) A policy agenda for pro-poor agricultural growth. World Dev 32(1):73–89. https://doi.org/10.1016/j.worlddev.2003.06.012

Farrington J (2002) Recent and future challenges in agricultural extension, 6–8

Fuglie K, Gautam M, Goyal A, Maloney WF (2020) Harvesting prosperity: technology and productivity growth in agriculture

Gebremedhin B, Hoekstra D, Tegegne A (2006a) Commercialization of Ethiopian agriculture: extension service from input supplier to knowledge broker and facilitator IPMS (improving productivity and market success) of Ethiopian farmer's project. https://cgspace.cgiar.org/bitstream/handle/10568/564/CommercialPMSWorkingPaper_1.pdf?sequence=1&isAllowed=y

Gebremedhin B, Hoekstra D, Tegegne A (2006b) Improving productivity and market success of Ethiopian farmers commercialization of Ethiopian agriculture: extension service from input supplier to knowledge broker and facilitator

Gebremedhin B, Jemaneh S, Hoekstra D, Anandajayasekeram P (2012) A guide to market-oriented extension services with special reference to Ethiopia. www.cgiar.org
Gebreselassie S, Sharp K (2008) Commercialisation of smallholder agriculture in selected *Tef*-growing areas of Ethiopia. Ethiop J Econ 16(1). https://doi.org/10.4314/eje.v16i1.39824
GFRAS (2012) Module 1: introduction to the new extensionist
GoM (2016) National agriculture policy
Gwandu T, Mtambanengwe F, Mapfumo P, Chikowo R, Nezomba H (2013) Factors influencing access to integrated soil fertility management information and knowledge and its uptake among smallholder farmers in Zimbabwe. J Agric Edu, 37–41. https://doi.org/10.1080/1389224X.2012.757245
Hamilton A, Hudson J (2017) The perceived impact of agricultural advice in Ethiopia. J Agric Edu Ext 23(2):159–173. https://doi.org/10.1080/1389224X.2016.1245151
Islam MR, Madsen JB (2018) Knowledge diffusion and agricultural development. Agric Econ 49(2):265–276. https://doi.org/10.1111/agec.12414
Israel GD (2013) Determining sample size 1. April 2009, 1–5
Jensen PF, Prowse M, Larsen MN (2019) Smallholders' demand for and access to private-sector extension services: a case study of contracted cotton producers in northern Tanzania. J Agrar Change 19(1):122–134. https://doi.org/10.1111/joac.12262
Jin Y, Huffman WE (2016) Measuring public agricultural research and extension and estimating their impacts on agricultural productivity: new insights from U.S. evidence. Agric Econ 47(1):15–31. https://doi.org/10.1111/AGEC.12206
Kaaria S, Sanginga P, Njuki J, Delve R, Chitsike C, Best R (2004) Enabling rural innovation in Africa: an approach for empowering smallholder farmers to access market opportunities for improved livelihoods. OECD, 1–9
Kahan D (2014) Global trends and challenges in extension: a farm management
Kelly M (2013) NGOs, pluralism and advisory services—Timor Leste. 8622. https://doi.org/10.1080/1389224X.2012.741527
Knorr J, Gerster-bentaya M, Hoffmann V (2007) The history of agricultural extension in Malawi (Issue January 2002)
Knowles MS (1980) The modern practice of adult education: from pedagogy to andragogy. Retrieved 4 Dec 2021, from https://pdfs.semanticscholar.org/8948/296248bbf58415cbd21b36a3e4b37b9c08b1.pdf
Lemma M, Mekonnen F, Hoekstra D, Gebremedhin B (2014) Innovative approaches and tools for market- oriented extension services: examples from LIVES project
Lindsjö K, Mulwafu W, Andersson Djurfeldt A, Joshua MK (2021) Generational dynamics of agricultural intensification in Malawi: challenges for the youth and elderly smallholder farmers. Int J Agric Sustain 19(5–6):423–436. https://doi.org/10.1080/14735903.2020.1721237
Loki O, Aliber M, Sikwela MM (2021) Assessment of socio-economic characteristics that determine farmers' access to agricultural extension services in Eastern Cape, South Africa. S Afr J Agric Ext 49(1):198–209. https://doi.org/10.17159/2413-3221/2021/V49N1A10908
Masangano C, Mthinda C (2012) Pluralistic extension system in Malawi. IFPRI Discussion paper 01171 (Issue April). http://www.ifpri.org/publication/pluralistic-extension-system-malawi
Mbo'o-Tchauawou M, Colverson K (2014) Increasing access to agricultural extension and advisory services: how effective are new approaches in reaching women farmers in rural areas? ILRI project report
Mgalamadzi LM, Matita M, Chimombo M, Chinsinga B, Chirwa EW, Kaiyatsa S, Mazalale J (2021) Understanding gender and social differentiation in the context of agricultural commercialisation and implications for livelihoods in rural Malawi. IDS working paper
Mgalamadzi LM, Matita M, Chimombo M (2024) The gendered nature of household decision making and expenditure choices in the context of smallholder agricultural commercialisation in Malawi. CABI Agric Biosci 5:65
Msofi L (2023) Agricultural extension and commercialisation in rural Malawi: implications for livelihoods, class and gender differentiation. Ph.D. dissertation, University of the Western Cape

Mudege NN, Chevo T, Nyekanyeka T, Kapalasa E, Demo P (2016) Gender norms and access to extension services and training among potato farmers in Dedza and Ntcheu in Malawi. J Agric Edu Ext 22(3):291–305. https://doi.org/10.1080/1389224X.2015.1038282
Mudege NN, Mdege N, Abidin PE, Bhatasara S (2017) The role of gender norms in access to agricultural training in Chikwawa and Phalombe, Malawi. Gend Place Cult 24(12):1689–1710. https://doi.org/10.1080/0966369X.2017.1383363
Mur R, Oonk L, Bitzer V (2016) Quality of content in agricultural extension, 1–12
Muricho GS (2015) Determinants of agricultural commercialization and its impacts on welfare among smallholder farmers in Kenya. University of Nairobi, 1–118.
Muyanga M, Jayne TS (2006) Agricultural extension in Kenya: practice and policy lessons
Nakano Y, Tsusaka TW, Aida T, Pede VO (2018) Is farmer-to-farmer extension effective? The impact of training on technology adoption and rice farming productivity in Tanzania. World Dev 105:336–351. https://doi.org/10.1016/j.worlddev.2017.12.013
Nordin M, Höjgård S (2017) An evaluation of extension services in Sweden. Agric Econ 48(1):51–60. https://doi.org/10.1111/AGEC.12294
NPC (2020) Malawi 2063: an inclusively wealthy and self-reliant nation
Olagunju FI, Adesiji GB (2013) Impact of agricultural extension services on cocoyam production in Ogun State, Nigeria. J Agric Food Inf, 37–41. https://doi.org/10.1080/10496505.2011.588937
Onwuegbuzie AJ (2009) A qualitative framework for collecting and analysing data in focus group research, 1–21.
Padel S (2001) Conversion to organic farming: a typical example of the diffusion of an innovation? Sociol Rural 41(1):40–61. https://doi.org/10.1111/1467-9523.00169
Phiri MAR, Chilonda P, Manyamba C (2012) Challenges and opportunities for raising agricultural productivity in Malawi. Int J Agric for 2(5):210–224. https://doi.org/10.5923/j.ijaf.20120205.04
Poulton C (2017) What is agricultural commercialisation, why is it important, and how do we measure it? November
Ragasa C, Kaima E (2017) Capacity and accountability in the agricultural extension system in Malawi insights from a survey of service providers in 15 districts
Ragasa C, Mazunda J (2018) The impact of agricultural extension services in the context of a heavily subsidized input system: the case of Malawi. World Dev 105:25–47. https://doi.org/10.1016/j.worlddev.2017.12.004
Ragasa C, Berhane G, Tadesse F, Taffesse AS (2012) Gender differences in access to extension services and agricultural productivity. ESSP working paper
resources/Pluralistic%20Extension%20Service%20in%20Malawi.pdf
Rogers EM (1983) Diffusion of innovations, 3rd edn. The Free Press. 82-70998
Scott JC (1998) Seeing like a state: how certain schemes to improve the human condition have failed. Yale University Press
Suvedi M, Ghimire R, Kaplowitz M (2017) Farmers' participation in extension programs and technology adoption in rural Nepal: a logistic regression analysis. J Agric Edu Ext 23(4):351–371. https://doi.org/10.1080/1389224X.2017.1323653
Swanson BE, Rajalahti R (2010) strengthening agricultural extension and advisory systems: procedures for assessing, transforming, and evaluating extension systems. http://www.worldbank.org/rural (44)
Tefera T (2015) Extension programme participation and smallholder' s livelihoods: evidence from Awassa Zuria District. J Agric Ext Rural Dev 7(5):150–155. https://doi.org/10.5897/JAERD2013.0534
Umeta G, Lemecha F, Mume T (2011) Survey on women access to agricultural extension services at selected districts of Mid Rift Valley. J Agric Ext Rural Dev 3(March):51–63
Van den Ban AW, Samanta RK (2006) Changing roles of agricultural extension in Asian Nations
Varshney D, Joshi PK, Dubey SK (2019) Direct and spill over effects of agricultural advisory services: evidence from the farm science centre in Uttar Pradesh, India. IFPRI Discussion paper 01850. https://doi.org/10.2499/P15738COLL2.133316

Von Braun J, Kennedy E (1994) Agricultural commercialisation, economic development, and nutrition

Waddington H, Snilstveit B, White H, Anderson JR (2010) The impact of agricultural extension services

Wossen T, Abdoulaye T, Alene A, Haile MG, Feleke S, Olanrewaju A, Manyong V (2017) Impacts of extension access and cooperative membership on technology adoption and household welfare. J Rural Stud 54:223–233. https://doi.org/10.1016/j.jrurstud.2017.06.022

Zhou S, Minde I, Mtigwe B (2013) Smallholder agricultural commercialisation for income growth and poverty alleviation in Southern Africa: a review. Afri J Agric Res 8(22):2599–2608. https://doi.org/10.5897/AJAR11.1040

Open Access This chapter is licensed under the terms of the Creative Commons Attribution 4.0 International License (http://creativecommons.org/licenses/by/4.0/), which permits use, sharing, adaptation, distribution and reproduction in any medium or format, as long as you give appropriate credit to the original author(s) and the source, provide a link to the Creative Commons license and indicate if changes were made.

The images or other third party material in this chapter are included in the chapter's Creative Commons license, unless indicated otherwise in a credit line to the material. If material is not included in the chapter's Creative Commons license and your intended use is not permitted by statutory regulation or exceeds the permitted use, you will need to obtain permission directly from the copyright holder.

What Determines the Continued Use of Improved Maize Varieties? Evidence from Smallholder Farmers in Kilosa District Council, Tanzania

Nicholaus Nchembi Shokela, Charity Chanza, Isaac Mambo, and Abdi Khalil Edriss

Abstract The adoption of advanced agricultural technologies among maize farmers has been widely encouraged to enhance productivity. In response, the Tanzanian government has implemented various initiatives and policies to improve maize varieties, aiming for higher yields and long-term sustainability. This study explores the key factors influencing farmers' continued use of improved maize varieties in Kilosa District Council, Tanzania. It focuses on farmers who have utilized these varieties for at least three years, examining why some persist while others discontinue their use. Farmers initially decide to adopt improved varieties, and over time, they determine whether to continue based on various influencing factors. Data were collected through questionnaire surveys from 286 farmers across eight wards using a multistage purposive sampling technique. Additionally, focus group discussions (FGDs) and key informant interviews (KIIs) provided deeper insights. The study employed unordered multinomial logistic regression, the Logistic Regression Model, and content analysis to objectively assess the data. The findings reveal that high yield, early maturity, farming experience, household labour, and hired labour significantly contribute to the continued use of improved maize varieties. Conversely, factors such as farm size, education level, pest and disease resistance, and access to agricultural extension services negatively affect sustained adoption. Content analysis from FGDs and KIIs further supports these results, with high yield and early maturity identified as the primary drivers of continuous use. These findings highlight that farmers' perceptions

N. N. Shokela (✉)
Tanzania Agricultural Research Institute (TARI) Ilonga Centre, Kilosa Morogoro, Tanzania
e-mail: nicholaus.shokela@tari.go.tz

C. Chanza · I. Mambo
Extension Department, Bunda College, Lilongwe University of Agriculture and Natural Resources, Lilongwe, Malawi
e-mail: cchanza@luanar.ac.mw

A. K. Edriss
Department of Agricultural and Applied Economics, Bunda College, Lilongwe University of Agriculture and Natural Resources, Lilongwe, Malawi

© The Author(s) 2026

B. R. Singh et al. (eds.), *Sustainable Food Systems in Sub-Saharan Africa*,
https://doi.org/10.1007/978-3-032-11288-0_24

of technological attributes, individual characteristics, and institutional factors play a crucial role in determining the sustained adoption of improved maize varieties.

Keywords Adoption · Continued use · Discontinued use · Improved maize varieties and technological attributes

1 Introduction

Growth of the agricultural sector is a key driver in achieving food security and reducing poverty across Sub-Saharan Africa (World Bank, 2008). To address these challenges, significant efforts have been made to equip farmers with appropriate technologies that enhance productivity. By increasing agricultural output, farmers can reduce food insecurity and improve household incomes through the sale of their produce.

1.1 Agriculture's Contribution to Food Security

A substantial portion of the population in Sub-Saharan Africa, including Tanzania, faces food insecurity and malnutrition. Addressing food shortages remains a major concern for many developing nations (Food and Agriculture Organization of the United Nations [FAO], 2015). The crisis is particularly severe in Sub-Saharan Africa, where food security is closely linked to low agricultural productivity, limited production, and other contributing factors (FAO, 2012). Research indicates that agricultural growth is up to four times more effective in reducing poverty than expansion in other sectors, such as the oil industry (United Nations [UN], 2008). Consequently, scaling up initiatives that improve agricultural productivity, especially among smallholder farmers, is critical to strengthening food security and ensuring efficient food distribution systems.

Responding to challenges of food security and poverty in agricultural-based economies requires significant initiative to enhance agricultural output and efficiency (World Development Report [WDR], 2008). In Tanzania, the government has implemented various policies that position agriculture as a cornerstone of the economy. These initiatives prioritize food security while also fostering export growth to support economic diversification.

1.2 Enhanced Maize Varieties in Kilosa District Council

The adoption of improved maize varieties in Kilosa District Council dates back to the early 1960s, when the Ilonga Composite White was first developed at the

Ilonga Research Centre. Over the years, several open-pollinated and hybrid varieties have been introduced. Despite this, some farmers continue to cultivate local maize varieties (KDC, 2015; Magembe et al., 2016). Farmers do not immediately abandon traditional varieties when adopting new technologies. Many follow a safety-first approach, gradually integrating improved varieties while retaining local ones for familiar consumption and production traits (Ramsom et al., 2003; Smale & Olwande, 2013). As a result, complete replacement does not always occur. The most commonly grown improved maize varieties in Kilosa District Council include STAHA, SITUKA M1, TMV1, DK 8033, SEEDCO, and PANNAR (KDC, 2015).

2 Methodology

2.1 *Study Area*

Kilosa District Council, located in Tanzania's Morogoro Region, borders Tanga and Arusha to the north, Mvomero District to the east, Dodoma and Iringa to the west, and Kilombero District to the south. It lies between 6° South and 8° North latitude and 36° 30' East and 38° West longitude (NBS, 2012). Spanning 14,918 square kilometers, Kilosa District Council had a population of 438,175 across 105,635 households, with an average household size of 5 (NBS, 2012). The council experiences a semi-humid climate, receiving an annual average rainfall of 800 mm over eight months (October–May), peaking between February and March. Temperatures range from 18°C in the highlands to 30°C in the lowlands.

Although Kilosa District Council has two rainy seasons, its rainfall pattern supports only one staple crop harvest per season. The early rains occur between November and January, followed by heavier rainfall from March to June. A prolonged dry season (June–October) follows, with diverse landscapes ranging from seasonally flooded plains to mountainous regions exceeding 2,000 m in altitude.

The council was selected for this study because agriculture is the primary livelihood for most residents, with maize as the dominant crop (KDC, 2015). The study focused on smallholder farmers who adopted three major improved maize varieties SITUKA M1, STAHA, and TMV1 developed by the Agricultural Research Institute (ARI) Ilonga in 1980, 1983, and 1987, respectively. These farmers provided critical insights into the factors influencing the sustained use of improved maize varieties in Kilosa District Council.

2.2 *Sampling Procedures*

The study adopted a multi-stage purposive sampling technique, which is well-suited for examining large and diverse populations (Kothari, 2004).

1. First stage – Four divisions within the district were purposively selected based on the cultivation of improved maize varieties: STAHA, SITUKA M1, and TMV1.
2. Second stage – Two wards from each division were purposively selected, resulting in a total of eight wards. Selection was based on the prevalence of farmers who had adopted the three improved maize varieties for at least three years.
3. Third stage – A simple random sampling (SRS) approach was used to select 286 farmers proportionally from the eight wards.

To determine the representative sample size, the following formula was applied:

$$n = \frac{z^2(1-p)p}{e^2} \tag{1}$$

where:

- n = Desired sample size
- z = Z-score, representing the confidence level
- p = Estimated population proportion
- e = Margin of error

Kilosa District Council has a total population of 438,175 (NBS, 2012), with 77% of farmers adopting improved maize varieties (KDC, 2015, Magembe et al., 2016). The sample size was calculated accordingly using a 95% confidence level (z = 1.96) and a 5% margin of error.

$$n = \frac{1.96^2(1-0.77)0.77}{0.05^2} = 272$$

The final sample size was adjusted to 286 smallholder farmers to account for potential non-respondents. The number of households selected from each ward was determined using the following formula:

$$n_1 = \frac{P_{1*n}}{P_2} \tag{2}$$

where:

- n = Total sample (286)
- n_1 = Expected sub-sample
- P_1 = Population of the ward
- P_2 = Total number of households across the eight sampled wards

2.3 Data Collection Methods

The study adopted a mixed-methods approach, integrating both qualitative and quantitative data collection techniques. Information was gathered through focus group discussions (FGDs), household surveys, and key informant interviews (KIIs). Five FGDs were conducted, each comprising 10 participants, with representation from both male and female farmers. Discussions were digitally recorded and transcribed into Word files for analysis. A semi-structured questionnaire was administered to 286 respondents to collect individual level data. Eighteen KIIs were conducted with community experts possessing specialized knowledge to provide deeper insights into the challenges faced and potential solutions. KIIs were chosen for their ability to complement FGDs and household surveys (Lokot, 2021).

2.4 Data Analysis

Data analysis involved both qualitative and quantitative approaches. Content analysis was applied to FGDs and KIIs, while unordered multinomial logistic (MNL) and binary Logistic Regression Model were used to assess determinants influencing farmers' choices and the continued or discontinued use of improved maize varieties. The analysis was structured into two decision stages:

1. MNL regression model – Identified the key factors influencing farmers' selection of improved maize varieties.
2. Binary Logistic Regression Model – Evaluated farmers' decisions regarding the continued or discontinued use of a selected variety after three or more years of adoption.

2.4.1 Multinomial Logistic Model

The multinomial logistic (MNL) model assumes that error terms are independently and identically distributed (Greene, 2003). This study employs an unordered MNL model, which is well-suited for analyzing categorical dependent variables with multiple choices. The model estimates the impact of individual factors on the probability of selecting a specific improved maize variety.

The improved maize varieties were categorized, and the most commonly preferred varieties (decision categories) were identified. These decision categories in the multinomial logit model were denoted as follows:

- $j = 1 \rightarrow$ Farmer chooses STAHA (STA1) variety
- $j = 2 \rightarrow$ Farmer chooses SITUKA M1 (STU1) Variety
- $j = 3 \rightarrow$ Farmer chooses TMV1 (TM1) Variety
- $j = 4 \rightarrow$ Farmer chooses STAHA and SITUKA M1 (STA1STU1) varieties
- $j = 5 \rightarrow$ Farmer chooses SITUKA M1 and TMV1 (STU1TM1) varieties
- $j = 6 \rightarrow$ Farmer chooses STAHA and TMV1 (STA1TM1) varieties
- $j = 7 \rightarrow$ Farmer chooses STAHA, SITUKA M1, and TMV1 (STA1STU1TM1) varieties

- j = 8 → Farmer chooses other varieties (STA0STU0TM0) beyond the selected maize varieties

Based on the principle of discrete choice utility theory (Train, 2003, Green, 2003), farmers select the improved maize variety that maximizes utility, considering both financial and non-financial costs. This decision is influenced by factors such as the perceived attributes of the technology, institutional conditions, and socioeconomic characteristics of the farmer.

However, among the eight possible variety combinations, only a small number of farmers selected multiple varieties (e.g., STAHA and SITUKA M1, STAHA and TMV1, SITUKA M1 and TMV1, or all the three varieties). Due to this low adoption rate for multiple-variety choices, the study focused only on individual maize variety choices that provided statistically significant results. Consequently, the final dependent variables in the model were STAHA, SITUKA M1, TMV1, and other varieties (STA0STU0TM0), treated as categorical variables.

$$Pr(Y = 1) = \frac{e\beta_j{}'x_i}{\sum e\beta_k{}'x_i}, j = 0, 1, 2, 3 \tag{3}$$

The estimated equation defines the probabilities for a farmer (i), selecting a maize variety (j) from four available options: STAHA, SITUKA M1, TMV1, and other varieties (STA0STU0TM0). The vector X_i represents perceived technology attributes, institutional factors, and individual socioeconomic characteristics, while β_j denotes the vector coefficients associated with each maize variety (j) (Greene, 2003). The probability of choosing an improved maize variety is given by:

$$Pr(y = 0) = \frac{1}{1 + e\sum e^{\beta_k{}'x_i}} \tag{4}$$

Normalization results from setting $\beta_0 = 0$, as presented in Eq. 4. This allows for the estimation of β_j for each variety, except for one designated as the normalized alternative (reference category). The model's estimated coefficients reflect the effect of X_i on the probability of selecting an improved maize variety (j) relative to the reference category. Under these circumstances, TMV1 serves as the reference outcome.

The impact of a unit change in any explanatory variable (X) on the probability of a farmer (i) selecting a particular maize variety (j) is given by the marginal effect statistic (Greene, 2003), expressed as

$$\Delta P_j/\Delta X_i = P_j\left[P_j - \sum_{k-1}^{m} Pk\beta k\right] \tag{5}$$

Based on theoretical foundations, empirical literature, and researchers' understanding of the contextual setting, nine explanatory variables were identified and tested. The following hypotheses were formulated:

a. High yield has a positive relationship with the choice of improved maize varieties.
b. Early maturity has a positive relationship with the choice of improved maize varieties.
c. Farmers Age has a positive or negative relationship with the choice of improved maize varieties.
d. The farm land size owned has a positive or negative relationship with the choice of improved maize varieties.
e. Farmers' cooperative affiliation has a positive relationship with the choice of improved maize varieties.
f. Available household labour has a positive or negative relationship with the choice of improved maize varieties.
g. Available hired labour has a positive or negative relationship with the choice of improved maize varieties.
h. Access to agricultural extension services has a positive relationship with the choice of improved maize varieties.
i. The level of education has a positive or negative relationship with the choice of improved maize varieties.

2.4.2 Logistic Regression Model (MNL)

Farmers initially decide whether to adopt an improved maize variety, and over time, they determine whether to continue using it. This decision-making process occurs in two distinct stages, each represented by a separate model. The MNL regression model previously discussed, identified the factors influencing farmers' selection of improved maize varieties. In contrast, the binary logistic regression model examines the continued or discontinued use of these varieties after at least three years of adoption.

To determine the key factors influencing farmers' decisions to either continue or discontinue using selected improved maize varieties, the Logistic Regression Model was applied. When the dependent variable is binary, taking only two values continued use (1) or discontinued use (0), Logistic Regression Model serves as the standard analytical method (Lang, 2008). One advantage of this model is that it does not require a normality assumption for predictors, instead directly estimating the probability of an event occurring or not occurring (Muchabaiwa, 2013). To assess the impact of various factors including high yield, early maturity, resistance to pests and diseases, farming experience, farmland size, household labour, hired labour, education level, and agricultural extension services three regressions were conducted using the binary logistic model, specified as follows:

$$Logit[\varnothing(x)] = log[\varnothing(x)/1 - \varnothing(x)] = \alpha + \beta_1 x_1 + \beta_2 x_2 + \bullet\bullet\bullet\bullet\ \beta_i x_i \bullet\bullet\bullet\ e \quad (6)$$

where:

- $\varnothing(x)/phi(x)$ represents:the probability of success (case)
- $1 - \varnothing(x) 1 - /phi(x)$ represents the probability of failure (non- case)

- $\alpha/alpha$ is the intercept (constant of the equation
- $\beta/beta$ represents the coefficients of the predictors variables
- xx represents the explanatory variables affecting the dependent variable
- e is the error term

Equation 6 can be simplified as:

$$Y_1 = In(P/1 - P) = f\left(x_{1,}x_{2,}x_{3,}x_{4,}x_{5,}x_{6,}x_{7,}x_{8,}x_9\right) + e \tag{7}$$

The simplified equation represents the logistic regression model, where the log-odds of the probability (P) of continued use versus discontinuation of improved maize varieties is a function of several explanatory variables.

Where:

- Y_1 represents the log odds of the probability (continued vs. discontinued)
- P is the probability of continued use, while *1-P* is the probability of discontinuation
- $f(x_1, x_2, x_3, \ldots\ldots x_9)$ represents the explanatory variables which may include high yield, early maturity, pest and disease resistance, farming experience, household labour, hired labour, education level agricultural extension services e is the error term, accounting for unobserved factors

This equation enables the estimation of how each independent variable influences the likelihood of continued adoption of improved maize varieties. This setup clearly defines the dependent and explanatory variables used in the binary logistic regression model, ensuring a structured analysis of the factors influencing farmers' continued adoption of improved maize varieties. By incorporating high yield, early maturity, resistance to pests and diseases, farmland size, farming experience, household labour, hired labour, education level, and Agricultural extension services contact, the model effectively examines the key drivers of sustained maize variety use over time. Building on theoretical frameworks and empirical studies, the researchers formulated the following hypotheses to examine factors influencing the continued or discontinued use of improved maize varieties:

a. High yield has a positive relationship with the continued use of improved maize varieties.
b. Early maturity has a positive relationship with the continued use of improved maize varieties.
c. Resistance to pests and diseases has a positive relationship with the continued use of improved maize varieties.
d. Farm size has a positive or negative relationship with the continued use of improved maize varieties.
e. Farming experience has a positive relationship with the continued use of improved maize varieties.
f. Availability of household labour has a positive or negative relationship with the continued use of improved maize varieties.
g. Availability of hired labour has a positive or negative relationship with the continued use of improved maize varieties.

h. Access to extension service contact has a positive relationship with the continued use of improved maize varieties.
i. The level of education has a positive or negative relationship with the continued use of improved maize varieties.

3 Results and Discussion

3.1 Farmers' Selection of Improved Maize Varieties

This section focuses on the significant variables influencing farmers' selection of improved maize varieties, as identified in the multinomial logistic regression (MNL) regression model. As shown in Table 1, the varietal attributes of high yield and early maturity play a crucial role in farmers' decision-making. Specifically, the probability of selecting STAHA increases by 30.8% compared to TMV1, given an equivalent unit of high yield. This suggests that farmers tend to adopt maize varieties that demonstrate superior yield performance relative to other available options. These findings align with Ransom et al. (2003), who noted that a high-yielding variety serves as a strong incentive for farmers to adopt a particular maize variety. Additionally, Akhassan (2015) found that dis-adoption occurs when improved varieties fail to show a significant yield advantage over traditional alternatives.

Interestingly, farmers who perceived STAHA as an early-maturing variety were 25.5% less likely to choose it compared to TMV1. This suggests that STAHA may take longer to mature than TMV1, making early-maturing varieties preferable for drought escape rather than drought tolerance (Cavatassi, 2011).

Table 1 Factors influencing the selection of STAHA Maize Variety

Explanatory Variable	Coefficient	Marginal effect	P-value
High yield	1.6882	0.3084	0.000***
Early maturity	-0.9264	-0.2557	0.000***
Farmers Age	0.0047	0.0044	0.045**
Farmland	0.1478	0.0199	0.517
Affiliation to Cooperatives	0.0370	-0.0707	0.229
Agricultural Extension services	-0.8144	-0.1751	0.012**
Hired labour	0.2138	0.1229	0.207
Household labour	0.2416	0.1398	0.042**
Education level	0.0014	-0.0145	0.287

Model summary; Number of observation = 267 Likelihood Ratio LR chi^2 (27) = 139.1 Prob > chi^2 = 0.0000*** Log likelihood = -289.03 Pseudo R^2 = 0.1940 Significance Levels: *** = 1%, ** = 5%, * = 10%, Baseline Outcome: TMV1

Additionally, an increase in the household head's age by one unit raised the probability of choosing STAHA by 0.04% compared to TMV1. Nwakor et al. (2011) found that age significantly influences farming technology choices, though Langyintuo & Mungoma (2008) reported mixed findings regarding its effect on improved technology adoption. Defrancesco et al. (2008) suggest that young farmers, who tend to be more innovative and less risk-averse, are more likely to adopt complex and labour-intensive technologies such as improved maize varieties. Conversely, older farmers may become more conservative and resistant to new ideas over time (Tshikala et al., 2015).

Access to agricultural extension services was found to decrease the probability of selecting STAHA by 17.5%, relative to choosing TMV1. This may be due to the greater emphasis on other crops in extension programs, leading to inefficiencies in the dissemination of information about STAHA (Kassie et al., 2012). However, Haggblade & Tembo (2003) and Mavunganidze et al. (2013) argued that agricultural extension services increase the likelihood of adopting agricultural technologies, highlighting the importance of quality extension programs.

Furthermore, access to labour increased the probability of selecting STAHA by 13.9% compared to TMV1. Households with more available labour were more likely to choose the variety, aligning with studies that found certain improved maize varieties to be labour-intensive (Haggblade & Tembo, 2003; Nyanga et al., 2011). Kafle (2010) concluded that when farmers meet the labour requirements of an improved maize variety, their likelihood of adoption increases significantly.

The results in Table 2 indicate that high yield negatively influences the choice of SITUKA M1 by 30.2% relative to TMV1. This finding suggests that higher perceived yield reduces the probability of selecting SITUKA M1, contradicting previous studies (Tura et al., 2010; Kudi et al., 2011), which reported that farmers generally prefer high-yielding varieties. Since yield directly measures seed performance, varieties with higher yields typically enhance output and gross earnings (Kudi et al., 2011).

However, when farmers perceived SITUKA M1 as an early-maturing variety, their probability of selecting it increased by 38.6% compared to TMV1. Early maturity is a critical trait for farmers, particularly in drought-prone areas, as it enables drought escape rather than drought tolerance. The shorter the growing period, the lower the risk of drought exposure. Cavatassi (2011) found that when farmers choose maize varieties in drought-affected regions, they prioritize attributes such as early maturity and drought tolerance.

The findings indicate that younger farmers are more likely to choose SITUKA M1, as a unit increase in age significantly decreases the probability of selecting this variety by 0.06% compared to TMV1. Simtowe et al. (2007) suggest that younger farmers are more willing to take risks, whereas older farmers tend to stick to traditional production techniques, making them less open to adopting new technologies (Tshikala et al., 2015).

An increase in agricultural land size inversely affects the likelihood of choosing SITUKA M1, with a 9.5% decrease compared to TMV1. This suggests that as farmers expand their landholdings, they tend to reduce their production of SITUKA M1.

Table 2 Factors influencing the selection of SITUKA M1 Maize Variety

Explanatory Variable	Coefficient	Marginal effect	P-value
High yield	-0.9224	-0.3021	0.000***
Early maturity	1.7546	0.3864	0.000***
Farmers Age	-0.0329	-0.0057	0.040**
Farmland	-0.2525	-0.0955	0.039**
Affiliation to Cooperatives	0.8193	0.1502	0.051*
Agricultural Extension services	0.5477	0.1838	0.005***
Hired labour	-0.6253	-0.1029	0.207
Household labour	-1.0710	-0.2067	0.002***
Education level	0.1724	0.0332	0.042**

Model summary; Number of observation = 267 Likelihood Ratio LR chi^2 (27) = 139.1 Prob > chi^2 = 0.0000*** Log likelihood = -289.03 Pseudo R^2 = 0.1940 Significance Levels: *** = 1%, ** = 5%, * = 10%, Baseline Outcome: TMV1

Affiliation with cooperatives positively influences the selection of SITUKA M1, increasing the probability by 15% compared to TMV1. Farmers who leave cooperatives are less likely to choose the variety, likely due to reduced access to production knowledge, input-sharing benefits, and economies of scale in farming practices such as harvesting, transportation, and market access. Several studies (Kassie et al., 2011; Ghimire et al., 2015; Olalekan & Simeoni, 2015) support the notion that cooperative participation significantly enhances the adoption of improved maize varieties, as cooperatives facilitate knowledge exchange and technology consolidation. Tura et al. (2010) suggest that exposure to multiple information sources enables farmers to assess risks and benefits, thus encouraging adoption.

Access to agricultural extension services increases the probability of choosing SITUKA M1 by 18.4% compared to TMV1, reinforcing the role of agricultural advisory services in promoting maize variety adoption (Tura et al., 2010). Conversely, limited access to agricultural extension services may significantly reduce the likelihood of adopting improved maize varieties. Several studies (Asfaw et al., 2012; Mariano et al., 2012; Ghimire et al., 2015) emphasize that extension services enhance technology uptake by reducing uncertainty and improving farmers' confidence in innovation.

Dependence on household labour negatively impacts the selection of SITUKA M1, reducing the probability by 20.6% compared to TMV1. This suggests that farmers prefer alternative sources of labour rather than relying solely on family members, whose availability is often affected by rural–urban migration and off-farm work commitments (Oluyole et al., 2013). Finally, a unit increase in schooling years results in a 3.3% increase in the probability of choosing SITUKA M1 over TMV1. This reinforces findings that higher education levels enhance farmers' ability to evaluate the advantages and disadvantages of agricultural technologies (Johannes et al., 2010; Abunga et al., 2012; Tshikala et al., 2015.

Table 3 indicates that when farmers perceive other maize varieties as early-maturing, their likelihood of selecting them decreases by 11.7% compared to TMV1. This suggests that farmers may still opt for these varieties despite their longer than others maturation period. Research highlights that certain varietal attributes hold more significance in farmers' decision-making. Belay and Wale (2021) found that farmers prioritize yield and overall benefits of a variety over its maturation period, whether short, intermediate, or long. As a result, farmers may prefer a long-maturing variety if it offers other desirable characteristics that outweigh its maturity duration.

A unit increase in farmland size raised the probability of farmers selecting other improved maize varieties by 9.1% compared to TMV1. This suggests that when farmers do not expand their cultivated area, they are more likely to continue growing STAHA, SITUKA M1, and TMV1. Access to larger farmland positively influences the adoption of alternative improved maize varieties (Tura et al., 2010; Mariano et al., 2012).

Farmers with limited land ownership may struggle to maximize returns from new technology investments, making them less inclined to adopt new seed varieties (Tura et al., 2010). This finding aligns with earlier research (Mendola, 2007; Kassie et al., 2011; Mariano et al., 2012; Ghimire et al., 2015), which highlights that when farmers gain additional land, they tend to shift production towards other maize varieties rather than continue growing previously adopted ones.

Table 3 Factors influencing the selection of other maize varieties (STA0SITU0TM0)

Explanatory Variable	Coefficient	Marginal effect	P-value
High yield	-0.0601	-0.0093	0.879
Early maturity	-0.4740	-0.1172	0.064*
Farmers Age	-0.0234	-0.0021	0.390
Farmland	0.4737	0.0908	0.001**
Affiliation to Cooperatives	0.3560	0.0004	0.995
Agricultural Extension services	-0.3591	-0.0448	0.498
Hired labour	-0.6620	-0.0817	0.264
Household labour	-0.3867	-0.0125	0.843
Education level	0.0462	-0.0036	0.784

Model summary; Number of observation = 267 Likelihood Ratio LR chi^2 (27) = 139.1 Prob > chi^2 = 0.0000*** Log likelihood = -289.03 Pseudo R2 = 0. 1940 Significance Levels: *** = 1%, ** = 5%, * = 10% Baseline Outcome TMV1

3.2 *Factors Influencing to Farmers' Continued or Discontinued Use of Adopted Improved Maize Varieties*

Table 4 presents the maximum likelihood estimates from the logistic regression model, examining the factors influencing farmers' continued or discontinued use of the STAHA maize variety.

Results in Table 4 indicate that farmers who perceive STAHA as a high-yielding variety are 61.5% more likely to continue using it. This suggests that yield performance is a key factor driving sustained adoption. Ghimire et al. (2015) found that farmers who viewed a variety as superior in yield tended to maintain its use over time.

Conversely, if farmers perceive STAHA as early maturing, the likelihood of continued use declines by 26.1%, indicating that early maturity does not positively influence long-term adoption. Additionally, an increase in land size reduces the probability of continued STAHA use by 5.3%, suggesting that farmers with larger landholdings may prefer other maize varieties. However, Tura et al. (2010) reported that having sufficient land generally supports the sustained adoption of improved seed technologies. The decision to adopt and continue using a technology is complex, influenced by policy, institutional dynamics, environmental conditions, household resources, farm type, and the technology itself(Tura et al., 2010).

Farmers' experience in farming significantly influences STAHA's continued use, with each additional year of experience increasing the probability by 0.7%. This suggests that greater farming experience reinforces continued reliance on STAHA. Previous studies (Tura et al., 2010; Alhassan et al., 2015) affirm that experience in agriculture positively impacts the decision to sustain improved maize cultivation.

Table 4 Logistic Regression Model Results for Farmers' Continued or Discontinued Use of STAHA Maize Variety

Explanatory Variable	Coefficient	Marginal effect	P-value
High yield	3.0175	0.6151	0.000***
Early maturity	-1.6885	-0.2612	0.004***
Resistance to pests and diseases	0.8916	0.1468	0.146
Farmland size	-0.3394	-0.0530	0.010**
Farming Experience	0.0467	0.0073	0.087*
Household labour	1.4858	0.2208	0.025**
Hired labour	1.3420	0.1630	0.045**
Education level	-0.2823	-0.0440	0.097*
Agricultural Extension service contact	-0.4380	-0.0683	0.012**

Model Summary; Number of observation = 105 Likelihood Ratio LR chi2 (9) = 48.91 Prob > chi2 = 0.0000***Log likelihood = -39.26 Pseudo R2 = 0.38 Significance Levels: *** = 1%, ** = 5%, * = 10%

Both household labour and hired labour play a crucial role in the sustained adoption of STAHA. The availability of household labour increases the probability of continued use by 22.1%, while hiring external labour raises the probability by 16.3%. This suggests that access to adequate labour resources enables farmers to sustain their use of the variety post-adoption. Household labour is often preferred because it is cost-effective(Omotesho et al., 2014). Additionally, according to Bedemo et al. (2013), the decision to utilize both household and hired labour is influenced by the seasonality of agricultural technologies and various socioeconomic factors.

Maintaining other covariates at a constant level, a unit increase in the household head's education level led to a 4.4% decrease in the likelihood of continuing to use STAHA. This indicates that more educated household heads are less inclined to sustain STAHA cultivation over time. This finding is consistent with Tura et al. (2010), who noted that literate-headed households in Central Ethiopia were less likely to continue using improved maize varieties. As individuals accumulate human capital through formal education, they gain access to off-farm employment opportunities, making agriculture a less attractive livelihood option (Uematsu & Mishra, 2010). On the other hand, Johannes et al. (2010), Abunga et al. (2012), and Tshikala et al. (2015) found that education positively influences the sustained use of improved varieties, as educated farmers can effectively assess the benefits and drawbacks of available technologies.

Interestingly, an increase in the frequency of extension agent visits resulted in a 6.8% reduction in the probability of continuing to use STAHA. This suggests that more frequent visits do not necessarily lead to greater adoption, reinforcing the idea that the quality of extension services plays a crucial role in technology utilization (Kassie et al., 2012). Extension agents provide diverse information and alternatives, which affect farmers differently based on prevailing farming conditions. Farmers are expected to choose options that best suit their needs (Baethgen et al., 2003). However, Tura et al. (2010) reported that more frequent farmer-agent interactions can positively influence the continued use of improved maize varieties.

When all other variables are held constant, the findings in Table 5 indicate that farmers' perception of SITUKA M1 as a high-yielding variety lowers the probability of continued use by 18.2%. This suggests that farmers are less likely to persist with the variety if they perceive an increase in its yield. The result implies that yield performance alone does not drive long-term adoption, and that farmers may prioritize other key attributes over yield when deciding whether to continue using the variety. This finding contrasts with the results of Ransom et al. (2003) and Kudi et al. (2011), who observed that an increase in yield positively influenced continuous use.

On the other hand, if farmers perceive SITUKA M1 as an early-maturing variety, the probability of continued use increases by 44.6%, highlighting the importance of early maturity in the sustained adoption of the variety. Raphael (2014) notes that early-maturing maize varieties are particularly valuable in marginal areas, as they help farmers cope with harsh conditions and severe food shortages. The popularity of a maize variety among farmers depends on how well it meets their specific needs (Jaleta et al., 2013) and how they evaluate the total benefits accrued within a growing season (Sinja et al., 2004).

Table 5 Logistic Regression Model Results for Farmers' Continued or Discontinued Use of SITUKA M1 Maize Variety

Explanatory Variable	Coefficient	Marginal effect	P-value
High yield	-1.1046	-0.1816	0.023**
Early maturity	2.1016	0.4457	0.000***
Resistance to pests and diseases	-0.8341	-0.1442	0.078*
Farmland size	-0.0292	-0.0050	0.872
Farming Experience	-0.0186	-0.0032	0.373
Household labour	-1.4745	-0.2734	0.011**
Hired labour	-0.7570	-0.1457	0.333
Education level	-0.0587	-0.0100	0.636
Agricultural Extension service contact	0.0182	0.0031	0.895

Model Summary; Number of observation = 133 Likelihood Ratio LR chi2 (9) = 38.48 Prob > chi2 = 0.0000***Log likelihood = -59.40 Pseudo R2 = 0.24 Significance Levels: *** = 1%, ** = 5%, * = 10%

Furthermore, if farmers perceive SITUKA M1 as resistant to pests and diseases, the probability of continued use decreases by 14.4%. This suggests that pest and disease resistance does not significantly drive sustained adoption, contrasting with Ghimire et al. (2015), who found that farmers tend to continue using varieties perceived as disease- or pest-resistant.

Additionally, sole reliance on household labour reduces the likelihood of continued SITUKA M1 use by 27.3%, indicating that farmers who depend entirely on family labour are less likely to sustain the variety over time. However, Jaleta et al. (2013) found that greater availability of household labour generally supports the long-term use of improved maize varieties.

The findings in Table 6 indicate that farmers who perceive TMV1 as a high-yielding variety are 61.2% more likely to continue using it. This confirms that yield potential plays a crucial role in the sustained adoption of TMV1, aligning with Kafle (2010), who noted that farmers prefer varieties that demonstrate high productivity.

Conversely, if farmers consider TMV1 an early-maturing variety, the probability of continued use decreases by 26.7%, highlighting the importance of attribute preference in variety selection. This suggests that early maturity alone does not strongly influence long-term adoption, contrasting with the findings of Katengeza et al. (2012) and Raphael (2014), who reported that farmers prefer early-maturing maize varieties for their adaptability to specific farming conditions.

An increase in agricultural land by one hectare reduces the likelihood of continued TMV1 use by 5.2%, indicating that as farmers expand their cultivated area, they may shift to alternative crop varieties. This observation is consistent with Etoundi & Dia (2008) and Muchangi (2016), who found that farmers allocating larger land areas tend to diversify their crops. However, Langyintuo & Mekuria (2008) and Tura et al. (2010) argue that households with sufficient land to sustain their families are more likely to maintain improved maize varieties over time.

Table 6 Logistic Regression Model Results for Continued or Discontinued Use of TMV1 Maize Variety

Explanatory Variable	Coefficient	Marginal effect	P-value
High yield	3.0895	0.6124	0.000***
Early maturity	-1.8623	-0.2672	0.004***
Resistance to pests and diseases	1.0521	0.1635	0.111
Farmland size	-0.3621	-0.0520	0.007***
Farming Experience	0.0330	0.0047	0.243
Household labour	1.0321	0.1439	0.150
Hired labour	1.0398	0.1217	0.162
Education level	-0.1656	-0.0238	0.335
Agricultural Extension service contact	-0.5472	-0.0786	0.003***
Model summary; Number of observation = 101 Likelihood Ratio LR chi2 (9) = 49.42 Prob > chi2 = 0.0000***Log likelihood = -35.84 Pseudo R2 = 0.41 Significance Levels: *** = 1%, ** = 5%, * = 10%			

Interestingly, an increase in extension agent visits lowers the probability of continued TMV1 use by 7.9%, suggesting that more frequent interactions with extension agents may not necessarily lead to technology adoption. This implies potential issues in information dissemination or barriers preventing farmers from implementing new technologies (Muchangi, 2016). Some researchers suggest that extension services may be ineffective if the messages are impractical or irrelevant to a large portion of farmers, limiting their impact on sustained technology use. In contrast, Tura et al. (2010) found that increased extension visits enhance the continued adoption of improved maize varieties.

4 Conclusion

The study findings highlight notable differences in farmers' variety choices based on specific attributes. In terms of high yield, farmers who opted for SITUKA M1 did not prioritize yield as a positive selection factor, unlike those who chose TMV1. However, early maturity had a strong positive influence on farmers' preference for SITUKA M1 over TMV1.

Additionally, factors such as age, cooperative affiliation, access to agricultural extension services, household labour, hired labour, and education level exhibited statistically significant effects on farmers' selection of maize varieties.

The logistic regression results for continued use of STAHA and TMV1 suggest that high yield consistently played a significant role in influencing farmers' sustained adoption of these varieties. However, despite its initial appeal, high yield did not positively affect farmers' long-term use of SITUKA M1.

Furthermore, farmers' continued adoption of maize varieties was influenced by pest and disease resistance, total farm size, farming experience, household labour availability, hired labour, education level, and agricultural extension service contact, each contributing variably to sustained maize variety utilization.

5 Recommendations and Implications

This section outlines key recommendations and implications aimed at ensuring farmers continue utilizing adopted improved maize seed technology, while also identifying potential areas for future research.

To enhance the effectiveness of improved maize seed technology, breeders and researchers should actively consider farmers' perspectives within their local agricultural context during the technology design process. The study highlights the need for an efficient and effective extension system, ensuring farmers receive comprehensive and accessible information to support long-term technology utilization.

5.1 Policy and Institutional Improvements

The government of Tanzania should reform extension services to make them more demand-driven, aligning with the specific needs of farmers. Additionally, farmers should be educated on the importance of financing extension services, enhancing their active participation in agricultural knowledge dissemination.

While this study was conducted in Kilosa District Council, the findings are broadly applicable to other developing countries where government-led extension institutions face challenges related to limited resources and operational constraints. Strengthening agricultural extension frameworks is crucial to addressing these gaps and ensuring effective technology adoption.

5.2 Future Research Directions

Any future agricultural interventions should incorporate farmers' needs, characteristics, and perspectives. Policy makers, researchers, breeders, and extension workers must consider farmers' practical experiences to enhance the successful implementation of improved maize seed technology.

References

Abunga M, Emelia A, Samweli G, Dadzie K (2012) Adoption of modern agricultural production technologies by farm households in Ghana: What factors influence their decision? Journal of Biology Agriculture and Healthcare 2(3):1–13
Alhassan ASH, Salifu K (2015) Determinants of farmers' adoption of improved maize varieties in the Wa Municipality. Am Int J Contemp Res 5(4):1–9
Asfaw S, Shiferaw B, Simtowe F, Lipper L (2012) Impact of modern agricultural technologies on smallholder welfare: Evidence from Tanzania and Ethiopia. Food Policy 37(3):283–295
Baethgen, W.E., Meinke, H., & Gimenez, A. (2003). Adaptation of agricultural production systems to climate variability and climate change: Lessons learned and proposed research approach. In *Climate Adaptation Net Conference – Insights and Tools for Adaptation: Learning from Climate Variability*, 18–20.
Bedemo A, Getnet K, Kassa B (2013) Determinants of household demand for and supply of farm labour in rural Ethiopia. Australian Journal of Labour Economics 16(3):351–367
Belay BF, Wale MF (2021) Participatory on-farm evaluation of improved sorghum varieties in North Gondar areas of Ethiopia. Cogent Food & Agriculture 7(1):1–20
Cavatassi R, Lipper L, Narloch U (2011) Modern variety adoption and risk management in drought-prone areas: Insights from sorghum farmers in eastern Ethiopia. Agric Econ 42(3):279–292
Defrancesco E, Gatto P, Runge F, Trestini S (2008) Factors affecting farmers' participation in agri-environmental measures: A Northern Italian perspective. J Agric Econ 59(1):114–131
Etoundi, N.S.M., & Kamgnia, D.B. (2008). Determinants of the adoption of improved varieties of maize in Cameroon: Case of CMS 8704. Available at: https://mpra.ub.uni-muenchen.de/37783/ [Accessed on May 9, 2015].
Food and Agriculture Organization (FAO) of the United Nations. (2012). *The State of Food Insecurity in the World*, 1–63.
Food and Agriculture Organization (FAO) of the United Nations. (2015). *Regional Overview of Food Security in Africa: African Food Insecurity Prospects Brighter Than Ever*. FAO: Accra, 5–20.
Ghimire R, Huang W, Poudel PM (2015) Adoption intensity of agricultural technology: Empirical evidence from smallholder maize farmers in Nepal. International Journal of Agriculture Innovations and Research 4(1):139–146
Greene WH (2003) Econometric Analysis. Pearson Education, India, pp 6–12
Haggblade, S., & Tembo, G. (2003). Development, diffusion, and impact of conservation farming in Zambia. Lusaka, Zambia: *Food Security Research Project*.
Harrell, F. (2015). *Regression Modeling Strategies: With Applications to Linear Models, Logistic and Ordinal Regression, and Survival Analysis*. Springer. Available at: https://www.springer.com/us/book/9783319194240 [Accessed on April 13, 2017].
Jaleta, M., Yirga, C., Kassie, M., de Groote, H., & Shiferaw, B. (2013). Knowledge, adoption, and use intensity of improved maize technologies in Ethiopia. In *Proceedings of the 4th International Conference of the African Association of Agricultural Economists*, **AAAE**, Hammamet, Tunisia, 22–25.
Johannes T, Vabi M, Malaa D (2010) Adoption of maize and cassava production technologies in the forest-savannah zone of Cameroon: Implications for poverty reduction. World Applied Science Journal 11(2):196–209
Kabunga NS, Dubois T, Qaim M (2012) Heterogeneous information exposure and technology adoption: The case of tissue culture bananas in Kenya. Agric Econ 43(5):473–486
Kafle B (2010) Determinants of adoption of improved maize varieties in developing countries: A review. International Research Journal of Applied Basic Science 1(1):1–7
Kassie M, Shiferaw B, Muricho G (2011) Agricultural technology, crop income, and poverty alleviation in Uganda. World Dev 39(1):1784–1795
Kassie, M., Jaleta, M., Shiferaw, B., Mmbando, F., & De Groote, H. (2012). Improved maize technologies and welfare outcomes in smallholder systems: Evidence from parametric and non-parametric approaches. *Selected Paper, IAAE Triennial Conference*, Foz do Iguaçu, Brazil, 18–24.

Katengeza S, Mangisoni JH, Kassie GT, Sutcliffe C, Langyintuo AS, La Rovere R, Mwangi WM (2012) Drivers of improved maize variety adoption in drought-prone areas of Malawi. J Dev Agric Econ 4(14):393–403

Kilosa District Council (KDC). (2015). *The three-year development plan 2014–2017 for Kilosa District Council* Morogoro, Tanzania.

Kothari CR (2004) Research Methodology: Methods and Techniques, 2nd edn. New Delhi, India, Wiley Eastern Ltd

Kudi TM, Bolaji M, Akinola MO, Nasa DH (2011) Analysis of the adoption of improved maize varieties among farmers in Kwara State, Nigeria. International Journal of Peace and Development Studies 1(3):8–12

Lang MJ (2008) Categorical data analysis: Away from ANOVAs (transformation or not) and towards logit mixed models. National Institutes of Health 59(4):434–446

Langyintuo A, Mekuria M (2008) Assessing the influence of neighborhood effects on the adoption of improved agricultural maize in developing countries. African Journal of Agricultural Research Economics 2(1):151–169

Lokot M (2021) Whose voices? Whose knowledge? A feminist analysis of the value of key informant interviews. Int J Qual Methods 20:1–8

Magembe, K.S., Mwatawala, M.W., Mamiro, D.P., & Chingonikaya, E.E. (2016). Assessment of awareness of mycotoxin infections in stored maize (*Zea mays* L.) and groundnut (*Arachis hypogaea* L.) in Kilosa District Council, Tanzania. *International Journal of Food Contamination*, 3(12), 3–8.

Mariano MJ, Villano R, Fleming E (2012) Factors influencing farmers' adoption of modern rice technologies and good management practices in the Philippines. Agric Syst 110:41–53

Mazvimavi K, Twomlow S (2009) Socioeconomic and institutional factors influencing the adoption of conservation farming by vulnerable households in Zimbabwe. Agric Syst 101(1):20–29

Mendola M (2007) Agricultural technology adoption and poverty reduction: A propensity-score matching analysis for rural Bangladesh. Food Policy 32(3):372–393

National Bureau of Statistics (NBS). (2012). *National sample census of agriculture: Smallholder agriculture crop sector national report, Volume 11*. NBS: Dar es Salaam, Tanzania, pp. 2–50

Nwakor FN, Ifenkwe GE, Okoye BC, Onummdu FN, Anyaegbunam HN, Ekedo TO, Onyia CE (2011) Socio-economic factors affecting the adoption of improved cassava varieties among farmers in Abia State. Journal of Agriculture and Social Research (JASR) 11(1):20–32

Nyanga PH (2012) Factors influencing adoption and area under conservation agriculture: A mixed methods approach. Sustainable Agriculture Research 1(2):27

Olalekan WA, Simeoni AB (2015) Discontinued use of improved maize varieties in Osun State, Nigeria. Journal of Development and Agriculture 7(3):85–91

Oluyole KA, Usman JM, Oni OA, Oduwole OO (2013) Input use efficiency of cocoa farmers in Ondo State, Nigeria. Journal of Finance and Economics 1(1):8–10

Omotesho FK, Muhammad-Lawal A, Ismaila E (2014) Assessment of hired labour use and food security among rural farming households in Kwara State. Nigeria. Journal of Agricultural Sciences 59(3):353–361

Ransom JK, Paudyal K, Adhikari K (2003) Adoption of improved maize varieties in the hills of Nepal. Agric Econ 29(3):299–305

Raphael D (2014) Determinants of adoption of early maturing maize varieties in Nzega District. Thesis, Sokoine University of Agriculture, Tanzania, Tabora Region. M.Sc

Sinja, J., Karugia, J.T., Baltenweck, I., Waithaka, M.M., Miano, M.D., Nyikal, R.A., & Romney, D. (2004). Farmer perception of technology and its impact on technology uptake: The case of fodder legumes in Central Kenya highlands. In *2004 Inaugural Symposium, December 6–8, 2004* (No. 9543). *African Association of Agricultural Economists (AAAE), Nairobi, Kenya*, 2–10.

Smale M, Olwande J (2013) Demand for maize hybrids and hybrid change on smallholder farms in Kenya. Agric Econ 45(2014):409–420

Train, E.K. (2003). *Discrete Choice Methods with Simulation* (2nd Ed.). The Press Syndicate of the University of Cambridge: The Pitt Building, Trumpington Street, Cambridge, United Kingdom, 1–343.

Tshikala KS, Fonsah GE, Kostandin G, Ames G (2015) Technology adoption behaviors: Evidence from maize producers in drought-prone regions of Eastern Kenya. African Journal of Agricultural Economics and Rural Development 3(2):203–213

Tura M, Aredo D, Tsegaye W, Rovere RL, Tesfahun G, Mwangi W, Mwabu G (2010) Adoption and continued use of improved maize seeds: Case study of Central Ethiopia. Afr J Agric Res 5(17):2350–2358

Uematsu, H., & Mishra, A.K. (2010). Can education be a barrier to technology adoption? *Selected Paper* prepared for presentation at the Agricultural & Applied Economics Association, Annual Meeting, AAEA, CAES, & WAEA: Denver, Colorado, 25–27.

Open Access This chapter is licensed under the terms of the Creative Commons Attribution 4.0 International License (http://creativecommons.org/licenses/by/4.0/), which permits use, sharing, adaptation, distribution and reproduction in any medium or format, as long as you give appropriate credit to the original author(s) and the source, provide a link to the Creative Commons license and indicate if changes were made.

The images or other third party material in this chapter are included in the chapter's Creative Commons license, unless indicated otherwise in a credit line to the material. If material is not included in the chapter's Creative Commons license and your intended use is not permitted by statutory regulation or exceeds the permitted use, you will need to obtain permission directly from the copyright holder.

Enhancing Attitudes Towards Implementing Sustainable Land Management Practices Through Farmer Field Schools in Rural Malawi: A Snapshot of the New Ecological Paradigm (NEP) Scale

Austin Tibu, Emmanuel Enokye, and Bishal K. Sitaula

Abstract Addressing the challenges of land degradation in Sub-Saharan Africa (SSA) requires giving greater attention to farmers' human-nature perspectives, in addition to providing agriculture extension. The concept of transformation provides an opportunity to evaluate the way farmer-to-farmer extension approaches align with farmers' evolving perspectives. These perspectives can be reflected in smallholder farmers' human-nature attitudes, offering insights into their values, beliefs, and motivations. However, there is limited understanding of farmers' attitudes and their contribution to sustainable land management implementation. This chapter presents the results of an exploratory study that examined smallholder farmers' environmental attitudes and perceptions regarding sustainable land management practices in Dowa district, Malawi. The attitudes and perceptions were assessed using the modified New Ecological Paradigm (NEP) scale. The typical scale uses fifteen questions to assess perceptions of the environmental attitudes of individual farmers. For this study, these questions were modified into ten questions that were judged to be more relevant to smallholder farmers in Malawi. Data were collected from 80 smallholder farmers in two purposely selected Extension Planning Areas (EPA), using a semi-structured questionnaire. The presence of Farmer Field Schools (FFS) was the primary criterion for selecting the two EPAs. Findings indicate that respondents view landscape and farmland degradation as significant threats to smallholder agricultural production. However, farmers' perceptions exhibit a multidimensional view of environmental attitudes towards conserving their farmland. The study concludes with recommendations for future strategies to support behavior change in farmer-to-farmer extension approaches for sustainable land management in Malawi. Recognizing farmers'

A. Tibu (✉) · E. Enokye · B. K. Sitaula
Department of International Development and Environment, NMBU, As, Norway
e-mail: austin.tibu@nmbu.no

A. Tibu
Depatment of Environmental Sciences and Management, Lilongwe University of Agriculture and Natural Resources, Lilongwe, Malawi

© The Author(s) 2026

B. R. Singh et al. (eds.), *Sustainable Food Systems in Sub-Saharan Africa*,
https://doi.org/10.1007/978-3-032-11288-0_25

attitudes will facilitate guidance to integrate human dimensions of land degradation that support the development and upscaling of locally adapted sustainable land management practices.

Keywords Environmental attitude · Farmer behavior · Ability · Sustainable livelihoods · Motivation

1 Introduction

The adoption of farmer-to-farmer approaches, where farmers act as extension agents in their communities, is becoming increasingly crucial in addressing land degradation. Agricultural extension needs to evolve to support sustainable practices sensitive to environmental degradation (Bongaarts 2019). This transition involves harnessing the human potential to care for the land and effectively implement sustainable management programmes. (O'Brien and Sygna 2013). Understanding the attitudes embedded in smallholder farmer-to-farmer approaches is essential for transitioning to environmentally sensitive agricultural production.

Sustainable land management offers a promising approach to tackle environmental degradation and enhance food production. The selection and implementation of sustainable land management practices is often based on the available biophysical information. To inform the choices, new tools for assessing land degradation are increasingly becoming available, providing spatial and temporal data on different forms of degradation. However, broader biophysical and socioeconomic determinants alone are insufficient to support the adoption of sustainable land management practices. Further, the extent and nature of land degradation across regions remain unclear (Place 2023). Context-specific determinants for smallholder farming systems are generally unavailable, especially regarding individual farmers' attributes (Burnham et al. 2023). Consequently, the various rates and drivers of land degradation and how to sustain agricultural production continue to be debatable (Place 2023). Furthermore, several studies suggest a list of reasons for the lower-than-expected adoption of sustainable land management practices (Chinseu et al. 2019; Chomba et al. 2020; Giller et al. 2021; Bouwman et al. 2021; Mgomezulu et al. 2023).

Farmers' attitudes have the potential to drive behavior change toward sustainable land management practices, independent of whether implementation of these practices has been approached through individually driven adoption, incentive-based approaches, or adoption in compliance with legislation (Burnham et al. 2023; Mulwafu and Kamchedzera 2024). The Farmer Field School (FFS) approach is one of the learning approaches that promote individually driven adoption of sustainable land management practices. The approach was introduced in Sub-Saharan Africa by the Food and Agriculture Organization of the United Nations (FAO) in the mid-1990s (Waddington et al. 2014; van den Berg et al. 2020b). Initially focused on adult learning principles and experiential learning promoting integrated pest management (IPM), the FFS approach has been shown to reduce the overuse of chemical pesticides and improve farmers' knowledge and decision-making abilities (Pemsl et al. 2006; Ali and Sharif 2012). The FFS approach aims to reduce environmental degradation by educating farmers on ecological farm management. However, there is limited research on the role of individual FFS members in driving deliberate actions on sustainable land management (van den Berg et al. 2020a; van den Berg et al. 2020b) despite farmers being viewed as value-driven actors working collectively to achieve individual goals on their farmlands (Bakker et al. 2023). Thus, understanding the role of farmers' environmental attitudes in driving the FFS approach can go a long way in supporting the implementation of sustainable land management practices.

Lack of awareness or knowledge is cited as one of the constraints to adopting certain sustainable land management practices (Mango et al. 2017), which include soil conservation and fertility measures, tree planting, and rainwater harvesting practices, among others (Nkonya et al. 2016). Learning sustainable land management practices through farmer-to-farmer approaches is becoming popular (Martini et al. 2023). Conversely, exploring how farmers' attitudes influence the learning process can contribute to the successful upscaling of sustainable land management (Ogunyiola et al. 2022; Bakker et al. 2023). However, current research is limited to the role of the attitudes of farmer trainers (Martini et al. 2023) and the wealth characteristics of the member farmers (Tchuwa et al. 2022). Few studies explore the farmers' environmental attitudes in farmer-to-farmer approaches. This research gap builds on the general pattern that the current understanding of smallholder farmers' behavior in rural settings is limited, particularly in the global South, including Malawi (Meijer et al. 2016; Bakker et al. 2023).

Based on this background, the study examines the environmental attributes of smallholder farmers in Malawi and their potential to support sustainable agricultural practices. It aims to assess the environmental attitudes of smallholder farmers and to compare their attitudes and perceptions towards sustainable land management in selected Farmer Field Schools (FFS). The study focuses on the potential of (FFS) programs in transforming farmers' attitudes to implement sustainable land management practices on their farmland. The study understands that other factors can influence farmers' adoption capacity. Some of these factors are the quality of farmer-trainers, technological complexity, communication methods, and external barriers (Maertens et al. 2021). Understanding these barriers can help design effective interventions to support farmers in adopting sustainable practices and promoting behavior change towards sustainable land management practices in Malawi. The research formulated the following hypotheses:

1. Farmers' membership in FFS characterizes their environmental attitudes, and
2. Farmers' environmental attitudes are positively correlated with the implementation of sustainable land management practices.

1.1 Theoretical Framework

The study integrated O'Brien and Sygna's (2013) transformation framework with the learning theories outlined in Stam et al. (2023). By maintaining the concept of a 'farmer's attitude,' the study applied the interaction between discursive and reflexive learning processes at the FFS level. This interaction was evaluated within the personal sphere of transformation to uncover the potential of FFS as a tool for transformation. As a result, the framework aided in comprehending smallholder farmers' environmental attitudes towards sustainable land management practices. The environmental attitudes were analyzed from farmers' responses documented during the household survey. The study further investigated the responses regarding the implementation of sustainable land management practices on their farmland. According to Stam et al. (2023), learning is a non-linear, iterative process of acquiring and generating new knowledge, insights, and meaning-making through developing, experimenting, and pursuing ideas and possibilities. From this perspective, a learning process becomes discursive when knowledge and meaning exchange among actors can lead to a shared understanding of the problem and practical solutions. The reflexive learning process occurs when actors modify their practices due to their action processes and reflections. The study chose to structure its questions on attitudes in relation to learning expectations, as it is challenging to directly inquire about behavior (Neaman et al. 2024). Regarding FFS, the potential for transforming farmers can be explored by understanding how the FFS approach empowers communities to enhance their capacity for informed decision-making (FAO 2016). The capacity can be explored by understanding whether the FFS approach can include environmental attitudes.

To understand farmers' environmental attitudes toward change, learning expectations must be considered in three areas: politics, practical, and personal transformation (O'Brien and Sygna 2013). The political sphere includes legal, cultural, and economic systems. The practical sphere involves behaviors, technical solutions, and practices. The personal sphere encompasses individual and collective beliefs, values, and worldviews. The study considers the personal sphere as a starting point for understanding the potential for transforming smallholder agriculture (Bakker et al. 2023).

Learning in the personal sphere can focus on self-reflection of the farmers' frames of reference or the attitudes of the individual farmers. The frame of reference can be based on the environmental attitudes held among the farmers. The practical sphere is driven by what farmers do in the personal sphere. Thus, there are no clear boundaries between the practical and the two other spheres. Figure 1 provides an overview of how the study related to the concepts.

The study cautiously noted that the transformation framework is not comprehensive and lacks specificity in describing decision-making and attitudes toward sustainable land management practices. However, it has the potential to help identify the underlying mechanisms for agricultural transformation. The learning process plays a crucial role in establishing a transformation process. Therefore, understanding the

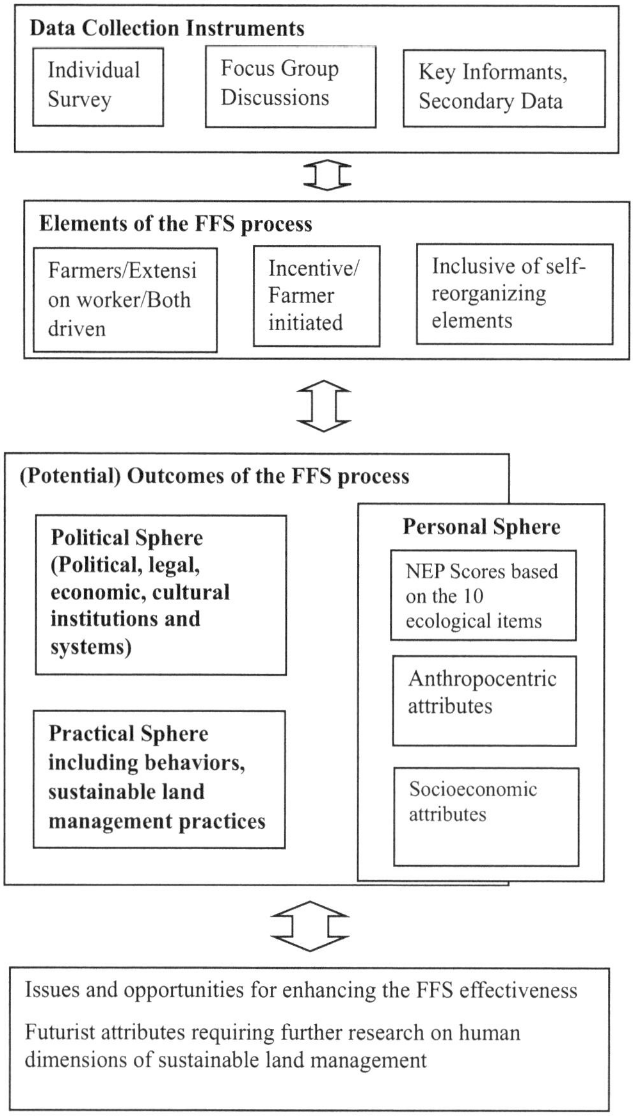

Fig. 1 Conceptual Framework for Attitudes and Implementing Sustainable Land Management practices (adapted from O'Brien and Sygna, 2013)

learning process can reveal the transformative potential of smallholder farmers by capturing their perspectives on engaging in the FFS approach.

Key informant interviews and Focus Group Discussions indicated that the established FFS focused on the practical sphere, emphasizing sustainable land management options such as manure application, pit planting, water harvesting, and improved crop varieties. An informant mentioned that these skills are acquired from national trainers as part of the FFS process supported by the Food and Agricultural Organization. The study then utilized the New Environmental Paradigm (NEP) scale (Dunlap and Van Liere 1978) and a list of elements compiled by Woiwode et al. (2021) to assess the inclusion of elements that have the potential to strengthen farmers' environmental attitudes within the FFS process. We also explored the socioeconomic characteristics of the respondents in the study to create an overview of the study households.

Sustainable land management is crucial for addressing environmental degradation and increasing food production. New tools for assessing land degradation are becoming available, but data on smallholder farming systems remains inadequate. The debate continues on different rates and drivers of land degradation, and why the adoption of sustainable land management practices is lower than expected. Lack of awareness or knowledge is one of the constraints to adopting sustainable land management practices. Learning sustainable land management practices through farmer-to-farmer approaches is becoming popular. Understanding how farmers' environmental attitudes influence the learning process can support the successful upscaling of sustainable land management.

This chapter is divided into four sections. Section one introduces the study and its objectives. Section two provides the theoretical basis for the study's conceptual framework. This is preceded by a description of the data collection methods and a brief description of the study areas. Section three outlines the data analyzed and the results. The respective results are discussed in this section. Section four concludes the chapter by providing insights and future research direction that may support behavior change in transitioning agriculture in Malawi.

2 Materials and Methods

The study addresses these questions by conducting an exploratory case study on the attitudes of smallholder farmers participating in two Farmer Field Schools as part of a sustainable food systems project in Malawi. In this project, the FFS approach is seen as a transformative learning strategy to promote sustainable food systems. Achieving sustainable food systems is contingent upon conserving natural resources on smallholder farmlands (Fanzo et al. 2021). In this regard, we view the FFS as a farmer-to-farmer approach that supports farmers' environmental attitudes in conserving their farmlands. The study highlights the importance of individual attributes in driving the transformation of food systems by managing natural resources on their farmland. Therefore, the study contributes to the discourse on the influence of individual and collective actions in sustainable land management and the potential of farmers' internal attributes to drive broader transitions (World Bank 2024).

Additionally, the study adds to the ongoing discussions about the role of research on internal and external transformation in sustainable development (Wamsler et al. 2021). The study is a snapshot of smallholder farmers' environmental attitudes based on the sample farmers interviewed in February–March 2023. Therefore, the study acknowledges the importance of time for environmental attitudes to be institutionalized (van Poeck et al. 2020). In addition, the study focuses on the learning expectations from the smallholder farmers' perceptions while acknowledging the need for further research and longitudinal benefits of the FFS approach.

2.1 Study Methods and Data Collection

This study investigates the potential of Farmer Field School (FFS) initiatives in the transition to sustainable food systems in Malawi. Specifically, it examines the role of FFS in knowledge transfer and environmental attributes of smallholder farmers. Given the significant food insecurity and environmental challenges in Malawi, understanding the potential of FFS in promoting sustainable agricultural practices is crucial and timely. The study utilizes an exploratory mixed methods approach, combining qualitative and quantitative data collection and analysis methods in two selected case studies in Dowa district in Malawi. By focusing on the two FFS, the study aims to provide an in-depth contextual examination of individual members. The exploratory approach allows for a comprehensive understanding of the notion of environmental attitudes by offering various perspectives and angles of analysis (Flyvbjerg 2006). The study's primary focus is on FFS members as its study population. However, an equal number of farmers were surveyed in the villages covered by the study FFS.

Through detailed questionnaire survey and Focused Group Discussions (FGD), the researchers gathered information on the experiences, challenges, and benefits received from FFS initiatives from two Agricultural Extension Planning Areas (EPAs) of Mndolera and Chibvala under Dowa District Agricultural Office (Fig. 2). The EPA is one of Malawi's lowest sub-district planning entities for agriculture programs.

The average annual temperature of the Dowa district, according to Sibale et al. (2021), ranges from 19.4°C to 21.3 °C, which is in the humid subtropical climatic zone. These EPAs receive 868 mm and 886 mm of rainfall, respectively, on average each year. Dowa district lies in central Malawi and has a population of 411,387 (Sibale et al. 2021). The majority ethnic group is called Chewas. Farming communities make up the majority of these two groups.

The two case study Dzoole and Mbalame FFS Groups, were established as part of the TRANSFORM program.[1] The FFS approach was recommended as a complementary approach to the Lead Farmer model in previous Norwegian-supported programs. Consequently, the two case study FFS were established in 2022 under the TRANSFORM program. According to the key informants, TRANSFORM had established 24 FFS in target districts across Malawi, with each FFS consisting of a maximum of 30 farmers. The primary focus of the two case studies was to analyze and understand

[1] TRANSFORM is a Norwegian funded program implementing Climate Smart Agriculture (CSA), climate adapted villages, support Farmer Field Schools (FFS) and the Lead Farmer (LF) approach to test developed innovations in agricultural transformation in Malawi.

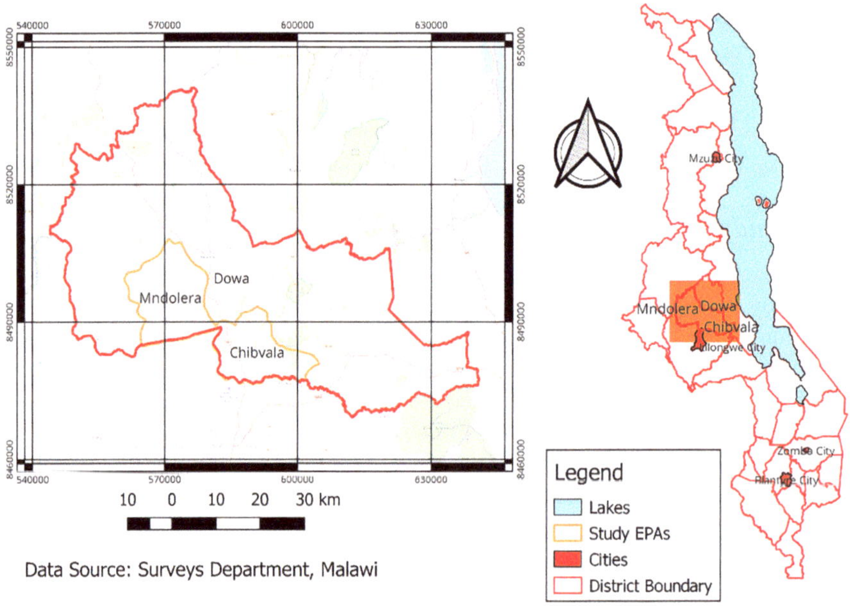

Fig. 2 Map of Malawi and Dowa district showing study sites

the local farming systems through regular Agroecosystem Analysis (AESA) and by considering the existing capacities of the farmers. The Dzoole FFS is situated in Mndolera EPA, characterized by aflat and gentle terrain, while the Mbalame FFS is located close to Dowa Hills.

The study interviewed eighty respondents, including forty FFS members and forty non-FFS members in targeted Extension Planning Areas (EPAs). To ensure consistency, households in villages around established FFS were surveyed. Membership in the FFS was a primary criterion for participant selection, but non-FFS members in the vicinity of established FFS were also interviewed. Non-FFS members were selected using the snowball sampling method (Bhardwaj et al. 2024). Three research assistants were trained over two days, followed by pilot testing of a structured questionnaire in one FFS in Mndolera EPA before full deployment.

The survey interview lasted for 60 min. The questionnaire examined general socioeconomic characteristics, disorienting dilemmas (such as shocks), and coping strategies, as well as FFS members' expectations in developing environmental attitudes, connectedness, and attitudes toward nature (Maertens et al. 2021; Woiwode et al. 2021). The NEP scale is a tool used to measure the relationship between humans and the environment by assessing environmental concerns (Hawcroft and Milfont 2010). The NEP scale was initially developed by Dunlap in the 1970s (Dunlap and Van Liere 1978) and later revised by Dunlap et al. (2000). It has been effectively utilized to explore people's perceptions regarding controlling nature, the finite nature of ecosystem resources, and the detrimental effects of human behavior on nature.

The scale determines the degree to which individuals value nature. It does this by assigning a score to each respondent, reflecting their overall perception, ranging from conservation to exploitation of natural resources. Higher scores on the NEP scale indicate a stronger inclination toward nature conservation, while lower scores indicate the opposite (Dunlap et al. 2000; Dunlap 2008). Consequently, the study employs NEP values as indicators of farmers' environmental attitudes, which can be considered a potential benefit of the FFS process.

Participants rated their responses on a scale ranging from strongly disagree (1) to strongly agree (5). They also rated Woiwode's elements for the self-organizing sustainable field of action on a scale from not at all integrated (1) to fully integrated (5). In addition, the survey included questions about perceptions of sustainable land management practices such as conservation agriculture, intercropping, agroforestry, and the use of improved seeds, among others (Pangapanga-Phiri et al. 2022). Participants provided self-reported yes/no responses regarding their knowledge of, and engagement in, sustainable land management on smallholder farmland (Maertens et al. 2021).

Prior to commencing the survey, a single FGD was held with farmers from each FFS to gather their perspectives on land restoration practices, attitudes towards the agricultural environment and climate, and their vision for the future of their agricultural land. Additionally, five key informant interviews were carried out with extension workers and FFS coordinators to further enrich the understanding of FFS's potential and challenges in facilitating knowledge exchange and collaborative learning among smallholder farmers in Malawi. Furthermore, secondary data from the Department of Agricultural Extension Services was collected to assess the effectiveness of extension services provided to smallholder farmers in rural areas. The analysis aimed to evaluate the impact of these services on smallholder farmers, particularly in terms of knowledge transfer and the adoption of sustainable agricultural practices. Additionally, four field observations were conducted to directly witness the implementation of FFS practices and their influence on farming activities and community dynamics.

2.2 Analytical Framework

The study's analysis was conducted in three stages: (1) summarizing descriptive statistics for the respondents in the two case studies; (2) presenting a summary of the NEP scores calculated from responses to Likert scale on smallholder farmers' attitudes in the two case studies; and (3) providing a brief discussion on the potential of the FFS approach in developing farmers' environmental attitudes. General socioeconomic characteristics, disorienting dilemmas (e.g., shocks), and coping strategies data were analyzed using the gtsummary package in R. Descriptive statistics were tabulated to display means and frequencies. Chi-squared statistical tests based on the p-values were included in the tables. The 5-scale responses for the NEP and self-organizing elements were analyzed using MASS and psych packages in R to test significant differences using the Chi-test. The mean or total scores of all the 10

NEP items were calculated, with possible mean values ranging from 1 to 15 or total values ranging from 10 to 50. Higher scores in this study indicate a stronger attitude towards conserving smallholder farmland. The results were presented in tables and graphs.

Constant Comparative Analysis (CCA) was used to analyze qualitative data. This involved systematically comparing with individual interview data to develop conceptual frameworks and identify relationships. We also used a comparative approach to relate statistically gathered information. In the quantitative analysis, we conducted coding, tabulation, and descriptive analysis to categorize and statistically treat the information, along with frequency analysis to compare responses.

3 Study Results and Discussion

3.1 Household Characteristics of the Respondents

Table 1 presents a summary of the socioeconomic and household characteristics of the respondents who participated in the case studies. The total number of respondents was 80, with 40 from Mndolera and 40 from Chibvala EPA. Out of the 80 respondents, 55 were female, and the average age was 41.77 years. 79% of the respondents were married, and 70% had attended primary education (0 to 8 years of education). In terms of agricultural production, 51% practiced mixed-cropping systems on smallholder agricultural land. The majority of the respondents (64%) were cultivating two or more plots, often scattered around their villages. 58% of the respondents owned goats and chickens, while cattle ownership was relatively low. Additionally, 49% were members of the FFS, and 61% reported being visited by an agricultural extension agent during the 2022/23 growing season.

In terms of respondents' demographics, such as age, sex, marital status, household size, FFS membership, number of plots cultivated in the 2022/2023 season, major agricultural systems practiced, livestock ownership, and shocks experienced in 2022/2023, there were no significant differences between the two EPA. However, there were notable variations in climate shocks experienced, crops grown in the 2022/2023 season, how respondents secured their agricultural plots, and their level of education. These findings align with the NSO (2019) report, which indicates, for instance, a higher female population in Dowa and a literacy level of 66% compared to the national average of 68% (NSO 2019). In low-literacy agricultural systems like those in these case studies, the role of agricultural extension is significant. For instance, Ragasa and Mazunda (2018) found that only 28% of households could access extension services in Malawi in 2018. The study revealed that 50% of surveyed respondents were FFS members, indicating a potential increase in access to extension services. The study specifically focused on FFS members to gain insights into the FFS learning process and its potential personal benefits.

Table 1 Descriptive Statistics of the Respondents (N = 80)

Socioeconomic Variable	**Overall** N = 80[1]	**Mndolera** N = 40[1]	**Chibvala** N = 40[1]	**p-value**[2]
Age (Mean)	42	40	43	0.4
Sex, n (%)				0.8
Female	55 (69%)	28 (70%)	27 (68%)	
Male	25 (31%)	12 (30%)	13 (33%)	
Marital Status, n (%)				0.7
Divorced	6 (7.5%)	4 (10%)	2 (5.0%)	
Married	63 (79%)	32 (80%)	31 (78%)	
Separated	3 (3.8%)	1 (2.5%)	2 (5.0%)	
Widowed	8 (10%)	3 (7.5%)	5 (13%)	
Household size (Mean)	5.30	5.70	4.90	0.1
Education Level (Mean)	6.6	7.7	5.6	0.006
FFS Membership (Yes), n (%)	39 (49%)	19 (48%)	20 (50%)	0.8
Major Cropping System, n (%)				0.13
Intercropping	4 (5.1%)	4 (10%)	0 (0%)	
Mixed cropping	40 (51%)	18 (45%)	22 (56%)	
Mono-cropping	35 (44%)	18 (45%)	17 (44%)	
Weather shock experienced in 2022, n (%)				< 0.001
None	54 (68%)	20 (50%)	34 (85%)	
Too much rain or floods	23 (28%)	19 (48%)	4 (10%)	
Drought/dry spell	2 (2.5%)	0 (0%)	2 (5.0%)	
Both drought/floods	1 (1.3%)	1 (2.5%)	0 (0%)	

[1]Mean; n (%)
[2]Wilcoxon rank sum test; Pearson's Chi-squared test; Fisher's exact test

In relation to the FFS approach, the analysis found no significant differences between the sex, marital status, household size, education level, crop production practices, and shocks experienced during the 2022/2023 growing season among FFS members and non-members. However, there were notable distinctions in age, agricultural systems, and livestock ownership between FFS members and non-members. It was observed that FFS members tended to be older and had a more significant number of goats compared to non-members. This indicates that younger farmers were less likely to participate in the FFS approach, despite its potential benefits to strengthen farmer-to-farmer learning.

The study revealed that there were no significant differences between sources of agricultural extension across the two study EPAs in terms of farmer-to-farmer learning. However, significant differences were found between information sources for FFS members and non-members. Over 82% of FFS members had access to Lead Farmers, whereas only 51% of non-members did. Non-FFS members seemed to rely significantly on friends for innovation and technical knowledge of sustainable agricultural practices. This finding aligns with Martini et al. (2023), who suggested that the farmer-to-farmer approach leverages existing platforms where farmers traditionally communicate with each other. Furthermore, the study indicated that agricultural extension workers visited FFS members more frequently than non-members. Additionally, FFS members actively shared technical knowledge and innovations. It is evident that the FFS approach is gaining popularity as a format for participatory extension.

3.2 Use of Sustainable Land Management Practices on Farmland

The study conducted a series of inquiries to assess awareness regarding various sustainable land management practices among smallholder farmers for the upcoming 2022/2023 growing season. Survey results indicated that 96.3% of farmers were familiar with conservation agriculture, 85.0% with agroforestry, and 98.8% with crop rotation. However, only 57.7% practiced conservation agriculture, 30.0% practiced agroforestry, and 71.3% practiced crop rotation. Additionally, the study revealed that 80.0% of respondents were aware of different climate-smart agriculture practices, but only 18.8% implemented these options on their farmlands. These findings differ from those of Mango et al. (2017), who reported lower awareness of soil and water conservation, mulching, and rainwater harvesting practices in the Chinyanja triangle. The variance in findings can be attributed to this study's focus on FFS members, who may have been exposed to information on sustainable land management practices during the FFS learning process. Consequently, there were notable discrepancies in the adoption of conservation agriculture, agroforestry, climate-smart agriculture, and crop rotation between FFS members and non-members. According to Pangapanga-Phiri et al. (2022), access to information plays a crucial role in influencing the decision to adopt sustainable land management-related practices.

In Martini et al.'s (2023) study, it was noted that Farmer Field Schools (FFS) are shifting from focusing solely on technology transfer to a more collaborative approach at the farm level. At the same time, Burnham et al. (2023) emphasized the increasing importance of motivational and cognitive factors in driving behavior changes that support land conservation. Their papers provide an overview of the role of farmer-to-farmer learning in strengthening farmers' attitudes. This overview can help to gain a deeper understanding of how FFS have and could continue to facilitate the transition to a human-centered FFS process within the framework of sustainable food systems. To analyze the integration of human-centered elements into FFS, we utilized the self-organizing sustainable fields of action list developed by Woiwode et al. (2021). The findings, including the elements and their descriptive statistics, are summarized in Table 2.

Table 2 Self-organizing sustainable fields of action included in the FFS Approach (from Woiwode et al., 2020)

Self-Organizing Element	Overall	Farmer Field School Member		
		No	**Yes**	**p-value**[2]
	N = 80[1]	N = 40[1]	N = 40[1]	
Nurturing mindfulness and self-reflection during training and practical sessions during the season				< 0.001
Not at all integrated	30 (38%)	29 (70%)	0 (0%)	
Fully integrated	43 (54%)	10 (25%)	33 (85%)	
Somewhat integrated	7 (8.8%)	2 (4.9%)	5 (13%)	
Embracing diversity, building trust, and clarifying shared vision				< 0.001
Not at all integrated	33 (42%)	28 (70%)	0 (0%)	
Fully integrated	42 (53%)	10 (25%)	32 (82%)	
Somewhat integrated	5 (6.3%)	3 (7.3%)	2 (5.1%)	
Contributing to social integration and cohesion and enriching life during meetings, trainings, and discussions				< 0.001
Not at all integrated	45 (56%)	32 (78%)	13 (34%)	
Fully integrated	17 (34%)	5 (12%)	22 (57%)	
Somewhat integrated	8 (10%)	4 (9.8%)	4 (10%)	
Preserving natural resources, including soils, water, and forests				< 0.001
Not at all integrated	28 (35%)	28 (67%)	0 (0%)	
Fully integrated	47 (59%)	12 (30%)	35 (90%)	
Somewhat integrated	5 (6.3%)	1 (2.4%)	4 (10%)	

[1]n (%)

[2]Fisher's exact test

The study revealed significant differences in the self-organizing elements and the FFS process between the two case study EPAs. These differences could be attributed to the context-specific nature of such attributes. In line with the study's findings, Tam and Chan (2017) cautioned that managing global environmental issues may be influenced by personal views and general orientations rooted in worldviews. However, the study also found significant differences between FFS members and non-members. The FFS process seems to include elements related to human dimensions that could potentially enhance engagement in sustainable agricultural practices. Identifying avenues to promote behavior change is an emerging aspect of sustainability research, economics, and environmental behavior (Woiwode et al. 2021). For instance, Lal et al. (2021) suggested that education could strengthen the human connection to the soil. Thus, exploring the influence of environmental knowledge and attitudes might be important. Studies have used scales to measure soil knowledge, connection to nature, and soil conservation behavior (Burnham et al. 2023). In Malawi, a few studies explored farmers' decision-making on land restoration practices, agroforestry, and sustainable land management practices using some of these tools (Meijer et al. 2016; Jambo et al. 2019; Djenontin et al. 2020). While the existing literature may favor one measurement tool over another, it is advisable to adopt existing measurements and instruments for environmental attitudes (Cruz and Manata, 2020). Thus, we applied the NEP scale because it has been validated in various cultural and demographic contexts (Tonin and Benedetto 2024).

3.3 Farmers' Attitudes and Perceptions

Figure 3 shows the responses to the ten items of the New Ecological Paradigm (NEP) scale from a sample of 80 individuals. The NEP scores for each item were calculated for both FFS members and non-members. Higher scores indicate greater environmental concerns and attitudes toward sustainable land management on smallholder farmland.

The average NEP score in this study was 3.12 (median = 3.17), with a standard deviation of 0.495, ranging from 3.00 to 3.33. The average total NEP score for each respondent was 37.5, with a standard deviation of 5.94. The total NEP scores ranged from 36.0 to 46.0 from Q1 to Q3 (refer to Fig. 3). Research suggests that high NEP scores indicate a stronger environmental connection of farmers to their smallholder land. It is worth noting that the total NEP score in this study was slightly lower than the total score of 56.1 in Italy (Tonin and Benedetto 2024), and the average NEP score was lower than the NEP score of 3.60 computed for farmers in Serbia and New Zealand (Durpoix, 2010; Despotović et al., 2021).

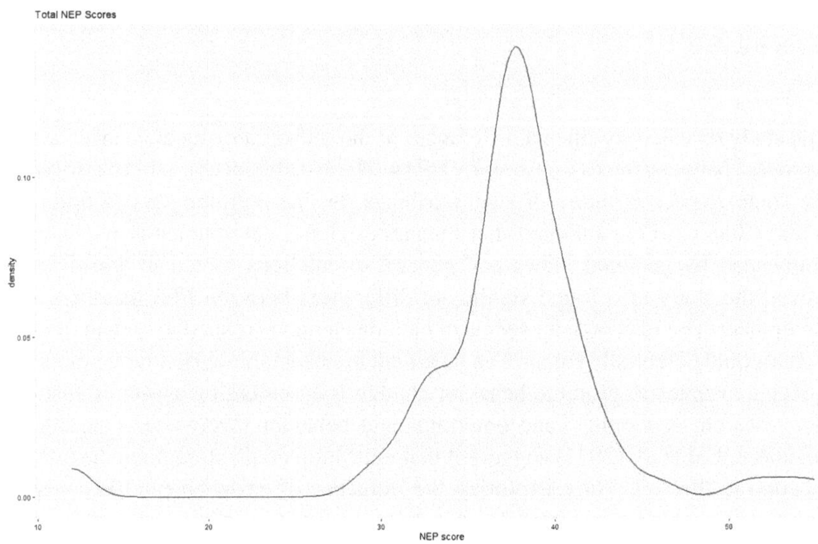

Fig. 3 Total NEP Scores for the two EPA study sites

There are several factors that may account for the variation in scores between previous studies and the present findings. Firstly, the NEP scale was developed based on the environmental perspectives of developed countries, and thus the theoretical concepts it employs, such as ecological and human domination perspectives, may have been defined according to Western environmental views (Tonin and Benedetto 2024). The smallholder farmers in the study EPAs may not easily differentiate between Western or anthropocentric dimensions of human domination. The two directions assumed by the NEP scale, whether eco-centric or anthropocentric, may be unstable in different contexts (Tonin and Benedetto 2024). For instance, the NEP scale may not exhibit clear patterns in certain contexts because environmental behavior is not always driven by eco-centric motives (Grúňová et al. 2019). Secondly, the sampling methods used in the studies should be taken into consideration. Additionally, socio-economic and sociocultural context characteristics may influence environmental attitudes, with socio-economic characteristics including age, gender, income level, and membership in farmers' groups.

There were significant variations in the NEP scores between the two study case EPAs, which could be attributed to the different contexts of the Mndolera and Chibvala EPAs. Additionally, significant differences in the NEP could be based on potential land availability for agricultural expansion. According to NSO (2019), the population density in Traditional Authority (TA) Nkukula covering Chibvala EPA was 309, compared to 276 in TA Dzoole covering Mndolera EPA. Our study also identified a similar trend, with 50% of the respondents expressing potential for land expansion in Mndolera EPA, compared to 35% in Chibvala EPA.

In addition, the study analyzed the environmental attitudes of smallholder farmers through the lens of opposing attitudes outlined by Ntanos et al. (2019). This perspective categorizes environmental attitudes into Dominance Social Paradigm (DSP) and the New Ecological Paradigm (NEP), based on individuals' perceptions of whether natural resources are inexhaustible and can be extensively utilized to meet human needs (anthropocentric approach) or the perception that degradation of farmland can have adverse effects on the entire ecosystem (pro-NEP) (Ntanos et al. 2019). Factor Analysis with varimax rotation was employed in this study to determine if the NEP scores of the respondents could be treated as a single dimension. Factor identification was based on the Kaiser Criterion for factor retention, using an Eigenvalue of 1.00 as the threshold. The analysis yielded two factors from the NEP scores, which collectively accounted for 99.0% of the total variance, with Eigenvalues of 2.28 and 1.20 (see Table 3) (Govindasamy et al. 2024).

The analysis categorized the 10 items into two dimensions: Factor 1 included the aspects of anthropocentric (2), rejection of human exceptionalism (4), recognition of limits to growth (6), and acknowledgment of the potential for an ecological crisis (10), as well as anthropocentric (12), rejection of human exceptionalism (14), and the potential for an ecological crisis (15). Factor 2 comprised nature's balance fragility (3), anthropocentricity (7), and rejection of human exceptionalism (9). Thus, the findings indicate that the NEP score in the two case studies was multidimensional, reflecting two dimensions of smallholder farmers' attitudes (see Table 4).

Table 3 Total Variance Explained by the 10 Factors of the Analysis

Factor	Eigenvalue	Difference	Proportion	Cumulative
Factor1	2.28809	1.08592	0.6515	0.6515
Factor2	1.20217	0.55342	0.3423	0.9938
Factor3	0.64875	0.47325	0.1847	1.1785
Factor4	0.17551	0.11990	0.0500	1.2285
Factor5	0.05560	0.04234	0.0158	1.2443
Factor6	0.01326	0.14091	0.0038	1.2481
Factor7	-0.12765	0.08222	-0.0363	1.2118
Factor8	-0.20986	0.04564	-0.0598	1.1520
Factor9	-0.25551	0.02288	-0.0728	1.0793
Factor10	-0.27839		-0.0793	1.0000
Number of observation = 80				

Table 4 The factor Analysis of the 10 Items of the Modified New Ecological Paradigm (n = 80)

Adapted NEP Item Assessed	F1	F2
Farmers have the right to modify our homes, fields, and our environment to suit our needs	0.69	
When humans interfere with nature, it often produces disastrous consequences for our food production systems		0.37
Human ingenuity will ensure that we do not make our environment unproductive for the current production systems	0.65	
Our farmlands have plenty of natural resources if we just learn how to develop them for increased production	0.62	
Plants and animals in our agricultural systems have as much right as humans to exist		0.56
Despite our special abilities to practice agriculture, humans are still subject to the laws of nature		0.65
The so-called ecological crisis facing our agricultural systems has been greatly exaggerated	0.39	
Farmers were meant to rule over the rest of nature for agricultural production	0.52	
Farmers will eventually learn enough about how nature works to be able to control it	0.66	
If things continue on their present course, we will soon experience a major crisis in agricultural production	0.26	

These findings align with numerous studies that have identified two dimensions in their analyses and the multifaceted nature of the NEP scores in Italy (Tonin and Benedetto 2024). Similar findings have identified multidimensional factors in New Zealand farmers' attitudes (Durpoix 2010). Consequently, the NEP scale can be used to generalize findings in areas that are underrepresented in the existing literature, similar to Malawi. Furthermore, the NEP score can demonstrate changes in attitudes toward the natural environment as individuals gain knowledge of their agricultural ecosystems (Ntanos et al. 2019). However, despite several studies using different scales and tools, the NEP has not been widely applied to comprehend farmers' attitudes in Malawi (Meijer et al. 2016; Bakker et al. 2023). In this context, the study investigated the patterns in attitudes among members of FFS.

3.4 Farmers' Attitudes and Worldviews in the FFS Extension Approach

The study found no significant differences in other variables such as sex, education level, marital status, and land holding between the two groups. However, there were significant disparities in the adoption of sustainable agricultural practices, with FFS members demonstrating a higher inclination towards sustainable practices compared to non-members. Additionally, the FFS process seems to have effectively integrated elements that could potentially foster the development of sustainable agriculture practices among smallholder farmers. For instance, the results indicate that the FFS process includes elements that can aid members in strengthening their connection to nature. However, it's worth noting that the NEP score for FFS members did not significantly differ from that of non-members.

The findings of this study have two important implications. Firstly, it is evident that there is no single defining dimension that characterizes the environmental attitudes of the respondents. The average NEP score of FFS members was slightly higher than that of non-members, indicating that FFS members in the study EPAs may have a greater concern for conserving their farmland. The study also sought to analyze the factor loadings for both FFS members and non-members. The Chi-square test revealed a weaker relationship with farmers' membership, suggesting that membership in the FFS did not necessarily influence environmental attitudes. However, it is important to note that this does not mean that the respondents in the study will prioritize human domination over pro-environmental behavior. Existing literature suggests that farmers may express positive attitudes towards the environment without necessarily translating them into action (Tonin and Benedetto 2024).

Nevertheless, the stronger relationship between the NEP score and certain sustainable land management practices, as well as the highly significant test on the inclusion of self-organizing elements outlined by Woiwode et al. (2021), indicate the potential for FFS to strengthen environmental attitudes. Martini et al. (2023) have suggested that farmer-to-farmer agricultural extension systems can serve as a platform for exchanging human-nature attitudes. This exchange can potentially lead to a locally grounded movement for conserving natural resources in practical and political spheres of transformation. For instance, environmental attributes are often linked to more significant points of intervention with a higher potential for impactful change (Woiwode et al. 2021).

Future research could delve into the potential benefits of integrating environmental attitudes and points of intervention within the FFS process. However, the scope of this study was limited in exploring this potential due to time and resource constraints.

Furthermore, van den Berg et al. (2020a) noted an increasing adoption of sustainable land management practices within Farmer Field Schools (FFS), suggesting improved access to information about sustainable land management among its members. In contrast, this study did not find a higher environmental attitude among FFS members partly due to its snapshot nature. Future research could explore the emerging temporal individual attributes that may help to establish a link between personal qualities and behavior change by influencing attitudes (Woiwode et al. 2021; Wamsler et al. 2021). These attributes have the potential to influence a sense of agency and bolster one's perception of identity, which was deemed crucial in sustainable land management (van Noordwijk et al. 2020). In this context, identity can be seen as encompassing skills that facilitate the transition to sustainable land management, such as creativity, critical thinking, and reflection (van Poeck et al. 2020). These skills represent underlying factors that influence human decision-making processes. According to Wamsler et al. (2021), these human qualities can be categorized into factors that impact people's awareness, connection, insights, purpose, and agency, referred to as transformative qualities. These qualities have the potential to drive greater action toward the adoption of sustainable land management practices, particularly within Farmer Field Schools in rural Malawi.

4 Conclusions and Potential Research Recommendations

The research shows that environmental attitudes are increasingly important factors that influence smallholder farmers' engagement in sustainable land management. The study found that farmers in Chibvala EPA had higher NEP scores than those in Mndolera EPA, indicating a context-specific nature of the NEP scale. The study suggests that there is a need to enhance the effectiveness of Farmer Field Schools in promoting sustainable agricultural practices among smallholder farmers in Malawi. It recommends further studies to analyze human-centered variables in characterizing farmers in Farmer Field Schools to assess the potential of FFS in contributing to agricultural transformation.

However, environmental attitudes are just one aspect of the determinants of sustainable land management. Current research in Malawi has focused on biophysical and socioeconomic drivers of sustainable land management, but few studies have integrated human dimensions of sustainable land management. This research has shown that understanding how humans interact with the environment might be crucial for addressing challenges like land degradation. The study recommends further research to measure Farmer Field School learning outcomes related to attitudes and worldviews across socioecological zones and over time to measure potential agricultural transformation through sustainable land management.

References

Ali A, Sharif M (2012) Impact of farmer field schools on adoption of integrated pest management practices among cotton farmers in Pakistan. J Asia Pac Econ 17(3):498–513. https://doi.org/10.1080/13547860.2012.694706

Bakker E, Hassink J, van Veluw K (2023) The 'inner' dimension of Dutch farmers' trajectories of change: drivers, triggers and turning points for sustained agroecological practices. Agroecol Sustain Food Syst 47(5):687–717. https://doi.org/10.1080/21683565.2023.2180563

Bhardwaj R, Agrawal U, Vashist P, Manna S (2024) Determination of sample size for various study designs in medical research: A practical primer. J Family Med Prim Care 13(7):2555–2561. https://doi.org/10.4103/JFMPC.JFMPC_1675_23

Bongaarts J (2019) IPBES, 2019. Summary for policymakers of the global assessment report on biodiversity and ecosystem services of the Intergovernmental Science-Policy Platform on Biodiversity and Ecosystem Services. Popul Dev Rev 45(3):680–681. https://doi.org/10.1111/PADR.12283

Bouwman TI, Andersson JA, Giller KE (2021) Adapting yet not adopting? Conservation agriculture in Central Malawi. Agric Ecosyst Environ 307:107224. https://doi.org/10.1016/J.AGEE.2020.107224

Burnham E, Zabel S, Navarro-Villarroel C, Ermakov DS, Castro M, Neaman A, Otto S (2023) Enhancing farmers' soil conservation behavior: Beyond soil science knowledge. Geoderma 437. https://doi.org/10.1016/j.geoderma.2023.116583

Chinseu E, Dougill A, Stringer L (2019) Why do smallholder farmers dis-adopt conservation agriculture? Insights from Malawi. Land Degrad Dev 30(5):533–543. https://doi.org/10.1002/LDR.3190

Chomba S, Sinclair F, Savadogo P, Bourne M, Lohbeck M (2020) Opportunities and Constraints for Using Farmer Managed Natural Regeneration for Land Restoration in Sub-Saharan Africa. Frontiers in Forests and Global Change 3:122. https://doi.org/10.3389/FFGC.2020.571679/BIBTEX

Despotović J, Rodić V, Caracciolo F (2021) Farmers' environmental awareness: Construct development, measurement, and use. J Clean Prod 295. https://doi.org/10.1016/j.jclepro.2021.126378

Djenontin INS, Zulu LC, Ligmann-Zielinska A (2020) Improving Representation of Decision Rules in LUCC-ABM: An Example with an Elicitation of Farmers' Decision Making for Landscape Restoration in Central Malawi. Sustainability 2020, Vol 12, Page 5380 12(13):5380. https://doi.org/10.3390/SU12135380

Dunlap RE (2008) The New Environmental Paradigm Scale: From Marginality to Worldwide Use. J Environ Educ 40(1):3–18. https://doi.org/10.3200/JOEE.40.1.3-18

Dunlap RE, Van Liere KD (1978) The "new environmental paradigm." Journal of Environmental Education 9(4):10–19. https://doi.org/10.1080/00958964.1978.10801875

Dunlap RE, Van Liere KD, Mertig AG, Jones RE (2000) New Trends in Measuring Environmental Attitudes: Measuring Endorsement of the New Ecological Paradigm: A Revised NEP Scale. J Soc Issues 56(3):425–442. https://doi.org/10.1111/0022-4537.00176

Durpoix D (2010) Farmers' attitudes and behavior towards the natural environment: a New Zealand case study

Fanzo J, Haddad L, Schneider KR, Béné C, Covic NM, Guarin A, Herforth AW, Herrero M, Sumaila UR, Aburto NJ, Amuyunzu-Nyamongo M, Barquera S, Battersby J, Beal T, Bizzotto Molina P, Brusset E, Cafiero C, Campeau C, Caron P, Cattaneo A, Conforti P, Davis C, DeClerck FAJ, Elouafi I, Fabi C, Gephart JA, Golden CD, Hendriks SL, Huang J, Laar A, Lal R, Lidder P, Loken B, Marshall Q, Masuda YJ, McLaren R, Neufeld LM, Nordhagen S, Remans R, Resnick D, Silverberg M, Torero Cullen M, Tubiello FN, Vivero-Pol JL, Wei S, Rosero Moncayo J (2021) Viewpoint: Rigorous monitoring is necessary to guide food system transformation in the countdown to the 2030 global goals. Food Policy 104

FAO (2016) Farmer Field School Guidance Document: Planning for quality programmes

Flyvbjerg B (2006) Five misunderstandings about case-study research. Qual Inq 12(2):219–245. https://doi.org/10.1177/1077800405284363

Giller KE, Hijbeek R, Andersson JA, Sumberg J (2021) Regenerative Agriculture: An agronomic perspective. Outlook Agric 50(1):13–25. https://doi.org/10.1177/0030727021998063

Govindasamy P, Isa NJM, Mohamed NF, Noor AM, Ma L, Olmos A, Green K (2024) A systematic review of exploratory factor analysis packages in R software. Wiley Interdiscip Rev Comput Stat 16

Grúňová M, Sané M, Cincera J, Kroufek R, Hejcmanová P (2019) Reliability of the new environmental paradigm for analysing the environmental attitudes of Senegalese pupils in the context of conservation education projects. Environ Educ Res 25(2):211–221. https://doi.org/10.1080/13504622.2018.1428942

Hawcroft LJ, Milfont TL (2010) The use (and abuse) of the new environmental paradigm scale over the last 30 years: A meta-analysis. J Environ Psychol 30(2):143–158. https://doi.org/10.1016/J.JENVP.2009.10.003

Jambo I, Groot J, … KD (2019) Motivations for the use of sustainable intensification practices among smallholder farmers in Tanzania and Malawi. Taylor & Francis

Lal R, Bouma J, Brevik E, Dawson L, Field DJ, Glaser B, Hatano R, Hartemink AE, Kosaki T, Lascelles B, Monger C, Muggler C, Ndzana GM, Norra S, Pan X, Paradelo R, Reyes-Sánchez LB, Sandén T, Singh BR, Spiegel H, Yanai J, Zhang J (2021) Soils and sustainable development goals of the United Nations: An International Union of Soil Sciences perspective. Geoderma Reg 25:e00398. https://doi.org/10.1016/J.GEODRS.2021.E00398

Maertens A, Michelson H, Nourani V (2021) How Do Farmers Learn from Extension Services? Evidence from Malawi. Am J Agric Econ 103(2):569–595. https://doi.org/10.1111/ajae.12135

Mango N, Makate C, Tamene L, Mponela P, Ndengu G (2017) Awareness and adoption of land, soil and water conservation practices in the Chinyanja Triangle, Southern Africa. International Soil and Water Conservation Research 5(2):122–129. https://doi.org/10.1016/J.ISWCR.2017.04.003

Martini E, Pagella T, Mollee E, van Noordwijk M (2023) Relational values in locally adaptive farmer-to-farmer extension: how important? Curr Opin Environ Sustain 65

Meijer SS, Sileshi GW, Catacutan D, Nieuwenhuis M (2016) Agroforestry and deforestation in Malawi: inter-linkages between attitudes, beliefs and behaviours. Agrofor Syst 90(4):645–658. https://doi.org/10.1007/S10457-015-9844-4/TABLES/10

Mgomezulu WR, Machira K, Edriss A-K, Pangapanga-Phiri I (2023). Modelling Farmers' Adoption Decisions of Sustainable Agricultural Practices under Varying Agro-Ecological Conditions: A New Perspective. https://doi.org/10.1016/j.igd.2023.100036

Mulwafu T, Kamchedzera G (2024) Land degradation neutrality and the weak avoid, reduce and reverse priorities in Malawi's soil laws. Soil Security 14:100134. https://doi.org/10.1016/J.SOISEC.2024.100134

Neaman A, Navarro-Villarroel C, Poblete-Ramos F, Lizardi N, Burnham E, Huerta-Salinas O, Zabel S, Ermakov DS, Castro M, Otto S (2024) Reconciling the soil stewardship paradox: Knowledge without care, care without knowledge. Geoderma Regional 37. https://doi.org/10.1016/j.geodrs.2024.e00794

Nkonya E, Mirzabaev A, Von Braun Editors J (2016) Economics of Land Degradation and Improvement-A Global Assessment for Sustainable Development

NSO (2019) 2018 Malawi Population and Housing Census: Main Report

Ntanos S, Kyriakopoulos G, Skordoulis M, Chalikias M, Arabatzis G (2019) An application of the new environmental paradigm (NEP) scale in a Greek context. Energies (Basel) 12(2). https://doi.org/10.3390/en12020239

O'Brien K, Sygna L (2013) Responding to Climate Change: The Three Spheres of Transformation. Proceedings transformation in a changing climate : International Conference in Oslo 19–21 June 2013. Oslo

Ogunyiola A, Gardezi M, Vij S (2022) Smallholder farmers' engagement with climate smart agriculture in Africa: role of local knowledge and upscaling. Climate Policy 22(4):411–426. https://doi.org/10.1080/14693062.2021.2023451
Pangapanga-Phiri I, Mungatana ED, Pangapanga L, Nkoka FS (2022) Understanding the impact of sustainable land-scape management practices on farm productivity under intensifying tropical cyclones: Evidence from Southern Malawi. Tropical Cyclone Research and Review 11(4):265–276. https://doi.org/10.1016/J.TCRR.2023.02.002
Pemsl DE, Waibel H, Witt R, Waibel H, Pemsl DE (2006) Diffusion of information among small-scale farmers in Senegal: the concept of Farmer Field Schools
Place F (2023) Sustaining Natural Resources in African Agriculture: What Have We Learned in the Past Two Decades?:139–150. https://doi.org/10.1007/978-981-19-5542-6_11
Ragasa C, Mazunda J (2018) The impact of agricultural extension services in the context of a heavily subsidized input system: The case of Malawi. World Dev 105:25–47. https://doi.org/10.1016/J.WORLDDEV.2017.12.004
Sibale D, Mwenelupembe G, Chikabvumbwa S, Chisale S, Sibale D, Mwenelupembe G, Chikabvumbwa S, Chisale S (2021) Evaluation of Water Delivery Performance of Nkhafi Irrigation Scheme in Dowa District, Malawi, Africa. Computational Water, Energy, and Environmental Engineering 10(3):95–107. https://doi.org/10.4236/CWEEE.2021.103007
Stam K, van Ewijk E, Chan PW (2023) How does learning drive sustainability transitions? Perspectives, problems and prospects from a systematic literature review. Environ Innov Soc Transit 48:100734. https://doi.org/10.1016/J.EIST.2023.100734
Tam KP, Chan HW (2017) Environmental concern has a weaker association with pro-environmental behavior in some societies than others: A cross-cultural psychology perspective. J Environ Psychol 53:213–223. https://doi.org/10.1016/J.JENVP.2017.09.001
Tchuwa F, Wellard K, Morton J, Kambewa D, Mkweu D, Tchuwa F, Wellard K, Morton J, Kambewa D, Mkweu D, Mhango W (2022). From Learning Plot to Main Field: Scaling-out Soil Health Innovations in Malawi. https://doi.org/10.3390/su14031532
Tonin S, Benedetto D (2024) Exploring Sustainability Concerns and Ecosystem Services: The Role of the New Ecological Paradigm Scale in Understanding Public Opinion. Sustainability (Switzerland) 16(5). https://doi.org/10.3390/su16051902
van den Berg H, Ketelaar JW, Dicke M, Fredrix M (2020a) Is the farmer field school still relevant? Case studies from Malawi and Indonesia. Wageningen Journal of Life Sciences 92(1):1–13. https://doi.org/10.1016/j.njas.2020.100329
van den Berg H, Phillips S, Dicke M, Fredrix M (2020b) Impacts of farmer field schools in the human, social, natural and financial domain: a qualitative review. Food Secur 12(6):1443–1459. https://doi.org/10.1007/S12571-020-01046-7/TABLES/2
van Noordwijk M, Gitz V, Minang PA, Dewi S, Leimona B, Duguma L, Pingault N, Meybeck A (2020) People-Centric Nature-Based Land Restoration through Agroforestry: A Typology. Land 2020, Vol 9, Page 251 9(8):251. https://doi.org/10.3390/LAND9080251
van Poeck K, Östman L, Block T (2020) Opening up the black box of learning-by-doing in sustainability transitions. Environ Innov Soc Transit 34:298–310. https://doi.org/10.1016/j.eist.2018.12.006
Waddington H, Snilstveit B, Hombrados J, Vojtkova M, Phillips D, Davies P, White H (2014) Farmer Field Schools for Improving Farming Practices and Farmer Outcomes: A Systematic Review. Campbell Syst Rev 10(1):i–335. https://doi.org/10.4073/CSR.2014.6
Wamsler C, Osberg G, Osika W, Herndersson H, Mundaca L (2021) Linking internal and external transformation for sustainability and climate action: Towards a new research and policy agenda. Glob Environ Chang 71:102373. https://doi.org/10.1016/J.GLOENVCHA.2021.102373
Woiwode C, Schäpke N, Bina O, Veciana S, Kunze I, Parodi O, Schweizer-Ries P, Wamsler C (2021) Inner transformation to sustainability as a deep leverage point: fostering new avenues for change through dialogue and reflection. Sustain Sci 16(3):841–858. https://doi.org/10.1007/S11625-020-00882-Y/FIGURES/2
World Bank (2024) Malawi WB 2024 Country Economic Monitor

Open Access This chapter is licensed under the terms of the Creative Commons Attribution 4.0 International License (http://creativecommons.org/licenses/by/4.0/), which permits use, sharing, adaptation, distribution and reproduction in any medium or format, as long as you give appropriate credit to the original author(s) and the source, provide a link to the Creative Commons license and indicate if changes were made.

The images or other third party material in this chapter are included in the chapter's Creative Commons license, unless indicated otherwise in a credit line to the material. If material is not included in the chapter's Creative Commons license and your intended use is not permitted by statutory regulation or exceeds the permitted use, you will need to obtain permission directly from the copyright holder.

Do Land Reforms Promote Equity in Land Rental Market Participation? Evidence from Malawi

Horace Happy Phiri, Maggie Munthali, Moses Chitete, Mark Malata, Lemekezani Chilora, Kennedy Machira, and Miru Muyanga

Abstract There is renewed interest in understanding the links between land reforms, land markets and poverty reduction in sub–Saharan Africa. Applying a three-fold extended Kitagawa–Oaxaca–Blinder decomposition approach to a two-wave panel data from Malawi, this study investigated the inequities between male and female-headed households in land rental market participation in sub–Saharan Africa. Using the amount of land rented-in, overall results reaffirm that rental market participation is higher among male-headed households, but the gap is on the decline. However, the contribution of the reforms, particularly the transfer from communal ownership to individual property rights is yet to play a role in addressing the gap between the gender groups. Rather, gains in literacy levels among female household heads seem to have been responsible for the reduction in disparities.

Keywords Land rental markets · Gender equity · Soil · Market participation · Sub-Saharan Africa

1 Introduction

Rural land markets provide an avenue through which groups that have limited access, for example youth and women, gain access to land (Abay et al. 2021; Ricker-Gilbert and Chamberlin 2008; Wineman and Tassie 2017). Potentially, land markets improve the gender equity by allowing female-headed households to access land outside non-market channels, such as inheritance (Lastarria-Cornhiel 2009). Welfare gains are recorded for those who rent in the land (Chamberlin and Ricker-Gilbert 2016). On the other hand, there are productivity gains by transferring land from less productive to more productive households (Holden et al. 2010).

H. H. Phiri (✉) · M. Chitete · M. Malata · K. Machira
Department of Agricultural and Applied Economics, Bunda College, Lilongwe, Malawi
e-mail: hphiri@luanar.ac.mw

M. Munthali · L. Chilora · M. Muyanga
MwAPATA Institute, Lilongwe, Malawi

© The Author(s) 2026

B. R. Singh et al. (eds.), *Sustainable Food Systems in Sub-Saharan Africa*,
https://doi.org/10.1007/978-3-032-11288-0_26

Land markets have rapidly increased over the decades due to increases in rural population densities, where off-farm employment opportunities are scarce, coupled with land policy reforms, especially well-defined land tenure and social security rights (Jayne et al. 2014). Land market transactions in SSA are both formal and informal, and land rental and sales markets are found within both formal and customary land tenure systems (Chamberlin and Ricker-Gilbert 2016; Holden et al. 2010; Ricker-Gilbert et al. 2022). Over the years, land markets have contributed to increased farm production and technical efficiency gains by transferring land from relatively less productive households to more productive households (Chamberlin and Ricker-Gilbert 2016; Holden et al. 2010).

Previous literature has shown that land market transactions in SSA are both formal and informal, and land rental and sales markets are found within formal and customary land tenure systems (Chamberlin and Ricker-Gilbert 2016; Holden et al. 2010; Ricker-Gilbert et al. 2022). In developing countries like Malawi, formal land markets rarely work; as such, land is acquired through informal processes such as inheritance from parents/relations, private donations and informal purchase e.g. "pinyolo" Jayne et al. (2021). Throughout SSA, land is recognized under various tenure regimes which include freehold, leasehold (private), and customary (traditional) tenure system. In the region, land transactions are popular in customary tenure systems, where land rights are tenuous or ambiguous and customary land tenure alone account for 78% of the land holding in Africa. Holden and Otsuka postulated that land transactions (both land rentals and sales markets) in SSA are affected and driven by risks, shocks and liquidity constraints. Despite the importance of informal land markets that operate outside of a statutory legal framework, these markets are not well understood and are often overlooked in political discourse (Chimhowu and Woodhouse 2006; Wineman and Tassie 2017).

Although markets for farmland are growing within the traditional tenure systems in Sub-Saharan Africa (SSA), there is evidence that this growth is uneven (Ricker-Gilbert and Chamberlin 2008). For instance, Wineman et al. (2017) found that female-headed households remain marginalized in terms of land market participation in Tanzania. Studies in gender market participation (Gebre et al. 2020; Marenya et al. 2017) suggest that gender differences in resource endowment, and its effects on the returns, result in disparities in market participation between male and female farmers.

One of the key drivers of vibrant land markets is tenure security. Tenure insecurity and weak enforcement of property rights increase the risk of losing land when it is rented out. This discourages potential landlords from participating in rental markets (Besley and Ghatak 2010). The imperfections in rental markets are more likely to disproportionally affect female landowners since they normally have weaker property rights (Menon et al. 2017). Land titles can be an efficient mechanism to correct this kind of distortion. Holden et al. (2011) and Xu and Du (2022) analyzed tenure insecurity, gender, low-cost land certification and land rental market in participation in Ethiopia. Land certification initially enhanced land rental market participation of (potential) tenant and landlord households, especially those that are headed by females.

Although there have been studies on the effects of titling/certification on tenure security and how the resultant security status affects land rental markets; gaps remain on whether the tenure security land reforms can contribute to equitable women participation in markets. It is important to understand that improving land rights might have differing outcomes, and also depends on the initial security of the landholding. The socially embedded nature of customary rights means the land rights of many women depend on social entitlements that can be eroded due to reforms that make land rights marketable. This result into a de facto transfer of a greater share of rights to (typically) male title holders, a case in point being the Kenyan tenure reform.

The African Development Bank (2020) gender profile for Malawi revealed that only 31.7% of women own land as compared to 68.3% of men. This has huge implications on women's economic empowerment, let alone food security attainment of a majority of women headed households in the country. However, synergies in the National Agriculture Policy (GoM 2016) and the National Gender Policy (GoM 2015) remain an earlier arrangement that has to be critically evaluated. For instance, the linkage of the gender policy to the agriculture policy is observed through its emphasis on advocating for gendered food utilization and dietary diversification. Nonetheless, the policies fail to recognize that issues of food security are to a larger extent explained by land ownership in rural Malawi, where over 80% of smallholder farmers live. The goal of the National Gender Policy remains to reduce gender inequality and hence ensure gender parity and women empowerment by enhancing participation of men, women, boys and girls in socioeconomic development processes. Considering that Malawi is an agrarian economy, it is imperative that policy makers understand gendered differences in access to land, and most importantly land market participation in Malawi.

Ironically, women constitute a majority of farmers in SSA contributing 60–80% of agricultural labour force, but their levels of participation in rental markets is disproportionately low (Lunduka et al. 2009). Ultimately, the weak participation of female farmers in land markets entails that the promise of improved household welfare that from land market transactions eludes this gender group. Although the factors that encourage or discourage participation on land rental markets have been widely documented, key questions remain on what sort of policy interventions would achieve more equitable participation. This study was carried out to address this knowledge gap by analyzing market participation amongst Malawian farmers, and how it has evolved overtime. The Malawi case offers a unique opportunity as the country Land Acts were reformed in 2016. The reforms, such as certification of customary land, were enacted in order to improve land tenure security (Benjamin 2020; Zuka 2019). Empirical evidence shows that tenure security enhances land rental market participation (Chamberlin and Ricker-Gilbert 2016; Lunduka et al. 2009). Therefore, a longitudinal study of the pre and post reform period in Malawi gives insight on land reforms, and can achieve more balanced participation in land markets.

This study seeks to make a contribution to literature on gender inclusive land policies in order to improve market-based land access by comparing market participation of gender groups in Malawi. The study hence adds to existing literature in two ways: First, it adds to the growing literature on gender equality and women empowerment by providing insights to policy makers on key policy factors that should be emphasized in order to improve women's access to land, in addressing gendered food utilization and food security considering that Malawi is an agrarian economy. Secondly, it provides era-segregated findings on gendered differences in access to land in Malawi by providing the key findings for the pre land reform era (2016) and the post land reform era (2019). Lastly, it provides novel empirical methods by employing the Kitagawa–Oaxaca–Blinder decomposition technique in land rental markets to provide robust empirical evidence of the extent of the gender gap in access to land in Malawi.

2 Materials and Methods

Data Sources

The study utilized panel data from the Integrated Household Panel Survey, 2016 (pre-reform year), and 2019 (post-reform). In 2019, the fourth round of the Integrated Household Panel Survey (IHPS 4) was implemented, targeting a national sample of 2508 households that were interviewed in 2016, and which could be traced back to the third cross-sectional survey (IHPS3) in 2011 but were interviewed again in 2013. A balanced panel of 1177 was extracted and used in this study. Data is available at https://microdata.worldbank.org/index.php/catalog/2939.

Analytical Techniques

We used a three-fold extended Kitagawa–Oaxaca–Blinder decomposition approach (1) to examine how gender differences in access to resources and returns from the mobilization of the resources contribute to the gender gap in land market participation. The decomposition is done to explain the average gender gap in the quantities of land rented in between the two gender groups (i.e. male, female). The outcome measure was the amount of land rented in. This was calculated using IHSP data from 2016 and 2019. The longitudinal nature of the variable means that a panel data B–O decomposition is applicable using "xtoaxaca" Stata command. The standard linear regression equation modelling for the relationship between the outcome variable (Y) and a set of predictors (X) for two comparable groups of a household is given as:

$$Y_{ig} = B_g X_{ig} + E_{ig};\, E\left(E_{ig}\right) = 0 \tag{1}$$

where Y is the natural log of the value of the outcome variable (land rented), g represents the gender group, such as male or female group, Xig is a vector of average values of observable characteristics, βg is a vector of coefficient estimates for gender g (including an intercept), and E_{ig} is the gender-specific random error term assumed to be independently and identically distributed with mean zero and variance, σ^2. The rationale behind the B–O decomposition approach is therefore to show how much of the average quantity gap between two groups (e.g., male and female groups). Following Jann (2) and Daymont and Andrisani (3), the mean gender gap of the quantity used by two groups can be written as:

$$Gap = \bar{Y}_m - \bar{Y}_F =$$

$$\underbrace{\{(\bar{x}_m - \bar{x}_F)\beta_f\}}_{\text{Endowment effect (E)}} + \underbrace{\{\bar{x}_F(\beta_m - \beta_f)\}}_{\text{Coefficient effect (C)}} + \underbrace{\{(\bar{x}_m - \bar{x}_F)(\beta_m - \beta_f)\}}_{\text{Interaction effect (CE)}} \quad (2)$$

where Ym and Yf denote the average value of the quantity of rented out or rented in by male and female headed households, X is a vector of average values of observable characteristics and is a vector of coefficient estimates for a male or female group. Equation (2) is a 'threefold' decomposition where the mean gender gap is divided into three components. The first component is the portion of the gap that is due to the gender differences in observable characteristics (called the "endowment effect"). The second component, the "structural or return effect", is the part of the gap emanating from differences in the coefficients of the observable characteristics; and the third component, the "interaction effect", is the portion of gap attributable to the joint effects of both observable characteristics and their estimated coefficients. Thus, gender differences in the quantities of fertilizers applied can be explained by these three factors. Using panel data, we also estimate β from a panel regression model. Since panel regressions model time-constant individual error terms, a decomposition using panel regression models must take empirical group differences in these time-constant, unobserved variables into account (1). As such, the time constant individual error term U_i becomes part of the decomposition of group-level differences.

$$Y_t^l = X_t^l B_t^l + u^l + \varepsilon_t^l \quad (3)$$

Taking the time-constant error terms into account adds the differences in the expectation of u^l as a fourth component U to the decomposition. This component is not time-dependent. It only comprises differences between groups in the time-constant error terms.

$$U = [E(u^m) - E(u^f)] \quad (4)$$

Accordingly, a decomposition using panel regression models attribute parts of the differences between groups to unobserved factors that do not change within the period of observation.

3 Results

3.1 Land Rental Market Participation and Its Determinants

It is crucial for women to have secure access to, use of and control over land and property rights in order for them to achieve economic empowerment and increase productivity and agricultural commercialization. When women secure land rights, they participate in land rental markets, thereby improving household food security and income diversification and alleviating poverty. Levels and degrees of participation in land rental markets vary across Malawi. Figure 1 shows how farm households participate in land rental markets by gender in 2016 and 2019. Results shows that higher levels of land market participation is reported among male-headed households than female-headed households. Most male and female-headed households from Central region rented-in land more than those from Northern and Southern Region. It is worth noting that female-headed households from Northern region (Chitipa, Karonga, Mzimba, Nkhatabay districts) and Southern Region (Balaka, Blantyre, Chikwawa, Chiradzulu, Mulanje, Neno, Nsanje, Phalombe and Thyolo districts) did not participate in land rental markets at all.

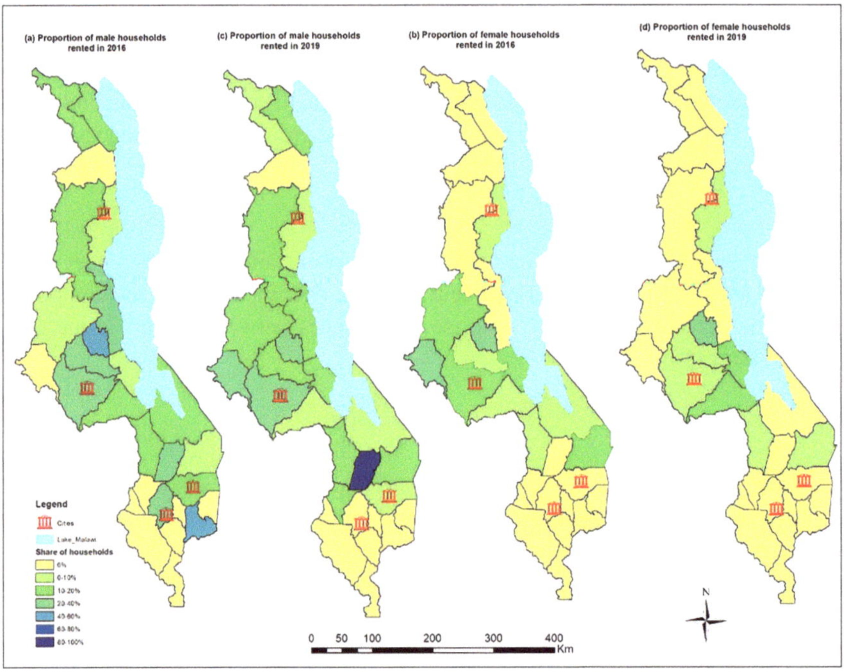

Fig. 1 Proportion of households that rented land by gender of decision maker

As mentioned in the methods section, mixed-level logit regression is used to examine the influence of gender and other variables in household land rental market participation. The results are presented in Table 1. We estimated a pooled model, and men and women models separately to investigate the gender effects on the probability of household positions in the market. If a household did not participate in the rental markets i.e., autarchic the outcome is zero (0) while the lessee was assigned a value of one (1). The model is significant at a 1% level, showing that the explanatory variables taken together explain the land market position of households. In the pooled sample estimation, we included the sex of the household head as a gender indicator variable. The results reaffirm that the gender of the household head significantly influences participation as a lessee i.e., rent in land, but its effect on being a lessor was found to be insignificant.

The age of the household head carries a negative coefficient and was significant in the overall model and male only model, but insignificant in the female only model. This implies that older male farmers have a lower probability of participating in rent markets. Similarly, more educated farmers were more likely to rent in land compared to otherwise. The relationship between market participation and livestock ownership was found to be positive among male household heads. Livestock is a proxy of the wealth status of an individual, households with livestock are generally considered more well off than those without. Relatively wealthy households are more likely to afford inputs and thus be able to demand more land. Strangely, male-headed households with more land had a higher probability of rent in land than those with smaller farms.

Access to credit, agricultural or otherwise, boosts the ability of the household to pay rental fees and purchase the required inputs. A positive and significant relationship was observed in both the male and female models. Marriage systems also significantly affected participation in rental markets for both male and female heads. The marriage systems in Malawian societies are at three levels patrilineal, matrilineal or mixed society. The results showed that those residents in mixed societies had a higher probability of rent in land for both gender groups.

Despite the shocks being insignificant influencers in the overall and male only models, irregular rainfall had a positive effect on the decision to rent in land among female decision-makers. The result suggests that more land is demanded in this category of farmers to offset the reduced farm output arising from irregular rainfall. Likewise, land certificates and other forms of title documents did not significantly affect the probability to rent in land.

Table 1 Factors influencing net land rented in by gender of decision maker

	(1)	(2)	(3)
Variables	Overall	Male only	Female only
Age	– 0.266***	– 0.254**	– 0.159
	(0.0914)	(0.105)	(0.207)
Education	0.173***	0.184***	0.212
	(0.0654)	(0.0710)	(0.204)
Livestock ownership	0.489***	0.508***	0.520
	(0.173)	(0.195)	(0.420)
Family size	0.0970	0.111	0.154
	(0.121)	(0.136)	(0.290)
Landholding size	0.317***	0.298***	0.311
	(0.0971)	(0.107)	(0.273)
Access to credit	0.495***	0.401**	1.052***
	(0.163)	(0.183)	(0.404)
FISP beneficiary	– 0.0423	– 0.0189	– 0.377
	(0.183)	(0.201)	(0.486)
Asset value	0.158	0.113	0.701
	(0.290)	(0.298)	(1.494)
Land_title	0.109	0.0795	0.406
	(0.454)	(0.508)	(1.097)
Drought	0.136	0.0700	0.288
	(0.170)	(0.190)	(0.423)
Floods	– 0.0923	– 0.0132	0.176
	(0.202)	(0.230)	(0.466)
Crop_Pests	– 0.109	– 0.121	– 0.258
	(0.183)	(0.202)	(0.467)
Irregular_rains	– 0.0301	– 0.190	0.925*
	(0.168)	(0.184)	(0.505)
Patrilineal		– 0.259	– 1.498
		(0.356)	(1.280)
Mixed		0.593***	0.861*
		(0.201)	(0.497)
Density	0.000573	– 4.17e–06	– 0.00102
	(0.000475)	(0.000684)	(0.00184)
City	– 1.404	– 0.375	1.044
	(1.026)	(1.496)	(4.176)
Sex	– 0.394*		
	(0.214)		
Constant	– 3.049***	– 2.894**	– 5.626**

(continued)

Table 1 (continued)

	(1)	(2)	(3)
Variables	Overall	Male only	Female only
	(1.024)	(1.147)	(2.828)
Observations	1.603	1.176	427

**** p-value < 0.01, ** p-value < 0.05, p-value <0.1*

3.2 *Gender Disparities in Land Rental Market Participation*

To better understand the disparities between male and female decision-makers in participation in land rental markets, we utilize the extended Kitagawa–Oaxaca–Blinder Decomposition. We compare the net land rented in among male and female decision-makers to quantify the gap between the two gender groups. Table 2 shows that the mean land rented for females was 0.46 acres in 2016, and it declined to 0.39 acres in 2019. In the same period, the average for males moved from 0.16 acres to 0.39 acres. Although, the amount of land rented in has declined for both gender groups, a steeper decline [25%] was observed among males compared to females [17%]. Consequently, the difference between the two groups which constitute the gender gap narrowed from 0.11 acres in 2016 to 0.80 in 2019. The differences in our control variables explain 67.6% of the net land rented in gap in 2016 and 32% in 2019. Differences between the gender groups in terms of the proportion that has never been to school accounted for 40.8%.

4 Discussion

The findings from this study show that land market participation is decreasing despite tenure security-enhancing land reforms enacted in 2016. The age of the household head, education, landholding size, access to credit and prevalent marriage system were found to significantly influence market participation for the lessee. However, access to credit and marriage systems were the only factors that universally affected both gender groups. This result implies that '*one size fits all*' policy intervention aimed at promoting market participation is less desirable; rather, tailor-made solutions addressing specific bottlenecks face by a gender group would be more appropriate.

The decline in levels of participation in both gender groups in the post reform period is contrary to previous studies (4–6).It suggests that the envisaged benefits of reforms have not manifested, and households still have a lower propensity to venture onto markets. This can be attributed to two reasons. First, one of the key changes expected to improve tenure security and promote market participation is conversion of customary land to the individual property rights system through issuance

Table 2 Oaxaca decomposition of factors influencing net land rented

		2016			2019			Change		
	Male	0.16***			0.12***					
	Female	0.046			0.039					
	Difference	0.115			0.08			– 0.034		
		[0.033]			[0.021]			[0.014]		
	Endowments	0.17*** (67.6%)			0.017			0.051* (– 25.8%)		
		[0.031]			[0.036]			[0.024]		
	Coefficients	– 0.012			0.005			– 0.19* (96.52%)		
		[0.051]			[0.030]			[0.095]		
	Interaction	0.1			0.03			– 0.056*** (28.5%)		
		[0.080]			[0.059]			[0.017]		
	RE	– 0.006			– 0.008			– 0.001		
		[0.026]			[0.031]			[0.006]		
		End	*Coef*	*Int*	*End*	*Coef*	*Int*	*End*	*Coef*	*Int*
Age	2	0	0.013	0	– 0.007	0.016	0.003	– 0.002	0.006	– 0.005
		[– 0.004]	[– 0.042]	[– 0.009]	[– 0.012]	[– 0.037]	[– 0.014]	[– 0.003]	[– 0.067]	[– 0.006]
	3	0.009	– 0.014	– 0.012	– 0.008	– 0.005	– 0.004	– 0.001	0.001	0
		[– 0.013]	[– 0.017]	[– 0.019]	[– 0.028]	[– 0.018]	[– 0.023]	[– 0.002]	[– 0.031]	[– 0.002]
	4	– 0.002	– 0.094	0.024	0.024	0.028	– 0.009	– 0.001	0.107***	0.008
		[– 0.017]	[– 0.076]	[– 0.024]	[– 0.031]	[– 0.083]	[– 0.027]	[– 0.004]	[– 0.032]	[– 0.007]

(continued)

Table 2 (continued)

		2016			2019			Change		
Education	0	0.102***	− 0.006	0.004	0.003	− 0.011	0.006	0.061***	− 0.101***	
		[− 0.015]	[− 0.007]	[− 0.004]	[− 0.008]	[− 0.011]	[− 0.006]	[− 0.012]	[− 0.011]	[− 0.009]
	2	0.002	0.001	0	− 0.002	0.017*	0.005	0.002	0.017	− 0.002
		[− 0.005]	[− 0.017]	[− 0.005]	[− 0.004]	[− 0.007]	[− 0.005]	[− 0.002]	[− 0.019]	[− 0.002]
	3	− 0.002	0.003	0.005	0.002	− 0.001	− 0.003	− 0.001	− 0.009	0.001
		[− 0.007]	[− 0.005]	[− 0.007]	[− 0.01]	[− 0.003]	[− 0.011]	[− 0.001]	[− 0.013]	[− 0.001]
	4	0.016***	0.006	0.015	0.022	− 0.004	− 0.016	0.002	− 0.038*	0
		[− 0.005]	[− 0.003]	[− 0.011]	[− 0.026]	[− 0.004]	[− 0.02]	[− 0.002]	[− 0.017]	[− 0.002]
Livestock ownership	Yes	0.01	− 0.017	− 0.008	0.011	0.006	0.003	0.001	0.036	− 0.001
		[− 0.011]	[− 0.03]	[− 0.016]	[− 0.008]	[− 0.027]]	− 0.011	[− 0.002]	[− 0.051]	[− 0.001]
Landholding size	1	− 0.168	− 1.037	0.16	− 0.124	− 1.049	0.126	0.005	− 0.004	− 0.002
		[− 0.136]	[− 0.794]	[− 0.13]	[− 0.098]	[− 0.764]	− 0.104	[− 0.038]	[− 0.079]	[− 0.006]
	2	− 0.154	− 0.757	0.142	− 0.14	− 0.761	0.138	0.007	− 0.001	0
		[− 0.095]	[− 0.565]	[− 0.088]	[− 0.088]	[− 0.49]	− 0.088	[− 0.055]	[− 0.087]	[− 0.005]
	3	0.302	− 0.22	− 0.243	0.217*	− 0.234	− 0.204	− 0.023	− 0.043*	0.007
		[− 0.161]	[− 0.246]	[− 0.158]	[− 0.108]	[− 0.248]	− 0.115	[− 0.059]	[− 0.018]	[− 0.006]
	4	0.041	0	0	0.008	0	0	0	− 0.033*	0
		[− 0.027]	[0]	[0]	[− 0.009]	[.]	0	[− 0.013]	[− 0.017]	[− 0.007]
Credit access	1	0.001	0	0	0.003	− 0.003	− 0.001	0.001	− 0.002	0.001
		[− 0.001]	[− 0.025]	[− 0.002]	[− 0.003]	[− 0.031]	− 0.006	[− 0.002]	[− 0.062]	[− 0.003]

(continued)

Table 2 (continued)

		2016			2019			Change		
Marriage system	1	− 0.001	0.01	0.001	0	− 0.008	0	0	− 0.02	0.001
		[− 0.002]	[− 0.009]	[− 0.002]	[− 0.002]	[− 0.009]	− 0.004	[− 0.001]	[− 0.014]	[− 0.003]
	3	0.005	0.047	0.009	0.002	− 0.008	− 0.001	− 0.002	− 0.067	0.002
		[− 0.009]	[− 0.034]	[− 0.005]	[− 0.002]	[− 0.012]	− 0.001	[− 0.002]	[− 0.041]	[− 0.001]
City resident	1	0.007	− 0.001	0	0.005	0.017	− 0.005	0.001	0.012	− 0.002
		[− 0.004]	[− 0.018]	[− 0.003]	[− 0.004]	[− 0.016]	− 0.004	[− 0.003]	[− 0.022]	[− 0.002]
Land title	1	0	0.07	0.001	0	− 0.021	0	0	− 0.092	0
		[− 0.003]	[− 0.342]	[− 0.003]	[− 0.001]	[− 0.039]	− 0.001	[− 0.001]	[− 0.371]	[− 0.002]

of land certificates. By 2019, only about 3% of study participants had title documents, suggesting sluggish implementation of the reforms. Secondly, as Lunduka, Holden (7) observes, the movement from communal rights to individualized property rights do not necessarily translate into more secure tenure. For example, in patrilocal societies women often obtain usufruct rights to family land, but do not possess in-heritance rights (8). Although land certificates confer ownership of land, its immovability means that inheritances rules still have a huge influence on how secure landowners feel, thereby limiting their propensity to participate in markets. The significance of the "type of marriage system" supports this notion. Households resident in areas where mixed marriage systems are practiced i.e., matrimonial and patrimonial systems are both dominant, were more likely to participate in rental markets than otherwise. Using land reforms to promote rental market participation will thus require effective implementation, complemented by laws that address the negative influence of inheritance customs.

Although the post reform period has seen a decline in participation in rental markets, the gap between the gender groups has reduced. The results from Extended B–O decomposition carried out in this study show that out of the socio demographics, farm traits and institution factors tested, only changes to levels of education significantly contributed to the more equitable participation of female headed households. Declining levels of illiteracy amongst female headed households, as shown by the endowment effect, result in reduction of the gap. At the same time, the coefficient effect, which accounts for differences in outcomes between the gender groups, despite having the same level of endowments, shows that it contributed to a widening gap. This entails that female headed households face inherent factors that limit participation in rental markets despite having the same level of endowment. Policy intervention to address these gender specific constraints is vital to achieving a more balanced participation.

5 Concluding Remarks

Using panel data from the Integrated Household Panel Survey from 2016 (the year before the reform) and 2019 (the year after the reform), the study evaluated the role of gender on market participation in the context of land reforms. We applied the extended Kitagawa–Oaxaca–Blinder Decomposition to analyze the differences between male and female decision-makers' participation in land rental markets. To measure the difference between the two gender groups, we compared the net land rented in by male and female decision-makers. To examine the influence of gender on the likelihood of household positions in the market, we independently estimated models for men and women, as well as a pooled model.

Results indicate that households with male heads of family report higher levels of land market activity than do households with female heads of household. Compared to homes in the Northern and Southern regions, more male and female-headed households from the Central area rented land. Furthermore, results from the extended Kitagawa–Oaxaca–Blinder Decomposition show that the land rented for females declined between 2016 and 2019, whereas males experienced an increase in land rented in the same period.

The findings confirm that the household head's gender greatly affects involvement as a lessee, or paying land rent, while its impact on being a lessor was shown to be negligible. The household's ability to pay rent and buy the necessary inputs is improved by having access to credit, whether it is agricultural or otherwise. In both the male and female models, there was a noticeable and favorable link. For both male and female heads, marriage systems had a major impact on involvement in the rental market. Depending on the dominant system in the area where the farm households were located, there were three degrees of measurement for the marriage systems. A patrilineal, matrilineal, or mixed society was home to the households. The findings indicated that residents of mixed societies had a higher likelihood of renting out land to both gender categories.

Although it has been determined that females have participated less in the land rent market than male counterparts since the implementation of land reform, the disparity is reducing. Therefore, the study suggests that the National Agricultural Policy adopts measures to encourage women's participation in land markets. More changes to land tenure laws should be tailored to the promotion of participation of women in land rent markets as lessors and lessees.

References

Abay KA, Chamberlin J, Berhane G (2021) Are land rental markets responding to rising population pressures and land scarcity in sub-Saharan Africa? Land Use Policy. https://doi.org/10.1016/j.landusepol.2020.105139

African Development Bank (2020) Malawi gender profile country report. Lilongwe, Malawi

Benjamin EO (2020) Smallholder agricultural investment and productivity under contract farming and customary tenure system: a Malawian perspective. Land, MDPI 9(8):1–17

Besley T, Ghatak M (2010) Property rights and economic development. In: Rodrik D, Rosen M (eds) Handbook of development economics, vol 5. North Holland, pp 4525–4595. https://doi.org/10.1016/B978-0-444-52944-2.00006-9

Chamberlin J, Ricker-Gilbert A (2016) Participation in rural land rental markets in Sub-Saharan Africa: who benefits and by how much? Evidence from Malawi and Zambia. Am J Agric Econ 98. https://doi.org/10.1093/ajae/aaw021

Chimhowu A, Woodhouse P (2006) Customary versus private property rights? Dynamics and trajectories of vernacular land markets in Sub-Saharan Africa. J Agrarian Change 6(3):346–371. https://doi.org/10.1111/j.1471-0366.2006.00125.x

Gebre GG, Isoda H, Rahut DB, Amekawa Y, Nomura H (2020) Gender gaps in market participation among individual and joint decision-making farm households: evidence from Southern Ethiopia. Eur J Dev Res 33:649–683. https://doi.org/10.1057/s41287-020-00289-6

GoM (2015) Malawi national gender policy. Government Press, Lilongwe, Malawi

GoM (2016) Malawi national agriculture policy. Government Press, Lilongwe, Malawi

Holden ST, Deininger K, Ghebru H (2011) Tenure insecurity, gender, low-cost land certification and land rental market participation in Ethiopia. J Dev Stud 47(1):31–47. https://doi.org/10.1080/00220381003706460

Holden ST, Otsuka K, Place FM (2010) The emergence of land markets in Africa: impacts on poverty, equity, and efficiency. Routledge

Jayne TS, Chamberlin J, Headey DD (2014) Land pressures, the evolution of farming systems, and development strategies in Africa: a synthesis. Food Policy 48:1–17

Jayne TS, Sanchez PA (2021) Agricultural productivity must improve in sub-Saharan Africa. Science 372(6546):1045–1047

Lastarria-Cornhiel S (2009) Module 4: gender issues in land policy and admin-istration. Gender in Agriculture Sourcebook. Food and Agriculture Organization (FAO), The International Bank for Reconstruction and Development/The World Bank, International Fund for Agricultural Development (IFAD), Rome, pp 125–171

Lunduka R, Holden ST, Øygard R (2009) Land rental market participation and tenure security in Malawi. In: Holden ST, Otsuka K, Place FM (eds) The emergence of land markets in Africa: impacts on poverty, equity and efficiency. Resources for the Future Press, pp 112–130

Marenya P, Kassie M, Jaleta M, Rahut D (2017) Maize market participation among female and male-headed households in Ethiopia. J Dev Stud 53(4):481–494

Menon N, Rodgers M, Kennedy AR (2017) Land reform and welfare in Vietnam: why gender of the land-rights holder matters. J Int Dev 29(4):454–472

Ricker-Gilbert J, Chamberlin J (2008) Transaction costs, land rental markets and their impact on youth access to agriculture in Tanzania. Land Econ 94(4)

Ricker-Gilbert J, Chamberlin J, Kanyamuka J, Jumbe CB, Lunduka R, Kaiyatsa S (2019) How do informal farmland rental markets affect smallholders' well-being? Evidence from a matched tenant–landlord survey in Malawi. Agric Econ 50(5):595–613

Ricker-Gilbert J, Jayne TS, Chamberlin J (2022) Absentee tenants and farmland transfers in sub-Saharan Africa: evidence from Malawi. Dev Pract 32(3):375–393

Wineman A, Tasie L (2017). Land markets and land access among female-headed households in Northwestern Tanzania. World Dev 100:108–122. https://doi.org/10.1016/j.worlddev.2017.07.027

Xu L, Du X (2022) Land certification, rental market participation, and household welfare in rural China. Agric Econ 53. https://doi.org/10.1111/agec.12688

Zuka SP (2019) Customary land titling and inter-generational wealth transfer in Malawi: will secondary land rights holders maintain their land rights? Land Use Policy 81(C):680–688

Open Access This chapter is licensed under the terms of the Creative Commons Attribution 4.0 International License (http://creativecommons.org/licenses/by/4.0/), which permits use, sharing, adaptation, distribution and reproduction in any medium or format, as long as you give appropriate credit to the original author(s) and the source, provide a link to the Creative Commons license and indicate if changes were made.

The images or other third party material in this chapter are included in the chapter's Creative Commons license, unless indicated otherwise in a credit line to the material. If material is not included in the chapter's Creative Commons license and your intended use is not permitted by statutory regulation or exceeds the permitted use, you will need to obtain permission directly from the copyright holder.

Prospects of Green Revolution and Research Perspectives

Road Map to Promoting Green Revolution in Sub-Saharan Africa

Agnes Mwangwela, Liveness Banda, Rattan Lal, and Bal Ram Singh

Abstract There is an urgent need to translate the known and proven technologies into strategies for improving soil health, increasing crop productivity for food and nutritional security, environmental sustainability, economic viability, and resilience to climate change and thus promoting Green Revolution in SSA. SSA is most vulnerable to climate change and changes in the distribution and magnitude of extreme rainfall events and drought, and changes in land use/cover are the major causes of soil degradation, aridity, and land desertification. Food insecurity and undernourishment situations are on the rise in Africa, being most serious in SSA. The rising African population, coupled with the changing food preferences, imply that food production must more than double than that in 2020 to meet the food needs in 2050. Therefore, policies to incentivize farmers to adopt proven technologies which are essential to protecting, restoring, and sustainably managing finite and fragile soil resources in harsh environments are needed, and practices of regenerative agriculture based on sound principles of agroecology must be encouraged. Such policies must be pro-nature, pro-agriculture, and pro-farmers. Furthermore, upscaling of management practices, e.g. conservation agriculture, agroforestry, precision/digital agriculture for fertilizers and irrigation use in SSA, are needed to meet the growing demands for food and nutritional security of the growing SSA population.

Keywords Climate change · Green revolution · Food and nutritional security · Traditional food crops

---

A. Mwangwela · L. Banda
Lilongwe University of Agriculture and Natural Resources, Bunda Campus, Lilongwe, Malawi

R. Lal
CFAES Rattan Lal Center for Carbon Management and Sequestration, The Ohio State University, Columbus, USA

B. R. Singh (✉)
Faculty of Environmental Sciences and Natural Resource Management, Norwegian University of Life Sciences, Ås, Norway
e-mail: balram.singh@nmbu.no

© The Author(s) 2026

B. R. Singh et al. (eds.), *Sustainable Food Systems in Sub-Saharan Africa*,
https://doi.org/10.1007/978-3-032-11288-0_27

1 Introduction

The Green Revolution, achieved in Asia and Latin America, resulted in increased food supply, and thereby meeting ongoing needs for rural development (Townsend et al. 2021) and avoiding persistent widespread undernutrition (FAO, 2022). There is an urgent need to also bring about food system transformation in Sub Saharan Africa (SSA) for meeting the food demands of a growing and more affluent population in the coming decades. However, Sub-Saharan Africa (SSA) is vast and diverse area consisting of 49 countries of all sizes, ranging from the largest—Democratic Republic of the Congo to small island countries such as Seychelles. Also, climate and topography of SSA countries vary widely, with some containing desert areas (e.g. Sahara and Kalahari) while others containing rainforests in the center of the continent. Thus, bringing such transformations in SSA must build upon the advances in food systems made in the past, and at the same time alleviate the many shortcomings, for example, greenhouse gas (GHG) emission (Tubiello et al., 2021) and ground water depletion (Bhattarai et al., 2021). Food systems involve the entire value chain from production to consumption and nutrition, along with the disposal of waste products. Thus, a convergence of agricultural, health, nutritional, economic, and environmental goals is critical. Therefore, we must also consider potentially conflicting priorities such as environmental sustainability, economic viability, nutrition and human health, as well as resilience to climate change and other environmental and socio-political disruptions (Davis et al., 2022). Although technologies and management practices for more efficient and sustainable production systems are available and can be effective, distorted policy incentives have limited their adoption and use (Springmann and Freund, 2022). There is also an urgent and critical need to translate proven technologies into action, which necessitates strong cooperation between policy makers, the private sector and researchers. In comparison with Asia and South America, there is a lack of policies which are pro-agriculture and pro-farmers. Such policies are needed to incentivize farmers in adoption of proven technologies which are essential for the protection, restoration, and sustainably managing the finite and fragile soil resource in harsh environments prevailing in SSA.

In addition to subsidies, incentivization involves making payments to farmers and land managers for the provision of ecosystem services such as sequestration of carbon in land-based sinks (soil, trees, wetlands, mine lands, and degraded soils and ecosystems), improving quality and renewability of water resources, and strengthening of above-and-below ground biodiversity. Payments to land managers are critical to widespread adoption of practices under the overall category of regenerative agriculture based on sound principles of agroecology. The goal is to make agriculture a part of the solution to address problems of anthropogenic climate change, depletion and contamination of water resources, purification of polluted air, and enrichment of biodiversity. Adoption of regenerative practices, in some soils and ecoregions, may lead to loss of productivity (low agronomic yields), especially during the initial period of 3–5 years. Thus, payments to farmers are essential to ensure societal benefits for human well-being and nature conservancy. Such policies are lacking in SSA.

Private sector can also play a critical role in translating science into action by activities such as: (1) increasing access of essential inputs to land managers, (2) making payments for strengthening of ecosystem services through adoption of innovative practices (i.e. improved varieties, micro-irrigation, nano-fertilizers, digital agriculture involving artificial intelligence and robots), (3) providing funding support for research and education, and (4) enhancing value addition of agricultural produce, which increases shelf life and reduces food waste. Indeed, private sector initiatives (both small and large scale) are important to modernize agriculture in SSA where policy initiatives are lagging.

Educational and extension institutions are also essential to translating science into action. Education about the importance of sustainable management of soil and agriculture to the protection and restoration of the environment is needed at all levels from primary school to graduate levels. The curricula must be revised at all levels to include environmental education, with a focus on sustainable management of natural resources (soil, water, air, biodiversity) right from the kindergarden through primary, middle and high school and college. Education of farmers, policy makers and the public about the importance of soil-centric strategy to Green Revolution (rather than the seed-centric approach) is also necessary, because of the widespread problems of soil degradation in SSA.

Education of farmers requires effective and strong extension services in SSA. In this context, adoption of the model of Land Grant University (LGU) system of USA may be relevant in SSA. The Green Revolution in India was facilitated by the development of LGUs during the early 1960s, and a similar strategy may be needed in SSA. Above all, there is a strong need and urgency for cooperation and synergism between the policy makers, private sector, and education institutions to bring about the transformation of the traditional food systems in SSA.

This chapter presents aspects related to climate and soil resources, food production and future demands, farming systems used, factors affecting crop yield and breaking the vicious cycle of low productivity and food insecurity, which may bring about sustainability of food production, leading to the much-needed Green Revolution in SSA.

The information presented in this chapter is entirely based on the review of literature of the available information published in peer review journals and from chapters of books by well-known and easily accessible publishers. The data were collated from academic journals focused on agricultural science in general and soil science in particular. The primary sources of academic journal articles were Web of Science and Google. The data were also collated from research publications by the Food and Agriculture Organization (FAO), the World Bank, studies from the Alliance for a Green Revolution in Africa (AGRA), U.S. Agency for International Development and research conducted by universities, regional/national l research institutions, CGIAAR Centers and those that focused on specific countries or agro-ecological zones within SSA. Specific articles and book chapters cited in this chapter are listed in the reference list.

2 Geographical Region

SSA is the part of Africa south of the Sahara Desert. The region excludes Algeria, Djibouti, Egypt, Libya, Morocco, Somalia, Sudan, and Tunisia. Yet, it consists of 49 countries which have similar cultural agricultural issues because of the subsistence farming practices.

2.1 Climate

The climate of SSA is highly diverse and driven by a range of distinct climate systems, depending upon the position of the continent in relation to the major circulation patterns, the seasonal migration of the Inter-Tropical Convergence Zone (ITCZ), the topography, the existence of large lakes, and the surrounding oceans. Temperatures are highest in the northern Sahara Desert, with cooler areas over high altitude plateaus and mountainous regions. Rainfall varies considerably across the continent, with the northern and southern extremes of the continent experiencing winter rainfall associated with the passage of mid-latitude cyclones (Davis- Reddy 2018). Furthermore, SSA is most vulnerable to climate change, and changes in the distribution and magnitude of extreme rainfall events observed in many parts of the African continent are associated with both climate change and climate variability (Williams et al., 2010). Drought, resulting from changes in land use/cover and climate, is considered as a key cause of land degradation, aridity, and desertification, further producing immense socio-economic impacts such as famine and large-scale migration in the semi-arid sub-humid regions of SSA (Masih et al., 2014).

2.2 Soil

Most soils of SSA are inherently poor in plant nutrients, highly acidic, and low in soil organic carbon (SOC), clay contents, and cation exchange capacity (CEC). It is estimated that more than 70% of African soils are inadequately fertile or depleted by agricultural practices, human, and animal pressure (Bado and Bationo, 2018). Despite the development of many technologies and concepts to improve soil fertility over the years, the paradigm of external inputs to start a "Green Revolution", as achieved in Asia and Latin America, has not been experienced in SSA, due to many reasons, severe soil-related constraints being one of them. Soils of SSA are highly diverse with different physical, chemical, and agronomic characteristics; 40% of soils have low nutrient capital reserves, 25% have aluminum toxicity, 18% have high leaching potential, and 9% have high P-fixation capacity (Bationo et al., 2006). In addition, land degradation caused by soil water erosion (46%), wind erosion (36%), loss of nutrients (9%), physical deterioration (4%), and salinization (3%) is a serious threat to food production, food security, and natural resource conservation in Africa, particularly for the poor and vulnerable population of the drylands (Bado and Bationo, 2018). Under the hot climate of SSA, SOC content decreases quickly, and the land management following conversion of forest to agriculture also impacts SOC status, soil fertility, and agricultural sustainability—a point underlined by Lal (2006) and Zivin and Lipper (2008). The other threats to soil health, productivity, and ecosystem services in SSA include accelerated soil erosion by water or wind, loss of SOC stock, soil nutrient depletion, dwindling of soil biodiversity, soil contamination, soil acidification, salinization and sodification, waterlogging; and compaction, crusting and sealing/capping (FAO, 2015).

2.3 *Vegetation*

SSA has experienced rapid deforestation of about 3.4 million ha/year between 2000 and 2010, with Malawi having the highest deforestation rate (Mapulanga and Naito 2019). With increasing rates of deforestation in SSA, the SOC storage and sequestration capacity of trees reduces, causing increased concentration of atmospheric CO_2 and higher rates of warming events. Reduction in vegetation by deforestation or other vegetation degrading processes expose the land to soil erosion and the loss of available plant nutrients and moisture necessary for increased vegetation production. Variability in climate plays an important role in the growth of vegetation across the terrestrial ecosystems of SSA; therefore, understanding the response of the vegetation to both climate change and climate variability provides the basis for developing a comprehensive analysis of the impacts of climate change on vegetation (Davis-Reddy, 2018).

2.4 *Land Use*

Changes in land use/cover and climate have led to observable changes in the distribution and dynamics of different types of terrestrial ecosystems, e.g. grasslands and shrublands, savannas and woodlands, and forests (Field et al., 2014). Increases in woody vegetation in SSA have been documented (Stevens et al., 2016), and modeling studies aimed at vegetation responses to climate change project that the extent of woody vegetation is likely to continue to increase, resulting in a variety of biome shifts (Higgins and Scheiter, 2012). Rural livelihoods in SSA are frequently affected by changes in land use/cover. These impacts are commonly intensified by the dependence of rural households on agricultural activities that are significantly reliant on the climate (Antwi-agyei et al., 2019). Drought, resulting from changes in land use/cover and climate, is a key cause of land degradation, aridity, and desertification (Masih et al., 2014).

3 Population and Projections for 2100

Sub Saharan Africa (SSA) started the nineteenth century with a population of 140 million (M), and had 9% of the world population (Stanley, 2023). Since the turn of the century in 1900s, Africa's population has continued to grow at an unprecedented rate, from 1.2% in the period ranging from 1900 to 1950, to 2.8% in the period from 1950 to the close of the century (Population Division ESA/UN 2011). The population growth rate is on the decline at 2.47%, and the current population is 1.45Billion (B). During the 2005–2010 period, the global population growth rate continued to steadily decelerate at 1.16% per year in 2010 and is further projected to decline to 0.34% by mid-century (2050).

During this same period, the SSA population growth rate reached 2.45% per year, and the rate is projected to decline to 1.68% by 2050. This implies that despite the deceleration in global population growth, there is continuing population growth in SSA, caused by the general improvement in health care, reduced mortality coupled with slow decline in total fertility rate. The UN total fertility rate (TFR) estimates for SSA in the 2015–2020 period was at 4.7 births per woman, which is more than double that in other regions of the world. Hence Africa's population is expected to grow from 1 B in 2015 to more than 2 B in 2050 and approximately 4 B in 2100 (United Nations, 2019).

It is projected that Africa's relative proportion to the global human population will rise from its current 15% to an astonishing 35% of the global population by 2100, while that of Europe will decline from its current 11–8% during the same period. It is important to note that the population growth in Africa is not homogenous. There is a two-tiered growth pattern, with most SSA countries, for example, populations of Chad, Mali, Niger, Nigeria, and South Sudan being forecasted to continue to grow even beyond 2100. These countries, according to the Institute for Health Metrics and Evaluation (2024) projections, may witness a doubling, tripling, or even as high as eightfold increase by 2100 compared to their current population, while the whole SSA population is forecasted to triple during that period.

The projected tripling of the population in SSA over the next 76 years demands a need to triple all existing resources and infrastructure (health, education, housing, energy, and so on) over that period, just to maintain existing inadequate levels of basic services and amenities (Ezeh et al., 2020). Sub-Saharan Africa should recognize its own unique population trajectory and should also, as a matter of urgency, engage in serious discussions at the highest level about how current and projected rates of population growth will affect development prospects in the region.

The conditions for population growth in Africa were favorable in the twentieth century, and they are predicted to be even better in the twenty-first century. In the situation of positive changes such as declines in fertility, improved access to safe and nutritious food, increasing education opportunities, and land reform, population growth may not be a significant problem for food security. Massive population growth on its own may not necessarily represent a major challenge, but the situation for SSA is characterized with a significantly youthful population, leading to a high dependency ratio, urbanization and decline in available productive resources. As a result, Africa will have a larger percentage of the world's population to feed. This is because the population growth rate in SSA is not in tandem with the capacity to feed itself; and thus, the growth is a major challenge for the relationship between population size and food security.

4 Food Demand

4.1 Present and Future

Agricultural productivity growth in SSA has been slow and has remained significantly below what could be achieved by using the best available practices and technology. According to a study of the FAO (2019), the low levels of productivity and competitiveness are the result of poor infrastructure, lack of productive technologies, lack of access to inputs and weak institutions.

SSA has the highest total population growth and urban population growth rates in the world. This urbanization and modernization in SSA has resulted in the adoption of Western lifestyles and trends that involve diverting from traditional food practices, food ingredients and eating patterns (Wentzel-Viljoen et al., 2018). Such lifestyle changes can lead to increased risk of nutrient deficiency because the refined wheat products that are now consumed are rich in starch, fat, and sugars but low in proteins, micronutrients and particularly in dietary fiber. Replacing these starch-rich staples like refined wheat, rice, maize, and cassava with climate-resilient crops (CRCs) will provide a more nutritious diet rich in proteins, dietary fiber, essential fatty acids, vitamins, minerals, and phytochemicals. Therefore, the transition of SSA agriculture to produce CRCs, such as cereals (e.g., sorghum, teff and finger millet), pseudocereals (amaranth), roots and tubers (cassava and sweet potato), pulses (Phaseolus beans, cowpeas, chickpeas, pigeon peas and Bambara groundnuts) and oilseed legumes (soya beans and peanuts), can be a good alternative for meeting food demands (Noort et al., 2022).

During the last six decades (from 1960 to 2020s), agricultural productivity in SSA could not keep pace with the increasing internal demands due to rapid population growth (OECD, 2016). Population in Africa is rapidly rising (estimated to be 1.48 billion by the end of 2023), and this is especially so in SSA; and this is in addition to the changing food preferences. This means that food production must more than double to meet the food needs in 2050 (FAO, 2017).

4.2 Food Deficit and Nutritional Insecurity

Although persistent efforts have been made to increase the productivity and quality of food resources, food insecurity still exists in SSA, and nearly 33% of the population in the region is malnourished (Fraval et al., 2019). Many SSA countries continue to struggle with undernourishment and the food security situation in SSA is deteriorating due to changes in consumer habits, population growth, and global economic upheavals (Onwujekwe et al., 2021).

The food security and undernourishment situation is on the rise in Africa as a whole, and it is most serious in SSA (Table 1). The number of people facing food insecurity in SSA increased from 170 million in 2014 to 294 million in 2021, while the number of undernourished people increased from 174 million in 2005 to 261milliom in 2021 (Table 1). In the last decade, the number of malnourished people in Sub-Saharan Africa has risen from 5.5 million to 30 million, resulting in the death of over 3.5 million children under the age of five per year owing to insufficient food intake (Drammeh et al., 2019). Even though most countries in SSA have begun to progress towards the goal of reversing the malnutrition trends, most efforts are not progressing at the appropriate rate.

Over the past 50–60 years, there have been profound changes in the types of staple food products consumed in SSA. Consumption of refined wheat breads is increasing rapidly and displacing traditional meals based on local crops, for example, porridges made from sorghum, millets, and tubers (Noort et al., 2022). This dietary transition from traditional foods to wheat-based bread products is attributed to a significant change in consumer attitudes which results from circumstantial factors such as changing lifestyles, demographics, urbanization, increased product availability, accessibility, and technological changes (Mehmeti et al., 2014).

Food insecurity in SSA has increased since 2015 and thus the SDG2 of the 2030 Agenda for Sustainable Development to end hunger and malnutrition may not be achieved. SSA has the highest level of hunger in the world according to the Global Hunger Index (GHI) value 27.1 in the year 2022 and is experiencing a stagnation in the fight against hunger (GHI, 2022). People with low income are the most food insecure and they also eat least diverse food. For example, in SSA, some populations rely on maize or cassava as their only staple. In Lesotho, for instance, the average maize

Table 1 Number of food insecure and undernourished people in Africa and SSA

Indicator	2005	2014	2016	2018	2021
Severely food insecure—Africa	–	192.1	–	246.8	322
Severely food insecure—SSA	–	169.7	–	224.8	293.8
Undernourished—Africa	189.9	–	198	–	278
Undernourished—SSA	174.3	–	185.8	–	260.6

Modified and extracted from FAO (2022)

consumption is 328 g maize per person per day for the country's entire population compared to an average consumption of 141 g/day across SSA (Ranum et al. 2014).

5 Farming Systems

A farming system is a dynamic framework that integrates available resources, management practices, and interactions between ecosystems and human activities to achieve agricultural production and sustainability goals (FAO, 2022). These systems address interconnected processes—such as food production, resource conservation, and the application of innovative technologies—while considering environmental and economic factors (Pretty et al., 2018; Adger et al., 2019). They are shaped by internal factors, including farming practices, land resources, input resources and external drivers like population growth, climate change, and policy reforms (Van Ittersum et al., 2016; Nhamo et al., 2021). In the African context, the increasing pressures from urbanization, population dynamics, and climate variability necessitate adaptive farming systems that enhance resilience and sustainability (Njenga et al., 2021; Nhliziyo & Mushunje, 2024).

The projected population growth in Africa, which may reach 5.8 billion by 2100, will limit per capita cropland to only 450 m^2, highlighting the urgent need for circular and efficient agricultural systems (van der Esch et al., 2017). Africa's rapid population growth presents both challenges and opportunities, necessitating a shift towards circular economy principles. This approach promotes sustainable resource management and waste reduction, addressing the increasing demand for food, water, and energy while ensuring environmental resilience (Sekabira et al., 2022). Implementing circular economy practices can enhance resource efficiency and stimulate economic growth in the context of Africa's demographic changes (Binsuwadan et al. 2023; Elroi et al., 2023). Other innovations to address these challenges include off-farm biomass utilization for food, energy, and organic fertilizers (Chew et al., 2019). Beyond optimizing traditional practices, future systems must leverage nonedible biomass and focus on integrating agroecological approaches to achieve environmental sustainability. Using renewable resources such as residues of agricultural biomass, food processing waste and municipal and industrial sewage sludge for additional purposes within the agricultural production system is another central principle of circularity in agriculture. Renewable resources such as the previous season's crop biomass returned to the soil, for example, will reverse or at least reduce the rate of nutrient mining, leading in turn, to a reduction in external inputs, such as chemical fertilizers (Tindwa et al. 2024).

5.1 Cropping Systems

Crop diversification in relation to crop combinations and production methods is a key to the future, at local, regional and global levels. Land is a critical resource in agriculture production, and the projected population growth will continue to exert enormous pressure on land use. The future of agriculture will therefore have to balance between food and industrial crops, take into account the use of marginal lands for agricultural production, as well as the detrimental effect on soil carbon sequestration/storage, biodiversity and water resource. Therefore, Africa will have to embrace innovative technologies that offer land and water use efficiency, such as sustainable exploitation of water resources for efficient irrigation systems, aeroponics and hydroponics coupled with vertical farming. These technologies would allow multiple production cycles in a year, defying the limitation of seasonality (Lakhiar et al., 2020).

It is expected that small holder farmers will continue to dominate agriculture production in SSA, hence the need to optimize smallholder multicrops and rainfed systems by incorporating climate adapted plants such as millets, sorghum, cassava and cowpeas.

5.2 Traditional Food Crops and Crop Yield

Traditional food crops have been part of the sub-Saharan agriculture and diets with minimal investment in production. SSA is endowed with wide variety of cereal grains, pulses, roots, tubers, nuts and oils, fruits and vegetables, some of which have been identified by the Vision for Adapted Crops (VACS), Forum for Agricultural Research in Africa (FARA) and African Union Development Agency (AUDA) to have potential for development in Africa (Table 2).

Most of the food crops and plants in Table 1 have not been integrated into human diets despite their nutritional value and as food security resilience crops. There is a need for these foods to transition to be an integral part of a functional and sustainable food system. One of the barriers to this transition is their low productivity when compared with other conventional crops. Low productivity is a very big concern, because of the declining agricultural land, draughts and unreliable rains, and the ever-growing population that require increased food productivity. These crops need significant investment in research and development to enhance productivity and uptake by farmers.

Crop improvement is one of the primary areas of investment to maximize traits such as drought tolerance, pest resistance, or nutrient and water use efficiency, which will lead to better crop performance and improve yields (Mayes, 2012), and thus address the double challenges of uncertain future climates and the need to increase

Table 2 Indigenous and traditional crops grown in Africa categorized by VACS as high to medium potential

Crop type	Examples of indigenous and traditional crops
Cereals	Finger millet (*Eleusine coracana*), Pearl millet (*Pennisetum glaucum*), Tef (*Eragrostis tef*), Fonio (*Digitaria exilis*), Sorghum (*Sorghum bicolor*), Barley (*Hordeum vulgare*), African rice (*Oryza glaberrima*), Oats (*Avena sativa*), Grain amaranth (*Amaranthus cruentus*)
Roots and tubers	Cassava (*Manihot esculenta*), Yams (*Dioscorea spp.*), Cocoyam (*Xanthosoma sagittifolium*), Taro (*Colocasia esculenta*), Enset (*Ensete ventricosum*), Sweet potato (*Ipomoea batatas*)
Legumes	Cowpea (*Vigna unguiculata*), Bambara groundnut (*Vigna subterranea*), Pigeon pea (*Cajanus cajan*), African yam bean (*Sphenostylis stenocarpa*), Lablab (*Lablab purpureus*), Kersting's groundnut (*Macrotyloma geocarpum*), Lupin (*Lupinus albus*), Fava bean (*Vicia faba*), Lentils (*Lens culinaris*), Mung bean (*Vigna radiata*), Chickpea (*Cicer arietinum*), Grass pea (*Lathyrus sativus*), Peas (*Pisum sativum*)
Fruits and tree crops	Baobab (*Adansonia digitata*), Tamarind (*Tamarindus indica*), Custard apple (*Annona reticulata*), Bush mango (*Irvingia gabonensis*), Desert date (*Balanites aegyptiaca*), Wild loquat (*Uapaca kirkiana*), African jujube (*Ziziphus mauritiana*), Miracle berry (*Synsepalum dulcificum*), Jackfruit (*Artocarpus heterophyllus*), Breadfruit (*Artocarpus altilis*), Cooking banana (*Musa paradisiaca*), Plantain (*Musa balbisiana*), Moringa (*Moringa oleifera*), Locust bean (*Parkia biglobosa*)
Vegetables and leafy crops	Roselle (*Hibiscus sabdariffa*), Jute mallow (*Corchorus olitorius*), Spiderplant (*Cleome gynandra*), Amaranth leaves (*Amaranthus tricolor*), African eggplant (*Solanum anguivi*), African nightshade (*Solanum americanum*), Cassava leaves (*Manihot esculenta*), Black jack (*Bidens pilosa*)
Nuts and oilseeds	Shea tree (*Vitellaria paradoxa*), Sesame (*Sesamum indicum*), Groundnut (*Arachis hypogaea*), Cashew (*Anacardium occidentale*), Flax (*Linum usitatissimum*), Safflower (*Carthamus tinctorius*), Macadamia (*Macadamia ternifolia*), Allanblackia (*Allanblackia floribunda*)

Source Adapted and modified from FARA (2023)

food production with fewer inputs. There is need to leverage modern breeding techniques that have already been optimized for conventional crops such as genetic transformation through genome editing (like CRISPR-Cas9) and marker-assisted selection (MAS) to select plants with desired traits more efficiently to develop improved varieties (Ahmar et al., 2020; Zhang et al., 2023). Furthermore, adoption of biotechnological tools such as tissue culture will enable production of propagation materials effectively and of high quality.

Proper agronomic practices for monoculture and multi-crop systems, precision agriculture techniques for optimized fertilizer application and integrated pest management among others can be tailored for traditional crops to ensure optimal yields. In addition to agronomical practices, there is a need for postharvest handling and value chain development to create a demand pull for the traditional crop products and ingredients. The demand pull is essential for farmers to have a sustainable market, and to encourage them to grow traditional crops. The new generation of farmers taking up traditional crops require appropriate training and support systems.

6 Causes of Low Yields

6.1 Soil Health

Soil health refers to the capacity of the soil to function as a biological system which enhances the life-support processes of the pedosphere. Use of the term "soil health" implies that soil is a living entity. Indeed, a healthy soil is characterized by high activity and species diversity of fauna and flora (macro, meso and micro). However, the biotic component of soil health must be supported by physical, chemical and ecological properties and processes. Therefore, a biologically healthy soil (with optimal level of SOC content, live biomass of organisms, respiration quotient, enzyme activity) must also have optimal properties and processes for physical (e.g., structure, strength, porosity, aeration, water retention and transmission, temperature regime, rooting depth),chemical (pH, electrical conductivity, CEC, availability of macro and micro-nutrients) and ecological (such as vulnerability to compaction/ crusting, accelerated erosion, leaching of plant nutrients, acidification, alkalization, gaseous emissions) parameters. Indeed, soil health is a complex parameter and a major challenge to quantify.

6.2 Fertilizer Use and Use Efficiency

Annual rate of fertilizer use in SSA is estimated at about 20 kg/ha in 2023 compared with the global average rate of about 135 kg/ha. With arable land area of 20% of the world cropland, the fertilizer use in Africa is merely 2% of the global fertilizer use. In addition to the low rate, the fertilizer use efficiency is low and may be 20–30% for nitrogenous fertilizers. The low fertilizer use efficiency is due to a range of interacting factors such as soil degradation, use of conventional systems of soil and crop management, and harsh climate characterized by drought, unfavorable soil temperatures, and high losses of nutrients by erosion, volatilization (of nitrogen) and leaching. Low level of SOC content (often less than 0.25% in the root zone of 0–30 cm depth) is a major determinant of low fertilizer use efficiency. In addition to restoring SOC content to an optimal range (~ 1.1 to 1.5% in 0–30 cm depth), adoption of 4 R concept of fertilizer use is also important to enhancing the use efficiency. The term 4 R refers to right rate, time, method and formulation of fertilizers.

6.3 Irrigation Use

Low fertilizer use efficiency in SSA is also aggravated by the extent and severity of drought stress on cropland in SSA. Indeed, low water productivity and frequent drought stress are important determinants of low agronomic yield in SSA. Thus, improving soil water conservation (by reducing losses of water by surface runoff and evaporation) is critical to improving and sustaining agronomic yield. Conservation agriculture and mulch farming techniques must be scaled up to major agro-ecoregions of SSA. Furthermore, water productivity is also vulnerable to changing and uncertain climate. Like the low rate of use of fertilizers, there is also low use of supplemental irrigation in SSA. In comparison with cropland area of 37–40% equipped for irrigation in Asia (e.g., South Asia, China), only 6–7% of the cropland area is equipped for irrigation in SSA. Thus, there is an urgency and strong need to expand the land area equipped for irrigation and adopting conservation agriculture and mulch farming techniques.

6.4 Poverty Trap

There is a predominance of resource-poor and small land holders in SSA, and the majority of these are women. These resource-poor farmers are gripped in the poverty trap which is not easy to break. Poverty trap is a vicious cycle that is hard to get out of. It is set in motion by degradation of soil, which leads to low productivity, low income, vulnerability to hunger and malnutrition, lack of education, and failure to access opportunities to get out of the poverty trop. The poverty feeds on itself, and its victims give up the hope and accept the situation as it is. Just as the poverty trap is set-in-motion by the onset of soil degradation, the process can also be reversed by onset of soil restoration through conversion to a restorative land use and adoption of recommended management practices, which lead to increase in productivity and improvement in nutritional quality of the food produced.

7 Breaking the Vicious Cycle

7.1 Restoring Soil Health

Degraded and depleted soils of Africa must be restored, and the principal degradative processes reversed. Increasing SOC content is critical to restoring degraded and depleted soils. The later can be accomplished by adopting land use and management practices which create a positive soil-ecosystem carbon budget. Enhancing SOC content is critical to improving soil structure, increasing water retention in the root zone, decreasing risks of accelerated erosion by water and wind, increasing the cycling of nutrients and water, strengthening of the activity and species diversity of biota, and improving agronomic yield. Restoring soil health is the critical step needed to break the vicious cycle of poverty which has trapped the farmers of Africa and have forced them to live under sub-optimal conditions.

7.2 Soil and Water Conservation

Most cropland soils of Africa are prone to degradation due to erosion by water and wind. Soil erosion is a selective process which preferentially removes SOC and clay contents.

Adoption of system-based conservation agriculture is the most cost-effective and scientifically proven strategy of protecting the fragile soils against erosivity of water and wind. However, effectiveness of conservation agriculture to controlling soil erosion deepens on the ability of small land holders and resource-poor farmers to recycle the crop residues and leave these on the soil surface as mulch. Therefore, payments to farmers for provisioning of ecosystem services is critical to upscaling of the conservation agriculture. Data from field experiments of the usefulness of CA in SSA are available since early 1970s, yet the adoption has lagged for 50 years or more because of the weak extension services and lack of access to input needed for the adoption of innovative technologies.

7.3 Carbon Farming

The term carbon farming implies growing carbon in land-based sinks (soil, wooded biomass, degraded and desertified ecosystems) so that it can also create an income stream for farmers. The goal is to reward farmers financially for increasing carbon stocks in soil and vegetation. Farmers must be rewarded financially through payments at US$50 per credit of carbon (one Mg of CO_2). This cost is based on societal value of C and not on the market forces of demand and supply. Estimates of the rate of C sequestration in soil and wooded vegetation can be obtained by remote sensing techniques based on land use and adoption of recommended soil/crop/trees/animal management practices. However, a protocol for payments must be developed so that farmers are rewarded fairly, transparently and justly. The protocol adopted must ensure that any cost of doing business must not exceed 10% of the payments to be made to the farmers.

7.4 Policies and Their Implementation

Lack of appropriate policies at the Continental level is one of the many reasons that the Green Revolution of the 1960s by-passed Africa. There is a lack of political will as a common resource which has led to neglect of the farming communities across the continent. Policies to be implemented must be pro-nature, pro-agriculture, and pro-farmers and implemented in such a manner so that the respect and dignity of the farming profession is enhanced and sustained in perpetuity. Neglect of the farming communities by policy makers has stagnated agronomic productivity and made Africa dependent for food import. This trend must be reversed through policies which are farmer-centric and focused on their well-being so that they can translate science into action and make Africa the foodbasket of the world.

7.5 Putting Sustainable Development Goals on Tracks

Like the Millennium Development Goals (2000 to 2015), the Sustainable Development Goals or SDGs (2015–2030) are also not on track for accomplishment by 2030. This is especially the case for Africa where the agenda 2030 of the United Nations will not be realized. Just as the Green Revolution of the 1960s by-passed Africa, similarly SDGs are not on track because soil health and its management are ignored. Indeed, the word "soil "does not appear in the primary language of any of the SDGs. With 5 years remaining, SDGs can still be put on the track if soil health is part of the agenda. In this context, Declaration of COP28 (Dubai, UAE), with focus on agriculture and specific mention of the term soil health, is a step in the right direction. There is still an opportunity to make progress in several SDGs if soil restoration and sustainable management of soil health are given the priority they deserve.

8 Conclusions

There is strong need to translate the known and proven science of soil health and strategies of its management into action in Africa so that the trends of declining agronomic productivity and tightening grip of the poverty trap can be reversed. Ending poverty (SDG#1), zero hunger (SDG#2), clean water and sanitation (SDG #6), climate action (SDG #13), life on land (SDG#15), and cooperation among institution (SDG #17) can be put on track in Africa and elsewhere by focusing on restoration and sustainable management of soil health in agroecosystems of Africa. This can be done by implementing policies which motivate farmers and encourage them to adopt recommended and proven management practices. Payments to farmers for ecosystem services, such as for sequestration of carbon in soil and increase in renewability and quality of water, are critical to translating science into action. In this context, private sector can also play a critical role by improving access to inputs, promoting extension and education, and making payments to farmers for strengthening of ecosystem services. Strengthening of extension services is also a step in the right direction. Thus, a close cooperation between researchers, policy makers, private sector and farmers' organization is needed to adopt a farmer-oriented and a soil-centric approach to improving agronomic productivity. The so-called Cerrado miracle must also happen in Africa. With upscaling of recommended management practices (e.g., conservation agriculture, agroforestry, precision/digital agriculture for fertilizers and irrigation), Africa can be a foodbasket of the world and SDGs of the U.N. can be put on the track by restoring and managing soil health.

References

Adger WN, Brown K, Surminski S (2019) Adaptation to climate change: insights into emerging issues. Annu Rev Environ Resour 44(1):1–26. https://doi.org/10.1146/annurev-environ-101718-033026

Ahmar S, Saeed S, Khan MHU, Ullah Khan S, Mora-Poblete F, Kamran M, Faheem A, Maqsood A, Rauf M, Saleem S, Hong WJ, Jung KH (2020) A revolution toward gene-editing technology and its application to crop improvement. Int J Mol Sci 21:5665. https://doi.org/10.3390/ijms21165665

Antwi-agyei P, Kpenekuu F, Hogarh JN, Obiri-danso K, Abaidoo RC, Jeppesen E, Andersen MN (2019) Land use and land cover changes in the Owabi reservoir catchment, Ghana: implications for livelihoods and management. Geosciences 9:286. https://doi.org/10.3390/geosciences9070286

Bado VB, Bationo A (2018) Integrated management of soil fertility and land resources in Sub-Saharan Africa: involving local communities. Adv Agron # 2018. ISSN 0065-2113. https://doi.org/10.1016/bs.agron.2018.02.001

Bationo A, Hartemink A, Lungu O, Naimi PF, Okoth P, Smaling E, Thiombiano, L (2006) African soils: their productivity and profitability of fertilizer use. In: Background paper prepared for the African fertilizer summit. vol 2006. Abuja, Nigeria, June 9–13, p 25

Bhattarai N, Pollack A, Lobell DB, Fishman R, Singh B, Dar A, Jain M (2021) The impact of groundwater depletion on agricultural production in India. Environ Res Lett 16:085003

Binsuwadan J, Yousif G, Abdulrahim H, Alofaysan H (2023) The role of the circular economy in fostering sustainable economic growth in the GCC. Sustainability 15:15926. https://doi.org/10.3390/su152215926

Davis KF, Dalin C, Kummu M, Marston L, Prabhu Pingali P, Tuninetti M (2022) Beyond the green revolution: a roadmap for sustainable food systems research and action. Environ Res Lett 17:100401. https://doi.org/10.1088/1748-9326/ac9425

Davis-Reddy C (2018) Assessing vegetation dynamics in response to climate variability and change across sub-Saharan Africa. Stellenbosch University. https://scholar.sun.ac.za

Drammeh W, Hamid NA, Rohana AJ (2019) Determinants of household food insecurity and its association with child malnutrition in Sub-Saharan Africa: a review of the literature. Curr. Res. Nutr. Food Sci. J. 7(3):610–623

Elroi H, Zbigniew G, Agnieszka WC, Piotr S (2023) Enhancing waste resource efficiency: circular economy for sustainability and energy conversion. Front Environ Sci 11. https://www.frontiersin.org/journals/environmental-science/articles/10.3389/fenvs.2023.1303792

Ezeh A, Kissling F, Singer P (2020) Why sub-Saharan Africa might exceed its projected population size by 2100. The Lancet 396

FAO (2015) Status of the world's soil resources. Main report. Food and Agriculture Organization Of The United Nations, Rome
FAO (2017) The future of food and agriculture—trends and challenges. FAO, Rome
FAO (2022) The state of food and agriculture 2022: leveraging automation in agriculture for transforming food systems. Food and Agriculture Organization of the United Nations. Available at: https://www.fao.org (Accessed 9 Oct 2024)
FAO, IFAD, UNICEF, WFP, WHO (2019) The state of food security and nutrition in the world 2019. In: Safeguarding against economic slowdowns and downturns. FAO, Rome, Italy
FAO, IFAD, UNICEF, WFP and WHO (2022) The State of food security and nutrition in the world 2022. In: Repurposing food and agricultural policies to make healthy diets more affordable. Rome, FAO. https://doi.org/10.4060/cc0639en
FARA (2023) The vision of VACS: considering an approach for advancing the African Agriculture and Food Systems Transformation. Available at: https://www.faraafrica.org/2023/12/07 (Accessed 11 Nov 2024)
Field CB, Barros VR, Mach K, Mastrandrea, M (2014) Climate change 2014: impacts, adaptation, and vulnerability. In: Working Group II contribution to the IPCC 5th assessment report-technical summary, Cambridge University Press, UK
Fraval S, Hammond J, Bogard JR, Ng'Endo M, van Etten J, Herrero M, Oosting SJ, de Boer IJM, Lannerstad M, Teufel NC, Todd S, Rosenstock, TS, Pagella T, Bernard Vanlauwe B, Dontsop-Ngueze, PM, David Baines D, Carpen P, Njingulula P, Okafor C, Wichern J, Ayantunde A, Bosire C, Kihoro E, Rao· EJO, Skirrow T, Steinke J, Yameogo· V, Mark T, van Wijk, MT(2019) Food access deficiencies in sub-Saharan Africa: Prevalence and implications for agricultural interventions. Front Sustain Food Syst 3:104
Higgins SI, Scheiter S (2012) Atmospheric CO_2 forces abrupt vegetation shifts locally, but not globally. Nature 488(7410):209–212
GHI (2022) Global regional and national trend Available online: https://www.globalhungerindex.org/trends.html.
Institute for Health Metrics and Evaluation (2024) Population of Africa (2024). https://www.Worldometer, Population of Africa (Accessed 10 Nov 2024)
IPCC (2012) Managing the Risks of Extreme Events and Disasters to Advance Climate Change Adaptation. A Special Report of Working Groups I and II of the Intergovernmental Panel on Climate Change, Field CB, Barros V, Stocker TF, Qin D, Dokken DJ, Ebi KL, Mastrandrea MD, Mach KJ, Plattner G-K, Allen SK, Tignor M, Midgley PM (eds). Cambridge University Press: Cambridge, UK and New York, NY.
Lal R (2006) Enhancing crop yields in the developing countries through restoration of the soil organic carbon pool in agricultural lands. Land Degrad Dev 17:197–209
Lakhiar IA, Gao JM, Syed TN, Chandio FA, Tunio MH, Ahmad F, Solangi KA (2020) Overview of the aeroponic agriculture –An emerging technology for global food security. Int J Agric & Biol Eng 13(1):1–10
Leclerc C, Mwongera C, Camberlin P, Moron V (2014) Cropping system dynamics, climate variability, and seed losses among East African smallholder farmers: a retrospective survey. Weather, Climate, and Society 6(3):354–370. https://doi.org/10.1175/WCAS-D-13-00035.1. hal-01056221
Masih I, Maskey S, Trambauer P (2014) A review of droughts on the African continent: a geospatial and long-term perspective. Hydrol Earth Syst Sci 18:3635–3649. https://doi.org/10.5194/hess-18-3635-2014
Mayes S, Massawe FJ, Alderson PG, Roberts JA, Azam-Ali SN, Hermann M (2012) The potential for underutilized crops to improve security of food production. J Exp Bot 63(3):1075–1079. https://doi.org/10.1093/jxb/err396
Mehmeti G, Xhoxhi O (2014) Future food trends. Ann. Food Sci. Technol. 15:392–400
Nhliziyo N, Mushunje A (2024) Determinants of Small-Scale Farmers' Participation in Social Capital Networks to Enhance Adoption of Climate Change Adaptation Strategies in OR Tambo District, South Africa. Agriculture 14(3):441

Nhamo L, Rwizi L, Mpandeli S, Botai J, Magidi J, Tazvinga H, Sobratee N, Liphadzi S, Naidoo D, Modi AT, Slotow R, Mabhaudhi T (2021) Urban nexus and transformative pathways towards a resilient Gauteng City-Region, South Africa. Cities 116:103266. https://doi.org/10.1016/j.cities.2021.103266

Njenga MW, Mugwe JN, Mogaka H., Nyabuga, G, Kiboi M, Ngetich F, Mucheru-Muna M, Sijali I, Mugendi D (2021) Communication factors influencing adoption of soil and water conservation technologies in the dry zones of Tharaka-Nithi County, Kenya. *Heliyon*, *7*(10).

Noort MWJ, Renzetti S, Linderhof, du Rand, GE,Marx-Pienaar JMM,de Kock L, Magano N, Taylor JRN (2022) Towards Sustainable Shifts to Healthy Diets and Food Security in Sub-Saharan Africa with Climate-Resilient Crops in Bread-Type Products: A Food System Analysis. Foods 11:135. https://doi.org/10.3390/foods11020135

OECD, FAO, OECD (2016) FAO Agricultural Outlook 2016–2025; OECD-FAO agricultural outlook; OECD: Paris, France; FAO: Rome, Italy,2016; ISBN 9789264253223

Onwujekwe EC, Ezemba CC (2021) Food security and safety: Africans perspectives a review. Arch Curr Res Int 21:14–20

OAU, ECA (1994) Population and development in Africa. https://www.unfpa.org/sites/default/files/resource-pdf/POPULATI0N_AND_DEVELOPMENT_IN_AFRICA_OAU_ECA.pdf

Pingali P (2012) Green revolution: impacts, limits, and the path ahead. Proc Natl Acad Sci 109:12302–12308

Pretty J, Benton TG, Bharucha ZP, Dicks LV, Flora CB, Godfray HCJ, Goulson D, Hartley S, Lampkin N, Morris C, Pierzynski G, Prasad PVV, Reganold J, Rockström J, Smith P, Thorne P, Wratten S (2018) Global assessment of agricultural system redesign for sustainable intensification. Nat Sustain 1(8):441–446. https://doi.org/10.1038/s41893-018-0114-0

Ranum P, Peña-Rosas JP, Garcia-Casal MN (2014) Global maize production, utilization, and consumption. Ann N Y Acad Sci 1312:105–112

Sekabira H, Nijman E, Späth L, Krütli P, Schut M, Vanlauwe B, Wilde B, Kintche K, Kantengwa S, Feyso A (2022) Circular bioeconomy in African food systems: what is the status quo? Insights from Rwanda, DRC, and Ethiopia. PLoS ONE 17(10):e0276319

Shrivastava SR, Shrivastava PS, Ramasamy J (2017) The global public health challenge of malnutrition: ensuring trend reversal. Ann Trop Med Publ Health 10:1375–1376

Springmann M, Freund F (2022) Options for reforming agricultural subsidies from health, climate, and economic perspectives. Nat Commun 13:82

Stanley A (2023) A demographic transformation in Africa has the potential to alter the world order. Finance Dev. https://www.imf.org/en/Publications/fandd/issues/2023/09/PT-african-century#:~:text=Africa's%20population%2C%20estimated%20to%20be,its%20share%20has%20since%20doubled

Stevens N, Erasmus BF, Archibald S, Bond WJ (2016) Woody encroachment over 70 years in South African savannahs: overgrazing, global change or extinction aftershock? Philos Trans R Soc Lond Ser B, Biol Sci 371(1703). https://doi.org/10.1098/rstb.2015.0437

Tesfa D, Tiruneh SA, Gebremariam AD, Azanaw MM, Engidaw MT, Kefale B, Abebe B, Dessalegn T, Tiruneh M (2022) The pooled estimate of the total fertility rate in sub-Saharan Africa using recent (2010–2018) demographic and health survey data. Front Public Health 10:1053302. https://doi.org/10.3389/fpubh.2022.1053302

Tindwa HJ, Semu, E, Singh, BR (2024) Circular regenerative agricultural practices in Africa: techniques and their potential for soil restoration and sustainable food production. agronomy 14:2423. https://doi.org/10.3390/agronomy14102423

Tubiello FN, Cynthia Rosenzweig C, Conchedda G, Karl1 K, Gütschow J, Xueyao P, Obli-Laryea G, Wanner N, Qiu SY, De Barros J, Flammini A, Mencos-Contreras E, Souza L, Quadrelli R, Heiðarsdóttir HH, Benoit P, Hayek M, Sandalow D (2021) Greenhouse gas emissions from food systems: building the evidence base. Environ Res Lett 16:065007

van der Esch S, ten Brink B, Stehfest E, Bakkenes M, Sewell A (2017) Exploring future agricultural production and land use under climate change scenarios: impacts on biodiversity and food security in Africa. PBL Netherlands Environmental Assessment Agency, The Hague

Van Ittersum MK, Van Bussel LG, Wolf J, Grassini P, Van Wart J, Guilpart N, Claessens L, De Groot H, Wiebe K, Mason-D'Croz D, Yang H (2016) Can sub-Saharan Africa feed itself? Proc Natl Acad Sci 113(52):14964–14969

Wentzel-Viljoen E, Lee S, Laubscher R, Vorster HH (2018) Accelerated nutrition transition in the North West Province of South Africa: results from the prospective urban and rural epidemiology (PURE-NWP-SA) cohort study, 2005 to 2010. Public HealthNutr 21:2630–2641

Williams CJR, Kniveton DR, Layberry R (2010) Assessment of a climate model to reproduce rainfall variability and extremes over Southern Africa. Theor Appl Climatol 99:9–27. https://doi.org/10.1007/s00704-009-0124-y

Zhang F, Neik TX, Thomas WJW, Batley J (2023) CRISPR-based genome editing tools: an accelerator in crop breeding for a changing future. Int J Mol Sci 24:8623. https://doi.org/10.3390/ijms24108623

Zivin JG, Lipper L (2008) Poverty, risk, and the adoption of soil carbon sequestration. Environ Dev Econ 13:353–373

Open Access This chapter is licensed under the terms of the Creative Commons Attribution 4.0 International License (http://creativecommons.org/licenses/by/4.0/), which permits use, sharing, adaptation, distribution and reproduction in any medium or format, as long as you give appropriate credit to the original author(s) and the source, provide a link to the Creative Commons license and indicate if changes were made.

The images or other third party material in this chapter are included in the chapter's Creative Commons license, unless indicated otherwise in a credit line to the material. If material is not included in the chapter's Creative Commons license and your intended use is not permitted by statutory regulation or exceeds the permitted use, you will need to obtain permission directly from the copyright holder.

Research Gaps and Future Perspectives

Bal Ram Singh, Agnes Mwangwela, Liveness Banda, Ruth Haug, and Rattan Lal

Abstract Increasing population, dwindling soil and water resources, severe soil degradation, resource poor and subsistence farming by small holders in SSA has created a serious food and nutritional security problem and the climate change impacting food productivity has aggravated this problem further. Agriculture sector programs are primarily focused on increased production, ignoring nutritional issues; and therefore, nutrition-sensitive agriculture investments and guidance for program formulation are needed. Successful management of drought requires adoption of conservation agriculture, including integrated soil fertility management, agroforestry, and integration of crops with trees and live stocks. Furthermore, incentivizing resource-poor small land holders through payments for ecosystem services (e.g., carbon sequestration in soil) is critical and policies which are pro-nature, pro-agriculture and pro-farmers are needed. Research and development priorities on technologies such as climate -resilient and nutrition-sensitive agriculture, irrigation and water management, integrated soil fertility management, protection and restoration of soil health and policy formulation through constructive integration of technological knowledge with societal knowledge are highly needed.

B. R. Singh (✉)
Faculty of Environmental Sciences and Natural Resource Management, Norwegian University of Life Sciences (NMBU), Ås, Norway
e-mail: balram.singh@nmbu.no

A. Mwangwela · L. Banda
Lilongwe University of Agriculture and Natural Resources (LUANAR), Lilongwe, Malawi

R. Haug
Department of International Environment and Development Studies, Faculty of Landscape and Society, (NMBU), Ås, Norway

R. Lal
CFAES Rattan Lal Center for Carbon Management and Sequestration, The Ohio State University, Columbus, OH, USA

© The Author(s) 2026

B. R. Singh et al. (eds.), *Sustainable Food Systems in Sub-Saharan Africa*,
https://doi.org/10.1007/978-3-032-11288-0_28

1 Introduction

Agriculture is the mainstay of the economies of sub-Saharan Africa (SSA), and agri-food systems are faced with a lot of challenges, despite the unrelenting efforts made by the scientific community and the agricultural programs initiated by governments of SSA region (Adjei-Nsiah et al., 2022). Agriculture in SSA is predominately rainfed and is managed by resource-poor smallholder farmers who cannot afford the recommended inputs. Cereal yields in SSA, for example, are extremely low and average about 1.6 Mg ha^{-1} compared to the global average of 3.9 Mg ha^{-1} (FAOSTATS 2020). A combination of increasing population, dwindling soil and water resources, severe soil degradation, resource poor and subsistence farming by small holders in SSA has created a serious food and nutritional security problem in the region. SSA countries are susceptible to climate change impacts, leading to agricultural productivity decline and the inability to mitigate climate change. Despite the availability of research-driven and proven technologies to resolve the above-mentioned problems, small farmers and especially women smallholders in SSA still lack such technologies. The FoodMa and TRANSFORM -Programs by NMBU in collaboration with LUANAR were initiated in the late 2010s to extend and implement the already researched pro-poor technologies and value-chains to small land-holder farmers. This book is based on some outputs of these programs. The book, containing 28 chapters, covers the major issues related to plant food systems and related innovative technologies, nutrient dense and biofortified food crops, soil management for climate change and soil carbon sequestration, policy formulations and their implementation for smallholders. It also explores the prospects of achieving the Green Revolution in SSA countries. The focus of this chapter is to provide a brief synthesis of the topics covered and identify the research gaps.

2 Resource Degradation and Plant Productivity

Land and Soil Degradation in SSA

Agricultural operations remain the primary driver of land and soil degradation in Sub-Saharan Africa (SSA), with an estimated 75% of agricultural land affected by some form of degradation (FAO 2015). The main contributors include unsustainable farming practices, deforestation, and overgrazing. These activities have led to the loss of soil fertility, erosion, salinity, and the depletion of biodiversity. Agricultural expansion continues to encroach on forested lands, with rates estimated at 0.8% annually, further exacerbating biodiversity loss and groundwater depletion (FAO 2020).

Types and Impacts of Soil Degradation

Soil degradation in SSA is characterized by several processes:

1. **Erosion**: Water and wind erosion leads to the loss of fertile topsoil, significantly affecting crop productivity.
2. **Physical Degradation**: Compaction and reduced infiltration caused by heavy machinery or overgrazing.
3. **Chemical Degradation**: Nutrient depletion, acidification, and salinity.
4. **Biological Degradation**: Decline in organic matter and microbial activity due to over-cultivation and poor management practices (Girija et al. 2020).

These forms of degradation are compounded by suboptimal agricultural practices among smallholder farmers, including:

- Inappropriate use of chemical fertilizers (<15 kg NPK nutrients per hectare annually).
- Limited organic nutrient inputs.
- Low adoption of climate-smart agricultural practices, such as conservation agriculture, integrated nutrient management, and the use of leguminous cover crops.

Agricultural Total Factor Productivity (ATFP)

Soil degradation is a significant factor contributing to SSA's low agricultural total factor productivity (ATFP), defined as the ratio of agricultural outputs to inputs. Despite the yield potential of cereal crops exceeding 5 $Mgha^{-1}$, actual yields remain stagnant at 1.5 $Mgha^{-1}$ over the past five decades. Similarly, legume yields average below 1 $Mgha^{-1}$, despite their potential to exceed 2 $Mgha^{-1}$ (Girija et al. 2020; Wollburg et al. 2023). This stagnation highlights the critical role of soil health in achieving agricultural productivity gains.

Interventions and Challenges

Although soil and nutrient management technologies, such as climate-smart agriculture, fertilizers, and organic manure, have been disseminated among smallholder farmers, their adoption faces significant barriers. Key challenges include:

1. **Economic Constraints**: Farmers prefer interventions that provide returns within a single growing season. Technologies requiring long-term investment, such as biochar and perennial crops, are less appealing.
2. **Land Tenure**: Insecure land ownership discourages long-term soil fertility investments.
3. **Limited Knowledge**: Many farmers lack access to training and resources to effectively adopt advanced soil management practices.

Pathways to Sustainable Land Management

To address these challenges, the following interventions should be prioritized:

1. **Policy and Technical Support**: Governments and NGOs must reduce barriers to adopting sustainable land management practices through subsidies, training programs, and infrastructure development.

2. **Soil Amendments**: Promoting the use of lime, biochar, and other sustainable soil amendments to enhance fertility and structure.
3. **Short-Term Economic Incentives**: Encourage adoption of economically viable practices, such as legume cover crops, which provide immediate income and soil restoration benefits.
4. **Crop Breeding Programs**: Develop and disseminate nutrient- and drought-efficient crop genotypes that are capable of thriving in degraded soils.
5. **Agroecological Intensification**: Explore the use of deep-rooted legumes and green manure genotypes for marginal soils.

Research and Knowledge Gaps

Despite progress, significant knowledge gaps remain, including the following:

- Technologies that reduce the time required for soil management interventions to yield results.
- Efficient methods for rehabilitating smallholder lands with minimal costs.
- Better integration of crop breeding, soil fertility management, and agroecological approaches to achieve sustainable intensification.

3 Climate-Resilient and Nutrition-Sensitive Agriculture

The need for agriculture to support better nutrition and health has been recognized globally as discussed in the United Nations' 2030 Agenda for Sustainable Development (United Nations 2017). Regionally in SSA, several initiatives to integrate nutrition interventions into their agricultural investment plans have been discussed (Rampa and van Seters 2013). Nutrition-sensitive agriculture includes several approaches such as biofortification of crops, homestead food production systems, livestock transfer programs, value chains for nutritious foods, and irrigation programs (Ruel et al. 2018). Among these, genetic biofortification, (i.e., the breeding of staple crops that are richer in essential micronutrients than traditional varieties) (Bouis and Saltzman 2017) and agronomic biofortification (i.e., use of mineral fertilizers through soil and foliar application to increase the concentration and bio-accessibility of specific nutrients in food crops) (Cakmak and Kutman, 2018) have been shown to be feasible and cost-effective approaches to addressing deficiencies of vitamin A, iron, and zinc. Another approach to improved nutrition and health is through diversifying diets by cultivating and consuming more legumes and vegetables, including neglected and underutilized crops. Access to improved seeds is an important factor in enhancing dietary diversification (Haug et al. 2023).

Although the focus on nutrition-sensitive agriculture over the past decade has increased, there are still knowledge gaps in research, development, and dissemination of obtained results to smallholders in SSA countries. The main gaps identified are as follows.

- Focus of agriculture sector programs is primarily on increased production, and nutritional issues are often ignored.
- There is a need to expand this work by looking into sustainability, scaling-up cost effectiveness, and promoting broader agriculture policy and investment decisions (Ruel et al. 2018).
- Thus, designing of nutrition-sensitive agriculture investments and guidance for program formulation is needed.
- Agrarian, agro-pastoral and pastoral communities in SSA have limited awareness on dietary practices and requirements, storage, food handling, processing of animal and plant-based foods.
- Most SSA countries have rain-fed agriculture with limited uptake of the nutrition-sensitive technologies mentioned above.
- With the rich set of ongoing studies on nutrition-sensitive agriculture, a greater understanding of what agriculture can and cannot do to contribute to nutrition improvements and achieving the Sustainable Development Goals for good health and wellbeing can be attained (SDG3) (Ruel et al. 2018).
- Poverty and inequality are the drivers of poor nutrition. Globally, FAO (2020) estimate that 42% of the population cannot afford a healthy diet. To improve nutrition and health in SSA, there is a need to reduce poverty and inequality. In order to do this, there is need to improve access to employment opportunities with decent salary levels, as well as addressing the question of access to farmland of sufficient size.

4 Policy Formulation and Research Gaps

Policy formulation is based upon several factors such as public opinion and interests of different groups in society as well as upon knowledge related to science and technology. Available knowledge in the field of food and agriculture might go in different directions e.g., regarding sustainable intensification or agro-ecology; using or not using input subsidizes; formal or informal seed system development; and biofortification or dietary diversification etc. Policymakers would need to assess the politics of various options, prioritize, and make decisions on policy directions. In the policy formulation process, research gaps might surface, and questions arise in relation to areas where more knowledge would be needed. In food and agriculture, there is a wealth of technological knowledge, but how to put this knowledge to good use is an area where more knowledge appears to be needed. Through constructive integration of technological knowledge with societal knowledge, more enabling environments for men and women farmers to improve their income and food security could be established. Accordingly, the focus of this chapter is to suggest some research gaps that might be of relevance for policymakers in their efforts to formulate the so-called "right" policy:

- How to close the yield gap in a nature-friendly way, and introduce methods that are not dependent on inputs that African smallholders can't afford or labor they do not have access to.
- How should African smallholders cope with land scarcity and plots becoming smaller in many African countries? To what degree is land reform possible or should there be more emphasis on off-farm employment opportunities for those who cannot make a livelihood from the farm?
- How will possible feminization of agriculture contribute towards changes in African agriculture? Increasingly, men are seeking jobs outside of the farm and women are left behind with an increased work burden.
- How to balance the call by urban population for cheap food and the call by smallholder farmers for higher farm gate prices? To what degree might more efficient value chains ensure that both farmers get a decent price for their products and consumers get access to affordable and healthy food?
- How to reach what is often called the "last-mile-smallholders" with agricultural services such as extension as well as market opportunities? In what ways could new approaches relating to mobile phones, joint transport and better storage facilities enable better reach of these groups?
- How could Governments, through agricultural sector policy formulation and implementation, contribute towards advancing equitable livelihoods and poverty reduction in African rural areas, in accordance with the UN Food System Summit of 2021?

5 Enhancing Irrigation and Water Management

Frequency and severity of agronomic and pedological droughts are aggravated by soil degradation and the current and projected climate change. The drought-soil degradation-anthropogenic climate change -nexus is strong and these three processes are mutually reinforcing. Soil degradation aggravates the drought, and both are reinforced by the anthropogenic climate change and have severe and adverse impacts on Earth's planetary functions and the processes governing them. Soil degradation is a serious problem in Africa, because accelerated erosion, nutrient mining and long-term use of extractive practices have depleted the soil organic carbon (SOC) content and stocks. Severe depletion of SOC content in the root zone, whose optimal level is 1.1–1.5% depending on soil type and ecoregional characteristics, is often as low as 0.1%. Severe depletion of SOC content degrades soil structure, reduces water infiltration and retention in the root zone, decreases plant available water capacity or the green water storage, accentuates diurnal and seasonal changes in soil temperature regimes with high levels surpassing the supra-optimal range (25–35 °C) and often reaching as high as 50 °C at 1 cm depth during the afternoon. Both pedological and agronomic droughts are aggravated by the prevalence of supra-optimal soil temperature regimes with adverse effects on plant growth and yield.

Successful management of drought involves storage of water in the root zone by adopting conservation agriculture (no-till, residue retention as mulch, complex rotations, cover cropping and integrated soil fertility management), agroforestry and establishment of contour ridges of perennials, and integration of crops with trees. The strategy is to maintain soil cover for decreasing losses by surface runoff evaporation, uptake by weeds by improving soil structure which leads to increase in storage of plant-available water in the root zone. Increasing SOC content is critical to improving soil structure and the latter is of critical importance to minimizing risks of accelerated soil erosion by water and wind.

In addition to in-situ conservation of water in the root zone, supplemental irrigation is another option to manage pedological and agronomic drought. Only 6% of the cropland area in Africa is equipped for irrigation, as compared with 37% in Asia (e.g., China, India, Pakistan).Thus, expanding irrigation by implementing small scale irrigation projects, including use of solar-powered pumps to lift the ground water, is an important strategy. Rather than flood and furrow or sprinkler irrigation, the preferred mode of using water is through micro-irrigation such as drip sub-fertigation.

Among agronomic management, choice of drought tolerant species and improved varieties is important to reduce risks of drought. Time of seeding and choice of agronomic practices, seeding geometry (e.g., clump planting) and use of mixed farming systems can avoid droughts and optimize agronomic yield. The strong interaction between water and plant nutrients (N, P, micronutrients) demands judicious management of fertilizers etc. Integrated soil fertility management (ISFM) can also enhance water use efficiency and reduce risks of drought during the growing season.

6 Integrated Soil Fertility Management for Nutritious Food and Sustainable Productivity

Integrated soil fertility management (ISFM) is a strategy of using fertilizer, organic inputs, and improved germplasm adapted to local conditions and for maximizing agronomic use efficiency of the applied nutrients and improving crop productivity. This is also one of the ways that may tackle climate change impacts in low-input systems practiced in SSA. Use of ISFM can enhance food security through improved productivity, adaptation, and mitigation of GHG emissions, and contribution to several pillars of climate-smart agriculture (CSA).

ISFM maximizes the use of organic resources and improved germplasm, minimizes nutrient losses, allows timely and judicious use of inorganic fertilizers based on need and economic viability, and maintains and enhances beneficial soil organisms and biological processes (Oladele and Braimoh 2014). Thus, ISFM strategies stimulate biomass production, which determines the amount of C stored in the plant biomass and the inputs of detritus to the soil, leading to soil carbon sequestration and assisting in climate mitigation (Kennedy and Singh 2020).

Even though ISFM addresses a suite of soil fertility-enhancing strategies, adaptation to climate change, and greenhouse gas mitigation, it is poorly documented and faces many challenges described below.

- The adoption of ISFM practices by smallholders in SSA is very low because of low awareness and common disbeliefs about the benefits of soil fertility management (Lambrecht et al. 2015), weak social networks and pervasive distrust (Wossen et al. 2015), lack of information about soil fertility and rainfall forecasts (Maro et al. 2013), and scarcity of organic residues and competition for residues with livestock (Rufino et al. 2011) among others.
- To scale out ISFM across smallholder farming systems in SSA, research and dissemination of practices at local, national, and international levels must be intensified.
- Also, high-resolution information on customized practices for soil fertility to maximize the benefits of ISFM is needed.
- As stated above, the contribution of ISFM to different pillars of CSA, such as increased productivity and food security, adaptation to climate change, mitigation of GHG emission must be disseminated to smallholders.

7 Green Revolution, Climate Change and Agroecological Prospects

This is the right time to usher in the soil -centric Green Revolution in SSA. The benefits of improved varieties and of appropriate species can only be realized if these are grown in soils with optimal physical, chemical, and biological properties. Even the elite varieties cannot extract water and nutrients (P, Ca, Mg, Fe, Zn, Mo etc) from a soil where they do not exist. The soil-centric Green Revolution is appropriate because it enhances eco-efficiency of inputs by reducing losses (or leakage into the environment) and improving uptake by crops. Protection, restoration and improvement of soil health, through re-carbonization of the terrestrial biosphere and buildup of SOC content in the root zone at an optimal level (1.1–1.5%), leads to accruing benefits from three basic principles of sustainable management of natural resources:

1. **The One Health Concept** which states that health of soil, plants, animals, people and the environment is one and indivisible. Thus, restoration and sustainable management of soil health is critical to producing safe, healthy and nutritious food while improving the environment and restoring the planetary processes.
2. **Protection and Restoration of Soil Health** leads to producing more from less. The objective is to produce more food from less land, less water, less use of fertilizers, less pesticides, and less emission of greenhouse gases into the atmosphere. Rather than bringing more land under production, the goal is to return some marginal land back to nature. Returning some land and water back to nature is the best strategy to make agriculture a solution to environmental issues including the adaptation and mitigation of climate change.

3. **The Law of Return** states that whatever is taken from soil must be replaced to the same site in one form or the other so that soil productivity and functionality can be maintained for ever. Soils of Africa have been suffering from nutrient mining (negative nutrient budget) since 1960s, and this trend must be reversed.

The traditional seed—centric Green Revolution of the 1960s, which was successful on alluvial soils with irrigation facilities, needs to be revisited to ensure that adverse effects on environment and soil degradation are minimized. With already degraded and depleted soils of Africa, it is important to adopt a soil-centric Green Revolution, which is based on protection, restoration and sustainable management of soil and other natural resources. In this regards, incentivizing resource-poor small land holders through payments for ecosystem services (e.g., carbon sequestration in soil) is critical. Africa, as anywhere else, needs policies which are from nature, pro-agriculture, and pro-farmers. Researchers, policy makers and private sector should work together to adopt the Soil Health Act, which promotes adoption of soil-centric Green Revolution and rewards farmers for protection, restoration and sustainable management of soil and other natural resources.

References

Adjei-Nsiah S, Opoku A, Frimpong KA, Danso I (2022) Editorial: soil fertility management for sustainable food production in sub-Saharan Africa. Front Sustain Food Syst 6:959604. https://doi.org/10.3389/fsufs.2022.959604

Bouis HE, Saltzman A (2017) Improving nutrition through biofortification: a review of evidence from HarvestPlus, 2003 through 2016. Glob Food Secur 12:49–58. https://doi.org/10.1016/j.gfs.2017.01.009

Cakmak I, Kutman UB (2018) Agronomic biofortification of cereals with zinc: a review. Eur J Soil Sci 69:172–180

FAO (2015) Status of the world soil resources. Food and Agriculture Organization of the United Nations, Rome. https://openknowledge.fao.org/items/f16010ce-1874-4108-bd03-a6a592e2e53a

FAO (2020) The state of food and agriculture 2020: overcoming water challenges in agriculture. Food and Agriculture Organization of the United Nations, Rome

FAOSTATS (2020) Food and agriculture organisation of the United Nations. FAO, Rome. http://www.fao.org/faostat/en/#home. Accessed 8 Oct2020

Girija Veni V, Srinivasarao C, Sammi Reddy K, Sharma KL, Rai A (2020) Soil health and climate change. In: Prasad MNV, Pietrzykowski M (eds) Climate change and soil interactions. Elsevier, pp 751–767. https://doi.org/10.1016/B978-0-12-818032-7.00026-6

Haug R, Hella JP, Mulesa TH, Kakwera MN, Westengen OT (2023) Seed systems development to navigate multiple expectations in Ethiopia. Malawi and Tanzania. World Develop Sustain 3:100092. https://doi.org/10.1016/j.wds.2023.100092

Lambrecht I, Vanlauwe B, Maertens M (2015) Integrated soil fertility management: from concept to practices in eastern DR Congo. Int J Agricult Sustain. https://doi.org/10.1080/14735903.2015.1026047

Maro G, Mrema J, Msanya B, Teri J (2013) Farmers' perception of soil fertility problems and their attitudes towards integrated soil fertility management for coffee in Northern Tanzania. J Soil Sci Environ Manag 5:93–99

Oladele O, Braimoh A (2014) Potential of agricultural land management activities for increased soil carbon sequestration in Africa: a review. Appl Ecol Environ Res 12(3):741–751
Rampa, F, van Seters, J, (2013) Toward the development and implementation of CAADP regional compacts and investment plans: the state of play. European Center for Development Policy Management (ECDPM), Maastricht
Ruel MT, Agnes R, Quisumbing AR, Mysbah-Balagamwala M (2018) Nutrition-sensitive agriculture: what have we learned so far? Glob Food Sec 17:128–153
Rufino M, Dury J, Tittonell P, Van Wijk M, Herrero M, Zingore S, Mapfumo P, Giller K (2011) Competing use of organic resources, village-level interactions between farm types and climate variability in a communal area of NE Zimbabwe. Agricult Syst 104:175–190
United Nations (2017) The sustainable development goals report 2017. United Nations, New York
Were K, Singh BR (2020) Effects of climate change on agricultural soils. In: Deryng D (ed) Climate change and agriculture, Burleigh Dodds Science Publishing Limited, pp 43–53
Wollburg P, Bentze T, Lu Y, Udry C, Gollin D (2023) Agricultural productivity growth in Africa: new evidence from microdata. World Bank working paper
Wossen T, Berger T, Di Falco S (2015) Social capital, risk preference and adoption of improved farm land management practices in Ethiopia. Agricult Econ 46:81–97

Open Access This chapter is licensed under the terms of the Creative Commons Attribution 4.0 International License (http://creativecommons.org/licenses/by/4.0/), which permits use, sharing, adaptation, distribution and reproduction in any medium or format, as long as you give appropriate credit to the original author(s) and the source, provide a link to the Creative Commons license and indicate if changes were made.

The images or other third party material in this chapter are included in the chapter's Creative Commons license, unless indicated otherwise in a credit line to the material. If material is not included in the chapter's Creative Commons license and your intended use is not permitted by statutory regulation or exceeds the permitted use, you will need to obtain permission directly from the copyright holder.

MIX
Papier aus verantwortungsvollen Quellen
Paper from responsible sources
FSC® C105338

If you have any concerns about our products,
you can contact us on
ProductSafety@springernature.com

In case Publisher is established outside the EU,
the EU authorized representative is:
Springer Nature Customer Service Center GmbH
Europaplatz 3, 69115 Heidelberg, Germany

Printed by Libri Plureos GmbH
in Hamburg, Germany